Controlled Fusion and Plasma Physics

Series in Plasma Physics

Series Editor:

Steve Cowley, Imperial College, UK and UCLA, USA

Other books in the series:

Plasma Electronics: Applications in Microelectronic Device Fabrication
T Makabe, Z Petrović

An Introduction to Inertial Confinement Fusion
S Pfalzner

Aspects of Anomalous Transport in Plasmas
R Balescu

Non-Equilibrium Air Plasmas at Atmospheric Pressure
K H Becker, R J Barker and K H Schoenbach (Eds)

Magnetohydrodynamic Waves in Geospace: The Theory of ULF Waves and their Interaction with Energetic Particles in the Solar-Terrestrial Environment
A D M Walker

Plasma Physics via Computer Simulation (paperback edition)
C K Birdsall, A B Langdon

Plasma Waves, Second Edition
D G Swanson

Microscopic Dynamics of Plasmas and Chaos
Y Elskens and D Escande

Plasma and Fluid Turbulence: Theory and Modelling
A Yoshizawa, S-I Itoh and K Itoh

The Interaction of High-Power Lasers with Plasmas
S Eliezer

Introduction to Dusty Plasma Physics
P K Shukla and A A Mamun

The Theory of Photon Acceleration
J T Mendonça

Laser Aided Diagnostics of Plasmas and Gases
K Muraoka and M Maeda

Reaction-Diffusion Problems in the Physics of Hot Plasmas
H Wilhelmsson and E Lazzaro

The Plasma Boundary of Magnetic Fusion Devices
P C Strangeby

Series in Plasma Physics

Controlled Fusion and Plasma Physics

K Miyamoto
University of Tokyo, Japan

CRC Press
Taylor & Francis Group
Boca Raton London New York

CRC Press is an imprint of the
Taylor & Francis Group, an **informa** business
A TAYLOR & FRANCIS BOOK

CRC Press
Taylor & Francis Group
6000 Broken Sound Parkway NW, Suite 300
Boca Raton, FL 33487-2742

First issued in paperback 2019

ISBN-13: 978-1-58488-709-6 (hbk)
ISBN-13: 978-0-367-39008-2 (pbk)

Library of Congress Cataloging-in-Publication Data

Miyamoto, Kenro, 1931-
Controlled fusion and plasma physics / Kenro Miyamoto.
p. cm. -- (Series in plasma physics ; 21)
Includes bibliographical references and index.
ISBN 1-58488-709-5 (alk. paper)
1. Plasma (Ionized gases) 2. Controlled fusion. I. Title.

QC718.M585 2007
621.48'4--dc22 2006050480

Visit the Taylor & Francis Web site at
http://www.taylorandfrancis.com

and the CRC Press Web site at
http://www.crcpress.com

Author's Biography

In 1955, Dr. Kenro Miyamoto received his B.S. degree in physics from the University of Tokyo. He went on to earn a Ph.D. degree from the University of Rochester in 1961. From 1963 to 1979, Dr. Miyamoto worked at the Institute of Plasma Physics, Nagoya University (presently, reorganized as the National Institute of Fusion Science). He went on to teach in the physics department at the University of Tokyo from 1971 to 1992. Dr. Miyamoto then joined the faculty of the engineering department at Seikei University from 1992 to 2000. Currently, he is professor emeritus of the University of Tokyo.

Preface

The worldwide effort to develop the fusion process as a major new source of energy has been going on for about a half century. Now the construction stage of "International Tokamak Experimental Reactor," called ITER, has already started.

The primary objective of this book is to present the recent status of controlled fusion research and related plasma physics for graduate and advanced undergraduate students. I also hope that this textbook will be a useful compact reference for scientists and engineers working in the relevant fields.

This book consists of two parts. The first part describes the various activities of controlled fusion. Chapters 1 through 3 cover the basic knowledge of controlled fusion research, followed by tokamak (Chapter 4), reversed field pinch (RFP) (Chapter 5), stellarator (Chapter 6), and mirror and tandem mirror (Chapter 7). Emphasis has been given to tokamak, as it is the most successful and most studied concept. Critical issues such as magnetohydrodynamic (MHD) stabilities, empirical energy confinement scalings of L and H modes, and steady-state operation are described in detail. The design concept of ITER is explained. Trials to innovative tokamaks beyond ITER are also introduced. Experimental results of improved confinement by pulsed parallel current drive (PPCD), current profile control in RFP, activities on quasi-symmetric stellarators, and recent results of tandem mirror experiments are described.

The second part discusses plasma physics closely associated with controlled fusion. This part begins with ideal MHD instabilities (Chapter 8) and resistive instabilities (Chapter 9), which essentially limit the beta ratio of confined plasmas. Neoclassical tearing mode and resistive wall mode are also explained in Chapter 9. Boltzmann equation and Vlasov equation are described in Chapter 10, which are the fundamentals of kinetic theory and MHD theory. The concept of Landau damping is discussed in this chapter, which characterizes the interaction of charged particles and wave (or perturbation) in hot plasmas. In Chapter 11, dielectric tensors of cold and hot plasmas are derived from Vlasov equation. Characteristics of various waves are discussed based on dispersion relations. Dielectric tensors of hot plasma include the effect of Landau damping (amplification) as well as cyclotron damping (amplification). Drift instability is introduced as a simple example of velocity space instabilities. In Chapter 12, the physical mechanism of wave heating and non-inductive current drive are explained. Wave heating as well as neutral beam heating are effective tools to increase ion and electron temperatures, and non-inductive current drives are essential for steady-state operation of tokamak. The subject of plasma transport by turbulence is a most complex and challenging issue for study. The many trials of theories and computer simulation of gyrokinetic particle and full orbit particle models are introduced in Chapter 13. Basic knowledge of zonal flow is also described in this chapter.

This textbook aims to introduce the various interesting aspects of recent fusion research activities through the minimum access path and to describe important issues from the standpoint of fusion research in the first part. In the second part, each chapter is related to an important subject in fusion research and explains it in a clear and compact manner.

Kenro Miyamoto
Professor Emeritus
University of Tokyo

Contents

1

Introduction to Plasmas

1.1 Charge Neutrality and Landau Damping

As the temperature of a material is raised, its state changes from solid to liquid and then to gas. If the temperature is elevated further, an appreciable number of the gas atoms are ionized and become the high temperature gaseous state in which the charge numbers of ions and electrons are almost the same and charge neutrality is satisfied in a macroscopic scale.

When ions and electrons move, these charged particles interact with Coulomb force which is long range force and decays only in inverse square of the distance r between the charged particles. The resultant current flows due to the motion of the charged particles and Lorentz interaction takes place.

"What a plasma really is is a collection of a very large number of individual charged particles, all interacting with each other through mutual Coulomb forces and through the electric currents associated with their motion, while at the same time interacting with (and thereby modifying) any electromagnetic fields of external origin. It is this property of collective long-range interactions, of plasma with itself and of plasma with its electromagnetic environment, that gives rise to the great complexity of its behavior" (from R. F. Post's text [1.1]).

When the temperature of a gas is T(K), the average velocity of the thermal motion of a particle with the mass m, that is, thermal velocity $v_{\rm T}$ is defined by

$$mv_{\rm T}^2/2 = \kappa T/2 \tag{1.1}$$

where κ is Boltzmann constant $\kappa = 1.380658(12) \times 10^{-23}$ J/K and κT indicates the thermal energy. Therefore the unit of κT is Joule (J) in SI unit (Système International d'Unitès). In many fields of physics, one electron volt (eV) is frequently used as a unit of energy. This is the energy necessary to move an electron, charge $e = 1.60217733(49) \times 10^{-19}$ Coulomb, against a potential difference of 1 volt:

$$1\,\text{eV} = 1.60217733(49) \times 10^{-19}\,\text{J}.$$

The temperature corresponding to the thermal energy of 1 eV is 1.16×10^4 K($=$e$/\kappa$). From now on, the thermal energy κT is denoted by just T for simplicity and new 'T' is named by just temperature. The ionization energy of a hydrogen atom is 13.6 eV. Even if the thermal energy (average energy) of hydrogen gas is 1 eV, that corresponds to around 10^4 K, a small amount of electrons with energy higher than 13.6 eV exists and ionizes the gas to a hydrogen plasma.

Plasmas are found in nature in various forms. There exits the ionosphere in the heights of 70∼500 km (density $n \sim 10^{12}\,\mathrm{m}^{-3}$, $T \sim 0.2$ eV). Solar wind is the plasma flow originated from the sun with $n \sim 10^{6\sim7}\,\mathrm{m}^{-3}$, $T \sim 10\,\mathrm{eV}$. Corona extends around the sun and the density is $\sim 10^{14}\,\mathrm{m}^{-3}$ and the electron temperature is $\sim 100\,\mathrm{eV}$ although these values depend on the different positions. White dwarf, the final state of stellar evolution, has the electron density of $10^{35\sim36}\,\mathrm{m}^{-3}$. Active research in plasma physics has been motivated by the aim to create and confine hot plasmas in fusion research.

One of the fundamental properties of plasma is the charge neutrality. Plasmas shield the electric potential applied to the plasmas. When a probe is inserted into a plasma and positive (negative) potential is applied, the probe attracts (repulses) electrons and the plasma tends to shield the electric disturbance. Let us estimate the shielding length. Assume that heavy ions are in uniform density ($n_\mathrm{i} = n_0$) and there is small perturbation in electron density n_e and potential ϕ. Since the electrons are in Boltzmann distribution with the electron temperature T_e, the electron density n_e becomes

$$n_\mathrm{e} = n_0 \, \exp(e\phi/T_\mathrm{e}) \simeq n_0(1 + e\phi/T_\mathrm{e})$$

where ϕ is the electrostatic potential and $e\phi/T_\mathrm{e} \ll 1$ is assumed. The equation for the electrostatic potential is given by Maxwell equation as follows: @

$$\boldsymbol{E} = -\nabla\phi, \qquad \nabla(\epsilon_0 \boldsymbol{E}) = -\epsilon_0 \nabla^2 \phi = \rho = -e(n_\mathrm{e} - n_0) = -\frac{e^2 n_0}{T_\mathrm{e}}\phi$$

and

$$\nabla^2\phi = \frac{\phi}{\lambda_\mathrm{D}^2}, \qquad \lambda_\mathrm{D} = \left(\frac{\epsilon_0 T_\mathrm{e}}{n_\mathrm{e} e^2}\right)^{1/2} = 7.45 \times 10^3 \left(\frac{1}{n_\mathrm{e}}\frac{T_\mathrm{e}}{e}\right)^{1/2} (\mathrm{m}) \tag{1.2}$$

where ϵ_0 is the dielectric constant of a vacuum and $\boldsymbol{E}$ is the electric intensity. n_e is in m^{-3} and T_e/e is in eV. When $n_\mathrm{e} \sim 10^{20}\,\mathrm{cm}^{-3}$, $T_\mathrm{e}/e \sim 10$ keV, then $\lambda_\mathrm{D} \sim 75\,\mu$m. In spherically symmetric cases, Laplacian ∇^2 becomes $\nabla^2\phi = (1/r^2)(\partial/\partial r)\,(r^2 \partial\phi/\partial r)$ and the solution is

$$\phi = \frac{q}{4\pi\epsilon_0}\frac{\exp(-r/\lambda_\mathrm{D})}{r}.$$

It is clear from the foregoing formula that Coulomb potential $q/4\pi\epsilon_0 r$ of point charge is shielded out to a distance λ_D. This distance λ_D is called *Debye length.* When the plasma size is a and $a \gg \lambda_\mathrm{D}$ is satisfied, then plasma is considered neutral in charge. If $a < \lambda_\mathrm{D}$, in contrary, individual particles are not shielded electrostatically and this state is no longer plasma but an assembly of independent charged particles.

The number of electrons included in the sphere of radius λ_D is called *plasma parameter* and is given by

$$n_e \lambda_D^3 = \left(\frac{\epsilon_0}{e}\frac{T_e}{e}\right)^{3/2} \frac{1}{n_e^{1/2}}. \tag{1.3}$$

When the density is increased while keeping the temperature constant, this value becomes small. If the plasma parameter is less than say ~ 1, the concept of Debye shielding is not applicable because the continuity of charge density breaks down in the scale of Debye length. Plasmas in the region of $n_e \lambda_D^3 > 1$ are called classical plasma or *weakly coupled plasma*, since the ratio of electron thermal energy T_e and coulomb energy between electrons $E_{\rm coulomb} = e^2/4\pi\epsilon_0 d$ ($d \simeq n_e^{-1/3}$ is the average distance between electrons with the density n) is given by

$$\frac{T_e}{E_{\rm coulomb}} = 4\pi(n_e \lambda_D^3)^{2/3}$$

and $n_e \lambda_D^3 > 1$ means that coulomb energy is smaller than the thermal energy. The case of $n_e \lambda_D^3 < 1$ is called *strongly coupled plasma.*

Fermi energy of degenerated electron gas is given by $\epsilon_F = (h^2/2m_e) \times (3\pi^2 n_e)^{2/3}$ ($h = 6.6260755(40) \times 10^{-34}$ J · s is Planck's constant). When the density becomes very high, it is possible to become $\epsilon_F \geq T_e$. In this case quantum effect is more dominant than thermal effect. This case is called *degenerated electron plasma.* One example of this is the electron plasma in metal. Most of plasmas in experiments of magnetic confinements are classical weakly coupled plasma.

Let us consider the case where a small perturbation occurs in a uniform plasma and the electrons in the plasma move by the perturbation. It is assumed that ions do not move because the ion's mass is much more heavy than the mass of electron. Due to the displacement of electrons, electric charges appear and an electric field is induced. The electric field is given by

$$\epsilon_0 \nabla \cdot \boldsymbol{E} = -e(n_e - n_0).$$

Electrons are accelerated by the electric field:

$$m_e \frac{d\boldsymbol{v}}{dt} = -e\boldsymbol{E}.$$

Due to the movement of electrons, the electron density changes:

$$\frac{\partial n_e}{\partial t} + \nabla \cdot (n_e \boldsymbol{v}) = 0.$$

Denote $n_e - n_0 = n_1$ and assume $|n_1| \ll n_0$; then we find

$$\epsilon_0 \nabla \cdot \boldsymbol{E} = -en_1, \qquad m_e \frac{\partial \boldsymbol{v}}{\partial t} = -e\boldsymbol{E}, \qquad \frac{\partial n_1}{\partial t} + n_0 \nabla \cdot \boldsymbol{v} = 0.$$

For simplicity, the displacement is assumed only in the x direction and is sinusoidal with the angular frequency ω:

$$n_1(x,t) = n_1 \exp(ikx - i\omega t).$$

Time differential $\partial/\partial t$ is replaced by $-i\omega$ and $\partial/\partial x$ is replaced by ik. The electric field has only the x component E. Then,

$$ik\epsilon_0 E = -en_1, \qquad -i\omega m_e v = -eE, \qquad -i\omega n_1 = -ikn_0 v$$

so that we find

$$\omega^2 = \frac{n_0 e^2}{\epsilon_0 m_e}. \tag{1.4}$$

This wave is called *electron plasma wave* or *Langmuir wave* and its frequency is called *electron plasma (angular) frequency* Π_e:

$$\Pi_e = \left(\frac{n_e e^2}{\epsilon_0 m_e}\right)^{1/2} = 5.64 \times 10^{11} \left(\frac{n_e}{10^{20}}\right)^{1/2} \quad \text{rad/sec.}$$

When $n_e = 10^{20}\text{m}^{-3}$, then $\Pi_e/2\pi = 90\text{GHz}$. There is the following relation between the plasma frequency and Debye length λ_D:

$$\lambda_D \Pi_e = \left(\frac{T_e}{m_e}\right)^{1/2} = v_{Te} = 4.19 \times 10^5 \left(\frac{T_e}{e}\right)^{1/2} \quad \text{m/sec.}$$

Therefore, the inverse of Π_e is the thermal electron's transit time of the Debye length ($\Pi_e^{-1} = \lambda_D/v_{Te}$).

The other fundamental process of plasma is collective phenomena of charged particles. Waves are associated with coherent motions of charged particles. When the phase velocity v_{ph} of wave or perturbation is much larger than the thermal velocity v_T of charged particles, the wave propagates through the plasma media without damping or amplification. However, when the refractive index N of plasma media becomes large and plasma becomes hot, the phase velocity $v_{ph} = c/N$ (c is light velocity) of the wave and the thermal velocity v_T become comparable ($v_{ph} = c/N \sim v_T$), then the exchange of energy between the wave and the thermal energy of plasma is possible. The existence of a damping mechanism of wave was found by L. D. Landau. The process of Landau damping involves a direct wave-particle interaction in collisionless plasma without necessity of randamizing collision. This process is the fundamental mechanism in wave heatings of plasma (wave damping) and instabilities (inverse damping of perturbations). Landau damping will be described in Chpater 10 and Chapter 11.

1.2 Fusion Core Plasma

Progress in plasma physics has been motivated by the realization of fusion core plasma. Necessary conditions for fusion core plasma are discussed in this section. Nuclear fusion reactions are the fused reactions of light nuclides to heavier ones. When the sum of the masses of nuclides after a nuclear fusion is smaller than the sum before the reaction by Δm, we call it *mass defect*. According to the theory of relativity, the amount of energy $(\Delta m)c^2$ (c is light speed) is released by the nuclear fusion.

Nuclear reactions of interest for fusion reactors are as follows (D= deuteron, T= triton, He^3= helium-3, Li= lithium):

(1) $D+D \to T(1.01\,MeV)+p(3.03\,MeV)$
(2) $D+D \to He^3(0.82\,MeV)+n(2.45\,MeV)$
(3) $T+D \to He^4(3.52\,MeV)+n(14.06\,MeV)$
(4) $D+He^3 \to He^4(3.67\,MeV)+p(14.67\,MeV)$
(5) $Li^6+n \to T+He^4+4.8\,MeV$
(6) $Li^7+n(2.5\,MeV) \to T+He^4+n$

where p and n are proton (hydrogen ion) and neutron, respectively (1 MeV $=10^6$ eV). Since the energy released by the chemical reaction of $H_2 + (1/2)O_2 \to H_2O$ is 2.96 eV, fusion energy released is about one million times as large as the chemical one. A binding energy per nucleon is smaller in very light or very heavy nuclides and largest in the nuclides with atomic mass numbers around 60. Therefore, large amounts of the energy can be released when the light nuclides are fused. Deuterium exists aboundantly in nature; for example, it comprises 0.015 atom percent of the hydrogen in sea water with the volume of about $1.35 \times 10^9\,km^3$.

Although fusion energy was released in an explosive manner by the hydrogen bomb in 1951, controlled fusion is still in the stage of research development. Nuclear fusion reactions were found in the 1920s. When proton or deuteron beams collide with targets of light nuclide, the beams lose their energy by the ionization or elastic collisions with target nuclides and the probability of nuclear fusion is negligible. Nuclear fusion research has been most actively pursued by the use of hot plasma.

In fully ionized hydrogen, deuterium, and tritium plasmas, the process of ionization does not occur. If the plasma is confined in some specified region adiabatically, the average energy does not decrease by the processes of elastic collisions. Therefore, if very hot D-T plasmas or D-D plasmas are confined, the ions have velocities large enough to overcome their mutual coulomb repulsion so that collision and fusion take place.

Let us consider the nuclear reaction of D colliding with T. The *cross-section* of T nucleus is denoted by σ. This cross-section is a function of the kinetic energy E of D. The cross-section of D-T reaction at $E = 100\,keV$

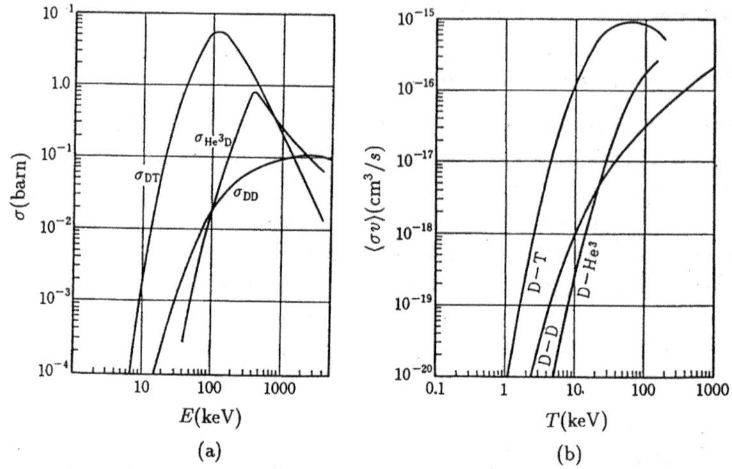

Figure 1.1 (a) The dependence of fusion cross-section σ on the kinetic energy E of colliding nucleous. $\sigma_{\rm DD}$ is the sum of the cross-sections of D-D reactions (1) (2). 1 barn $= 10^{-24}$ cm^2. (b) The dependence of fusion rate $\langle \sigma v \rangle$ on the ion temperature $T_{\rm i}$.

is 5×10^{-24} cm^2. The cross-sections σ of D-T, D-D, D-He^3 reaction versus the kinetic energy of colliding nucleous are shown in Figure 1.1(a) [1.2, 1.3]. The probability of fusion reaction per unit time in the case that a D ion with the velocity v collides with T ions with the density of $n_{\rm T}$ is given by $n_{\rm T}\sigma v$ (we will discuss the collision probability in more detail in Section 2.2.1). When a plasma is Maxwellian with the ion temperature of $T_{\rm i}$, it is necessary to calculate the average value $\langle \sigma v \rangle$ of σv over the velocity space. The dependence of $\langle \sigma v \rangle$ on ion temperature $T_{\rm i}$ is shown in Figure 1.1(b) [1.4]. A fitting equation of $\langle \sigma v \rangle$ of D-T reaction as a function of T in the unit of keV is [1.5]

$$\langle \sigma v \rangle ({\rm m}^{-3}) = \frac{3.7 \times 10^{-18}}{H(T) \times T^{2/3}} \exp\left(-\frac{20}{T^{1/3}}\right), \qquad (1.5)$$

$$H(T) \equiv \frac{T}{37} + \frac{5.45}{3 + T(1 + T/37.5)^{2.8}}.$$

Figure 1.2 shows an example of an electric power plant based on a D-T fusion reactor. Fast neutrons produced in fusion core plasma penetrate the first wall. A lithium blanket surrounding the plasma moderates the fast neutrons, converting their kinetic energy to heat. Furthermore, the lithium blanket breeds tritium due to reaction (5),(6). (Triton beta-decays to He^3 with the half-life of 12.3 years (T$\rightarrow$ He^3 + e($<$ 18.6keV)) and tritium does not exist as a natural resource). Lithium blanket gives up its heat to

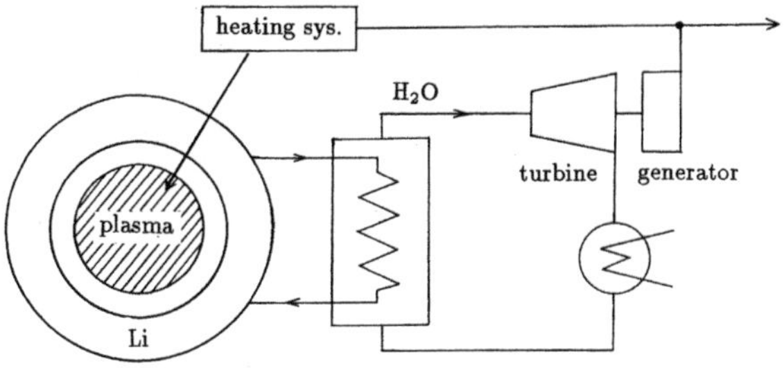

Figure 1.2 An electric power plant based on a D-T fusion reactor.

generate the steam by a heat exchanger; steam turbine generates electric power. A part of the generated electric power is used to operate the heating system of plasma. As alpha particles (He ions) are charged particles, alpha particles can heat the plasma by Coulomb collisions directly (see Section 2.6). The total heating power P_{heat} is the sum of α particle heating power P_α and the heating power P_{ext} by the external heating system. The necessary total heating power to sustain the plasma in steady state must be equal to the energy loss rate of fusion core plasma. Therefore, good energy confinement (small energy loss rate) of hot plasma is the most important issue.

The thermal energy of plasma per unit volume is given by $(3/2)n(T_{\mathrm{i}}+T_{\mathrm{e}})$. This thermal energy is lost by thermal conduction and convective losses. The notation P_{L} denotes these energy losses of the plasma per unit volume per unit time (power loss per unit volume). There is radiation loss R due to bremsstrahlung of electrons and impurity ion radiation in addition to P_{L}. The total energy confinement time τ_{E} is defined by

$$\tau_{\mathrm{E}} \equiv \frac{(3/2)n(T_{\mathrm{e}}+T_{\mathrm{i}})}{P_{\mathrm{L}}+R} \simeq \frac{3nT}{P_{\mathrm{L}}+R}. \tag{1.6}$$

The necessary heating input power P_{heat} to maintain the thermal energy of plasma is equal to $P_{\mathrm{L}}+R$.

In the case of D-T reaction, the sum of kinetic energies $Q_\alpha = 3.52\,\mathrm{MeV}$ of alpha particle and $Q_{\mathrm{n}} = 14.06\,\mathrm{MeV}$ of neutron is Q_{NF}=17.58 MeV per reaction ($Q_{\mathrm{n}} : Q_\alpha = m_\alpha : m_{\mathrm{n}}$=0.8 : 0.2 due to momentum conservation). Since the densities of D ions and T ions of equally mixed plasma are $n/2$, the number of D-T reaction per unit time per unit volume is $(n/2)(n/2)\langle\sigma v\rangle$ (refer to the discussion in Section 2.2.1), so that fusion output power per unit volume P_{NF} is given by

$$P_{\mathrm{NF}} = (n/2)(n/2)\langle\sigma v\rangle Q_{\mathrm{NF}}. \tag{1.7}$$

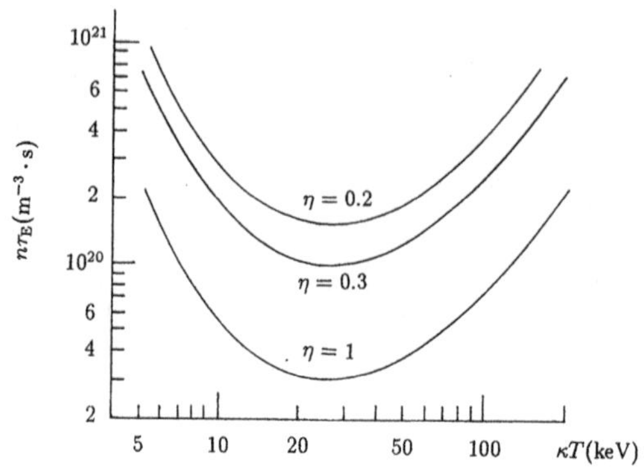

Figure 1.3 Condition of D-T fusion core plasma in $n\tau_E$ - T diagram in the case of $\eta = 0.3$, critical condition ($\eta = 1$), and ignition condition ($\eta = 0.2$).

When the fusion powers by the neutron and alpha particle are denoted by P_n and P_α, respectively, then P_n=0.8P_{NF} and P_α=0.2P_{NF}. Denote the thermal-to-electric conversion efficiency by η_{el} and heating efficiency (ratio of the deposit power into the plasma to the electric input power of heating device) by η_{heat}. When a part ($\gamma < 1$) of generated electric power is used to operate the heating system, then available heating power to plasma is

$$(0.8\eta_{el}\gamma\eta_{heat} + 0.2)P_{NF} = \eta P_{NF},$$

$$\eta \equiv 0.8\gamma\eta_{el}\eta_{heat} + 0.2.$$

The burning condition is

$$P_{heat} = P_L + R = \frac{3nT}{\tau_E} < \eta P_{NF} \tag{1.8}$$

that is

$$\frac{3nT}{\tau_E} < \eta \frac{Q_{NF}}{4} n^2 \langle \sigma v \rangle,$$

$$n\tau_E > \frac{12T}{\eta Q_{NF} \langle \sigma v \rangle}. \tag{1.9}$$

The right-hand side of (1.9) is the function of temperature T only. When $T = 10^4$ eV and $\eta \sim 0.3$ ($\gamma \sim 0.4$, $\eta_{el} \sim 0.4$, $\eta_{heat} \sim 0.8$), the necessary condition is $n\tau_E > 1.7 \times 10^{20}\,\mathrm{m}^{-3} \cdot$ sec. The burning condition of D-T fusion plasma in the case of $\eta \sim 0.3$ is shown in Figure 1.3. In reality, the plasma is hot in the core and is cold in the edge. For a more accurate discussion, we must take account of the profile effect of temperature and density, which will be analyzed in Section 4.8.

The ratio of the fusion output power due to α particles to the total is $Q_\alpha/Q_{\rm NF} = 0.2$. If the total kinetic energy (output energy) of alpha particles contributes to heat the plasma and alpha particle heating power can sustain the necessary high temperature of the plasma without heating from outside, the plasma is in the ignited state. The condition $P_\alpha = P_{\rm L} + R$ is called *ignition condition*, which corresponds to the case of $\eta = 0.2$ in (1.8).

The condition $P_{\rm heat} = P_{\rm NF}$ is called *break even condition*. This corresponds to the case of $\eta = 1$ in (1.8).

The ignition condition ($\eta = 0.2$) and break even condition ($\eta = 1$) are also shown in Figure 1.3.

Problems

1. Triple Product When ion temperature is near 10 keV, fusion rate $\langle \sigma v \rangle$ of D-T reaction is approximately given by $\langle \sigma v \rangle = \langle \sigma v \rangle_{10}(T_{\rm keV}/10)^2$, where $T_{\rm keV}$ is ion temperature in keV. Drive the more simplified burning condition than Equation (1.9) near $T_{\rm keV} \sim 10$. Calculate the necessary value of $n\tau_{\rm E}T_{\rm keV}$ in unit of $\rm m^3\,s\,keV$ in the case of $\eta \approx 0.3$. ($\langle \sigma v \rangle_{10} \approx 10^{-22}{\rm m}^3/s$ and $Q_{\rm NF}$=17.58MeV.)

2. Stability of Burning Plasma Time variation of temperature of burning plasma is given by $3({\rm d}(nT)/{\rm d}t) = -3nT/\tau_{\rm E} + \eta Q_{\rm NF} n^2 \langle \sigma v \rangle /4$. Let T_0 be the equilibrium temperature of fusion plasma, at which ${\rm d}(nT)/{\rm d}t = 0$. Is the reactor thermally stable or not, when the temperature changes to $T_0 + \delta T$ from the equilibrium? For simplicity, assume that n and η are constant. Derive the stability condition assuming $\tau_{\rm E} \propto T^{-\alpha}$, $\langle \sigma v \rangle \propto T^\beta$ near $T = T_0$.

2

Particle Orbit and Magnetic Configuration

2.1 Particle Orbit

2.1.1 Cyclotron Motion

The equation of motion of charged particle with mass m and charge q in an electric and magnetic field $\boldsymbol{E}$, $\boldsymbol{B}$ is given by

$$m\frac{\mathrm{d}\boldsymbol{v}}{\mathrm{d}t} = q(\boldsymbol{E} + \boldsymbol{v} \times \boldsymbol{B}). \tag{2.1}$$

When the magnetic field is homogenous and is in the z direction and the electric field is zero, the equation of motion becomes $\dot{\boldsymbol{v}} = (qB/m)(\boldsymbol{v} \times \boldsymbol{b})$ $(\boldsymbol{b} = \boldsymbol{B}/B)$ and

$$v_x = -v_\perp \sin(\Omega t + \delta),$$

$$v_y = v_\perp \cos(\Omega t + \delta),$$

$$v_z = v_{z0},$$

$$\Omega = -\frac{qB}{m}. \tag{2.2}$$

The solution of these equations is a spiral motion around the magnetic line of force with the angular velocity of Ω (see Figure 2.1). This motion is

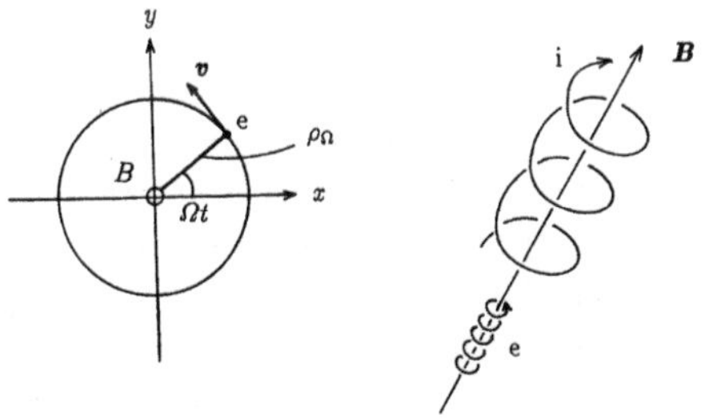

Figure 2.1 Larmor motion of charged particles in magnetic field.

called *Larmor motion*. The angular frequency Ω is called *cyclotron (angular) frequency*. Denote the radius of the orbit by ρ_Ω, then the centrifugal force is $mv_\perp^2/\rho_\Omega$ and Lorentz force is $qv_\perp B$. Since both forces must be balanced, we find

$$\rho_\Omega = \frac{mv_\perp}{|q|B}. \tag{2.3}$$

This radius is called *Larmor radius*. The center of Larmor motion is called *guiding center*. Electron's Larmor motion is right-handed rotation ($\Omega_e > 0$), and ion's Larmor motion is left-handed rotation ($\Omega_i < 0$). When $B = 1\,\text{T}$, $T = 100\,\text{eV}$, the values of Larmor radius and cyclotron freqency are given in Table 2.1.

Table 2.1 Mass, thermal velocities, Larmor radii, and cyclotron frequencies of electron and proton.

B=1T, T=100eV	electron	@ proton
mass (Kg)	$9.1093897(54) \times 10^{-31}$	$1.6726231(10) \times 10^{-27}$
v_T	4.2×10^6 m/s	9.8×10^4 m/s
ρ_Ω	23.8 μm	1.0 mm
Ω	1.76×10^{11} /s	-9.58×10^7 /s
$\Omega/2\pi$	28 GHz	−15.2 MHz

$v_T = (T/m)^{1/2}$ is thermal velocity, ρ_Ω is Larmor radius, Ω and $\Omega/2\pi$ are angular cyclotron frequency and cyclotron frequency, respectively.

2.1.2 Drift Velocity of Guiding Center, Magnetic Moment

When a uniform electric field $\boldsymbol{E}$ perpendicular to the uniform magnetic field is superposed, the equation of motion (2.1) is reduced to

$$m\frac{\mathrm{d}\boldsymbol{u}}{\mathrm{d}t} = q(\boldsymbol{u} \times \boldsymbol{B})$$

by introducing $\boldsymbol{u}$ and $\boldsymbol{u}_E$ defined by

$$\boldsymbol{v} \equiv \boldsymbol{u}_E + \boldsymbol{u}, \qquad \boldsymbol{u}_E \equiv \frac{\boldsymbol{E} \times \boldsymbol{b}}{B}. \tag{2.4}$$

Therefore, the motion of charged particles is superposition of Larmor motion and drift motion $\boldsymbol{u}_E$ of its guiding center. The direction of guiding center drift by $\boldsymbol{E}$ is the same for both ion and electron (Figure 2.2). When a gravitational field $\boldsymbol{g}$ is superposed, the force is $m\boldsymbol{g}$, which corresponds to $q\boldsymbol{E}$ in the case of electric field. Therefore, *the drift velocity of the guiding center due to the gravitation is given by*

$$\boldsymbol{u}_g = \frac{m}{qB}(\boldsymbol{g} \times \boldsymbol{b}) = -\frac{\boldsymbol{g} \times \boldsymbol{b}}{\Omega}. \tag{2.5}$$

Figure 2.2 Drift motion of guiding center in electric and gravitational field (conceptional drawing).

The directions of ion's drift and electron's drift due to gravitation are opposite each other and the drift velocity of the ion guiding center is much larger than that of the electron (see Figure 2.2). When the magnetic and electric fields change slowly and gradually in time and in space ($|\omega/\Omega| \ll 1, \rho_\Omega/R \ll 1$), the formulas of drift velocity are valid. However, because of the curvature of the field line of magnetic force, centrifugal force acts on the particle which runs along a field line with the velocity of $v_\parallel$. The acceleration of centrifugal force is

$$\boldsymbol{g}_{\text{curv}} = \frac{v_\parallel^2}{R}\boldsymbol{n}$$

where R is the radius of curvature of field line and $\boldsymbol{n}$ is the unit vector with the direction from the center of the curvature to the field line (Figure 2.3).

Furthermore, as is described later in this section, the resultant effect of Larmor motion in an inhomogeneous magnetic field is reduced to the acceleration of

$$\boldsymbol{g}_{\nabla B} = -\frac{v_\perp^2/2}{B}\nabla B.$$

Therefore, drift velocity $\boldsymbol{u}_{\text{G}}$ of the guiding center due to inhomogenous curved magnetic field is given by the *drift approximation* as follows:

$$\boldsymbol{u}_{\text{G}} = -\frac{1}{\Omega}\left(\frac{v_\parallel^2}{R}\boldsymbol{n} - \frac{v_\perp^2}{2}\frac{\nabla B}{B}\right) \times \boldsymbol{b}. \tag{2.6}$$

The first term is called *curvature drift* and the second term is called ∇B *drift.* Since $\nabla \times \boldsymbol{B} = \mu_0 \boldsymbol{j}$ ($\boldsymbol{j}$ is current density), the vector formula reduces

$$\frac{1}{2B}\nabla(\boldsymbol{B}\cdot\boldsymbol{B}) = (\boldsymbol{b}\cdot\nabla)\boldsymbol{B} + \boldsymbol{b}\times(\nabla\times\boldsymbol{B}) = \frac{\partial}{\partial l}(B\boldsymbol{b}) + \boldsymbol{b}\times\mu_0\boldsymbol{j}$$

$$= \frac{\partial B}{\partial l}\boldsymbol{b} + B\frac{\partial \boldsymbol{b}}{\partial l} - \mu_0\frac{\nabla p}{B} = \frac{\partial B}{\partial l}\boldsymbol{b} - B\frac{\boldsymbol{n}}{R} - \mu_0\frac{\nabla p}{B}$$

where p is plasma pressure and $\nabla p = \boldsymbol{j}\times\boldsymbol{B}$ is hold in equilibrium state (refer to Section 4.2.1). We used the following relation (see Figure 2.3):

$$\frac{\partial \boldsymbol{b}}{\partial l} = -\frac{\boldsymbol{n}}{R}$$

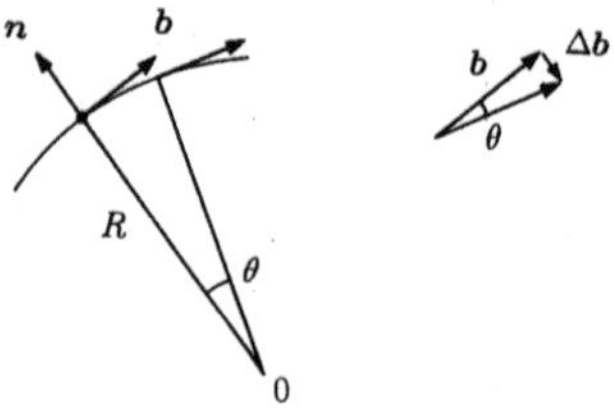

Figure 2.3 Radius of curvature of line of magnetic force.

where l is the length along the field line. Then we have

$$\frac{\boldsymbol{n}\times\boldsymbol{b}}{R} = -\left(\frac{\nabla B}{B} + \mu_0\frac{\nabla p}{B^2}\right)\times\boldsymbol{b}.$$

If ∇p is much smaller than $\nabla B^2/(2\mu_0)$, we find

$$\boldsymbol{u}_{\rm G} = -\frac{1}{\Omega}\frac{v_\parallel^2 + v_\perp^2/2}{R}(\boldsymbol{n}\times\boldsymbol{b}).$$

The parallel motion along the magnetic field is given by

$$m\frac{{\rm d}v_\parallel}{{\rm d}t} = qE_\parallel + mg_\parallel - \frac{mv_\perp^2/2}{B}\nabla_\parallel B. \tag{2.7}$$

Let us consider the effect of inhomogeneity of the magnetic field on a gyrating charged particle. The x component of Lorentz force $\boldsymbol{F}_{\rm L} = q\boldsymbol{v}\times\boldsymbol{B}$ perpendicular to the magnetic field (z direction) and the magnitude B of the magnetic field near the guiding center is

$$F_{{\rm L}x} = qv_yB = -|q|v_\perp\cos\theta B$$

$$B = B_0 + \frac{\partial B}{\partial x}\rho_\Omega\cos\theta + \frac{\partial B}{\partial y}\rho_\Omega\sin\theta.$$

The time average of x component of Lorentz force is given by $\langle F_{{\rm L}x}\rangle = \frac{1}{2}(\partial B/\partial x)(-|q|)v_\perp\rho_\Omega$ and the y component is also given the same way. We find

$$\langle\boldsymbol{F}_{\rm L}\rangle_\perp = -\frac{mv_\perp^2/2}{B}\nabla_\perp B.$$

Next it is necessary to estimate the time average of z component of Lorentz force. The equation $\nabla\cdot\boldsymbol{B} = 0$ near the guiding center in Figure 2.4 becomes $B_r/r + \partial B_r/\partial r + \partial B_z/\partial z = 0$ and we find

$$\langle F_{{\rm L}z}\rangle = -\langle qv_\theta B_r\rangle = |q|v_\perp\rho_\Omega\frac{\partial B_r}{\partial r} = -\frac{mv_\perp^2/2}{B}\frac{\partial B}{\partial z}$$

($r\sim\rho_\Omega$ is very small and $B_r/r \approx \partial B_r/\partial r$). Thus, the necessary expression of $\boldsymbol{g}_{\nabla B}$ is derived.

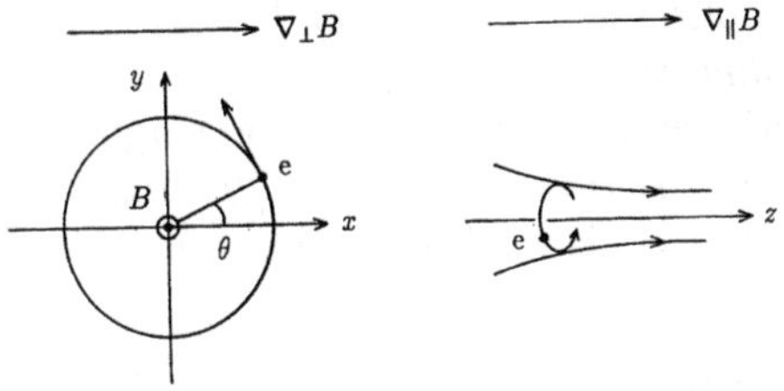

Figure 2.4 Larmor motion in inhomogeneous magnetic field.

A current loop with the current I encircling the area S has the magnetic moment of $\mu_{\rm m} = IS$. Since the current and encircling area of gyrating Larmor motion are $I = q\Omega/2\pi$ and $S = \pi\rho_\Omega^2$, respectively, it has the *magnetic moment* of

$$\mu_{\rm m} = \frac{q\Omega}{2\pi}\pi\rho_\Omega^2 = \frac{mv_\perp^2}{2B}.$$

This physical quantity is adiabatically invariant.

Let us check the invariance of $\mu_{\rm m}$ in the presence of a slowly changing magnetic field ($|\partial B/\partial t| \ll |\Omega B|$). Scalar product of $\boldsymbol{v}_\perp$ and the equation of motion is

$$m\boldsymbol{v}_\perp \cdot \frac{{\rm d}\boldsymbol{v}_\perp}{{\rm d}t} = \frac{\rm d}{{\rm d}t}\left(\frac{mv_\perp^2}{2}\right) = q(\boldsymbol{v}_\perp \cdot \boldsymbol{E}_\perp).$$

During one period $2\pi/|\Omega|$ of Larmor motion, the change $\Delta W_\perp$ of the kinetic energy $W_\perp = mv_\perp^2/2$ is

$$\Delta W_\perp = q\int(\boldsymbol{v}_\perp \cdot \boldsymbol{E}_\perp){\rm d}t = q\oint \boldsymbol{E}_\perp \cdot {\rm d}\boldsymbol{s} = q\int(\nabla \times \boldsymbol{E} \cdot \boldsymbol{n}){\rm d}S$$

where $\oint {\rm d}s$ is the closed line integral along Larmor orbit and $\int {\rm d}S$ is the surface integral over the encircled area of Larmor orbit. Since $\nabla \times \boldsymbol{E} = -\partial \boldsymbol{B}/\partial t$, $\Delta W_\perp$ is

$$\Delta W_\perp = -q\int\frac{\partial \boldsymbol{B}}{\partial t}\cdot \boldsymbol{n}{\rm d}S = |q|\pi\rho_\Omega^2\frac{\partial B}{\partial t}.$$

The change of magnetic field ΔB during one period of Larmor motion is $\Delta B = (\partial B/\partial t)(2\pi/|\Omega|)$, we find

$$\Delta W_\perp = \frac{mv_\perp^2}{2}\frac{\Delta B}{B} = W_\perp\frac{\Delta B}{B}$$

and

$$\mu_{\rm m} = \frac{W_\perp}{B} = \frac{mv_\perp^2/2}{B} = {\rm const.} \tag{2.8}$$

When a system is periodic in time, the action integral $\oint p{\rm d}q$, in terms of the canonical variables p, q is an adiabatic invariant in general. The action integral of Larmor motion is $J_\perp = (-m\rho_\Omega \Omega) 2\pi\rho_\Omega = -(4\pi m/q)\mu_{\rm m}$. $J_\perp$ is called *transversal adiabatic invariant.*

When the magnetic field changes slowly, the magnetic moment is conserved. Therefore, if B is increased, $mv_\perp^2/2 = \mu_{\rm m} B$ is also increased and the particles are heated. This kind of heating is called *adiabatic heating.*

2.1.3 Polarization Drift

Let us consider the case that $\boldsymbol{E} = E_0 \exp(-i\omega t)\hat{\boldsymbol{x}}$ in the x direction is time dependent but $\boldsymbol{B}$ is stationary and constant in the z direction. Then the equation of motion (2.1) is

$$\ddot{v}_x = \frac{q}{m}\dot{E}_x + \frac{q}{m}\dot{v}_y B = i\omega\Omega\frac{E_x}{B} - \Omega^2 v_x$$

$$\ddot{v}_y = -\frac{q}{m}\dot{v}_x B = -\Omega^2\frac{E_x}{B} - \Omega^2 v_y.$$

When we define

$$v_{\rm E} \equiv -\frac{E_x}{B}, \quad v_{\rm p} = i\frac{\omega}{\Omega}\frac{E_x}{B}$$

then the equation of motion is reduced to

$$\ddot{v}_x = -\Omega^2(v_x - v_{\rm p}), \quad \ddot{v}_y = -\Omega^2(v_y - v_{\rm E}).$$

When $\Omega^2 \gg \omega^2$, the solution is

$$v_x = -iv_\perp \exp(-i\Omega t) + v_{\rm p}, \quad v_y = v_\perp \exp(-i\Omega t) + v_{\rm E}.$$

This solution shows that the guiding center motion consists of the usual $\boldsymbol{E}\times\boldsymbol{B}$ drift (but slowly oscillating) and the new drift along $\boldsymbol{E}$. This new term is called *the polarization drift* and is expressed by

$$\boldsymbol{v}_{\rm p} = -\frac{1}{\Omega B}\frac{\partial \boldsymbol{E}}{\partial t}. \tag{2.9}$$

Since $\boldsymbol{v}_{\rm p}$ is in opposite directions for ions and electrons, there is a *polarization current*

$$\boldsymbol{j}_{\rm p} = en_{\rm e}(\boldsymbol{v}_{\rm pi} - \boldsymbol{v}_{\rm pe}) = \frac{n_{\rm e}(m_{\rm i} + m_{\rm e})}{B^2}\frac{\partial \boldsymbol{E}}{\partial t} = \frac{\rho_{\rm m}}{B^2}\frac{\partial \boldsymbol{E}}{\partial t}$$

where $\rho_{\rm m}$ is the mass density.

2.1.4 Pondromotive Force

The equation of motion of an electron in the electromagnetic wave $\boldsymbol{E}(\boldsymbol{r},t) = \widehat{\boldsymbol{E}}(\boldsymbol{r})\cos(\boldsymbol{k}\cdot\boldsymbol{r} - \omega t)$ is

$$m\frac{{\rm d}\boldsymbol{v}}{{\rm d}t} = -e(\boldsymbol{E} + \boldsymbol{v}\times\boldsymbol{B}).$$

Here we assume that $\widehat{\boldsymbol{E}}(\boldsymbol{r})$ varies slowly in the scale of the wavelength. Using the notation $\alpha \equiv \boldsymbol{k}\cdot\boldsymbol{r} - \omega t$, the magnetic field $\boldsymbol{B}$ is given as follows:

$$\frac{\partial \boldsymbol{B}}{\partial t} = -\nabla \times \boldsymbol{E} = \nabla \times \widehat{\boldsymbol{E}} \cos\alpha + \boldsymbol{k} \times \widehat{\boldsymbol{E}} \sin\alpha,$$

$$\boldsymbol{B} = \frac{\nabla \times \widehat{\boldsymbol{E}}}{\omega} \sin\alpha + \frac{\boldsymbol{k} \times \widehat{\boldsymbol{E}}}{\omega} \cos\alpha.$$

In the first order equations, we neglect the second order term $\boldsymbol{v} \times \boldsymbol{B}$ and we have

$$m\frac{\mathrm{d}\boldsymbol{v}_1}{\mathrm{d}t} = -e\boldsymbol{E}(\boldsymbol{r}_0, t) = -e\widehat{\boldsymbol{E}}(\boldsymbol{r}_0)\cos(\boldsymbol{k}\cdot\boldsymbol{r}_0 - \omega t),$$

$$\boldsymbol{v}_1 = \frac{e\widehat{\boldsymbol{E}}(\boldsymbol{r}_0)}{m\omega} \sin(\boldsymbol{k}\cdot\boldsymbol{r}_0 - \omega t),$$

$$\boldsymbol{r}_1 = \frac{e\widehat{\boldsymbol{E}}(\boldsymbol{r}_0)}{m\omega^2} \cos(\boldsymbol{k}\cdot\boldsymbol{r}_0 - \omega t),$$

where $\boldsymbol{r}_0$ is the initial position. We expand $\boldsymbol{E}(\boldsymbol{r}, t)$ about $\boldsymbol{r}_0$:

$$\boldsymbol{E}(\boldsymbol{r}, t) = \boldsymbol{E}(\boldsymbol{r}_0, t) + (\boldsymbol{r}_1 \cdot \nabla)\boldsymbol{E}(\boldsymbol{r}, t)|_{\boldsymbol{r}_0}$$

$$= \boldsymbol{E}(\boldsymbol{r}_0, t) + (\boldsymbol{r}_1 \cdot \nabla)\widehat{\boldsymbol{E}} \cos\alpha_0 - \widehat{\boldsymbol{E}}\,(\boldsymbol{r}_1 \cdot \boldsymbol{k}) \sin\alpha_0$$

where $\alpha_0 \equiv \boldsymbol{k}\cdot\boldsymbol{r}_0 - \omega t$.
In the second order, we must add the term $\boldsymbol{v}_1 \times \boldsymbol{B}$:

$$m\frac{\mathrm{d}\boldsymbol{v}_2}{\mathrm{d}t} = -e\big((\boldsymbol{r}_1 \cdot \nabla)\widehat{\boldsymbol{E}} \cos\alpha_0 - \widehat{\boldsymbol{E}}\,(\boldsymbol{r}_1 \cdot \boldsymbol{k}) \sin\alpha_0\big)$$

$$-e\boldsymbol{v}_1 \times \left(\nabla \times \widehat{\boldsymbol{E}} \sin\alpha_0 + \frac{\boldsymbol{k} \times \widehat{\boldsymbol{E}}}{\omega} \cos\alpha_0\right)$$

$$= -\frac{e^2}{m\omega^2}\Big((\widehat{\boldsymbol{E}} \cdot \nabla)\widehat{\boldsymbol{E}} \cos^2\alpha_0 + \widehat{\boldsymbol{E}} \times \nabla \times \widehat{\boldsymbol{E}} \sin^2\alpha_0\Big)$$

$$-\frac{e^2}{m\omega^2}\Big(-(\widehat{\boldsymbol{E}} \cdot \boldsymbol{k})\widehat{\boldsymbol{E}} + \widehat{\boldsymbol{E}} \times \boldsymbol{k} \times \widehat{\boldsymbol{E}}\Big) \sin\alpha_0 \cos\alpha_0$$

$$= -\frac{e^2}{2m\omega^2}\Big(\nabla\frac{\widehat{\boldsymbol{E}}^2}{2}(1 - \cos 2(\boldsymbol{k}\cdot\boldsymbol{r}_0 - \omega t)) + 2(\widehat{\boldsymbol{E}} \cdot \nabla)\widehat{\boldsymbol{E}} \cos 2(\boldsymbol{k}\cdot\boldsymbol{r}_0 - \omega t)$$

$$+\boldsymbol{k}\widehat{\boldsymbol{E}}^2 \sin 2(\boldsymbol{k}\cdot\boldsymbol{r}_0 - \omega t) - 2(\boldsymbol{k} \cdot \widehat{\boldsymbol{E}})\widehat{\boldsymbol{E}} \sin 2(\boldsymbol{k}\cdot\boldsymbol{r}_0 - \omega t)\Big). \tag{2.10}$$

We used the formula $\widehat{\boldsymbol{E}} \times (\nabla \times \widehat{\boldsymbol{E}}) = \nabla(\widehat{\boldsymbol{E}} \cdot \widehat{\boldsymbol{E}})/2 - (\widehat{\boldsymbol{E}} \cdot \nabla)\widehat{\boldsymbol{E}}$,
$\widehat{\boldsymbol{E}} \times (\boldsymbol{k} \times \widehat{\boldsymbol{E}}) = \boldsymbol{k}\widehat{\boldsymbol{E}}^2 - (\boldsymbol{k} \cdot \widehat{\boldsymbol{E}})\widehat{\boldsymbol{E}}$. In the case of transverse electromagnetic

wave, the terms $(\widehat{\boldsymbol{E}} \cdot \nabla)\widehat{\boldsymbol{E}}$ and $(\boldsymbol{k} \cdot \widehat{\boldsymbol{E}})\widehat{\boldsymbol{E}}$ are negligible and the terms due to Lorentz force are dominant. The time average of $m\mathrm{d}\boldsymbol{v}_2/\mathrm{d}t$ becomes

$$m\left\langle \frac{\mathrm{d}\boldsymbol{v}_2}{\mathrm{d}t} \right\rangle = -\frac{e^2}{4m\omega^2}\nabla\widehat{\boldsymbol{E}}^2.$$

This is the effective nonlinear force on a single electron. This nonlinear force on plasma per unit volume is

$$nm\left\langle \frac{\mathrm{d}\boldsymbol{v}_2}{\mathrm{d}t} \right\rangle = -\frac{\omega_\mathrm{p}^2}{\omega^2}\nabla\frac{\epsilon_0\widehat{\boldsymbol{E}}^2}{4} = -\frac{\omega_\mathrm{p}^2}{\omega^2}\nabla\frac{\epsilon_0\langle \boldsymbol{E}^2\rangle}{2} \tag{2.11}$$

where ω_p is electron plasma frequency. This force is called *ponderomotive force*. This force moves plasma out of the beam, so that electron plasma frequency Π_e is lower and the dielectric constant $\epsilon = (1 - \Pi_\mathrm{e}^2/\omega^2)$ (refer to Section 11.1) is higher inside the beam than outside; that is, the refractive index $N = \epsilon^{1/2}$ is larger inside the beam than outside. Then, the plasma acts as an optical fiber, focusing the beam to a small diameter. By the ponderomotive force, intense laser beam with Peta Watt (10^{15}W) can bore a hole and reach to the core of high density fuel pellet in inertial confinement and heat electrons by the oscillating components in (2.10). This concept is called *fast ignition* [2.1].

2.2 Coulomb Collision, Neutral Beam Injection

2.2.1 Coulomb Collision

The motions of charged particles were analyzed in the previous section without considering the effects of collisions between particles. In this section, phenomena associated with *Coulomb collisions* will be discussed. Let us start from a simple model. Assume that a sphere with the radius a moves with the velocity v in the region where spheres with the radius b are filled with the number density n (see Figure 2.5). When the distance between the two particles becomes less than $a + b$, collision takes place. The cross-section σ of this collision is $\sigma = \pi(a + b)^2$. Since the sphere a moves by the distance $l = v\delta t$ during δt, the probability of collision with the sphere b is

$$nl\sigma = n\sigma v\delta t$$

since nl is the possible number of the sphere b, with which the sphere a within a unit area of incidence may collide, and $nl\sigma$ is the total cross-section per unit area of incidence during the period of δt. Therefore, the collision time τ_coll, when the probability of collision becomes 1, is

$$\tau_\mathrm{coll} = (n\sigma v)^{-1}.$$

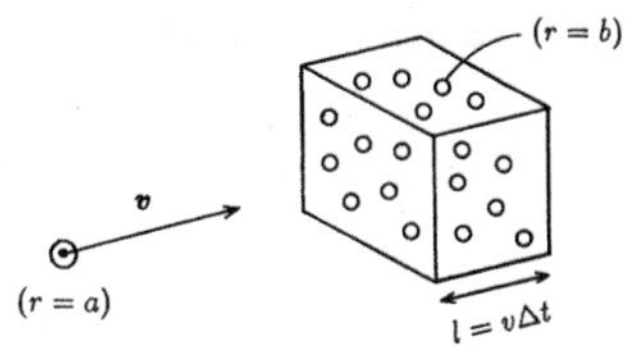

Figure 2.5 Probability of collision of a sphere a with sphere b.

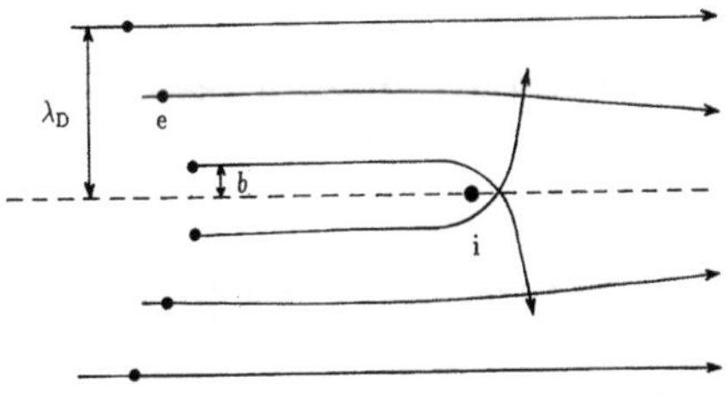

Figure 2.6 Coulomb collision of electron with ion.

In this simple case the cross-section σ of the collision is independent of the velocity of the incident sphere a. However, the cross-section is dependent on the incident velocity in general.

Let us consider strong Coulomb collision of an incident electron with ions with the charge Ze (see Figure 2.6) in which the electron is deflected strongly after the collision. Such a collision can take place when the magnitude of electrostatic potential of the electron at the closest distance b is the order of the kinetic energy of incident electron, that is,

$$\frac{Ze^2}{4\pi\epsilon_0 b} = \frac{m_e v_e^2}{2}.$$

The cross-section of the strong Coulomb collision is $\sigma = \pi b^2$. The inverse of the collision time $\tau_{\rm coll}$ of the strong Coulomb collision is

$$\frac{1}{\tau_{\rm coll}} = n_i \sigma v_e = n_i v_e \pi b^2 = \frac{n_i \pi (Ze^2)^2 v_e}{(4\pi\epsilon_0 m_e v_e^2/2)^2} = \frac{Z^2 e^4 n_i}{4\pi\epsilon_0^2 m_e^2 v_e^3}.$$

Since Coulomb force is long range interaction, a test particle is deflected by small angle even by a distant field particle, which the test particle does not become very close to. As is described in Section 1.1, the Coulomb field of a field particle is not shielded inside the Debye sphere with the radius of Debye length λ_D and there are many field particles inside the Debye sphere in the usual laboratory plasmas (weakly coupled plasmas). Accumulation of many collisions with small angle deflection results in large effect. When the effect of the small angle deflection is taken into account, the total Coulomb cross-section increases by the factor of *Coulomb logarithm*

$$\ln \Lambda \simeq \ln\left(\frac{2\lambda_{\rm D}}{b}\right) \simeq \int_{b/2}^{\lambda_{\rm D}} \frac{1}{r} {\rm d}r \simeq 15 \sim 20.$$

The time derivative of the momentum $p_{\|}$ parallel to the incident direction of the electron is given by use of the *collision time* $\tau_{\rm ei\|}$ as follows: [2.2], Chapter 4 of [2.3]

$$\frac{{\rm d}p_{\|}}{{\rm d}t} = -\frac{p_{\|}}{\tau_{\rm ei\|}},$$

$$\frac{1}{\tau_{\rm ei\|}} = \frac{Z^2 e^4 n_{\rm i} \ln \Lambda}{4\pi\epsilon_0^2 m_{\rm e}^2 v_{\rm e}^3} \tag{2.12}$$

where $\tau_{\rm ei\|}$ indicates the deceleration time of an electron by ions.

When a test particle with the charge q, the mass m, and the velocity v collides with the field particles with the charge q^*, the mass m^*, and the thermal velocity $v_{\rm T}^* = (\kappa T^*/m^*)^{1/2}$ in general, the collision time of the test particle is given by

$$\frac{1}{\tau_{\|}} = \frac{q^2 q^{*2} n^* \ln \Lambda}{4\pi\epsilon_0^2 m m_{\rm r} v^3} = \left(\frac{qq^*}{\epsilon_0 m}\right)^2 \frac{n^* \ln \Lambda}{4\pi (m_{\rm r}/m) v^3} \tag{2.13}$$

under the assumption of $v > v_{\rm T}^*$. $m_{\rm r}$ is the reduced mass $m_{\rm r} = mm^* /(m+m^*)$. Taking the average of $(m/2)v^2 = (3/2)T$, $1/\tau_{\|}$ becomes

$$\frac{1}{\tau_{\|}} = \frac{q^2 q^{*2} n^* \ln \Lambda}{3^{1/2} 12\pi\epsilon_0^2 (m_{\rm r}/m^{1/2}) T^{3/2}}. \tag{2.14}$$

The inverse of collision time is called *collisional frequency* and is denoted by ν. The *mean free path* is given by $\lambda = 3^{1/2} v_{\rm T}\, \tau$.

The collision frequency in the case of electron with ions is

$$\frac{1}{\tau_{\rm ei\|}} = \frac{Z^2 e^4 n_{\rm i} \ln \Lambda}{3^{1/2} 12\pi\epsilon_0^2 m_{\rm e}^{1/2} T^{3/2}}. \tag{2.15}$$

This electron-ion collision frequency is ~ 1.4 times the Spitzer result [2.4] of

$$\frac{1}{\tau_{\rm ei\|\ Spitzer}} = \frac{Z^2 e^4 n_{\rm i} \ln \Lambda}{9.3 \times 10\, \epsilon_0^2 m_{\rm e}^{1/2} T_{\rm e}^{3/2}}. \tag{2.16}$$

When an ion with the charge Z and the mass $m_{\rm i}$ collides with the same ions, the ion-ion collision frequency is given by

$$\frac{1}{\tau_{\rm ii\|}} = \frac{Z^4 e^4 n_{\rm i} \ln \Lambda}{3^{1/2} 6\pi\epsilon_0^2 m_{\rm i}^{1/2} T_{\rm i}^{3/2}}. \tag{2.17}$$

Electron-electron Coulomb collision frequency can be derived by substitution of $m_{\rm i} \to m_{\rm e}$ and $Z \to 1$ into the formula of $\tau_{\rm ii\|}$ as follows:

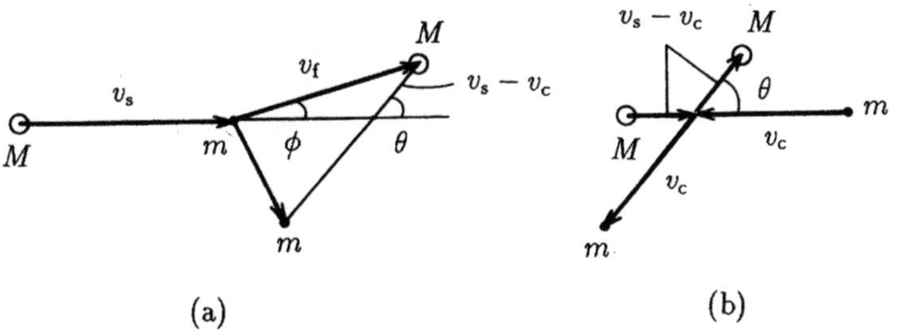

Figure 2.7 Elastic collision of test particle M and field particle m in laboratory system (a) and center-of-mass system (b).

$$\frac{1}{\tau_{\mathrm{ee}\|}} = \frac{n_e e^4 \ln \Lambda}{3^{1/2} 6\pi\epsilon_0^2 m_e^{1/2} T_e^{3/2}}. \tag{2.18}$$

However, the case of ion to electron Coulomb collision is more complicated to treat because the assumption $v_i > v_T^*$ no longer holds. Let us consider the case that a test particle with the mass M and the velocity v_s collides with a field particle with the mass m. In center-of-mass systems where the center of mass is rest, the field particle m moves with the velocity of $v_c = -Mv_s/(M+m)$ and the test particle M moves with the velocity of $v_s - v_c = mv_s/(M+m)$ (refer to Figure 2.7). Since the total momentum and total kinetic energy of two particles are conserved in the process of elastic collision, the velocities of the test particle and the field particle do not change and two particles only deflect their direction by the angle of θ in center-of-mass system. The velocity v_f and scattering angle ϕ of the test particle after the collision in laboratory systems are given by (see Figure 2.7)

$$v_f^2 = (v_s - v_c)^2 + {v_c}^2 + 2(v_s - v_c)v_c\cos\theta = v_s^2\frac{(M^2 + 2Mm\cos\theta + m^2)}{(M+m)^2},$$

$$\sin\phi = \frac{m\sin\theta}{(M^2 + 2Mm\cos\theta + m^2)^{1/2}}.$$

Denote the momentum and the kinetic energy of the test particle before and after the collision by p_s, E_s, and p_f, E_f, respectively; then we find

$$\frac{\Delta E}{E_s} \equiv \frac{E_f - E_s}{E_s} = -\frac{2Mm}{(M+m)^2}(1 - \cos\theta).$$

When the average is taken by θ, we obtain the following relations in the case of $m/M \ll 1$:

$$\langle\frac{\Delta E}{E_s}\rangle \simeq -\frac{2m}{M}, \qquad \langle\frac{\Delta p_\|}{p_s}\rangle \simeq -\frac{m}{M}.$$

From the foregoing discussion, the collision frequency $1/\tau_{\mathrm{ie}\|}$ where a heavy ion collides with light electrons is about m_e/m_i times the value of $1/\tau_{\mathrm{ei}\|}$ and is given by [2.2], ch.4 of [2.3]

$$\frac{1}{\tau_{\mathrm{ie}\parallel}} = \frac{m_\mathrm{e}}{m_\mathrm{i}} \frac{Z^2 e^4 n_\mathrm{e} \ln \Lambda}{(2\pi)^{1/2} 3\pi\epsilon_0^2 m_\mathrm{e}^{1/2} T_\mathrm{e}^{3/2}}. \tag{2.19}$$

When the parallel and perpendicular components of the momentum of a test particle are denoted by $p_\parallel$ and $p_\perp$, respectively, and the energy by E, there are the following relations

$$E = \frac{p_\parallel^2 + p_\perp^2}{2m},$$

$$\frac{\mathrm{d}p_\perp^2}{\mathrm{d}t} = 2m\frac{\mathrm{d}E}{\mathrm{d}t} - 2p_\parallel \frac{\mathrm{d}p_\parallel}{\mathrm{d}t}.$$

We define the *velocity diffusion time* $\tau_\perp$ in the perpendicular direction to the initial momentum and the *energy relaxation time* τ^ϵ by

$$\frac{\mathrm{d}p_\perp^2}{\mathrm{d}t} \equiv \frac{p_\perp^2}{\tau_\perp},$$

$$\frac{\mathrm{d}E}{\mathrm{d}t} \equiv -\frac{E}{\tau^\epsilon}$$

respectively. $1/\tau_\perp$ and $1/\tau^\epsilon$ are given by [2.2]

$$\frac{1}{\tau_\perp} = \frac{q^2 q^{*2} n^* \ln \Lambda}{2\pi\epsilon_0^2 v (mv)^2} = \frac{q^2 q^{*2} n^* \ln \Lambda}{2\pi\epsilon_0^2 m^2 v^3}, \tag{2.20}$$

$$\frac{1}{\tau^\epsilon} = \frac{q^2 q^{*2} n^* \ln \Lambda}{4\pi\epsilon_0^2 m^* v (mv^2/2)} = \frac{q^2 q^{*2} n^* \ln \Lambda}{2\pi\epsilon_0^2 m m^* v^3}. \tag{2.21}$$

In the case of electron to ion collision, we find

$$\frac{1}{\tau_{\mathrm{ei}\perp}} \simeq \frac{2}{\tau_{\mathrm{ei}\parallel}}, \qquad \frac{1}{\tau_{\mathrm{ei}}^\epsilon} \simeq \frac{m_\mathrm{e}}{m_\mathrm{i}} \frac{2}{\tau_{\mathrm{ei}\parallel}}. \tag{2.22}$$

In the case of electron to electron collision, and ion to ion collision, we find

$$\frac{1}{\tau_{\mathrm{ee}\perp}} \simeq \frac{1}{\tau_{\mathrm{ee}\parallel}} = \frac{e^4 n_\mathrm{e} \ln \Lambda}{3^{1/2} 6\pi\epsilon_0^2 m_\mathrm{e}^{1/2} T_\mathrm{e}^{3/2}} = \frac{\ln \Lambda}{32.6} \frac{\Pi_\mathrm{e}}{n_\mathrm{e} \lambda_\mathrm{D}^3} \tag{2.23}$$

$$\frac{1}{\tau_{\mathrm{ee}}^\epsilon} \simeq \frac{1}{\tau_{\mathrm{ee}\parallel}} \tag{2.24}$$

and

$$\frac{1}{\tau_{\mathrm{ii}\perp}} \simeq \frac{1}{\tau_{\mathrm{ii}\parallel}} = \frac{Z^4 e^4 n_\mathrm{i} \ln \Lambda}{3^{1/2} 6\pi\epsilon_0^2 m_\mathrm{i}^{1/2} T_\mathrm{i}^{3/2}} \tag{2.25}$$

$$\frac{1}{\tau_{\mathrm{ii}}{}^\epsilon} \simeq \frac{1}{\tau_{\mathrm{ii}\parallel}} \tag{2.26}$$

respectively. In the case of ion to electron collision we have the following relations [2.2]:

$$\frac{1}{\tau_{\mathrm{ie}\perp}} \simeq \frac{Z^2 e^4 n_{\mathrm{e}} \ln \Lambda}{(2\pi)^{3/2} \epsilon_0^2 m_{\mathrm{e}}^{1/2} E_{\mathrm{i}} T_{\mathrm{e}}^{1/2}} \frac{m_{\mathrm{e}}}{m_{\mathrm{i}}}, \tag{2.27}$$

$$\frac{1}{\tau_{\mathrm{ie}}{}^{\epsilon}} \simeq \frac{Z^2 e^4 n_{\mathrm{e}} \ln \Lambda}{4\pi \epsilon_0^2 m_{\mathrm{e}}^{1/2} T_{\mathrm{e}}^{3/2}} \frac{4}{3(2\pi)^{1/2}} \frac{m_{\mathrm{e}}}{m_{\mathrm{i}}} \simeq \frac{1}{\tau_{\mathrm{ie}\parallel}} \simeq \frac{m_{\mathrm{e}}}{m_{\mathrm{i}}} \frac{2.77}{\tau_{\mathrm{ei}\parallel}} \tag{2.28}$$

where $E_{\mathrm{i}} = (3/2)T_{\mathrm{i}}$ is the kinetic energy of the ion.

2.2.2 Neutral Beam Injection

High energy neutral particle beams can be injected into plasmas across strong magnetic fields. The neutral particles are converted to high-energy ions by means of charge exchange with plasma ions or ionization. The high energy ions (mass m_{b}, electric charge $Z_{\mathrm{b}}e$, energy E_{b}) running through the plasma slow down by Coulomb collisions with the plasma ions ($m_{\mathrm{i}}, Z_{\mathrm{i}}e$) and electrons ($m_{\mathrm{e}}, -e$) and the beam energy is thus transferred to the plasma. This method is called heating by *neutral beam injection* (NBI). The rate of change of the fast ion's energy, that is, the heating rate of plasma is [2.5]

$$\frac{\mathrm{d}E_{\mathrm{b}}}{dt} = -\frac{E_{\mathrm{b}}}{\tau_{\mathrm{bi}}^{\epsilon}} - \frac{E_{\mathrm{b}}}{\tau_{\mathrm{be}}^{\epsilon}}, \tag{2.29}$$

$$\frac{1}{\tau_{\mathrm{bi}}^{\epsilon}} = \frac{(Z_{\mathrm{b}}e)^2 (Z_{\mathrm{i}}e)^2 \ln \Lambda n_{\mathrm{i}}}{2\pi \epsilon_0^2 m_{\mathrm{i}} m_{\mathrm{b}} v_{\mathrm{b}}^3}$$

and

$$\frac{\mathrm{d}E_{\mathrm{b}}}{\mathrm{d}t} = -\frac{Z_{\mathrm{b}}^2 e^4 \ln \Lambda n_{\mathrm{e}}}{4\pi \epsilon_0^2 m_{\mathrm{e}} v_{\mathrm{b}}} \left(\sum_{\mathrm{i}} \frac{m_{\mathrm{e}}}{m_{\mathrm{i}}} \frac{n_{\mathrm{i}} Z_{\mathrm{i}}^2}{n_{\mathrm{e}}} + \frac{4}{3\pi^{1/2}} \left(\frac{m_{\mathrm{e}} E_{\mathrm{b}}}{m_{\mathrm{b}} T_{\mathrm{e}}} \right)^{3/2} \right) \tag{2.30}$$

when the beam ion's velocity v_{b} is much less (say 1/3) than the plasma electron thermal velocity and much larger (say 2 times) than the plasma ion thermal velocity. The first term in the right-hand side is due to beam-ion collisions and the second term is due to beam-electron collisions. A critical energy E_{cr} of the beam ion, at which the plasma ions and electrons are heated at equal rates, is given by

$$\frac{m v_{\mathrm{cr}}^2}{2} = E_{\mathrm{cr}} = 14.8 T_{\mathrm{e}} A_{\mathrm{b}} \left(\frac{1}{n_{\mathrm{e}}} \sum_{\mathrm{i}} \frac{n_{\mathrm{i}} Z_{\mathrm{i}}^2}{A_{\mathrm{i}}} \right)^{2/3} \tag{2.31}$$

where A_{b}, A_{i} are atomic weights of the injected ion and plasma ion, respectively. When the energy of the injected ion is larger than E_{cr}, the contribution to the electron heating is dominant. The slowing down time of the ion beam is given by

$$\tau_{\mathrm{slowdown}} = \int_{E_{\mathrm{cr}}}^{E_{\mathrm{b}}} \frac{-\mathrm{d}E_{\mathrm{b}}}{(\mathrm{d}E_{\mathrm{b}}/\mathrm{d}t)} = \frac{\tau_{\mathrm{be}}^{\epsilon}}{1.5} \ln\left(1 + \left(\frac{E}{E_{\mathrm{cr}}}\right)^{3/2}\right),$$

$$\frac{1}{\tau_{\mathrm{be}}^{\epsilon}} = \frac{Z^2 n_{\mathrm{e}} e^4 \ln \Lambda}{(2\pi)^{1/2} 3\pi\epsilon_0^2 m_{\mathrm{e}}^{1/2} T_{\mathrm{e}}^{3/2}} \frac{m_{\mathrm{e}}}{m_{\mathrm{b}}} \qquad (2.32)$$

where $\tau_{\mathrm{be}}^{\epsilon}$ is the energy relaxation time of beam ion with electrons.

2.2.3 Runaway Electron, Electric Resistivity

When a uniform electric field $\boldsymbol{E}$ is applied to a plasma, the motion of a test electron is

$$m_{\mathrm{e}} \frac{\mathrm{d}\boldsymbol{v}}{\mathrm{d}t} = -e\boldsymbol{E} - \frac{1}{\tau_{\mathrm{ee}}(v)} m_{\mathrm{e}} \boldsymbol{v},$$

$$\frac{1}{\tau_{\mathrm{ee}}} = n_{\mathrm{e}} \sigma v = \frac{e^4 \ln \Lambda}{2\pi\epsilon_0^2 m_{\mathrm{e}}^2 v^3}.$$

The deceleration term decreases as v increases and its magnitude becomes smaller than the acceleration term $|-e\boldsymbol{E}|$ at a critical value v_{cr}. When $v > v_{\mathrm{cr}}$, the test particle is accelerated. The deceleration term becomes smaller and the velocity starts to increase without limit. Such an electron is called a *runaway electron*. The critical velocity is given by

$$\frac{m_{\mathrm{e}} v_{\mathrm{cr}}^2}{2e} = \frac{e^2 n \ln \Lambda}{4\pi\epsilon_0^2 E}. \qquad (2.33)$$

The necessary electric field for a given electron velocity to be v_{cr} is called *Dreicer field*. Taking $\ln \Lambda = 20$, we find

$$\frac{m_{\mathrm{e}} v_{\mathrm{cr}}^2}{2e} = 5 \times 10^{-16} \frac{n}{E}$$

where SI unit is used. When $n = 10^{19}\,\mathrm{m}^{-3}$, $E = 1\,\mathrm{V/m}$, electrons with energy larger than 5 keV become runaway electrons.

When an electric field less than Dreicer field is applied to a plasma, electrons are accelerated and decelerated by collisions with ions to be an equilibrium state as follows:

$$\frac{m_{\mathrm{e}}(v_{\mathrm{e}} - v_{\mathrm{i}})}{\tau_{\mathrm{ei}}} = -eE.$$

The current density j induced by the electric field becomes

$$j = -en_{\mathrm{e}}(v_{\mathrm{e}} - v_{\mathrm{i}}) = \frac{e^2 n_{\mathrm{e}} \tau_{\mathrm{ei}}}{m_{\mathrm{e}}} E.$$

The *specific electric resistivity* defined by $\eta j = E$ is

$$\eta = \frac{m_e \nu_{ei}}{n_e e^2} \tag{2.34}$$

When the Spitzer result of e-i collision frequency (2.16) is used, the value of $\eta_{spitzer}$ is given by [2.4]

$$\eta_{spitzer} = \frac{m_e \nu_{ei\|spitzer}}{n_e e^2} = \frac{(m_e)^{1/2} Z^2 (n_i/n_e) e^2 \ln \Lambda}{9.3 \times 10 \epsilon_0^2} (T_e)^{-3/2}$$

$$= 5.2 \times 10^{-5} Z \ \ln \Lambda \left(\frac{T_e}{e} \right)^{-3/2} \quad (\Omega \mathrm{m}).$$

The specific resistivity of a plasma with $T_e = 1\,\mathrm{keV}$, $Z = 1$ is $\eta = 3.3 \times 10^{-8}\,\Omega\mathrm{m}$ and is slightly larger than the specific resistivity of copper at 20°C, $1.8 \times 10^{-8}\,\Omega\mathrm{m}$. When a current density of j is induced, the power ηj^2 per unit volume contributes to electron heating. This heating mechanism of electron is called *Ohmic heating.*

2.3 Time and Space Scales in Plasmas

Various kinds of plasma characteristics have been described in this chapter. Characteristic time scales are

- period of electron plasma frequency $2\pi/\Pi_e$
- electron cyclotron period $2\pi/\Omega_e$
- ion cyclotron period $2\pi/|\Omega_i|$
- electron to ion collision time τ_{ei}
- ion to ion collision time τ_{ii}
- electron-ion thermal energy relaxation time τ_{ei}^{ϵ}

Alfvén velocity v_A, which is a propagation velocity of magnetic perturbation, is $v_A^2 = B^2/(2\mu_0 \rho_m)$ (ρ_m is mass density)(refer to Section 12.4.1).

- Alfvén transit time $\tau_A = L/v_A$ is a typical magnetohydrodynamic time scale, where L is a plasma size. In a medium with the specific resistivity η, electric field diffuses with time scale of resistive diffusion time $\tau_R = \mu_0 L^2/\eta$ (refer to Chapter 9).

Characteristic scales in length are

- Debye length λ_D
- electron Larmor radius $\rho_{\Omega e}$
- ion Larmor radius $\rho_{\Omega i}$
- plasma size L

- electron-ion collision mean free path λ_{ei}

The relations between space and time scales are

$$\lambda_D \Pi_e = v_{Te}, \quad \rho_{\Omega e} \Omega_e = v_{Te}, \quad \rho_{\Omega i}|\Omega_i| = v_{Ti},$$

$$\lambda_{ei}/\tau_{ei} \simeq 3^{1/2} v_{Te}, \quad \lambda_{ii}/\tau_{ii} \simeq 3^{1/2} v_{Ti}, \quad L/\tau_A = v_A,$$

where v_{Te}, v_{Ti} are the thermal velocities

$$v_{Te}^2 = T_e/m_e, \quad v_{Ti}^2 = T_i/m_i.$$

The drift velocity of the guiding center is

$$v_{drift} \sim \kappa T/eBL = v_T(\rho_\Omega/L).$$

Parameters of a typical D fusion grade plasma with $n_e = 10^{20}\,\mathrm{m}^{-3}$, $T_e = T_i = 10\,\mathrm{keV}$, $B = 5\,\mathrm{T}$, $L = 1\,\mathrm{m}$ are as follows:

$2\pi/\Pi_e = 11.1\,\mathrm{ps}$ $\quad \Pi_e/2\pi = 89.8\,\mathrm{GHz}$
$2\pi/\Omega_e = 7.1\,\mathrm{ps}$ $\quad \Omega_e/2\pi = 140\,\mathrm{GHz}$
$2\pi/|\Omega_i| = 26\,\mathrm{ns}$ $\quad |\Omega_i|/2\pi = 38\,\mathrm{MHz}$
$\tau_A = 0.13\,\mu\mathrm{s}$, $\quad \tau_{\perp ei} = 0.12\,\mathrm{ms}$, $\quad \tau_{\perp ii} = 7.2\,\mathrm{ms}$
$\tau_{ei}^{\epsilon} = 0.3\,\mathrm{s}$, $\quad \tau_R = 1.2 \times 10^3\,\mathrm{s}$.
$\lambda_D = 74.5\,\mu\mathrm{m}$, $\quad \rho_{\Omega e} = 47.6\,\mu\mathrm{m}$, $\quad \rho_{\Omega i} = 2.88\,\mathrm{mm}$
$\lambda_{ei} = 8.6\,\mathrm{km}$, $\quad \lambda_{ii} = 8.6\,\mathrm{km}$.

The ranges of scales in time and space extend to $\tau_R \Pi_e \sim 10^{14}$, $\lambda_{ei}/\lambda_D \sim 1.6 \times 10^8$ and the wide range of scales suggests the variety and complexity of plasma phenomena. Equations for plasma parameters are listed in Table 2.2.

Table 2.2 Equations for plasma parameters (SI unit).

$\Pi_{\rm e} = (n_{\rm e}e^2/m_{\rm e}\epsilon_0)^{1/2} = 5.64 \times 10^{11}(n_{\rm e\,20})^{1/2}$
$\Omega_{\rm e} = eB/m_{\rm e} = 1.76 \times 10^{11}B$
$\lvert\Omega_{\rm i}\rvert = ZeB/m_{\rm i} = 9.58 \times 10^7(Z/A)B$
$\tau_{\rm A}^{-1} = a/v_{\rm A} = 0.459 \times 10^{-6}(An_{\rm i\,20})^{1/2}a/B$
$\nu_{\rm ei\perp} \equiv 1/\tau_{\rm ei\perp} = Z^2 n_{\rm i}e^4 \ln \Lambda/(3^{1/2}6\pi\epsilon_0^2 m_{\rm e}^{1/2}T_{\rm e}^{3/2})$ $= 8.41 \times 10^9 Z^2(T_{\rm e}/e)^{-3/2}n_{\rm i\,20}$
$\nu_{\rm ii\perp} \equiv 1/\tau_{\rm ii\perp} = Z^4 n_{\rm i}e^4 \ln \Lambda/(3^{1/2}6\pi\epsilon_0^2 m_{\rm i}^{1/2}T_{\rm i}^{3/2})$ $= 2.0 \times 10^8 \frac{Z^4}{A^{1/2}}(T_{\rm i}/e)^{-3/2}n_{\rm i\,20}$
$\tau_{\rm R}^{-1} = \eta/(\mu_0 a^2) = 41.4Z \ln \Lambda(T_{\rm e}/e)^{-3/2}a^{-2}$
$\nu_{\rm ei\parallel} \equiv 1/\tau_{\rm ei\parallel} = \nu_{\rm ei\perp}/2, \qquad \nu_{\rm ei\parallel spitzer} \approx 0.7\nu_{\rm ei\parallel}$
$\lambda_{\rm D} = (\epsilon_0 T/n_{\rm e}e^2)^{1/2} = 7.45 \times 10^{-7}(T_{\rm e}/e)^{1/2}n_{\rm e\,20}^{-1/2}$
$\rho_{\Omega \rm e} = v_{\rm Te}/\Omega_{\rm e} = 2.38 \times 10^{-6}(T_{\rm e}/e)^{1/2}(1/B)$
$\rho_{\Omega \rm i} = v_{\rm Ti}/\Omega_{\rm i} = 1.02 \times 10^{-4}(AT_{\rm i}/e)^{1/2}(1/ZB)$
$\lambda_{\rm ei} = (3T_{\rm e}/m_{\rm e})^{1/2}\tau_{\rm ei\parallel} = 1.73 \times 10^{-4}(T_{\rm e}/e)^2(n_{\rm e\,20})^{-1}$
$v_{\rm A} = (B^2/\mu_0 n_{\rm i}m_{\rm i})^{1/2} = 2.18 \times 10^6 B/(An_{\rm i\,20})^{1/2}$
$v_{\rm Te} = (T_{\rm e}/m_{\rm e})^{1/2} = 4.19 \times 10^5(T_{\rm e}/e)^{1/2}$
$v_{\rm Ti} = (T_{\rm i}/m_{\rm i})^{1/2} = 9.79 \times 10^3(T_{\rm i}/Ae)^{1/2}$
$\eta = m_{\rm e}\nu_{\rm ei\parallel\, spitzer}/n_{\rm e}e^2 = 5.2 \times 10^{-5}Z \ln \Lambda(T_{\rm e}/e)^{-3/2}$ (Ωm) in case of $Zn_{\rm i} = n_{\rm e}$)
$(v_{\rm Ti}/v_{\rm A})^2 = \beta_{\rm i}/2, \qquad (v_{\rm A}/c)^2 = (\lambda_{\rm D}/\rho_{\Omega \rm e})^2(m_{\rm e}n_{\rm e}/m_{\rm i}n_{\rm i})$
$(\Pi_{\rm e}/\nu_{\rm ei\perp}) = (32.6/Z \ln \Lambda)n_{\rm e}\lambda_{\rm D}^3, \quad \lambda_{\rm D}\Pi_{\rm e} = v_{\rm Te} \quad S_{\rm R} \equiv \tau_{\rm R}/\tau_{\rm A}$

$\ln \Lambda = 20$ is assumed. T/e in eV and $n_{20} \equiv n(\mathrm{m}^3)/10^{20}$.

2.4 Mirror

Let us consider a mirror field as is shown in Figure 2.8, in which the magnetic field is weak at the center and strong at both ends of the mirror field. For simplicity the electric field is assumed to be zero. Since Lorentz force is perpendicular to the velocity, the magnetic field does not contribute the change of kinetic energy and

$$\frac{mv_\parallel^2}{2} + \frac{mv_\perp^2}{2} = \frac{mv^2}{2} = E = \text{const.}, \tag{2.35}$$

where $v_\parallel$ and $v_\perp$ are parallel and perpendicular components of velocity to the magnetic field. Since the magnetic moment is conserved, we find

$$v_\parallel = \pm\left(\frac{2}{m}E - v_\perp^2\right)^{1/2} = \pm\left(v^2 - \frac{2}{m}\mu_\mathrm{m}B\right)^{1/2}.$$

When the particle moves toward the open ends, the magnetic field becomes large and $v_\parallel$ becomes small or even zero. Since the force along the parallel direction to the magnetic field is $-\mu_\mathrm{m}\nabla_\parallel B$, both ends of the mirror field repulse charged particles as a mirror reflects light. The ratio of magnitude of magnetic field at the open end to the central value is called the *mirror ratio*:

$$R_\mathrm{M} = \frac{B_\mathrm{M}}{B_0}.$$

Let us denote the parallel and perpendicular components of the velocity at the mirror center by $v_{\parallel 0}$ and $v_{\perp 0}$, respectively. The value $v_\perp^2$ at the position of maximum magnetic field B_M is given by

$$v_{\perp \mathrm{M}}^2 = \frac{B_\mathrm{M}}{B_0} v_{\perp 0}^2.$$

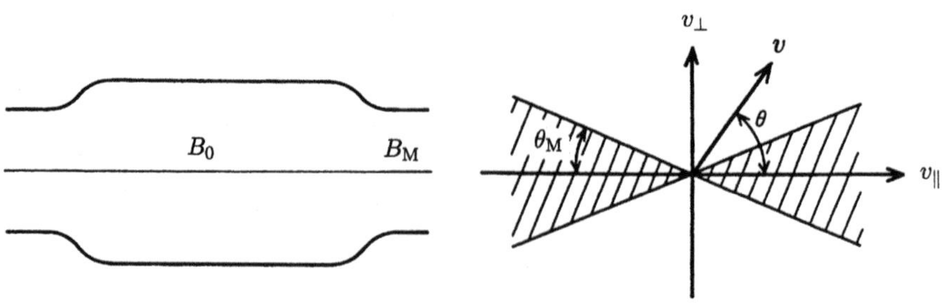

Figure 2.8 Mirror field and loss cone in $v_\parallel$ - $v_\perp$ space.

This value $v_{\perp M}^2$ cannot be larger than $v^2 = v_0^2$, so that the particle satisfying the following condition is reflected and is trapped in the mirror field:

$$\left(\frac{v_{\perp 0}}{v_0}\right)^2 > \frac{B_0}{B_M} = \frac{1}{R_M} \equiv \sin\theta_M. \tag{2.36}$$

Particles in the region where $\sin\theta \equiv v_{\perp 0}/v_0$ satisfies

$$\sin^2\theta \le \frac{1}{R_M}$$

are not trapped and the region is called *loss cone* in $v_{\parallel}$ - $v_{\perp}$ space (refer to Figure 2.8).

A particle trapped in a mirror field moves back and forth along the field line between both ends. The second action integral of this periodic motion

$$J_{\parallel} = m \oint v_{\parallel} \mathrm{d}l \tag{2.37}$$

is also another adiabatic invariant. $J_{\parallel}$ is called the *longitudinal adiabatic invariant.* As one makes the mirror length l shorter, $\langle v_{\parallel} \rangle$ increases (for $J_{\parallel} = 2m\langle v_{\parallel}\rangle l$ is conserved), and the particles are accelerated. This phenomena is called *Fermi acceleration.*

The line of magnetic force of the mirror is convex toward outside. The particles trapped by the mirror are subjected to curvature drift and gradient B drift, so that the trapped particles move back and forth, while drifting in θ direction. The orbit $(r,\ \theta)$ of the crossing point at $z = 0$ plane of back and forth movement is given by $J_{\parallel}(r, \theta, \mu_{\mathrm{m}}, E) = \mathrm{const.}$

2.5 Toroidal System

2.5.1 Magnetic Surface

A *line of magnetic force* satisfies the equations

$$\frac{\mathrm{d}x}{B_x} = \frac{\mathrm{d}y}{B_y} = \frac{\mathrm{d}z}{B_z} = \frac{\mathrm{d}l}{B} \tag{2.38}$$

where l is the length along a magnetic line of force $(\mathrm{d}l)^2 = (\mathrm{d}x)^2 + (\mathrm{d}y)^2 + (\mathrm{d}z)^2$. The *magnetic surface* $\psi(\boldsymbol{r}) = \mathrm{const.}$ is such that all magnetic lines of force lie upon that surface which satisfies the condition

$$(\nabla\psi(\boldsymbol{r})) \cdot \boldsymbol{B} = 0. \tag{2.39}$$

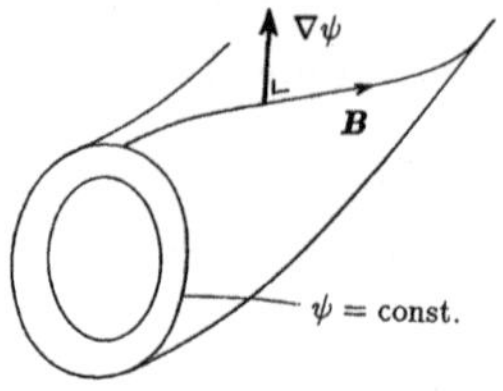

Figure 2.9 Magnetic surface $\psi =$ const., the normal $\nabla\psi$ and line of magnetic force.

The vector $\nabla\psi(\boldsymbol{r})$ is normal to the magnetic surface and must be orthogonal to $\boldsymbol{B}$ (see Figure 2.9).

In terms of cylindrical coordinates (r, θ, z) the magnetic field $\boldsymbol{B}$ is given by

$$B_r = \frac{1}{r}\frac{\partial A_z}{\partial \theta} - \frac{\partial A_\theta}{\partial z}, \quad B_\theta = \frac{\partial A_r}{\partial z} - \frac{\partial A_z}{\partial r}, \quad B_z = \frac{1}{r}\frac{\partial}{\partial r}(rA_\theta) - \frac{1}{r}\frac{\partial A_r}{\partial \theta}. \quad (2.40)$$

In the case of *axi-symmetric configuration* $(\partial/\partial\theta = 0)$,

$$\psi(r, z) = rA_\theta(r, z) \quad (2.41)$$

satisfies the condition (2.39) of magnetic surface; $B_r\partial(rA_\theta)/\partial r + B_\theta \cdot 0 + B_z\partial(rA_\theta)/\partial z = 0$.

The magnetic surface in the case of translational symmetry $(\partial/\partial z = 0)$ is given by

$$\psi(r, \theta) = A_z(r, \theta) \quad (2.42)$$

and the magnetic surface in the case of helical symmetry, in which ψ is the function of r and $\theta - \alpha z$ only, is given by

$$\psi(r, \theta - \alpha z) = A_z(r, \theta - \alpha z) + \alpha r A_\theta(r, \theta - \alpha z) \quad (2.43)$$

where α is helical pitch parameter.

2.5.2 Hamilonian Equation of Motion

The equation of motion of a particle with the mass m and the charge q in an electromagnetic field $\boldsymbol{E}$, $\boldsymbol{B}$ is

$$m\frac{\mathrm{d}^2\boldsymbol{r}}{\mathrm{d}t^2} = \boldsymbol{F} = q\left(\boldsymbol{E} + \frac{\mathrm{d}\boldsymbol{r}}{\mathrm{d}t} \times \boldsymbol{B}\right). \quad (2.44)$$

Since the Lorentz force of the second term in the right-hand side of Equation (2.44) is orthogonal to the velocity $\boldsymbol{v}$, the scalar product of Lorentz force and $\boldsymbol{v}$ is zero. The kinetic energy is given by

$$\frac{mv^2}{2} - \frac{m{v_0}^2}{2} = q\int_{t=t_0}^{t} \boldsymbol{E} \cdot \boldsymbol{v}\mathrm{d}t.$$

When the electric field is zero, the kinetic energy of the charged particle is conserved. When generalized coordinates q_i $(i = 1, 2, 3)$ are used, it is necessary to utilize the Lagrangian formulation. Lagrangian of a charged particle in the field with scalar and vector potentials ϕ, $\boldsymbol{A}$ is given by

$$L(q_i, \dot{q}_i, t) = \frac{mv^2}{2} + q\boldsymbol{v} \cdot \boldsymbol{A} - q\phi \tag{2.45}$$

where $\dot{q}_i$ is time derivetive of q_i.
Lagrangians in the orthogonal and cylindrical coordinates are given by

$$L(x, y, z, \dot{x}, \dot{y}, \dot{z}, t) = \frac{m}{2}(\dot{x}^2 + \dot{y}^2 + \dot{z}^2) + q(\dot{x}A_x + \dot{y}A_y + \dot{z}A_z) - q\phi,$$

$$L(r, \theta, z, \dot{r}, \dot{\theta}, \dot{z}, t) = \frac{m}{2}(\dot{r}^2 + (r\dot{\theta})^2 + \dot{z}^2) + q(\dot{r}A_r + r\dot{\theta}A_\theta + \dot{z}A_z) - q\phi$$

respectively. The Lagrange equation of motion is

$$\frac{\mathrm{d}}{\mathrm{d}t}\left(\frac{\partial L}{\partial \dot{q}_i}\right) - \frac{\partial L}{\partial q_i} = 0. \tag{2.46}$$

Canonical transformation is more general than the coordinates transformation. Hamiltonian equation of motion is conserved with respect to canonical transformation. In this formulation we introduce momentum coordinates (p_i), in addition to the space coordinates (q_i), defined by

$$p_i \equiv \frac{\partial L}{\partial \dot{q}_i} \tag{2.47}$$

and treat p_i as independent variables. Then we can express $\dot{q}_i$ as a function of (q_j, p_j, t) from (2.47) as follows:

$$\dot{q}_i = \dot{q}_i(q_j, p_j, t). \tag{2.48}$$

The Hamiltonian $H(q_i, p_i, t)$ is given by

$$H(q_i, p_i, t) \equiv -L(q_i, \dot{q}_i(q_j, p_j, t), t) + \sum_i p_i \dot{q}_i(q_j, p_j, t). \tag{2.49}$$

The x component of momentum p_x in the orthogonal coordinates and θ component p_θ in the cylindrical coordinates are written as examples as follows:

$$p_x = m\dot{x} + qA_x, \qquad \dot{x} = (p_x - qA_x)/m,$$

$$p_\theta = mr^2\dot{\theta} + qrA_\theta, \qquad \dot{\theta} = (p_\theta - qrA_\theta)/(mr^2).$$

Hamiltonian in the orthogonal coordinates is

$$H = \frac{1}{2m}((p_x - qA_x)^2 + (p_y - qA_y)^2 + (p_z - qA_z)^2) + q\phi(x, y, z, t)),$$

and Hamiltonian in the cylindrical coordinates is

$$H=\frac{1}{2m}\left((p_r-qA_r)^2+\frac{(p_\theta-qrA_\theta)^2}{r^2}+(p_z-qA_z)^2\right)+q\phi(r,\theta,z,t).$$

The variation of Lagrangian L is given by

$$\delta L=\sum_i\left(\frac{\partial L}{\partial q_i}\delta q_i+\frac{\partial L}{\partial \dot{q}_i}\delta\dot{q}_i\right)=\sum_i(\dot{p}_i\delta q_i+p_i\delta\dot{q}_i)=\delta(\sum_i p_i\dot{q}_i)+\sum_i(\dot{p}_i\delta q_i-\dot{q}_i\delta p_i)$$

and

$$\delta(-L+\sum_i p_i\dot{q}_i)=\sum_i(\dot{q}_i\delta p_i-\dot{p}_i\delta q_i), \qquad \delta H(q_i,p_i,t)=\sum_i(\dot{q}_i\delta p_i-\dot{p}_i\delta q_i).$$

Accordingly, the Hamiltonian equation of motion is reduced to

$$\frac{\mathrm{d}q_i}{\mathrm{d}t}=\frac{\partial H}{\partial p_i}, \qquad \frac{\mathrm{d}p_i}{\mathrm{d}t}=-\frac{\partial H}{\partial q_i}. \tag{2.50}$$

Equation (2.50) in the orthogonal coordinates is

$$\frac{\mathrm{d}x}{\mathrm{d}t}=\frac{p_x-qA_x}{m}, \qquad \frac{\mathrm{d}p_x}{\mathrm{d}t}=\frac{q}{m}\frac{\partial \boldsymbol{A}}{\partial x}\cdot(\boldsymbol{p}-q\boldsymbol{A})-q\frac{\partial\phi}{\partial x},$$

$$m\frac{\mathrm{d}^2x}{\mathrm{d}t^2}=\frac{\mathrm{d}p_x}{\mathrm{d}t}-q\frac{\mathrm{d}A_x}{\mathrm{d}t}=q\left[\left(\boldsymbol{v}\cdot\frac{\partial\boldsymbol{A}}{\partial x}\right)-\frac{\partial\phi}{\partial x}-\left(\frac{\partial A_x}{\partial t}+(\boldsymbol{v}\cdot\nabla)A_x\right)\right]$$
$$=q(\boldsymbol{E}+\boldsymbol{v}\times\boldsymbol{B})_x$$

and it was shown that (2.50) is equivalent to (2.44).

When H does not depend on t explicitly (when ϕ , $\boldsymbol{A}$ do not depend on t),

$$\frac{\mathrm{d}H(q_i,p_i)}{\mathrm{d}t}=\sum_i\left(\frac{\partial H}{\partial q_i}\frac{\mathrm{d}q_i}{\mathrm{d}t}+\frac{\partial H}{\partial p_i}\frac{\mathrm{d}p_i}{\mathrm{d}t}\right)=0,$$

$$H(q_i,p_i)=\text{const.} \tag{2.51}$$

is one integral of Hamiltonian equations. This integral expresses the conservation of energy.

When the electromagnetic field is axially symmetric, p_θ is constant due to $\partial H/\partial\theta=0$ as is seen in (2.50) and

$$p_\theta=mr^2\dot{\theta}+qrA_\theta=\text{const.} \tag{2.52}$$

This indicates conservation of the angular momentum. In the case of translational symmetry ($\partial/\partial z=0$), we have

$$p_z=m\dot{z}+qA_z=\text{const.} \tag{2.53}$$

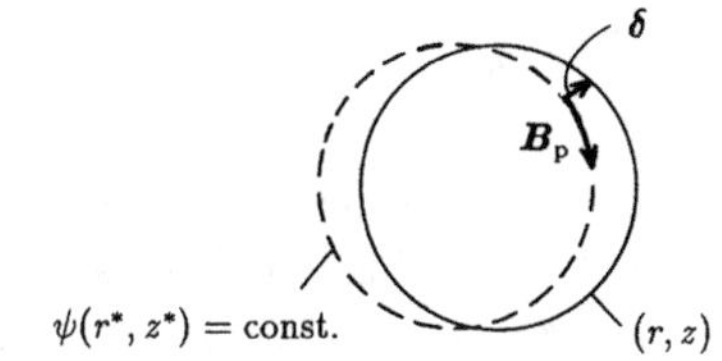

Figure 2.10 Magnetic surface (dotted line) and particle orbit (solid line).

2.5.3 Particle Orbit in an Axially Symmetric System

The coordinates (r^*, θ^*, z^*) on a magnetic surface of an axially symmetric field satisfy

$$\psi = r^* A_\theta(r^*, z^*) = c_{\mathrm{M}} \qquad (c_{\mathrm{M}} \text{ is constant}).$$

On the other hand, the coordinates (r, θ, z) of a particle orbit are given by the conservation of the angular momentum (2.52) as follows:

$$rA_\theta(r, z) + \frac{m}{q} r^2 \dot{\theta} = \frac{p_\theta}{q} = \text{const.}$$

If c_{M} is chosen to be $c_{\mathrm{M}} = p_\theta / q$, the relation between the magnetic surface and the particle orbit is reduced to

$$rA_\theta(r, z) - r^* A_\theta(r^*, z^*) = -\frac{m}{q} r^2 \dot{\theta}.$$

The distance $\boldsymbol{\delta}$ (Figure 2.10) between the magnetic surface and the orbit is given by

$$\boldsymbol{\delta} = (r - r^*)\boldsymbol{e}_r + (z - z^*)\boldsymbol{e}_z,$$

$$\boldsymbol{\delta} \cdot \nabla(rA_\theta) = -\frac{m}{q} r^2 \dot{\theta},$$

where $\boldsymbol{e}_r$ and $\boldsymbol{e}_z$ are unit vectors in the directions of r and z, respectively. Since $rB_r = -\partial(rA_\theta)/\partial z, \quad rB_z = \partial(rA_\theta)/\partial r$, this expression is reduced to

$$[-(z - z^*)B_r + (r - r^*)B_z] = -\frac{m}{q} r\dot{\theta}.$$

The left-hand side of the above equation is θ component of the vector product of $\boldsymbol{B}_{\mathrm{p}} = (B_r, B_z)$ and $\boldsymbol{\delta} = (r - r^*, z - z*)$ and

$$(\boldsymbol{B}_{\mathrm{p}} \times \boldsymbol{\delta})_\theta = -\frac{m}{q} r\dot{\theta}.$$

Denote the magnitude of poloidal component $\boldsymbol{B}_{\mathrm{p}}$ (component within (rz) plane) of $\boldsymbol{B}$ by B_{p}. Then we find the relation $-B_{\mathrm{p}}\delta = -(m/q)v_\theta \ (v_\theta = r\dot{\theta})$ and we have

$$\delta = \frac{mv_\theta}{qB_\text{p}} = \rho_{\Omega\text{p}}. \tag{2.54}$$

This value is equal to the Larmor radius corresponding to the magnetic field B_p and the tangential velocity v_θ. This is called *poloidal Larmor radius.* If c_M is chosen to be $c_\text{M} = (p_\theta - m\langle rv_\theta\rangle)/q$ ($\langle rv_\theta\rangle$ is the average of rv_θ), we find

$$\delta = \frac{m}{qB_\text{p}}\left(v_\theta - \frac{\langle rv_\theta\rangle}{r}\right).$$

When B_p is not zero in an axisymmetric system, all charged particles cannot deviate from the magnetic surface more than poloidal Larmor radius, if collisions do not occur.

2.5.4 Drift of Guiding Center in Toroidal Field (Circulating Particle and Banana)

Let us consider the drift of the guiding center of a charged particle in a simple toroidal field ($B_r = 0,\ B_\varphi = B_0R_0/R,\ B_z = 0$) in terms of cylindrical coordinates (R, φ, z). The φ component B_φ is called the toroidal field and B_φ decreases in the form of $1/R$ outward. The lines of magnetic force are circles around the z axis. The z axis is called the *major axis* of the torus. As was described in Section 2.4, the drift velocity of the guiding center is given by

$$\boldsymbol{v}_\text{G} = v_\parallel \boldsymbol{e}_\varphi + \frac{m}{qB_\varphi R}\left(v_\parallel^2 + \frac{v_\perp^2}{2}\right)\boldsymbol{e}_z$$

where $\boldsymbol{e}_\varphi$ is the unit vector in the φ direction (see Figure 2.11). Particles in this simple torus run fast in the toroidal direction and drift slowly in the z direction with the velocity of

$$v_\text{dr} = \frac{m}{qB_0R_0}\left(v_\parallel^2 + \frac{v_\perp^2}{2}\right) \sim \left(\frac{\rho_\Omega}{R_0}\right)v. \tag{2.55}$$

This drift is called *toroidal drift.* Ions and electrons drift in opposite direction along the z axis. As a consequence of the resultant charge

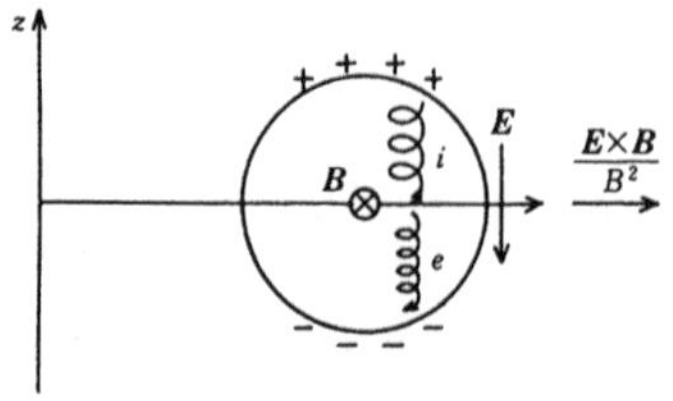

Figure 2.11 Toroidal drift.

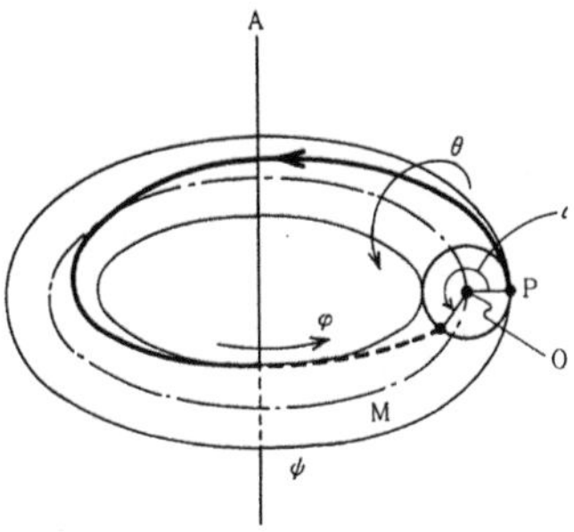

Figure 2.12 The major axis A, the minor axis M of toroidal field, and rotational transform angle ι.

separation, an electric field $\boldsymbol{E}$ is induced and both ions and electrons drift outward by $\boldsymbol{E} \times \boldsymbol{B}/B^2$ drift. Consequently, a simple toroidal field cannot confine a plasma (Figure 2.11), unless the separated charges are cancelled or short-circuited by an appropriate method. If lines of magnetic force connect the upper and lower regions as is shown in Figure 2.12, the separated charges can be short-circuited, as the charged particles can move freely along the lines of force. If a current is induced in a toroidal plasma, the component of magnetic field around the *magnetic axis* (which is also called the *minor axis*) is introduced as is shown in Figure 2.12. This component $\boldsymbol{B}_{\rm p}$ is called *poloidal magnetic field.* The radius R of the magnetic axis is called the *major radius* of torus and the radius a of the plasma cross-section is called the *minor radius.* Denote the radial coordinate in plasma cross-section by r. When a line of magnetic force circles the major axis of torus and come back to cross the plane P, the cross point rotates around the minor axis O by an angle ι in P plane, there is the following relation:

$$\frac{r\iota}{2\pi R} = \frac{B_{\rm p}}{B_\varphi}.$$

The angle ι is called the *rotational transform angle* and is given by

$$\frac{\iota}{2\pi} = \frac{R}{r}\frac{B_{\rm p}}{B_\varphi}. \qquad (2.56)$$

The inverse of $\iota/2\pi$ is called the *safety factor,*

$$q_{\rm s} \equiv \frac{2\pi}{\iota} = \frac{r}{R}\frac{B_\varphi}{B_{\rm p}}.$$

$A \equiv R/a$ is called the *aspect ratio* and $\epsilon \equiv a/R$ is called the *inverse aspect ratio.*

Guiding Center of Circulating Particles

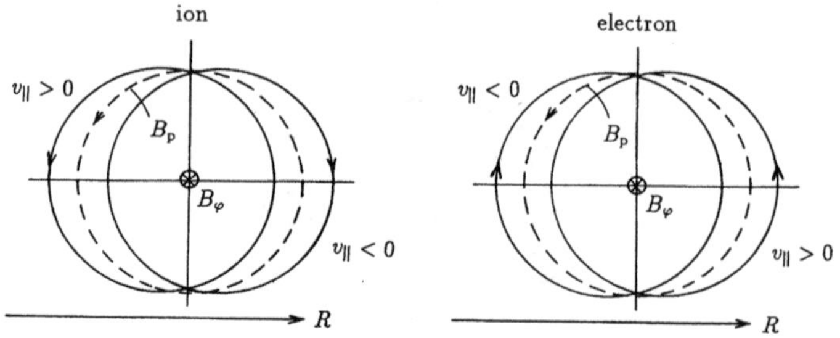

Figure 2.13 Orbits (solid lines) of guiding center of circulating ions, electrons and magnetic surfaces (dotted lines).

When a particle circulates torus with the velocity of $v_\parallel$, it takes $T \approx 2\pi R_0/v_\parallel$. Accordingly, the particle rotates around the minor axis with the angular velocity of

$$\omega = \frac{\iota}{T} = \frac{\iota v_\parallel}{2\pi R_0}$$

and drifts in z direction with the velocity of $v_{\rm dr}$. Introducing $x = R - R_0$ coordinate, the orbit of the guiding center of the particle is given by

$$\frac{{\rm d}x}{{\rm d}t} = -\omega z, \qquad \frac{{\rm d}z}{{\rm d}t} = \omega x + v_{\rm dr}.$$

The solution is

$$\left(x + \frac{v_{\rm dr}}{\omega}\right)^2 + z^2 = r^2.$$

If a rotational transform angle is introduced, the orbit becomes a closed circle and the center of orbit circle deviates from the center of magnetic surface by the amount of

$$\Delta = -\frac{v_{\rm dr}}{\omega} = -\frac{mv_\parallel}{qB_0}\frac{2\pi}{\iota}\left(1 + \frac{v_\perp^2}{2v_\parallel^2}\right), \tag{2.57}$$

$$|\Delta| \sim \rho_\Omega \left(\frac{2\pi}{\iota}\right) = \rho_\Omega q_{\rm s}$$

where ρ_Ω is Larmor radius. As is seen in Figure 2.13, the sign of the deviation is $\Delta < 0$ for the case of $v_\parallel > 0,\ q > 0$ (ion) since $v_{\rm dr} > 0,\ \omega > 0$ and the sign becomes $\Delta > 0$ for the case of $v_\parallel < 0$ (opposit to $v_\parallel > 0$) $q > 0$ (ion).

Guiding Center of Banana Particles

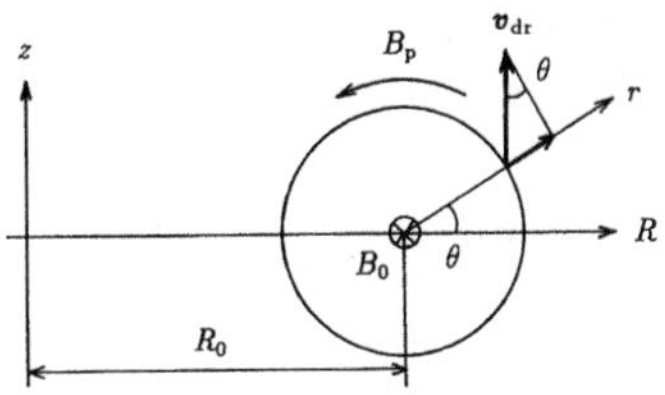

Figure 2.14 (r, θ) coordinates.

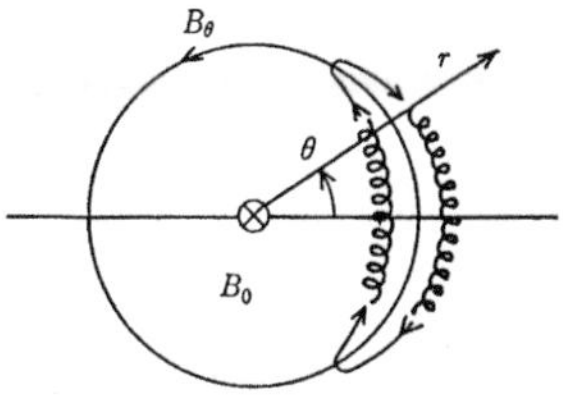

Figure 2.15 Projection of banana orbit of ion to poloidal plane (r, θ).

In the case of $|B_\varphi| \gg |B_{\rm p}|$, the magnitude of toroidal field is nearly equal to B_φ and

$$B = \frac{B_0 R_0}{R} = \frac{B_0}{1 + (r/R)\cos\theta} \simeq B_0 \left(1 - \frac{r}{R_0}\cos\theta\right).$$

Denote the length along the line of magnetic force by l, and denote the projection of a location on the magnetic line of force to (R, z) plane by the coordinates (r, θ) as is shown in Figure 2.14. Since the following relations

$$\theta = \frac{\varphi}{q_s} = \frac{l}{Rq_{\rm s}} = \left(\frac{B_{\rm p}}{rB_0}\right) l = \kappa l, \qquad (\theta - \frac{\varphi}{q_{\rm s}} = 0)$$

hold, we find

$$B = B_0 \left(1 - \frac{r}{R_0}\cos(\kappa l)\right).$$

If $v_\parallel$ (parallel component to magnetic field) is much smaller than $v_\perp$ component and satisfies the condition

$$\frac{v_\perp^2}{v^2} > 1 - \frac{r}{R}, \qquad \frac{v_\parallel^2}{v^2} < \frac{r}{R}, \tag{2.58}$$

the particle is trapped outside in the weak region of the magnetic field due to the mirror effect as is described in Section 2.4. (The mirror ratio is $(1/R)/(1/(R+r))$). This particle is called the *trapped particle.* The circulating particle that is not trapped is also called the *untrapped particle.*

Since $v_\parallel^2 \ll v_\perp^2$ for the trapped particle, the r component of the toroidal drift $v_{\rm dr}$ of the trapped particle is given by

$$\frac{{\rm d}r}{{\rm d}t} = v_{\rm dr}\sin\theta = \frac{m}{qB_0}\frac{v_\perp^2}{2R}\sin\theta.$$

The parallel motion of the guiding center is given by (refer to (2.7))

$$\frac{{\rm d}v_\parallel}{{\rm d}t} = -\frac{\mu_{\rm m}}{m}\frac{\partial B}{\partial l} = -\frac{\mu_{\rm m}}{m}\frac{r}{R}\kappa B_0\sin\kappa l = -\frac{v_\perp^2}{2R}\frac{B_{\rm p}}{B_0}\sin\theta.$$

The solution is

$$\frac{\rm d}{{\rm d}t}\left(r + \frac{m}{qB_{\rm p}}v_\parallel\right) = 0,$$

$$r - r_0 = -\frac{m}{qB_{\rm p}}v_\parallel. \tag{2.59}$$

Here $r = r_0$ indicates the radial coordinate of the turning point by the mirror effect. Since the orbit is of banana shape, the trapped particle is also called the *banana particle* (see Figure 2.15). The width of banana $\Delta_{\rm b}$ is given by

$$\Delta_{\rm b} = \frac{m}{qB_{\rm p}}v_\parallel \sim \frac{mv}{qB_0}\frac{v_\parallel}{v}\frac{B_0}{B_{\rm p}} \sim \frac{B_0}{B_{\rm p}}\left(\frac{r}{R}\right)^{1/2}\rho_\Omega \sim \left(\frac{R}{r}\right)^{1/2} q_{\rm s}\rho_\Omega. \tag{2.60}$$

2.5.5 Effect of Longitudinal Electric Field on Banana Orbit

In the tokamak configuration, a toroidal electric field is applied in order to induce the plasma current. The guiding center of a particle drift by $\boldsymbol{E}\times\boldsymbol{B}/B^2$, but the banana center moves in a different way. The toroidal electric field can be described by

$$E_\varphi = -\frac{\partial A_\varphi}{\partial t}$$

in (R, φ, z) coordinates. Since angular momentum is conserved, we can write

$$R(mR\dot\varphi + qA_\varphi) = {\rm const.}$$

Taking the average of the foregoing equation over a Larmor period, and using the relation

$$\langle R\dot\varphi\rangle = \frac{B_\varphi}{B}v_\parallel$$

we find

$$R\left(mv_\parallel\frac{B_\varphi}{B} + qA_\varphi\right) = {\rm const.} \tag{2.61}$$

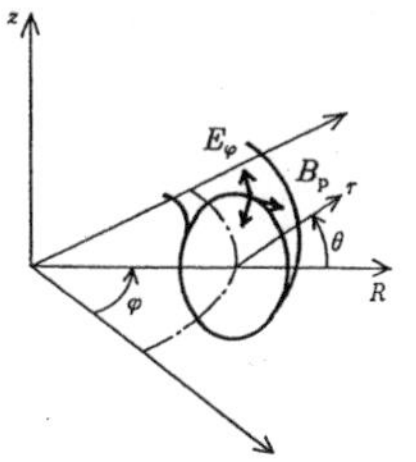

Figure 2.16 Coordinate system, in which Ware pinch is analyzed.

For particles in banana motion ($v_\parallel \ll v_\perp$), $v_\parallel$ becomes 0 at the turning points of the banana orbit. The displacement of a turning point (R, Z) per period Δt is obtained from

$$0 = \Delta(RA_\varphi(R,Z)) = \Delta r \frac{\partial}{\partial r} RA_\varphi + \Delta t \frac{\partial}{\partial t} RA_\varphi$$

where r is the radial coordinate of the magnetic surface. The differentiations of $RA_\varphi = \text{const.}$ with respect to φ and θ are zero, since $RA_\varphi = \text{const.}$ is the magnetic surface. By means of the relation

$$\frac{1}{R}\frac{\partial}{\partial r}(RA_\varphi) = \frac{1}{R}\left(\frac{\partial R}{\partial r}\frac{\partial(RA_\varphi)}{\partial R} + \frac{\partial Z}{\partial r}\frac{\partial(RA_\varphi)}{\partial Z}\right)$$

$$= \cos\theta B_Z - \sin\theta B_R = \boldsymbol{B}_\mathrm{p} \cdot \boldsymbol{e}_\theta = -B_\mathrm{p}$$

we obtaine the drift velocity

$$\frac{\Delta r}{\Delta t} = -\frac{E_\varphi}{B_\mathrm{p}}, \tag{2.62}$$

where $\boldsymbol{e}_\theta$ is the unit vector in the θ direction (refer to Figure 2.16). When the sign of B_p produced by the current induced by the electric field E_φ is taken into account ($B_\mathrm{p} > 0$, $E_\varphi > 0$ in the case of Figure 2.16), the sign of $\Delta r/\Delta t$ is negative and the banana center moves inward. Since $|B_\mathrm{p}| \ll |B_\varphi| \simeq B$, the drift velocity of the banana center is $(B/B_\mathrm{p})^2$ times as fast as the drift velocity $E_\varphi B_\mathrm{p}/B^2$ of the guiding center of the particle. This phenomena is called *Ware's pinch* [2.6].

2.5.6 Orbit of Guiding Center and Magnetic Surface

The velocity of guiding center was derived in Section 2.1.2 as follows:

$$\boldsymbol{v}_\mathrm{G} = v_\parallel \boldsymbol{b} + \frac{1}{B}(\boldsymbol{E} \times \boldsymbol{b}) + \frac{mv_\perp^2/2}{qB^2}(\boldsymbol{b} \times \nabla B) + \frac{mv_\parallel^2}{qB^2}(\boldsymbol{b} \times (\boldsymbol{b} \cdot \nabla)\boldsymbol{B}) \tag{2.63}$$

$$\mu_\mathrm{m} = mv_\perp^2/(2B) = \text{const.}$$

When the electric field $\boldsymbol{E}$ is static and is expressed by $\boldsymbol{E} = -\nabla\phi$, the conservation of energy

$$\frac{m}{2}(v_\parallel^2 + v_\perp^2) + q\phi = W$$

holds. Then $v_\parallel$ is expressed by

$$v_\parallel = \pm\left(\frac{2}{m}\right)^{1/2}(W - q\phi - \mu_{\rm m}B)^{1/2}.$$

Noting that $v_\parallel$ is a function of the coodinates, we can write

$$\nabla\times(mv_\parallel\boldsymbol{b}) = mv_\parallel\nabla\times\boldsymbol{b} + \nabla(mv_\parallel)\times\boldsymbol{b}$$

$$= mv_\parallel\nabla\times\boldsymbol{b} + \frac{1}{v_\parallel}(-q\nabla\phi - \mu_{\rm m}\nabla B)\times\boldsymbol{b}$$

and

$$\frac{v_\parallel}{qB}\nabla\times(mv_\parallel\boldsymbol{b}) = \frac{mv_\parallel^2}{qB}\nabla\times\boldsymbol{b} + \frac{1}{B}(\boldsymbol{E}\times\boldsymbol{b}) + \frac{mv_\perp^2/2}{qB^2}(\boldsymbol{b}\times\nabla B).$$

Then (2.63) for $\boldsymbol{v}_{\rm G}$ is reduced to [2.7]

$$\boldsymbol{v}_{\rm G} = v_\parallel\boldsymbol{b} + \left(\frac{v_\parallel}{qB}\nabla\times(mv_\parallel\boldsymbol{b}) - \frac{mv_\parallel^2}{qB}\nabla\times\boldsymbol{b}\right) + \frac{mv_\parallel^2}{qB^2}(\boldsymbol{b}\times(\boldsymbol{b}\cdot$$

$$= v_\parallel\boldsymbol{b} + \frac{v_\parallel}{qB}\nabla\times(mv_\parallel\boldsymbol{b}) - \frac{mv_\parallel^2}{qB}(\nabla\times\boldsymbol{b} - \boldsymbol{b}\times(\boldsymbol{b}\cdot\nabla)\boldsymbol{b}).$$

As the relation $\nabla(\boldsymbol{b}\cdot\boldsymbol{b}) = 2(\boldsymbol{b}\cdot\nabla)\boldsymbol{b} + 2\boldsymbol{b}\times(\nabla\times\boldsymbol{b}) = 0$ ($(\boldsymbol{b}\cdot\boldsymbol{b}) = 1$) holds, the third term in the right-hand side of the equation for $\boldsymbol{v}_{\rm G}$ becomes $\nabla\times\boldsymbol{b} - \boldsymbol{b}\times(\boldsymbol{b}\cdot\nabla)\boldsymbol{b} = (\nabla\times\boldsymbol{b}) - (\nabla\times\boldsymbol{b})_\perp = (\nabla\times\boldsymbol{b})_\parallel = (\boldsymbol{b}\cdot(\nabla\times\boldsymbol{b}))\boldsymbol{b}$. Accordingly, within accuracy of the 1st order of the Larmor radius/characteristic length of $\boldsymbol{b}$, the velocity of the guiding center is reduced to

$$\boldsymbol{v}_{\rm G} = \left(v_\parallel + \frac{mv_\parallel^2}{qB}(\boldsymbol{b}\cdot\nabla\times\boldsymbol{b})\right)\boldsymbol{b} + \frac{mv_\parallel}{qB}\nabla\times(v_\parallel\boldsymbol{b})$$

$$= \frac{1}{1 + (mv_\parallel/qB)\boldsymbol{b}\cdot\nabla\times\boldsymbol{b}}\left(v_\parallel\boldsymbol{b} + \frac{mv_\parallel}{qB}\nabla\times(v_\parallel\boldsymbol{b})\right). \tag{2.64}$$

The first factor in the right-hand side of (2.64) is necessary in order to conserve the phase space volume in the Lagrange-Hamiltonian formulation of the guiding center motion [2.8]. Since $\nabla\times\boldsymbol{B} = B\nabla\times\boldsymbol{b} + \nabla B\times\boldsymbol{b} = \mu_0\boldsymbol{j}$, we have $\boldsymbol{b}\cdot\nabla\times\boldsymbol{b} = \mu_0 j_\parallel/B$. The second term of the denominator is usually very small compared with 1 (zero in the case of $j_{\parallel} = 0$). If the second term of the denominator can be neglected, (2.63) for $\boldsymbol{v}_{\mathrm{G}}$ is reduced to [2.7]

$$\frac{\mathrm{d}\boldsymbol{r}_{\mathrm{G}}}{\mathrm{d}t} = \frac{v_{\parallel}}{B}\nabla \times \left(\boldsymbol{A} + \frac{mv_{\parallel}}{qB}\boldsymbol{B}\right). \tag{2.65}$$

The orbit of (2.64) and (2.65) are identical when the magnetic field does not depend on time.
The orbit of the guiding center is equal to the field line of the magnetic field $\boldsymbol{B}^* = \nabla \times \boldsymbol{A}^*$ with the vector potential

$$\boldsymbol{A}^* \equiv \boldsymbol{A} + \frac{mv_{\parallel}}{qB}\boldsymbol{B}.$$

By the reason analogous to that in Section 2.5.1, the *orbit surface of drift motion* of the guiding center is given by

$$rA^*_{\theta}(r, z) = \mathrm{const.} \tag{2.66}$$

Problems

1. Density of Gas at 0.1 Pascal and 0 °C Calculate the number density of ideal gas at 0.1 Pascal and 0 °C. (1 atmosphere = 760 mmHg = $13.5951 \times 76 \times 980.665 \times 10^{-1}$ Pa = 1.01325×10^5 Pa, N_{A}(Avogadro's number = 6.0221367×10^{23}) (0.1 Pa = 0.75 mtorr).

2. Electron Plasma Frequency and Debye Length Calculate the Debye length and the plasma electron plasma frequency in the case of fusion core plasma with the density of $n_{\mathrm{e}} = 10^{20}\mathrm{m}^{-3}$ and $T_{\mathrm{e}} = 10\mathrm{keV}$. Although the electron density and electron temperature of the ionosphere depend on the height and time, typical parameters are $n_{\mathrm{e}} \sim 10^{12}\mathrm{m}^{-3}$ and $T_{\mathrm{e}} \sim 0.2\mathrm{eV}$. Calculate the same parameters of plasma in ionosphere.

3. Larmor Radius Calculate the Larmor radius of fast deuteron D^+ in the magnetic field of $B = 5\mathrm{T}$ produced by charge exchange of fast neutral with the energy of 1 MeV (10^6 eV) injected by NBI in the direction at the angle 30° with magnetic field. Refer to (2.3).

4. Curvature Drift and ∇B Drift Confirm that drift velocities of curvature drift and ∇B drift are approximately $(\rho_{\Omega}/R)v_{\mathrm{T}}$, where R is the curvature of field line. Refer to (2.6).

5. Polarization Drift Polarization drift velocity is $\boldsymbol{v}_{\mathrm{p}} = (m/qB) \times(\partial \boldsymbol{E}/\partial t)/B$ and $\boldsymbol{v}_E$ is $E \times B$ drift velocity. Confirm the relation $(\partial/\partial t)(mv_E^2/2) - q\boldsymbol{v}_{\mathrm{p}} \cdot \boldsymbol{E} = 0$. Refer to (2.9).

6. Pondromotive Force Assume a laser beam with power of 1 Peta Watt = 10^{15} W is focused to a plasma with spot size of 0.02 mm radius. Estimate the light pressure $\epsilon\langle E^2\rangle/2$. The light intensity I_{L} is $I_{\mathrm{L}}(\mathrm{W/m^2}) = (\epsilon\langle E^2\rangle/2)c(\mathrm{J/m^3})\mathrm{m/s}$, where c is light velocity. Refer to (2.11).

7. Neutral Beam Injection (NBI) Fast neutral injected by NBI is charge exchanged and becomes fast ion. Fast ion loses the energy by Coulomb collisions with plasma ions and electrons. The rate of change of fast ion's energy, that is, heating rate of plasma by a fast ion is given by (2.30). Derive this formula.

8. A Period of Bounce Motion in Mirror Estimate a period of bounce motion of trapped particle with thermal velocity $v_{\rm T}$ in a mirror with the length L and the mirror ratio $R_{\rm M}$. Refer to (2.36).

9. A Period of Banana Motion Let the poloidal angle of turning point of banana particle be ± 0.5 radian in tokamak. Estimate a period of banana motion in this case. (Major radius: R, inverse aspect ratio: $\epsilon = A^{-1}$, safety factor: q, thermal velocity: $v_{\rm T}$). Refer to Section 2.5.4.

10. Ion Orbit and Magnetic Surface Calculate the maximum deviation of orbit of a deuteron with 10 keV from the magnetic surface with radius $r = 1\,$m of tokamak carrying the plasma current $I_{\rm p} = 15\,$MA with flat current profile with the minor radius of plasma $a = 2$m and the major radius $R = 6.2\,$m. Assume that collisons do not occur. Refer to (2.54).

3

Magnetohydrodynamics

3.1 Magnetohydrodynamic Equations for Two Fluids

Plasmas can be described as magnetohydrodynamic two fluids of ions and electrons with mass densities $\rho_{\rm mi}$, $\rho_{\rm me}$, charge density ρ, current density $\boldsymbol{j}$, flow velocities $\boldsymbol{V}_{\rm i}$, $\boldsymbol{V}_{\rm e}$, and pressures $p_{\rm i}$, $p_{\rm e}$. These physical quantities can be expressed by appropriate averages in velocity space by use of the phase space distribution functions $f_i(\boldsymbol{r}, \boldsymbol{v}, t)$ of ions and electrons. The number of ions $\mathrm{d}N_{\rm i}$ in a small phase space volume $\mathrm{d}x\mathrm{d}y\mathrm{d}z\mathrm{d}v_x\mathrm{d}v_y\mathrm{d}v_z$ is given by

$$\mathrm{d}N_{\rm i} = f_i(\boldsymbol{r}, \boldsymbol{v}, t)\mathrm{d}x\mathrm{d}y\mathrm{d}z\mathrm{d}v_x\mathrm{d}v_y\mathrm{d}v_z.$$

Phase space distribution function will be described in detail in Chapter 10. The number density of ion $n_{\rm i}$, the ion mass density $\rho_{\rm mi}$, and the ion flow velocity $\boldsymbol{V}_{\rm i}(\boldsymbol{r}, t)$ at (x, y, z, t) are expressed as follows:

$$n_{\rm i}(\boldsymbol{r}, t) = \int f_{\rm i}(\boldsymbol{r}, \boldsymbol{v}, t)\mathrm{d}\boldsymbol{v}, \tag{3.1}$$

$$\rho_{\rm mi}(\boldsymbol{r}, t) = m_{\rm i} n_{\rm i}(\boldsymbol{r}, t), \tag{3.2}$$

$$\boldsymbol{V}(\boldsymbol{r}, t) = \frac{\int \boldsymbol{v} f_{\rm i}(\boldsymbol{r}, \boldsymbol{v}, t)\mathrm{d}\boldsymbol{v}}{\int f_{\rm i}(\boldsymbol{r}, \boldsymbol{v}, t)\mathrm{d}\boldsymbol{v}} = \frac{1}{n_{\rm i}(\boldsymbol{r}, t)} \int \boldsymbol{v} f_{\rm i}(\boldsymbol{r}, \boldsymbol{v}, t)\mathrm{d}\boldsymbol{v}. \tag{3.3}$$

We have the same expressions for electrons as those for ions. Since magnetohydrodynamics will treat average quantities in the velocity space, phenomena associated with the shape of the phase space distribution function on velocity (refer to Chapter 10) will be neglected. However, the independent variables are $\boldsymbol{r}, t$ only and it is possible to analyze geometrically complicated configurations.

Magnetohydrodynamic equations are the following:

$$\frac{\partial n_{\rm e}}{\partial t} + \nabla \cdot (n_{\rm e}\boldsymbol{V}_{\rm e}) = 0, \tag{3.4}$$

$$\frac{\partial n_{\rm i}}{\partial t} + \nabla \cdot (n_{\rm i}\boldsymbol{V}_{\rm i}) = 0, \tag{3.5}$$

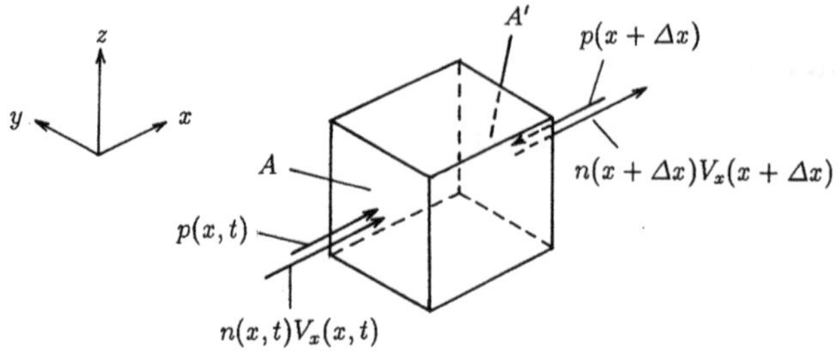

Figure 3.1 Particle flux and force due to pressure.

$$n_{\mathrm{e}}m_{\mathrm{e}}\frac{\mathrm{d}\boldsymbol{V}_{\mathrm{e}}}{\mathrm{d}t} = -\nabla p_{\mathrm{e}} - en_{\mathrm{e}}(\boldsymbol{E} + \boldsymbol{V}_{\mathrm{e}} \times \boldsymbol{B}) + \boldsymbol{R}, \tag{3.6}$$

$$n_{\mathrm{i}}m_{\mathrm{i}}\frac{\mathrm{d}\boldsymbol{V}_{\mathrm{i}}}{\mathrm{d}t} = -\nabla p_{\mathrm{i}} + Zen_{\mathrm{i}}(\boldsymbol{E} + \boldsymbol{V}_{\mathrm{i}} \times \boldsymbol{B}) - \boldsymbol{R}. \tag{3.7}$$

Here $\boldsymbol{R}$ denotes the rate of momentum (density) change of the electron fluid by the collision with the ion fluid. The rate of momentum change of the ion fluid due to the collision with electron fluid is $-\boldsymbol{R}$. The change of the number $n(x, y, z, t)\Delta x\Delta y\Delta z$ of particles within the region of $\Delta x\Delta y\Delta z$ is the difference between the incident particle flux $n(x, y, z, t)V_x(x, y, z, t)\Delta y\Delta z$ into the surface A in Figure 3.1 and outgoing particle flux $n(x + \Delta x, y, z, t)V_x(x + \Delta x, y, z, t)\Delta y\Delta z$ from the surface A′; that is,

$$(n(x, y, z, t)V_x(x, y, z, t) - n(x + \Delta x, y, z, t)V_x(x + \Delta x, y, z, t))\Delta y\Delta z$$

$$= -\frac{\partial(nV_x)}{\partial x}\Delta x\Delta y\Delta z.$$

When the particle fluxes of the other surfaces are taken into accout, we find the equations of continuity (3.4) and (3.5), that is

$$\frac{\partial n}{\partial t}\Delta x\Delta y\Delta z = -\left(\frac{\partial(nV_x)}{\partial x} + \frac{\partial(nV_y)}{\partial y} + \frac{\partial(nV_z)}{\partial z}\right)\Delta x\Delta y\Delta z.$$

The term $-\nabla p$ in (3.6) and (3.7) is the force per unit volume of plasma due to the pressure p and the second term in the right-hand side of (3.6) and (3.7) is Coulomb force and Lorentz force per unit volume. The third term is the collision term of electron-ion collision as is mentioned in Section 2.2 and is given by

$$\boldsymbol{R} = -n_{\mathrm{e}}m_{\mathrm{e}}(\boldsymbol{V}_{\mathrm{e}} - \boldsymbol{V}_{\mathrm{i}})\nu_{\mathrm{ei}} \tag{3.8}$$

where ν_{ei} is Coulomb collision frequency of the electron with the ion.

Let us consider the total time differential in the left-hand side of the equation of motion. The flow velocity $\boldsymbol{V}$ is a function of space coordinates $\boldsymbol{r}$ and time t. Then the acceleration of a small volume of fluid is given by

$$\frac{\mathrm{d}\boldsymbol{V}(\boldsymbol{r},t)}{\mathrm{d}t} = \frac{\partial \boldsymbol{V}(\boldsymbol{r},t)}{\partial t} + \left(\frac{\mathrm{d}\boldsymbol{r}}{\mathrm{d}t}\cdot\nabla\right)\boldsymbol{V}(\boldsymbol{r},t) = \frac{\partial \boldsymbol{V}(\boldsymbol{r},t)}{\partial t} + (\boldsymbol{V}(\boldsymbol{r},t)\cdot\nabla)\boldsymbol{V}(\boldsymbol{r},t).$$

Therefore, the equations of motion (3.6) and (3.7) are reduced to

$$n_\mathrm{e}m_\mathrm{e}\left(\frac{\partial \boldsymbol{V}_\mathrm{e}}{\partial t} + (\boldsymbol{V}_\mathrm{e}\cdot\nabla)\boldsymbol{V}_\mathrm{e}\right) = -\nabla p_\mathrm{e} - en_\mathrm{e}(\boldsymbol{E} + \boldsymbol{V}_\mathrm{e}\times\boldsymbol{B}) + \boldsymbol{R} \tag{3.9}$$

$$n_\mathrm{i}m_\mathrm{i}\left(\frac{\partial \boldsymbol{V}_\mathrm{i}}{\partial t} + (\boldsymbol{V}_\mathrm{i}\cdot\nabla)\boldsymbol{V}_\mathrm{i}\right) = -\nabla p_\mathrm{i} + Zen_\mathrm{i}(\boldsymbol{E} + \boldsymbol{V}_\mathrm{i}\times\boldsymbol{B}) - \boldsymbol{R}. \tag{3.10}$$

Conservation of particle (3.4) and (3.5), the equations of motion (3.9) and (3.10) can be derived from Boltzmann equation (refer to Chapter 11). Integration of Boltzmann equation over velocity space yields (3.4) and (3.5). Integration of Boltzmann equation multiplied by $m\boldsymbol{v}$ yields (3.9) and (3.10) (refer to [3.1]).

3.2 Magnetohydrodynamic Equations for One Fluid

Since the ion-to-electron mass ratio is $m_\mathrm{i}/m_\mathrm{e} = 1836\,A$ (A is atomic weight of the ion), the contribution of ions to the mass density of plasma is dominant. In many cases, it is more convenient to reorganize the equations of motion for two fluids to the equation of motion for one fluid and Ohm's law.

The total mass density of plasma ρ_m, the flow velocity of plasma $\boldsymbol{V}$, the electric charge density ρ, and the current density $\boldsymbol{j}$ are defined as follows:

$$\rho_\mathrm{m} = n_\mathrm{e}m_\mathrm{e} + n_\mathrm{i}m_\mathrm{i}, \tag{3.11}$$

$$\boldsymbol{V} = \frac{n_\mathrm{e}m_\mathrm{e}\boldsymbol{V}_\mathrm{e} + n_\mathrm{i}m_\mathrm{i}\boldsymbol{V}_\mathrm{i}}{\rho_\mathrm{m}}, \tag{3.12}$$

$$\rho = -en_\mathrm{e} + Zen_\mathrm{i}, \tag{3.13}$$

$$\boldsymbol{j} = -en_\mathrm{e}\boldsymbol{V}_\mathrm{e} + Zen_\mathrm{i}\boldsymbol{V}_\mathrm{i}. \tag{3.14}$$

From Equations (3.4) and (3.5), it follows that

$$\frac{\partial \rho_\mathrm{m}}{\partial t} + \nabla\cdot(\rho_\mathrm{m}\boldsymbol{V}) = 0, \tag{3.15}$$

$$\frac{\partial \rho}{\partial t} + \nabla \cdot \boldsymbol{j} = 0. \tag{3.16}$$

From Equations (3.9) and (3.10), we find

$$\rho_{\mathrm{m}} \frac{\partial \boldsymbol{V}}{\partial t} + n_{\mathrm{e}} m_{\mathrm{e}} (\boldsymbol{V}_{\mathrm{e}} \cdot \nabla) \boldsymbol{V}_{\mathrm{e}} + n_{\mathrm{i}} m_{\mathrm{i}} (\boldsymbol{V}_{\mathrm{i}} \cdot \nabla) \boldsymbol{V}_{\mathrm{i}}$$

$$= -\nabla(p_{\mathrm{e}} + p_{\mathrm{i}}) + \rho \boldsymbol{E} + \boldsymbol{j} \times \boldsymbol{B}. \tag{3.17}$$

The charge neutrality of the plasma allows us to write $n_{\mathrm{e}} \simeq Z n_{\mathrm{i}}$. Denote $\Delta n_{\mathrm{e}} = n_{\mathrm{e}} - Z n_{\mathrm{i}}$, then we have

$$\rho_{\mathrm{m}} = n_{\mathrm{i}} m_{\mathrm{i}} \left(1 + \frac{m_{\mathrm{e}}}{m_{\mathrm{i}}} Z\right), \quad p = p_{\mathrm{i}} + p_{\mathrm{e}}, \quad \boldsymbol{V} = \boldsymbol{V}_{\mathrm{i}} + \frac{m_{\mathrm{e}} Z}{m_{\mathrm{i}}} (\boldsymbol{V}_{\mathrm{e}} - \boldsymbol{V}_{\mathrm{i}}),$$

$$\rho = -e \Delta n_{\mathrm{e}}, \quad \boldsymbol{j} = -e n_{\mathrm{e}} (\boldsymbol{V}_{\mathrm{e}} - \boldsymbol{V}_{\mathrm{i}}).$$

Since $m_{\mathrm{e}}/m_{\mathrm{i}} \ll 1$, the second and third terms in the left-hand side of (3.17) can be written to be $(\boldsymbol{V} \cdot \Delta)\boldsymbol{V}$. Since $\boldsymbol{V}_{\mathrm{e}} = \boldsymbol{V}_{\mathrm{i}} - \boldsymbol{j}/e n_{\mathrm{e}} \simeq \boldsymbol{V} - \boldsymbol{j}/e n_{\mathrm{e}}$, (3.9) reduces to

$$\boldsymbol{E} + \left(\boldsymbol{V} - \frac{\boldsymbol{j}}{e n_{\mathrm{e}}}\right) \times \boldsymbol{B} + \frac{1}{e n_{\mathrm{e}}} \nabla p_{\mathrm{e}} - \frac{\boldsymbol{R}}{e n_{\mathrm{e}}} = \frac{m_{\mathrm{e}}}{e^2 n_{\mathrm{e}}} \frac{\partial \boldsymbol{j}}{\partial t} - \frac{m_{\mathrm{e}}}{e} \frac{\partial \boldsymbol{V}}{\partial t}. \tag{3.18}$$

By use of the expression of specific resistivity η, (refer to (2.34)) the collision term $\boldsymbol{R}$ is reduced to

$$\boldsymbol{R} = n_{\mathrm{e}} \left(\frac{m_{\mathrm{e}} \nu_{\mathrm{ei}}}{n_{\mathrm{e}} e^2}\right) (-e n_{\mathrm{e}}) (\boldsymbol{V}_{\mathrm{e}} - \boldsymbol{V}_{\mathrm{i}}) = n_{\mathrm{e}} e \eta \boldsymbol{j}. \tag{3.19}$$

Equation (3.18) corresponds to a generalized Ohm's law. Finally, the equation of motion for one fluid model and a generalized Ohm's law is given by

$$\rho_{\mathrm{m}} \left(\frac{\partial \boldsymbol{V}}{\partial t} + (\boldsymbol{V} \cdot \nabla) \boldsymbol{V}\right) = -\nabla p + \rho \boldsymbol{E} + \boldsymbol{j} \times \boldsymbol{B}, \tag{3.20}$$

$$\boldsymbol{E} + \left(\boldsymbol{V} - \frac{\boldsymbol{j}}{e n_{\mathrm{e}}}\right) \times \boldsymbol{B} + \frac{1}{e n_{\mathrm{e}}} \nabla p_{\mathrm{e}} - \eta \boldsymbol{j} = \frac{m_{\mathrm{e}}}{e^2 n_{\mathrm{e}}} \frac{\partial \boldsymbol{j}}{\partial t} - \frac{m_{\mathrm{e}}}{e} \frac{\partial \boldsymbol{V}}{\partial t} \simeq 0. \tag{3.21}$$

$$(|\omega/\Omega_{\mathrm{e}}| \ll 1)$$

The equation of continuity and Maxwell equations are

$$\frac{\partial \rho_{\mathrm{m}}}{\partial t} + \nabla \cdot (\rho_{\mathrm{m}} \boldsymbol{V}) = 0, \tag{3.22}$$

$$\frac{\partial \rho}{\partial t} + \nabla \cdot \boldsymbol{j} = 0, \tag{3.23}$$

$$\nabla \times \boldsymbol{E} = -\frac{\partial \boldsymbol{B}}{\partial t}, \tag{3.24}$$

$$\frac{1}{\mu_0}\nabla \times \boldsymbol{B} = \boldsymbol{j} + \frac{\partial \boldsymbol{D}}{\partial t} = \boldsymbol{j} + \epsilon_0 \frac{\partial \boldsymbol{E}}{\partial t}, \tag{3.25}$$

$$\nabla \cdot \boldsymbol{D} = \rho, \tag{3.26}$$

$$\nabla \cdot \boldsymbol{B} = 0. \tag{3.27}$$

From Equations (3.25) and (3.24), it follows $\nabla \times \nabla \times \boldsymbol{E} = -\mu_0 \partial \boldsymbol{j}/\partial t - \mu_0 \epsilon_0 \partial^2 \boldsymbol{E}/\partial t^2$. A typical phase velocity $v_{\rm ph} = \omega/k$ ($k = |\boldsymbol{k}|$) of magnetohydrodynamic wave ($\propto \exp i(\boldsymbol{k} \cdot \boldsymbol{r} - i\omega t)$) is Alfvén velocity $v_{\rm A} = B/(\mu_0 \rho_{\rm m})^{1/2}$ as is described in Section 8.2.2 and Section 11.4.1 and is much smaller than light velocity c; that is, $\omega/k \sim v_{\rm A} \ll c$. Since $|\nabla \times (\partial \boldsymbol{B}/\partial t)| = |\nabla \times \nabla \times \boldsymbol{E}| \sim k^2|\boldsymbol{E}|$, and $\mu_0\epsilon_0|\partial^2 \boldsymbol{E}/\partial t^2| \sim \omega^2|\boldsymbol{E}|/c^2$, the displacement current in vacuum $\partial \boldsymbol{D}/\partial t$ of (3.25) is negligible. The ratio of the first term $(m_{\rm e}/e)\partial \boldsymbol{j}/\partial t$ in the right-hand side of (3.21) to the term $(\boldsymbol{j} \times \boldsymbol{B})$ in the left-hand side is $\omega/\Omega_{\rm e}$, so that the first term can be neglected, if $|\omega/\Omega_{\rm e}| \ll 1$. The second term $(m_{\rm e}/e)\partial \boldsymbol{V}/\partial t$ in the right-hand side of (3.21) is of the order of $\omega/\Omega_{\rm e}$ times as large as the term $\boldsymbol{V} \times \boldsymbol{B}$ in the left-hand side. Therefore, we may set the right-hand side of (3.21) nearly zero. When the term $\boldsymbol{j} \times \boldsymbol{B}$ is eliminated by the use of (3.20), we find

$$\boldsymbol{E} + \boldsymbol{V} \times \boldsymbol{B} - \frac{1}{en_{\rm e}}\nabla p_{\rm i} - \eta \boldsymbol{j} = \frac{\Delta n_{\rm e}}{n_{\rm e}}\boldsymbol{E} + \frac{m_{\rm i}}{e}\frac{{\rm d}\boldsymbol{V}}{{\rm d}t}.$$

The ratio of $(m_{\rm i}/e){\rm d}\boldsymbol{V}/{\rm d}t$ to $\boldsymbol{V} \times \boldsymbol{B}$ is around $|\omega/\Omega_{\rm i}|$, and $\Delta n_{\rm e}/n_{\rm e} \ll 1$. When $|\omega/\Omega_{\rm i}| \ll 1$, we find

$$\boldsymbol{E} + \boldsymbol{V} \times \boldsymbol{B} - \frac{1}{en_{\rm e}}\nabla p_{\rm i} = \eta \boldsymbol{j}. \quad (\,|\omega/\Omega_{\rm i}| \ll 1) \tag{3.28}$$

When $|\omega/\Omega_{\rm i}| \ll 1$ and $|\omega/k| \ll c$, and if the ion pressure term $\nabla p_{\rm i}$ can be neglected in Ohm's law, magnetohydrodynamic equations are simplified as follows:

$$\boldsymbol{E} + \boldsymbol{V} \times \boldsymbol{B} = \eta \boldsymbol{j}, \tag{3.29}$$

$$\rho_{\rm m}\left(\frac{\partial \boldsymbol{V}}{\partial t} + (\boldsymbol{V} \cdot \nabla)\boldsymbol{V}\right) = -\nabla p + \boldsymbol{j} \times \boldsymbol{B}, \tag{3.30}$$

$$\nabla \times \boldsymbol{B} = \mu_0 \boldsymbol{j}, \tag{3.31}$$

$$\nabla \times \boldsymbol{E} = -\frac{\partial \boldsymbol{B}}{\partial t}, \tag{3.32}$$

$$\nabla \cdot \boldsymbol{B} = 0, \tag{3.33}$$

$$\frac{\partial \rho_{\mathrm{m}}}{\partial t} + (\boldsymbol{V} \cdot \nabla)\rho_{\mathrm{m}} + \rho_{\mathrm{m}} \nabla \cdot \boldsymbol{V} = 0. \tag{3.34}$$

We may add the adiabatic equation as an equation of state;

$$\frac{\mathrm{d}}{\mathrm{d}t}(p\rho_{\mathrm{m}}^{-\gamma}) = 0,$$

where the quantity γ is the ratio of specific heats and $\gamma = (2+\delta)/\delta$ (δ is the number of degrees of freedom) is 5/3 in the three-dimensional case $\delta = 3$. Combined with (3.34), the adiabatic equation becomes

$$\frac{\partial p}{\partial t} + (\boldsymbol{V} \cdot \nabla)p + \gamma p \nabla \cdot \boldsymbol{V} = 0. \tag{3.35}$$

Instead of this relation, we may use the more simple relation of incompressibility

$$\nabla \cdot \boldsymbol{V} = 0. \tag{3.36}$$

if $|(\mathrm{d}\rho_{\mathrm{m}}/\mathrm{d}t)/\rho_{\mathrm{m}})| \ll |\nabla \cdot V|$.

From Equations (3.31) and (3.32), the energy conservation law is given by

$$\frac{1}{\mu_0}\nabla \cdot (\boldsymbol{E} \times \boldsymbol{B}) + \frac{\partial}{\partial t}\left(\frac{B^2}{2\mu_0}\right) + \boldsymbol{E} \cdot \boldsymbol{j} = 0. \tag{3.37}$$

From Equation (3.29), the third term in the left-hand side of (3.37) becomes

$$\boldsymbol{E} \cdot \boldsymbol{j} = \eta j^2 + (\boldsymbol{j} \times \boldsymbol{B}) \cdot \boldsymbol{V}. \tag{3.38}$$

By use of (3.30) and (3.34), the Lorentz term in (3.38) is expressed by

$$(\boldsymbol{j} \times \boldsymbol{B}) \cdot \boldsymbol{V} = \frac{\partial}{\partial t}\left(\frac{\rho_{\mathrm{m}} V^2}{2}\right) + \nabla \cdot \left(\frac{\rho_{\mathrm{m}} V^2}{2}\boldsymbol{V}\right) + \boldsymbol{V} \cdot \nabla p.$$

From Equation (3.35), it follows that

$$-\nabla \cdot (p\boldsymbol{V}) = \frac{\partial p}{\partial t} + (\gamma - 1)p\nabla \cdot \boldsymbol{V}$$

and

$$\boldsymbol{V} \cdot \nabla p = \frac{\partial}{\partial t}\left(\frac{p}{\gamma - 1}\right) + \nabla \cdot \left(\frac{p}{\gamma - 1} + p\right)\boldsymbol{V}.$$

Therefore, the energy conservation law (3.37) is reduced to

$$\nabla \cdot (\boldsymbol{E} \times \boldsymbol{H}) + \frac{\partial}{\partial t}\left(\frac{\rho_{\mathrm{m}} V^2}{2} + \frac{p}{\gamma - 1} + \frac{B^2}{2\mu_0}\right) + \eta j^2 + \nabla \cdot \left(\frac{\rho_{\mathrm{m}} V^2}{2} + \frac{p}{\gamma - 1} + p\right)\boldsymbol{V}$$

$$= 0. \tag{3.39}$$

The substitution of (3.29) into (3.32) yields

$$\frac{\partial \boldsymbol{B}}{\partial t} = \nabla \times (\boldsymbol{V} \times \boldsymbol{B}) - \eta \nabla \times \boldsymbol{j} = \nabla \times (\boldsymbol{V} \times \boldsymbol{B}) + \frac{\eta}{\mu_0} \Delta \boldsymbol{B} \tag{3.40}$$

$$\frac{\partial \boldsymbol{B}}{\partial t} = -(\boldsymbol{V} \cdot \nabla)\boldsymbol{B} - \boldsymbol{B}(\nabla \cdot \boldsymbol{V}) + (\boldsymbol{B} \cdot \nabla)\boldsymbol{V} + \frac{\eta}{\mu_0} \Delta \boldsymbol{B}. \tag{3.41}$$

Here we used the vector formula for $\nabla \times (\boldsymbol{V} \times \boldsymbol{B})$ and $\nabla \times (\nabla \times \boldsymbol{B})$ (refer to Table 3.1). The quantity $\eta/\mu_0 = \nu_{\rm m}$ is called *magnetic viscosity.* The substitution of (3.31) into (3.30) yields

$$\rho_{\rm m} \frac{{\rm d}\boldsymbol{V}}{{\rm d}t} = -\nabla \left(p + \frac{B^2}{2\mu_0} \right) + \frac{1}{\mu_0} (\boldsymbol{B} \cdot \nabla) \boldsymbol{B}. \tag{3.42}$$

The equation of motion (3.42) and the equation of magnetic diffusion (3.41) are fundamental equations of magnetohydrodynamics. Equation (3.33), equation of continuity (3.34), and equation of state (3.35) or (3.36) are additional equations.

Table 3.1 Vector formulas

$\boldsymbol{a} \cdot (\boldsymbol{b} \times \boldsymbol{c}) = \boldsymbol{b} \cdot (\boldsymbol{c} \times \boldsymbol{a}) = \boldsymbol{c} \cdot (\boldsymbol{a} \times \boldsymbol{b})$

$\boldsymbol{a} \times (\boldsymbol{b} \times \boldsymbol{c}) = (\boldsymbol{a} \cdot \boldsymbol{c})\boldsymbol{b} - (\boldsymbol{a} \cdot \boldsymbol{b})\boldsymbol{c}$

$(\boldsymbol{a} \times \boldsymbol{b}) \cdot (\boldsymbol{c} \times \boldsymbol{d}) = (\boldsymbol{a} \cdot \boldsymbol{c})(\boldsymbol{b} \cdot \boldsymbol{d}) - (\boldsymbol{a} \cdot \boldsymbol{d})(\boldsymbol{b} \cdot \boldsymbol{c})$

$\nabla \cdot (\phi \boldsymbol{a}) = \phi \nabla \cdot \boldsymbol{a} + (\boldsymbol{a} \cdot \nabla)\phi$

$\nabla \times (\phi \boldsymbol{a}) = \nabla \phi \times \boldsymbol{a} + \phi \nabla \times \boldsymbol{a}$

$\nabla(\boldsymbol{a} \cdot \boldsymbol{b}) = (\boldsymbol{a} \cdot \nabla)\boldsymbol{b} + (\boldsymbol{b} \cdot \nabla)\boldsymbol{a} + \boldsymbol{a} \times (\nabla \times \boldsymbol{b}) + \boldsymbol{b} \times (\nabla \times \boldsymbol{a})$

$\nabla \cdot (\boldsymbol{a} \times \boldsymbol{b}) = \boldsymbol{b} \cdot \nabla \times \boldsymbol{a} - \boldsymbol{a} \cdot \nabla \times \boldsymbol{b}$

$\nabla \times (\boldsymbol{a} \times \boldsymbol{b}) = \boldsymbol{a}(\nabla \cdot \boldsymbol{b}) - \boldsymbol{b}(\nabla \cdot \boldsymbol{a}) + (\boldsymbol{b} \cdot \nabla)\boldsymbol{a} - (\boldsymbol{a} \cdot \nabla)\boldsymbol{b}$

$\nabla \times \nabla \times \boldsymbol{a} = \nabla(\nabla \cdot \boldsymbol{a}) - \nabla^2 \boldsymbol{a}$ (valid for x, y, z coordinates only)

$\nabla \cdot \boldsymbol{a} = (1/r)\partial(r a_r)/\partial r + (1/r)\partial a_\theta/\partial \theta + \partial a_z/\partial z$

$\nabla \times \boldsymbol{a} = [(1/r)\partial a_z/\partial \theta - \partial a_\theta/\partial z]\boldsymbol{i}_r + [\partial a_r/\partial z - \partial a_z/\partial r]\boldsymbol{i}_\theta$
$\quad + [(1/r)\partial(r a_\theta)/\partial r - (1/r)\partial a_r/\partial \theta]\boldsymbol{i}_z$

$\nabla^2 \phi = (1/r)\partial/\partial r(r \partial \phi/\partial r) + (1/r^2)\partial^2 \phi/\partial \theta^2 + \partial^2 \phi/\partial z^2$

The ratio of the first term to the second term of the right-hand side in (3.40), $R_{\rm m}$, defined by

$$\frac{|\nabla \times (\boldsymbol{V} \times \boldsymbol{B})|}{|\Delta \boldsymbol{B}(\eta/\mu_0)|} \approx \frac{VB/a}{(B/a^2)(\eta/\mu_0)} = \frac{\mu_0 V a}{\eta} \equiv S_{\rm R} \tag{3.43}$$

is called the *magnetic Reynolds number*. The notation a is a typical plasma size and its minor radius. Magnetic Reynolds number is equal to the ratio of magnetic diffusion time $\tau_{\rm R} = \mu_0 a^2/\eta$ to Alfvén transit time $\tau_{\rm A} = a/v_{\rm A}$ (it is assumed that $V \approx v_{\rm A}$); that is,

$$\tau_{\rm R} = \mu_0 a^2/\eta, \qquad \tau_{\rm A} = a/v_{\rm A}, \qquad S_{\rm R} = \tau_{\rm R}/\tau_{\rm A}. \tag{3.44}$$

When $S_{\rm R} \ll 1$, the magnetic field in a plasma changes according to the diffusion equation. When $S_{\rm R} \gg 1$, it can be shown that the lines of magnetic force are frozen in the plasma. Let the magnetic flux within the surface element ΔS be $\Delta\Phi$, and take the z axis in the $\boldsymbol{B}$ direction. Then $\Delta\Phi$ is

$$\Delta\Phi = \boldsymbol{B} \cdot \boldsymbol{n}\Delta S = B\Delta x\Delta y.$$

As the boundary of ΔS moves, the rate of change of ΔS is

$$\frac{\rm d}{{\rm d}t}(\Delta x) = \frac{\rm d}{{\rm d}t}(x + \Delta x - x) = V_x(x+\Delta x) - V_x(x) = \frac{\partial V_x}{\partial x}\Delta x,$$

$$\frac{\rm d}{{\rm d}t}(\Delta S) = \left(\frac{\partial V_x}{\partial x} + \frac{\partial V_y}{\partial y}\right)\Delta x\Delta y.$$

The rate of change of the flux $\Delta\Phi$ is

$$\frac{\rm d}{{\rm d}t}(\Delta\Phi) = \frac{{\rm d}B}{{\rm d}t}\Delta S + B\frac{\rm d}{{\rm d}t}(\Delta S) = \left(\frac{{\rm d}\boldsymbol{B}}{{\rm d}t} + \boldsymbol{B}(\nabla\cdot\boldsymbol{V}) - (\boldsymbol{B}\cdot\nabla)\boldsymbol{V}\right)_z \Delta S$$

$$= \frac{\eta}{\mu_0}\Delta B_z(\Delta S). \tag{3.45}$$

(refer to (3.41)). When $S_{\rm R} \to \infty$, $\eta \to 0$, the rate of change of the flux becomes zero, i.e., ${\rm d}(\Delta\Phi)/{\rm d}t \to 0$. This means the magnetic flux is frozen in the plasma.

Problems

1. Displacement Current What condition is necessary to be able to neglect the displacement current in vacuum of Maxwell equations in the frame of magnetohydrodynamics? Refer to Section 3.2.

2. Magnetic Reynolds Number Calculate Alfvén transit time $\tau_{\rm A} = a/v_{\rm A}$, resistive diffusion time $\tau_{\rm R} = \mu_0 a^2/\eta$, and the magnetic Reynolds number $S = \tau_{\rm R}/\tau_{\rm A}$ of D-T fusion core plasma with $n_{\rm e} = 10^{20}{\rm m}^{-3}$, $T_{\rm i} = T_{\rm e} = 10{\rm keV}$, $B = 5{\rm T}$, $a = 2\,{\rm m}$, effective $Z = 1.5$ and atomic weight of D-T fuel $= 2.5$. Refer to (3.44).

4

Tokamak

The word "tokamak" is said to be a contraction of the Russian words for current (т о к), vessel (к а м е р), magnet (м а г н и т), and coil (к а т у ш к а). Tokamaks are axisymmetric, with the plasma current itself giving rise to the poloidal field essential to the equilibrium of toroidal plasmas. In a tokamak, the toroidal field used to stabilize against MHD instabilities is strong enough to satisfy the Kruskal-Shafranov condition. This characteristic is quite different from that of reversed field pinch, with its relatively weak toroidal field. There are reviews and textbooks of tokamak experiments and theory [4.1],[4.2],[4.3] and diagnostics [4.4].

4.1 Tokamak Devices

The structure of the devices of Tokamak-T3, large tokamaks JET, JT60U, TFTR, and ITER are shown in Figures 4.1, 4.2, 4.3, 4.4, and 4.5, respectively as typical examples. The cross-section of ITER (International Tokamak Experimental Reactor) under construction is also displayed in Figure 4.6.

The toroidal field coils, equilibrium field coils (also called the poloidal field coils, which produce the vertical field and shaping field), ohmic heating coils (the primary windings of the current transformer), and vacuum vessel can be seen in the figures. Sometimes "poloidal field coils" means both the equilibrium field coils and the ohmic heating coils. By raising the current of the primary windings of the current transformer (ohmic heating coils), a current is induced in the plasma, which acts as the secondary winding. In the Tokamak-T3 and JET device, the current transformer is of the iron core type. The air-core type of current tranformer is utilized in JT60U, TFTR, and ITER. The vacuum vessel is usually made of thin stainless steel or inconel so that it has enough electric resistance in the toroidal direction. Therefore, the voltage induced by the primary windings can penetrate it. The thin vacuum vessel is called the liner. Before starting an experiment, the liner is outgassed by baking at a temperature of 150 to 400℃ for a long time under high vacuum. Furthermore, before running an experiment, a plasma is run with a weak toroidal field in order to discharge-clean the wall of the liner. Inside the liner there is a diaphragm made of tungsten, molybdenum, or graphite that limits the plasma size and minimizes the interaction of the plasma with the wall. This diaphragm is called a limiter. Recently a divertor configuration was introduced instead of the limiter. In this case the magnetic surface, including the separatrix point, determines the plasma boundary (refer to Section 4.5). A conducting shell surrounds the plasma outside the liner and is used to maintain the positional equilibrium or to stabilize MHD instabilities during the skin time scale. The magnitude of the vertical field is feedback controlled to keep the plasma at the center of the liner always. Many improvements have been made in tokamak devices over the years. Accuracy of the magnetic field is also important to improve the plasma performance in tokamak as well as other toroidal devices. The parameters of typical tokamak devices are listed in Table 4.1.

Table 4.1 Parameters of tokamaks.

	R	$a(\times b)$	R/a	B_t	I_p	Remarks
T-3	1.0	0.15	6.7	3.4	0.12	$\tau_E > 30\tau_B$ in 1968
T-10	1.5	0.39	3.8	5.0	0.65	
PLT	1.32	0.4	3.3	3.2	0.5	
TFTR	2.48	0.85	2.9	5.2	2.5@	Compact
JET	2.96	1.25(×2.1)	2.4	3.45	7	Noncircular
JT60U	3.4	1.1(×1.4)	3.1	4.2	6	Noncircular, JT60 upgraded
ITER	6.2	2.0(×3.4)	3.1	5.3	15	Aiming at Q$\sim$ 10

R, a, b in m, B_t in T, and I_p in MA.

4.2 Equilibrium

In order to maintain a hot plasma, we must confine and keep it away from the vacuum-container wall. The most promising method for such confinement of a hot plasma is the use of appropriate strong magnetic fields. An equilibrium condition must be satisfied for such magnetic confinement systems.

4.2.1 Equilibrium Equation

When a plasma is in the steady state and fluid velocity is zero ($\boldsymbol{V} = 0$), magnetohydrodynamic equation (3.30) yields the equilibrium equations

$$\nabla p = \boldsymbol{j} \times \boldsymbol{B}, \tag{4.1}$$

$$\nabla \times \boldsymbol{B} = \mu_0 \boldsymbol{j}, \tag{4.2}$$

$$\nabla \cdot \boldsymbol{B} = 0, \tag{4.3}$$

$$\nabla \cdot \boldsymbol{j} = 0. \tag{4.4}$$

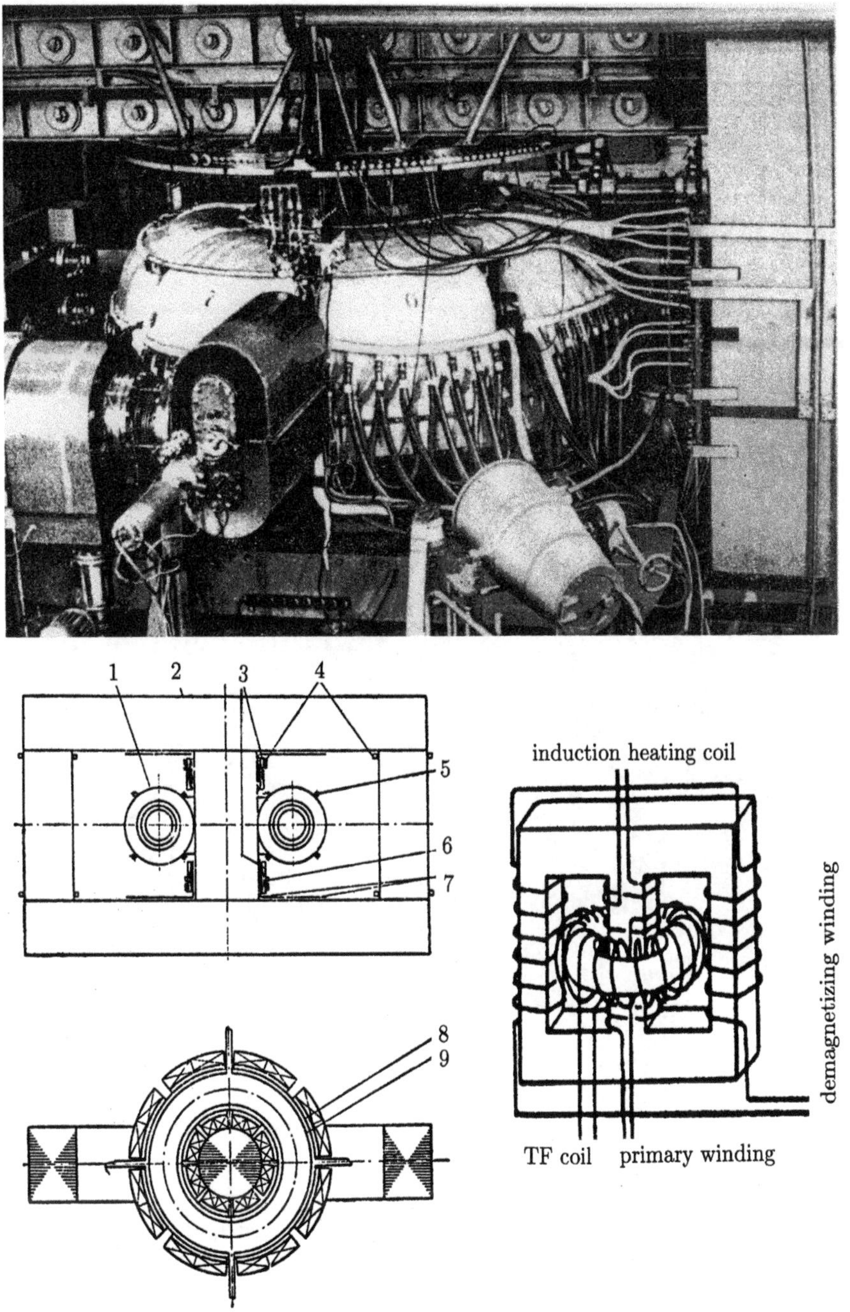

Figure 4.1 Photogragh (top), schematic drawing (lower left) and circuits of toroidal field coil and primary winding of current tranaformer, demagnetization winding, induction heating coil (lower right) of Tokamak-T3 device. The numbers in shematic drawing indicate (1) toroidal field coil, (2) current transformer, (3) primary winding, (4) demagnetization winding, (5) compensating coil, (6) induction heating coil, (7) screening for primary winding, (8) external chamber,

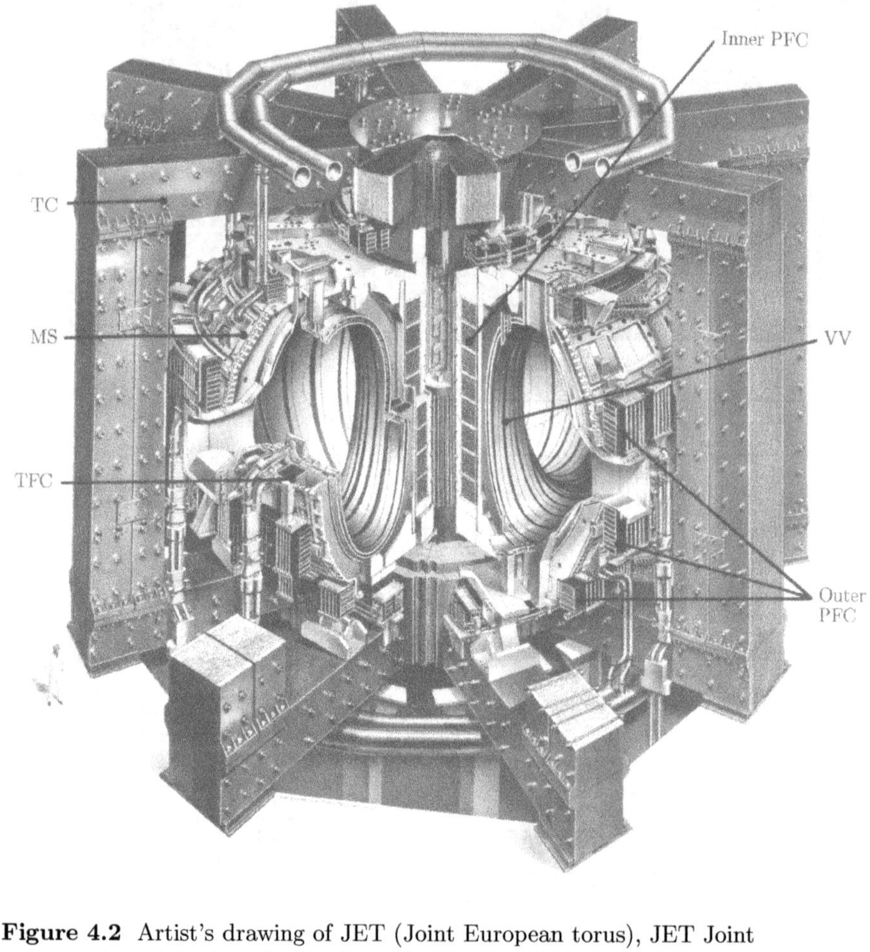

Figure 4.2 Artist's drawing of JET (Joint European torus), JET Joint Undertaking, Abingdon, Oxfordshire, England. The toroidal field coils (TFC) are arranged around the vacuum vessel (VV). The outer poloidal field coils (Outer PFC, equilibrium field coils) and inner poloidal field coils (Inner PFC, ohmic heating coils) are wound in the toroidal direction outside TFC. JET uses an ion-core current transformer (TC). The mechanical structures (MS) support the toroidal field coils against the large amount of torque due to the equilibrium field.

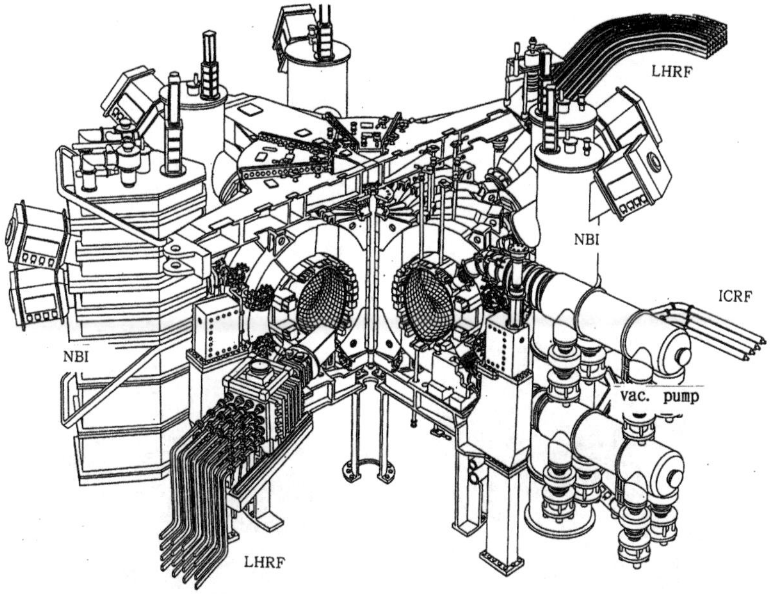

Figure 4.3 A birdview of JT60U, Japan Atomic Energy Research Institute.

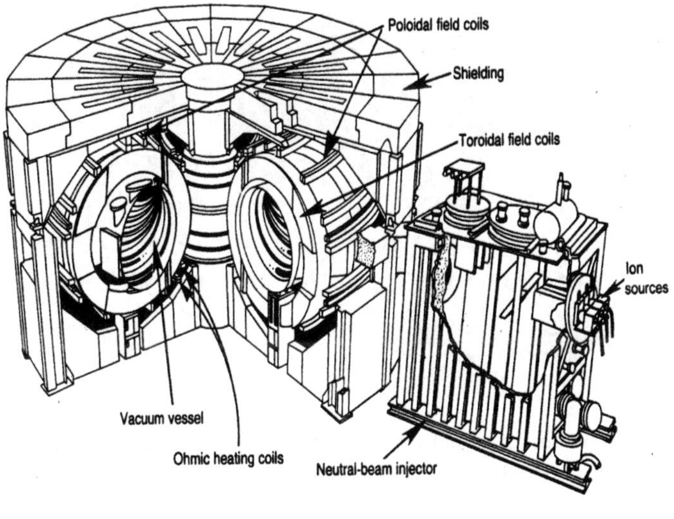

Figure 4.4 A birdview of TFTR (Tokamak Fusion Test Reactor), Plasma Physics Laboratory, Princeton University.

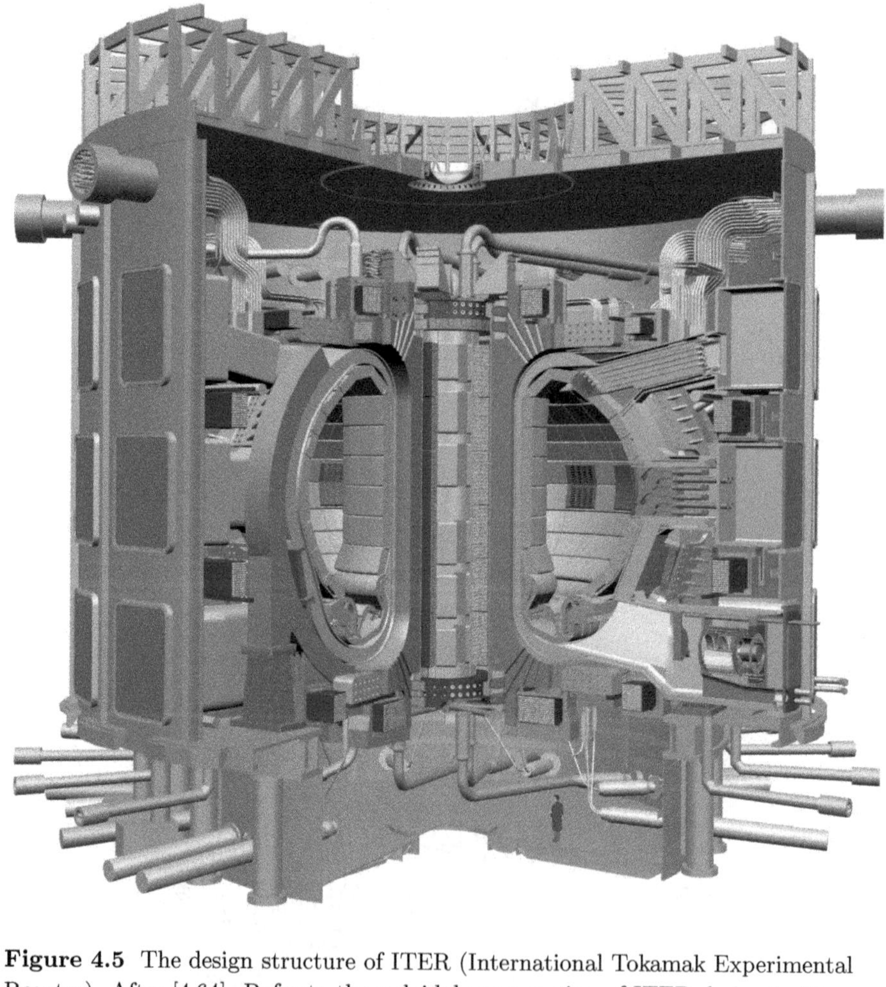

Figure 4.5 The design structure of ITER (International Tokamak Experimental Reactor). After [4.64]. Refer to the poloidal cross-section of ITER design in Figure 4.6.

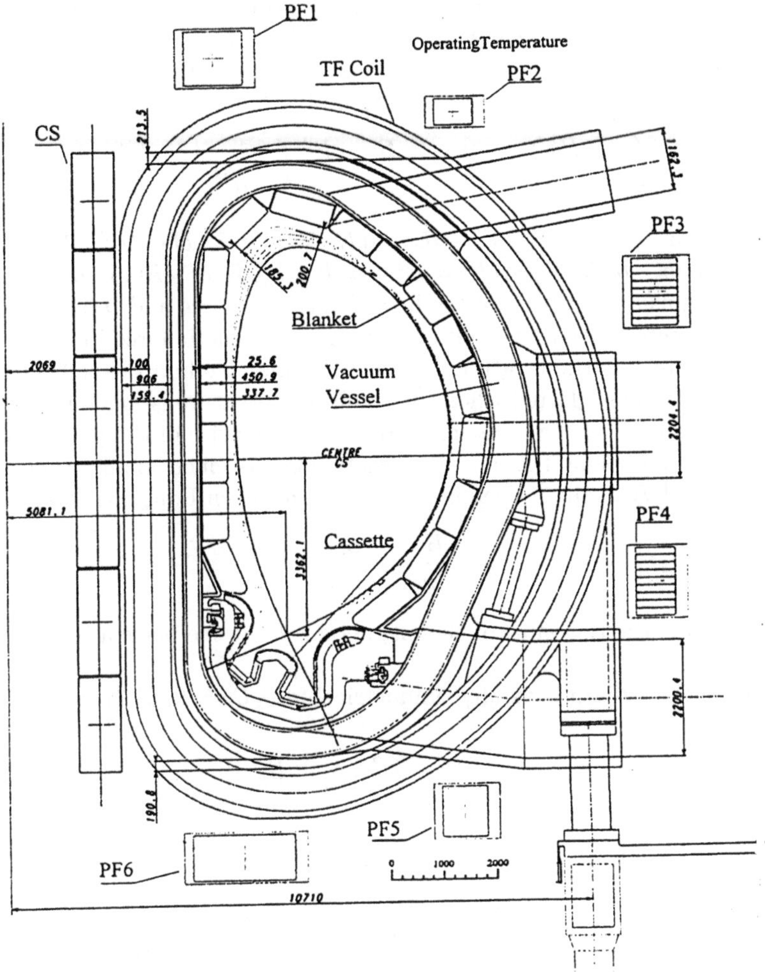

Figure 4.6 Poloidal cross-section of ITER outline design in 2000. After [4.64].

From (4.1), we have

$$\boldsymbol{B}\cdot\nabla p = 0, \tag{4.5}$$

$$\boldsymbol{j}\cdot\nabla p = 0. \tag{4.6}$$

Equation (4.5) indicates that $\boldsymbol{B}$ and ∇p are orthogonal, and the surfaces of constant pressure coincide with the magnetic surfaces. Equation (4.6) shows that the current-density vector $\boldsymbol{j}$ is everywhere parallel to the constant-pressure surfaces. Substitution of (4.2) into (4.1) yields

$$\nabla\left(p+\frac{B^2}{2\mu_0}\right) = (\boldsymbol{B}\cdot\nabla)\frac{\boldsymbol{B}}{\mu_0} = \frac{B^2}{\mu_0}\left(-\frac{1}{R}\boldsymbol{n}+\frac{\partial B/\partial l}{B}\boldsymbol{b}\right). \tag{4.7}$$

The vector relations $\boldsymbol{B}\times(\nabla\times\boldsymbol{B})+(\boldsymbol{B}\cdot\nabla)\boldsymbol{B}\equiv\nabla(\boldsymbol{B}\cdot\boldsymbol{B}/2)$, $1(\boldsymbol{B}\cdot\nabla)\boldsymbol{B} = B^2[(\boldsymbol{b}\cdot\nabla)\boldsymbol{b}+\boldsymbol{b}((\boldsymbol{b}\cdot\nabla)B)/B] = B^2[-\boldsymbol{n}/R+\boldsymbol{b}(\partial B/\partial l)/B]$ are used, where R is the radius of curvature of the line of magnetic force and $\boldsymbol{n}$ is the unit vector directed toward a point on the line of magnetic force from the center of curvature. l is the length along the field line. We find the right-hand side of (4.7) can be neglected when the radius of curvature is much larger than the scale length of the pressure gradient, i.e., the size of the plasma, and the variation of $\boldsymbol{B}$ along the line of magnetic force is much smaller than the variation of $\boldsymbol{B}$ in the perpendicular direction. Then Equation (4.7) becomes

$$p+\frac{B^2}{2\mu_0}\sim\frac{B_0^2}{2\mu_0}$$

where B_0 is the the value of the magnetic field at the plasma boundary ($p=0$).

When the system is axially symmetric and $\partial/\partial z = 0$, (4.7) exactly reduces to

$$\frac{\partial}{\partial r}\left(p+\frac{B_z^2+B_\theta^2}{2\mu_0}\right) = -\frac{B_\theta^2}{r\mu_0}. \tag{4.8}$$

By the multiplication of (4.8) by r^2 and the integration by parts we obtain

$$\left(p+\frac{B_z^2+B_\theta^2}{2\mu_0}\right)_{r=a} = \frac{1}{\pi a^2}\int_0^a\left(p+\frac{B_z^2}{2\mu_0}\right)2\pi r dr$$

and

$$\langle p\rangle+\frac{\langle B_z^2\rangle}{2\mu_0} = p_a+\frac{B_z^2(a)+B_\theta^2(a)}{2\mu_0}. \tag{4.9}$$

$\langle p\rangle$ is the volume average of p and p_a is the plasma pressure at the plasma boundary. As $B^2/2\mu_0$ is the pressure of the magnetic field, (4.9) is the equation of pressure equilibrium. The ratio of plasma pressure to the pressure of the external magnetic field B_0

$$\beta \equiv \frac{p}{B_0^2/2\mu_0} = \frac{n(T_e + T_i)}{B_0^2/2\mu_0} \tag{4.10}$$

is called the *beta ratio.* For a confined plasma, β is always smaller than 1, and is used as a figure of merit of the confining magnetic field. The ratio of plasma pressure to the pressure of poloidal field B_θ is called *poloidal beta.* When the pressure at the boundary is $p_a = 0$ and $|B_z(a) - B_z(r)| \ll |B_z(a)|$ in (4.9), the poloidal beta β_p is

$$\beta_p \equiv \frac{\langle p \rangle}{B_\theta^2(a)/2\mu_0} = 1 + \frac{B_z^2(a) - \langle B_z^2(r) \rangle}{B_\theta(a)^2} \approx 1 + \left(\frac{2B_z}{B_\theta^2}\right)_a \langle B_z(a) - B_z(r) \rangle . \tag{4.11}$$

$B_z(a)$ is the magnetic field in the direction of z in the case without plasma. In the case of $\beta_p > 1$, the magnitude of magnetic field $B_z(r)$ inside the plasma is smaller than that in the vacuum case ($B_z(r) < B_z(a)$). This indicates the *diamagnetism* of plasma. In the case of $\beta_p < 1$, $B_z(r)$ becomes larger than $B_z(a)$ ($B_z(r) > B_z(a)$). When the plasma current flows along a line of magnetic force, the current produces the poloidal magnetic field and a poloidal component of the plasma current appears and induces an additional z component of the magnetic field. This is the origin of *paramagnetism* of plasma.

Equilibrium Equation for Axially Symmetric Systems

Let us use cylindrical coordinates (r, φ, z) and denote the magnetic surface by ψ. The magnetic surface ψ in an axisymmetric system is given by (see (2.41)):

$$\psi = rA_\varphi(r, z).$$

The r and z components of the magnetic field are given by:

$$rB_r = -\frac{\partial \psi}{\partial z}, \qquad rB_z = \frac{\partial \psi}{\partial r}. \tag{4.12}$$

Therefore, ψ is also called the poloidal flux function. The relation $\boldsymbol{B} \cdot \nabla p = 0$ follows from the equilibrium equation and is expressed by:

$$-\frac{\partial \psi}{\partial z}\frac{\partial p}{\partial r} + \frac{\partial \psi}{\partial r}\frac{\partial p}{\partial z} = 0.$$

Accordingly p is a functon of ψ only, i.e.,

$$p = p(\psi). \tag{4.13}$$

Similarly, from $\boldsymbol{j} \cdot \nabla p = 0$ and $\nabla \times \boldsymbol{B} = \mu_0 \boldsymbol{j}$, we may write

$$-\frac{\partial p}{\partial r}\frac{\partial (rB_\varphi)}{\partial z} + \frac{\partial p}{\partial z}\frac{\partial (rB_\varphi)}{\partial r} = 0.$$

This means that rB_φ is a function of ψ only and

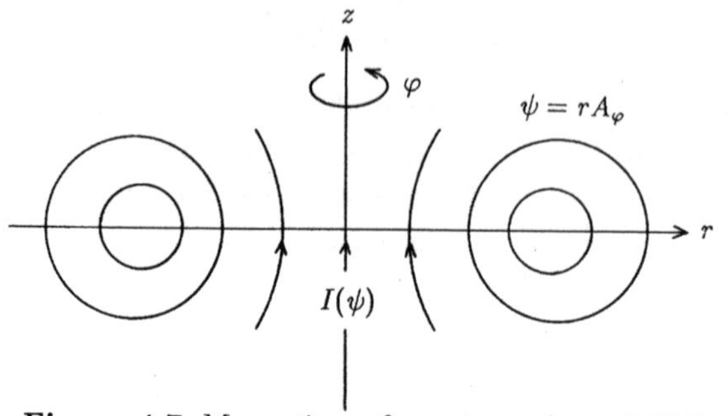

Figure 4.7 Magnetic surfaces $\psi = rA_\varphi$ and $I(\psi)$.

$$rB_\varphi = \frac{\mu_0 I(\psi)}{2\pi}. \tag{4.14}$$

Equation (4.14) indicates that $I(\psi)$ means the current flowing in the poloidal direction through the circular cross-section within $\psi = rA_\varphi$ (Figure 4.7). The r component of $\boldsymbol{j} \times \boldsymbol{B} = \nabla p$ leads to the equation on ψ:

$$L(\psi) + \mu_0 r^2 \frac{\partial p(\psi)}{\partial \psi} + \frac{\mu_0^2}{8\pi^2}\frac{\partial I^2(\psi)}{\partial \psi} = 0 \tag{4.15}$$

where

$$L(\psi) \equiv \left(r\frac{\partial}{\partial r}\frac{1}{r}\frac{\partial}{\partial r} + \frac{\partial^2}{\partial z^2} \right)\psi.$$

This equation is called the *Grad-Shafranov equation.* The current density is expressed in terms of the function of the magnetic surface as

$$j_r = \frac{-1}{2\pi r}\frac{\partial I(\psi)}{\partial z}, \qquad j_z = \frac{1}{2\pi r}\frac{\partial I(\psi)}{\partial r},$$

$$j_\varphi = \frac{-1}{\mu_0}\left(\frac{\partial}{\partial r}\frac{1}{r}\frac{\partial \psi}{\partial r} + \frac{1}{r}\frac{\partial^2 \psi}{\partial z^2} \right) = -\frac{L(\psi)}{\mu_0 r}$$

$$= \frac{1}{\mu_0 r}\left(\mu_0 r^2 p' + \frac{\mu_0^2}{8\pi^2}(I^2)' \right).$$

The prime (′) means differentiation by ψ. By use of (4.12) and (4.14), we have

$$\boldsymbol{j} = \frac{I'}{2\pi}\boldsymbol{B} + p' r \boldsymbol{e}_\varphi \tag{4.16}$$

$$L(\psi) + \mu_0 r j_\varphi = 0. \tag{4.17}$$

When the unit vectors with the directions of r, φ, z are denoted by $\boldsymbol{e}_r, \boldsymbol{e}_\varphi, \boldsymbol{e}_z$, respectively, then we have $\nabla\varphi = \boldsymbol{e}_\varphi/R$, $\boldsymbol{e}_r \times \boldsymbol{e}_\varphi = \boldsymbol{e}_z$, $\boldsymbol{e}_z \times \boldsymbol{e}_\varphi = -\boldsymbol{e}_r$. Therefore, $\boldsymbol{B}$ can be expressed from (4.12) and (4.14) as follows:

$$\boldsymbol{B} = \frac{\mu_0 I(\psi)}{2\pi}\nabla\varphi + \nabla\psi \times \nabla\varphi. \tag{4.18}$$

$p(\psi)$, $I^2(\psi)$ are arbitrary functions of ψ. When they are linear or quadratic functions of ψ, (4.15) becomes a linear differential equation. Let us consider a simple linear case of ψ. At the plasma boundary $\psi = \psi_{\mathrm{b}}$, we denote $p_{\mathrm{b}} = p(\psi_{\mathrm{b}})$ and $I_{\mathrm{b}}^2 = I^2(\psi_{\mathrm{b}})$; that is,

$$p(\psi) = p_{\mathrm{b}} - \frac{a}{\mu_0 R^2}(\psi - \psi_{\mathrm{b}}), \tag{4.19}$$

$$I^2(\psi) = I_{\mathrm{b}}^2 - \frac{8\pi^2}{\mu_0^2} b(\psi - \psi_{\mathrm{b}}). \tag{4.20}$$

Then (4.15) and (4.17) are reduced to

$$L(\psi) = a\frac{r^2}{R^2} + b = -\mu_0 r j_\varphi. \tag{4.21}$$

We set the position of the magnetic axis to $(R, 0)$. The following function

$$\psi - \psi_0 = \frac{b+a}{1+\epsilon}\left[\frac{1}{2}\left(1 + c\frac{r^2 - R^2}{R^2}\right)z^2 + \frac{\epsilon}{8R^2}(r^2 - R^2)^2 + \right.$$
$$\left. \frac{(1+\epsilon)b - (1-c)(b+a)}{24(b+a)R^4}(r^2 - R^2)^3\right] \tag{4.22}$$

is the solution of (4.21), which is correct up to the cubic of $(r - R), z$ [4.7],[4.8]. ϵ, c are constant and $\psi_0 = \psi(R, 0)$. When the coefficient of the third term of the right-hand side of (4.22) is 0, that is,

$$(1+\epsilon)b - (1-c)(b+a) = 0 \quad \rightarrow \quad \epsilon = -(c-1)(a/b) - c \tag{4.23}$$

(4.22) becomes the exact Solovev solution of the Grad-Shafranov equation (4.21) [4.7],[4.8]. When we set $c = R^2/(R^2 - R_{\mathrm{x}}^2)$, ϵ becomes $\epsilon = -(a/b + R^2/R_{\mathrm{x}}^2)R_{\mathrm{x}}^2/(R^2 - R_{\mathrm{x}}^2)$ due to (4.23) and then (4.22) is reduced to

$$\psi = \frac{b}{2}\left(1 - \frac{r^2}{R_{\mathrm{x}}^2}\right)z^2 + \frac{a + (R^2/R_{\mathrm{x}}^2)b}{8R^2}\left((r^2 - R^2)^2 - (R^2 - R_{\mathrm{x}}^2)^2\right). \tag{4.24}$$

(4.24) is an exact equilibrium solution in the interior region of plasma surrounded by the conductive wall specified by $\psi(r, z) = \psi_{\mathrm{b}}$. The surface $\psi(r, z) = 0$ is the separatrix surface (refer to Figure 4.8 and Section 4.4). The separatrix points X are located at $(R_{\mathrm{x}}, \pm Z_{\mathrm{x}})$, where Z_{x} is Z_{x}

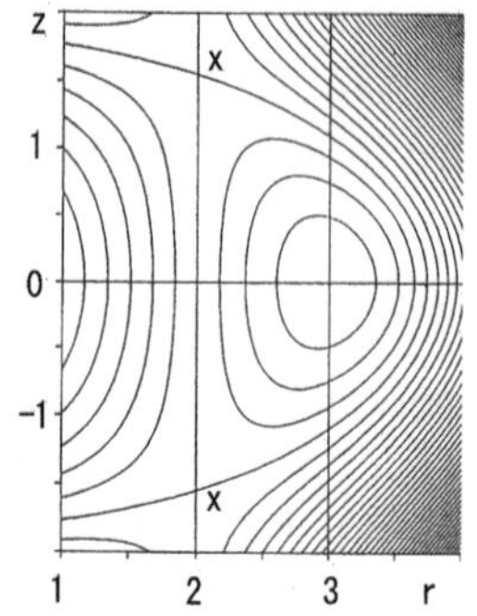

Figure 4.8 The contour (magnetic surface) of the flux function ψ of (4.24) in the case of $a/b = 4.4$, $R = 3$, $R_x = 2$. Xs are the separatrix points and the magnetic surface passing X points is the separatix surface.

$= [-(a/b + R^2/R_x^2)(1 - R_x^2/R^2)/2]^{1/2} R_x$. The maximum value R_{max} of r within the separatrix surface is $R_{max} = (2 - R_x^2/R^2)^{1/2} R$.
When we set the separatrix surface as the plasma boundary ($\psi_b = 0$), the aspect ratio A, elongation ratio κ_s, and central poloidal beta β_{p0} are

$$\frac{1}{A} = \frac{R_{max} - R_x}{2R} = \frac{(2 - R_x^2/R^2)^{1/2} - R_x/R}{2}$$

$$\kappa_s = \frac{2Z_x}{R_{max} - R_x} = \frac{AZ_x}{R}$$

$$\beta_{p0} \equiv \frac{p(R,0) - p_b}{B_z^2(R_x,0)/2\mu_0} = \frac{a}{a + (R^2/R_x^2)b}.$$

When A and κ_s are specified, β_{p0} is fixed. To avoid this inadequateness, Weening [4.9] added an additional particular solution $r^2 \ln(r^2/R_\alpha^2) - r^2$ to Solovev solution (4.24); that is,

$$\psi = \frac{b+d}{2}\left(1 - \frac{r^2}{R_x^2}\right) z^2 + \frac{a + (R^2/R_x^2)(b+d)}{8R^2}\left((r^2 - R^2)^2 - (R^2 - R_x^2)^2\right)$$

$$-\frac{d}{4}\left(r^2 \ln\frac{r^2}{R_x^2} - (r^2 - R_x^2)\right). \tag{4.25}$$

When the plasma boundary is chosen to be the separatrix $\psi(r,z) = 0$, the aspect ratio A, elongation ratio κ_s, and central poloidal beta β_{p0} are

$$\frac{Z_x^2}{R_x^2} = -\frac{1}{2}\left(\frac{a}{b+d} + \frac{R^2}{R_x^2}\right)\left(1 - \frac{R_x^2}{R^2}\right)$$

$$\frac{R_{\rm max}^2}{R^2} = \left(2 - \frac{R_{\rm x}^2}{R^2}\right) + \frac{2d[x\ln x/(x-1)-1]}{a+(R^2/R_{\rm x}^2)(b+d)}, \qquad x \equiv \frac{R_{\rm max}^2}{R_{\rm x}^2}$$

$$\frac{1}{A} = \frac{R_{\rm max}/R - R_{\rm x}/R}{2}, \qquad \kappa_{\rm s} = \frac{AZ_{\rm x}}{R}$$

$$\beta_{p0} = \frac{a}{a+(R^2/R_{\rm x}^2)(b+d)}\left[1+\frac{2d\big(\ln(R^2/R_{\rm x}^2)-(1-R^2/R_{\rm x}^2)\big)}{\big(a+(R^2/R_{\rm x}^2)(b+d)\big)(1-R_{\rm x}^2/R^2)}\right].$$

The magnetic surface ψ, the magnetic field $\boldsymbol{B}$ and the pressure p in translationally symmetric system ($\partial/\partial z = 0$) are given by

$$\psi = A_z(r,\theta),$$

$$B_r = \frac{1}{r}\frac{\partial\psi}{\partial\theta} \quad B_\theta = -\frac{\partial\psi}{\partial r} \quad B_z = \frac{\mu_0}{2\pi}I(\psi),$$

$$p = p(\psi).$$

The equilibrium equation is reduced to

$$\frac{1}{r}\frac{\partial}{\partial r}\left(r\frac{\partial\psi}{\partial r}\right) + \frac{1}{r^2}\frac{\partial^2\psi}{\partial\theta^2} + \mu_0\frac{\partial p(\psi)}{\partial\psi} + \frac{\mu_0^2}{8\pi^2}\frac{\partial I^2(\psi)}{\partial\psi} = 0, \tag{4.26}$$

$$\boldsymbol{j} = \frac{1}{2\pi}I'\boldsymbol{B} + p'\boldsymbol{e}_z, \quad \Delta\psi + \mu_0 j_z = 0. \tag{4.27}$$

It is possible to drive the similar equilibrium equation in the case of a helically symmetric system.

4.2.2 Vertical Field for Tokamak Equilibrium

The poloidal magnetic field produced by plasma current $I_{\rm p}$ inside the plasma ring is stronger than that outside the plasma ring. Therefore, it is necessary for the tokamak equilibrium to add a vertical field to reduce the poloidal field inside the ring and to increase the poloidal field outside the ring as is shown in Figure 4.9. Let us estimate the necessary vertical field $B_\perp$.

The *hoop force* by which the current ring of a plasma tends to expand is given by

$$F_{\rm h} = -\frac{\partial}{\partial R}\frac{L_{\rm p}I_{\rm p}^2}{2}\bigg|_{L_{\rm p}I_{\rm p}={\rm const.}} = \frac{1}{2}I_{\rm p}^2\frac{\partial L_{\rm p}}{\partial R}$$

where L_p is the self-inductance of the current ring:

$$L_p = \mu_0 R \left(\ln \frac{8R}{a} + \frac{l_i}{2} - 2 \right) \tag{4.28}$$

where $\mu_0 R(\ln(8R/a) - 2)$ is the inductance due to the magnetic field energy outside the plasma and $\mu_0 R l_i/2$ is the inductance due to the magnetic field energy inside the plasma where

$$l_i \equiv \frac{2\pi \int_0^a B_p^2(\rho)\rho d\rho}{\pi a^2 B_p^2(a)}. \tag{4.29}$$

Accordingly, the hoop force is

$$F_h = \frac{\mu_0 I_p^2}{2} \left(\ln \frac{8R}{a} + \frac{l_i}{2} - 1 \right).$$

The outward force F_p exerted by the plasma pressure is (Figure 4.10)

$$F_p = \langle p \rangle \pi a^2 2\pi.$$

The inward (contractive) force F_{B1} due to the tension of the toroidal field inside the plasma is

$$F_{B1} = -\frac{\langle B_\varphi^2 \rangle}{2\mu_0} 2\pi^2 a^2$$

and the outward force F_{B2} by the pressure due to the external magnetic field is

$$F_{B2} = \frac{B_{\varphi v}^2}{2\mu_0} 2\pi^2 a^2.$$

The force F_I acting on the plasma due to the vertical field $B_\perp$ is

$$F_I = I_p B_\perp 2\pi R.$$

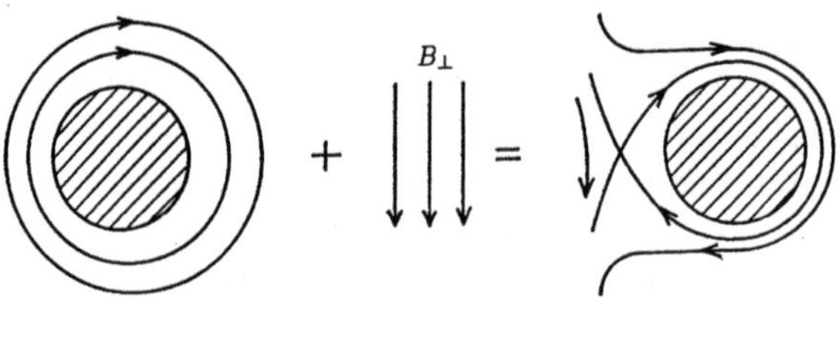

Figure 4.9 Poloidal magnetic field due to the combined plasma current and vertical field.

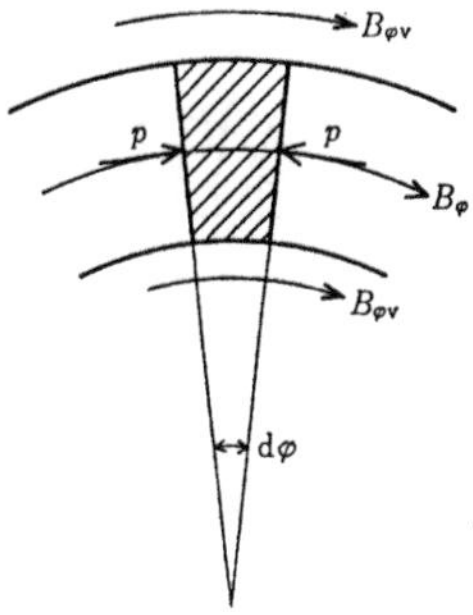

Figure 4.10 Equilibrium of forces acting on a toroidal plasma.

Balance of these forces gives

$$\frac{\mu_0 I_p^2}{2}\left(\ln\frac{8R}{a}+\frac{l_i}{2}-1\right)+2\pi^2a^2\left(\langle p\rangle+\frac{B_{\varphi v}^2}{2\mu_0}-\frac{\langle B_\varphi^2\rangle}{2\mu_0}\right)+2\pi R I_p B_\perp=0,$$

and the necessary amount of $B_\perp$ is

$$B_\perp=\frac{-\mu_0 I_p}{4\pi R}\left(\ln\frac{8R}{a}+\frac{l_i}{2}-1+\beta_p-\frac{1}{2}\right)=\frac{-\mu_0 I_p}{4\pi R}\left(\ln\frac{8R}{a}+\Lambda-\frac{1}{2}\right), \quad (4.30)$$

where

$$\beta_p=\frac{\langle p\rangle}{B_p^2(a)/2\mu_0}, \quad (4.31)$$

$$\Lambda=\beta_p+l_i/2-1. \quad (4.32)$$

Equation (4.9) is used for the derivation. The equilibrium of tokamak with circular cross-section is discussed in detail in References [4.3, 4.10].

Control of Plasma Position

If the vertical field $B_\perp$ is uniform in space, the equilibrium is neutral with regard to changes to plasma position in the horizontal direction. When the lines of the vertical field are curved, as shown in Figure 4.11, the plasma position is stable with regard to up and down motion. The z component F_z of the magnetic force applied to a plasma current ring with mass M is

$$F_z=-2\pi R I_p B_R.$$

From the relation $(\partial B_R/\partial z)-(\partial B_z/\partial R)=0$,

$$M\frac{d^2z}{dt^2}=-2\pi R I_p\frac{\partial B_R}{\partial z}z=2\pi I_p B_z\left(-\frac{R}{B_z}\frac{\partial B_z}{\partial R}\right)z. \quad (4.33)$$

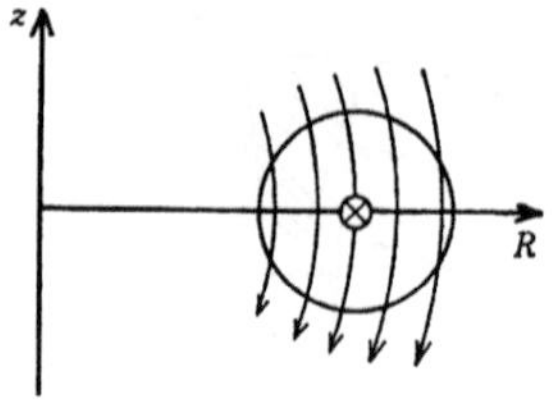

Figure 4.11 Vertical field for plasma equilibrium.

As $I_p B_z < 0$, the stability condition for decay index n for vertical movement is

$$n \equiv -\frac{R}{B_z}\frac{\partial B_z}{\partial R} > 0.$$

The horizontal component F_R of the magnetic force is

$$M\frac{d^2(\Delta R)}{dt^2} = F_R = 2\pi R I_p (B_z - B_\perp)\Delta R.$$

The amount of $B_\perp$ necessary for plasma equilibrium (refer to (4.30)) is

$$B_\perp = \frac{-\mu_0 I_p}{4\pi R}\left(\ln\frac{8R}{a} + \Lambda - \frac{1}{2}\right), \qquad \Lambda = \frac{l_i}{2} + \beta_p - 1.$$

When the plasma is ideally conductive, the magnetic flux inside the plasma ring is conserved and

$$\frac{\partial}{\partial R}(L_p I_p) + 2\pi R B_\perp = 0.$$

Here the self-inductance is $L_p = \mu_0 R(\ln(8R/a) + l_i/2 - 2)$. Therefore, the equation of motion is

$$M\frac{d^2(\Delta R)}{dt^2} = 2\pi I_p B_\perp \left(\frac{3}{2} - n\right)\Delta R \tag{4.34}$$

under the assumption $\ln(8R/a) \gg 1$. Then the stability condition for the horizontal movement is

$$\frac{3}{2} > n.$$

4.2.3 Limit of Poloidal Beta Ratio for Equilibrium

In the previous subsection, the necessary vertical field $B_\perp$ for plasma equilibrium is given by

$$B_\perp = B_a \frac{a}{2R}\left(\ln\frac{8R}{a} + \Lambda - \frac{1}{2}\right).$$

The direction of $B_\perp$ is opposite to that of B_ω produced by the plasma current inside the torus, so that the resultant poloidal field becomes zero at some points in the inside region of the torus and a separatrix is formed. When the plasma pressure is increased and β_p becomes large, the necessary amount of $B_\perp$ is increased and the separatrix shifts toward the plasma. For simplicity, let us consider a sharp-boundary model in which the plasma pressure is constant inside the plasma boundary, and in which the boundary encloses a plasma current I_p. Then, the pressure-balance equation is

$$\frac{B_\omega^2}{2\mu_0} + \frac{B_{\varphi\mathrm{v}}^2}{2\mu_0} \approx p + \frac{B_{\varphi\mathrm{i}}^2}{2\mu_0}, \tag{4.35}$$

where B_ω is ploidal field outside the plasma and $B_{\varphi\mathrm{v}}$, $B_{\varphi\mathrm{i}}$ are φ components of the field outside and inside the plasma boundary, respectively. $B_{\varphi\mathrm{v}}$ and $B_{\varphi\mathrm{i}}$ are proportional to $1/r$, according to (4.14). If the values of $B_{\varphi\mathrm{v}}$, $B_{\varphi\mathrm{i}}$ at $r = R$ are denoted by $B_{\varphi\mathrm{v}}^0$, $B_{\varphi\mathrm{i}}^0$, respectively, (4.35) may be written as

$$B_\omega^2 = 2\mu_0 p - ((B_{\varphi\mathrm{v}}^0)^2 - (B_{\varphi\mathrm{i}}^0)^2)\left(\frac{R}{r}\right)^2.$$

The upper limit of the plasma pressure is determined by the condition that the resultant poloidal field at $r = r_\mathrm{min}$ inside the torus is zero,

$$2\mu_0 p_\mathrm{max}\frac{r_\mathrm{min}^2}{R^2} = (B_{\varphi\mathrm{v}}^0)^2 - (B_{\varphi\mathrm{i}}^0)^2. \tag{4.36}$$

As r is expressed by $r = R + a\cos\omega$, (4.35) is reduced (with $(r_\mathrm{min} = R - a)$) to

$$B_\omega^2 = 2\mu_0 p_\mathrm{max}\left(1 - \frac{r_\mathrm{min}^2}{r^2}\right) = 8\mu_0 p_\mathrm{max}\frac{a}{R}\cos^2\frac{\omega}{2}.$$

Here $a/R \ll 1$ is assumed. From the relation $\oint B_\omega a\mathrm{d}\omega = \mu_0 I_\mathrm{p}$, the upper limit $\beta_\mathrm{p}^\mathrm{c}$ of the poloidal beta ratio is

$$\beta_\mathrm{p}^\mathrm{c} = \frac{\pi^2}{16}\frac{R}{a} \approx 0.5\frac{R}{a}. \tag{4.37}$$

Thus, the upper limit of $\beta_\mathrm{p}^\mathrm{c}$ is half of the aspect ratio R/a in this simple model. When the rotational transform angle ι and the safety factor $q_\mathrm{s} = 2\pi/\iota$ are introduced, we find that

$$\frac{B_\omega}{B_\varphi} = \frac{a}{R}\left(\frac{\iota}{2\pi}\right) = \frac{a}{Rq_\mathrm{s}}$$

so that β is

$$\beta = \frac{p}{B^2/2\mu_0} \approx \frac{p}{B_\omega^2/2\mu_0}\left(\frac{B_\omega}{B_\varphi}\right)^2 = \left(\frac{a}{Rq_\mathrm{s}}\right)^2\beta_\mathrm{p}.$$

Accordingly, the upper limit of the beta ratio is

$$\beta^{\rm c} = \frac{0.5}{q_{\rm s}^2}\frac{a}{R}. \tag{4.38}$$

Beta Limit of Elongated Plasma Cross-Section

The poloidal beta limit of a circular tokamak is given by $\beta_{\rm p} = 0.5R/a$, as was just derived. The same poloidal beta limit is derived by similar consideration for the elongated tokamak with horizontal radius a and vertical radius b. When the length of circumference along the poloidal direction is denoted by $2\pi aK$ for the elongated plasma, the average of poloidal field is

$$\bar{B}_{\rm p} = \mu_0 I_{\rm p}/(2\pi aK)$$

where K is approximately given by $K = [(1+(b/a)^2)/2]^{1/2}$. Since

$$\frac{\bar{B}_{\rm p}}{B_{\rm t}} = \frac{Ka}{Rq_{\rm I}}, \qquad q_{\rm I} = \frac{Ka}{R}\frac{B_{\rm t}}{\bar{B}_{\rm p}} \tag{4.39}$$

the beta limit of an elongated tokamak is

$$\beta \le 0.5K^2\frac{a}{Rq_{\rm I}^2} \tag{4.40}$$

and is K^2 times as large as that of a circular one.

In order to make the plasma cross-section elongated, the decay index n of the vertical field must be negative, and the elongated plasma is positionally unstable in the up-down motion. If there is not a conductive wall near the plasma boundary, the growth rate is the inverse of poloidal Alfvën transit time. Therefore, a (resistive) shell must be installed around the plasma to reduce the growth rate to the inverse of resistive skin time of the shell. Furthermore, feedback control of the variable horizontal field is necessary to keep the plasma position. (Refer to Problems 4 and 5 of this chapter.)

4.2.4 Pfirsch-Schlüter Current, Shafranov Shift

When the plasma pressure is isotropic, the current $\boldsymbol{j}$ in the plasma is given by (4.1) and (4.4) as

$$\boldsymbol{j}_\perp = \frac{\boldsymbol{b}}{B}\times\nabla p, \tag{4.41}$$

$$\nabla\cdot\boldsymbol{j}_\parallel = -\nabla\cdot\boldsymbol{j}_\perp = -\nabla\cdot\left(\frac{\boldsymbol{B}}{B^2}\times\nabla p\right) = -\nabla p\cdot\nabla\times\left(\frac{\boldsymbol{B}}{B^2}\right).$$

$\boldsymbol{j}_\perp$ is called diamagnetic current, since this current reduces the magnetic field inside the plasma. Then the parallel component of plasma current to the magnetic field $j_\parallel$ is

$$\nabla \cdot \boldsymbol{j}_{\parallel} = -\nabla p \cdot \left(\left(\nabla \frac{1}{B^2} \times \boldsymbol{B} \right) + \frac{\mu_0 \boldsymbol{j}}{B^2} \right) = 2\nabla p \cdot \frac{\nabla B \times \boldsymbol{B}}{B^3} \tag{4.42}$$

$$\frac{\partial j_{\parallel}}{\partial s} = 2\nabla p \cdot \frac{(\nabla B \times \boldsymbol{b})}{B^2} \tag{4.43}$$

where s is the length along a line of magnetic force. In the zeroth-order approximation, we can put $B \propto 1/R \propto (1 - (r/R_0 \cos\theta))$, $p = p(r)$, and $\partial/\partial s = (\partial\theta/\partial s)\partial/\partial\theta = (\iota/(2\pi R))\partial/\partial\theta$, where ι is the rotational transform angle. When s increases by $2\pi R$, θ increases by ι. Then, Equation (4.43) is reduced to

$$\frac{\iota}{2\pi R}\frac{\partial j_{\parallel}}{\partial \theta} = -\frac{\partial p}{\partial r}\frac{2}{RB}\sin\theta,$$

i.e.,

$$j_{\parallel} = \frac{2q_{\mathrm{s}}}{B}\frac{\partial p}{\partial r}\cos\theta. \tag{4.44}$$

This current is called the *Pfirsch-Schlüter current* [4.11]. These formulas are very important, and will be used to estimate the diffusion coefficient of a toroidal plasma in Section 4.5.1. The Pfirsch-Schlüter current is due to the short circuiting, along magnetic-field lines, of toroidal drift polarization charges. The resultant current is proportional to q_{s}.

Shafranov Shift

Pfirsh-Schlüter current produces the vertical field in the plasma, which brings a plasma current ring inward shift called the *Shafranov shift.* In the case of a tokamak with large aspect and circular plasma cross-section, Shafranov shift $\Delta(r)$ of the center of magnetic surface with the radius of r from the center of plasma boundary with the radius of a is given below as follows [4.12]:

$$\frac{\mathrm{d}\Delta}{\mathrm{d}r} = \frac{1}{RrB_\theta^2}\left(\beta_{\mathrm{p}}B_{\theta a}^2\int_0^r r^2\frac{\mathrm{d}}{\mathrm{d}r}\frac{p}{\langle p\rangle}\mathrm{d}r - \int_0^r rB_\theta^2\mathrm{d}r\right) \tag{4.45}$$

where B_θ is the magnitude of poloidal field at r and $B_{\theta a}$ is the magnitude of poloidal field at $r = a$. β_{p} is the poloidal beta and $\langle p\rangle$ is the volume average of pressure $\langle p\rangle = \int_0^a p2r\mathrm{d}r/a^2$. In the case of parabolic pressure profile, the pressure term of the right-hand side of (4.45) becomes $-(a/R)\beta_{\mathrm{p}}(B_{\theta a}/B_\theta)^2(r/a)^3$ and in the case of flat current profile, the second term of the right-hand side of (4.45) becomes $-(a/4R)(r/a)$. When the pressure profile is parabolic and the current profile is flat, the Shafranov shift is $\Delta/a = (a/2R)(\beta_{\mathrm{p}} + 1/4)(1 - (r/a)^2)$.

4.3 MHD Stability and Density Limit

A possible MHD instability in the low-beta tokamak is kink modes, which will be treated in Chapter 8. Kink modes can be stabilized by tailoring the current profile and by appropriate choice of the safety factor q_a. When the plasma pressure is increased, the beta value is limited by the ballooning modes (Section 8.6). This instability is a mode localized in the bad curvature region driven by a pressure gradient. The beta limit of a ballooning mode is derived to be $\beta_{\max} \sim 0.28(a/Rq_a)$ in Section 8.6. The β limit by kink and ballooning modes depends on the radial profile of the plasma current (shear) and the shape of the plasma cross-section.

Even if a plasma is ideally MHD stable, tearing modes can be unstable for a finite resistive plasma. When Δ' is positive at the rational surfaces (see Section 9.1) in which the safety factor $q_s(r)$ is rational $q_s(r) = 1, 3/2, 2$, tearing modes grow and magnetic islands are formed, as shown in Figure 4.12. When the profile of the plasma current is peaked, then the safety factor at the center becomes $q_s(0) < 1$ and the tearing mode with $m = 1,\ n = 1$ grows at the rational surface $q_s(r) = 1$, and the hot core of the plasma is pushed out when the reconnection of magnetic surfaces occurs (Figure 4.13) and the current profile is flattened. The thermal energy in the central hot core is lost in this way [4.13]. Since the electron temperature in the central part is higher than in the outer region and the resistance in the central part is smaller, the current profile is peaked again and the same process is repeated. This type of phenomenon is called *internal disruption* or *minor disruption.*

The stable operational region of a tokamak with plasma current I_p and density n_e is limited. With Greenward density defined by

$$n_G(10^{20}\mathrm{m}^{-3}) \equiv \frac{I_p(\mathrm{MA})}{\pi a(\mathrm{m})^2}, \tag{4.46}$$

an empirical scaling of the normalized Greenward density or Greenward-Hugill-Murakami parameter N_G

$$N_G \equiv \frac{\langle n_e \rangle}{n_G} < 1 \tag{4.47}$$

holds for most tokamak experiments [4.14], where n_{20} is the electron density in the unit of $10^{20}\mathrm{m}^{-3}$. N_G is expressed by the other form (refer to (4.39))

$$N_G = \frac{0.628}{K^2} \frac{\langle n_{20} \rangle}{B_t(\mathrm{T})/R(\mathrm{m})} q_I.$$

The upper limit of the electron density depends critically on the plasma wall interaction and tends to increase as the heating power increases, although

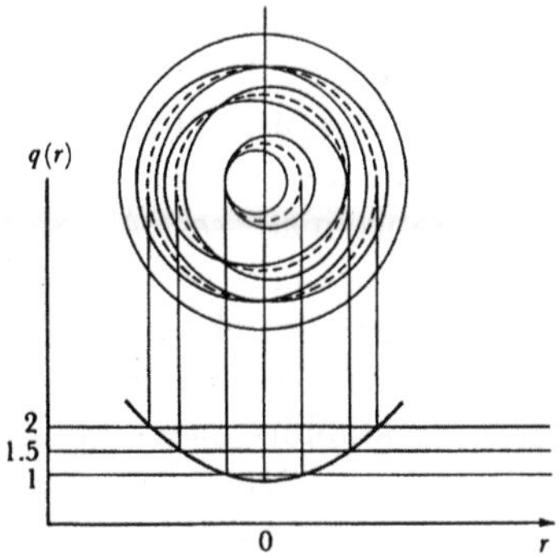

Figure 4.12 Magnetic islands of $m = 1, m = 3/2, m = 2$ modes appear at $q_s(r)$=1,3/2,2.

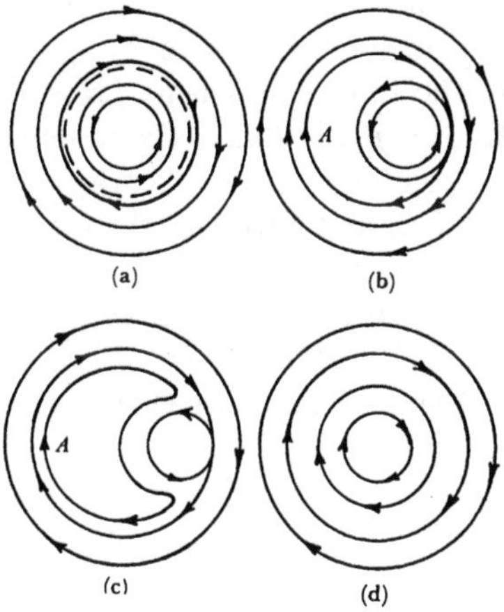

Figure 4.13 The hot core in the center is expelled by the reconnection of magnetic surfaces.

the scaling $N_G < 1$ does not reflect the power dependence. When hydrogen ice pellets are injected into a plasma for fueling from the high field side of ASDEX-U with advanced divertor [4.15], N_G becomes up to ~ 1.5. Therefore, there is the possibility to increase N_{GHM} further. The safety factor q_a at the plasma boundary is $q_a > 3$ in most cases. Beyond the stable region ($N_G < 1,\ 1/q_a < 1/2 \sim 1/3$), strong instability, called *disruptive instability*, occurs in usual operations. Negative spikes appear in the loop voltage due to the rapid expansion of the current channel (flattened current profile); that is, the rapid reduction of the internal inductance. The thermal energy of the plasma is lost suddenly. The electron temperature drops rapidly, and the plasma resistance increases. A positive pulse appears in the loop voltage. Then the plasma discharge is terminated rapidly. In some cases, the time scale of disruption is much faster than the time scale predicted by the resistive tearing mode (Section 10.1). For possible mechanisms of the disruptive instability, overlapping of the magnetic islands of $m = 2/n = 1\ (q_s(r) = 2)$ and $m = 3/n = 2\ (q_s(r) = 1.5)$ or the reconnection of $m = 2/n = 1, m = 1/n = 1$ magnetic islands are discussed. Reviews of the MHD instabilities of tokamak plasmas and plasma transport are given in [4.16]~[4.19].

MHD Stable Beta Limit of Elongated Tokamak

The output power density of nuclear fusion is proportional to $n^2\langle\sigma v\rangle$. Since $\langle\sigma v\rangle$ is proportional to T_i^2 in the region near $T_i \sim 10\,\text{keV}$, the fusion output power is proportional to the square of plasma pressure $p = nT$. Therefore, the higher the beta ratio $\beta = p/(B^2/2\mu_0)$, the more economical the possible fusion reactor. The average beta of $\langle\beta\rangle \sim 3\%$ was realized by NBI experiments in ISX-B, JFT-2, and PLT. All these tokamaks have a circular plasma cross-section.

In non-circular tokamak DIII-D, $\langle\beta\rangle$=11% was realized in 1990 [4.20], in which a=0.45 m,B_t=0.75 T, I_p=1.29 MA, I_p/aB_t=3.1 MA/Tm, $\beta_N \sim$3.6, κ_s=2.35 and R=1.43 m. Figure 4.14 shows the experimental data of DIII-D on the observed beta versus I_p/aB_t.

For the study of linear growth rate of MHD instability, the variational method of energy integral is used to evaluate the eigenvalue of ω^2 as follows:

$$\delta L = \delta \int \boldsymbol{\xi} \cdot \widehat{K}\boldsymbol{\xi}\mathrm{d}\boldsymbol{r} = 0 \tag{4.48}$$

under the constraint of $\int \rho_m \boldsymbol{\xi}^2 \mathrm{d}\boldsymbol{r}$; that is

$$\delta \int \rho_m \boldsymbol{\xi}^2 \mathrm{d}\boldsymbol{r} = 0.$$

Eigenvalue ω^2 is given by

$$\omega^2 = \frac{\int \boldsymbol{\xi} \cdot \widehat{K}\boldsymbol{\xi}\mathrm{d}\boldsymbol{r}}{\int \rho_m \boldsymbol{\xi}^2 \mathrm{d}\boldsymbol{r}}.$$

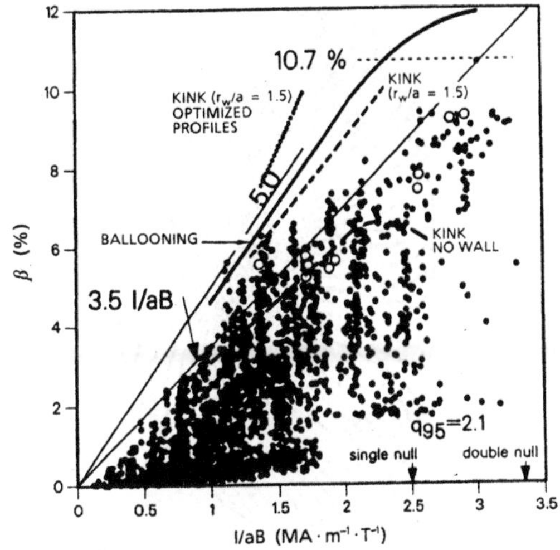

Figure 4.14 The observed beta versus I/aB for DIII-D. Various β limit calculations are summarized in the curves with different assumptions on the location of a conducting wall (r_w/a). After [4.20].

The energy integral is described in Section 8.4. The linear growth rate is $\gamma = (-\omega^2)^{1/2}$ when $\omega^2 < 0$ [$\boldsymbol{\xi}(\boldsymbol{r},t) = \boldsymbol{\xi}(\boldsymbol{r})\exp(-i\omega t)$]. To solve the eigenvalue problem, the "regular finite elements method" is used in ERATO code to avoid the introduction of spurious modes of numerical origin. Troyon et al. [4.21] evaluate the maximum growth rates of MHD instabilities in many cases of elongated tokamak plasma and derive the beta scaling on the upper limit of stable beta value under the optimized conditions as follows [4.21, 4.22]:

$$\beta_\mathrm{c}(\%) = \beta_\mathrm{N}\frac{I_\mathrm{p}\,(\mathrm{MA})}{a(\mathrm{m})B_t(\mathrm{T})}. \tag{4.49}$$

β_N is called the Troyon factor or normalized beta ($\beta_\mathrm{N} = 2 \sim 4$). Figure 4.15 shows the poloidal plasma flow associated with an unstable mode which develops when β exceeds the limit.

When the following definitions

$$\bar{B}_\mathrm{p} \equiv \frac{\mu_0 I_\mathrm{p}}{2\pi a K}, \qquad q_\mathrm{I} \equiv K\frac{a}{R}\frac{B_\mathrm{t}}{\bar{B}_\mathrm{p}} \tag{4.50}$$

are used, the critical beta is reduced to

$$\beta_\mathrm{c}(\%) = 5\beta_\mathrm{N}K^2\frac{a}{Rq_\mathrm{I}}. \tag{4.51}$$

where $2\pi Ka$ is the length of circumference of the plasma boundary and K is

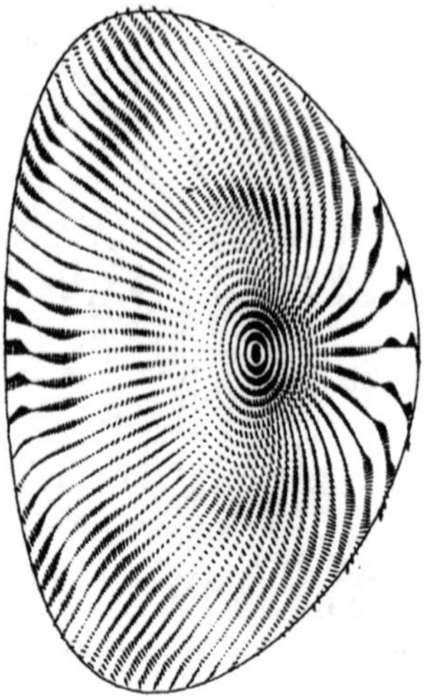

Figure 4.15 Unstable global $n = 1$ mode. After [4.21]. The singular surfaces $q = 2, 3$, and 4 are visible because of the peaked shear velocity on them. $q_0 = 1.35, \beta = 3\%$

approximately given by

$$K^2 \simeq (1 + \kappa_s^2)/2$$

and κ_s is the ratio of the vertical radius b to the horizontal radius a. q_I is called cylindrical safety factor, as q_I is the safety factor in the limiting case of large aspect ratio. In this case, there is the following relation between $\beta_c(\%)$ and β_p:

$$\beta_c(\%)\beta_p = 0.25\beta_N^2 K^2$$

The safety factor q_ψ at a magnetic surface ψ is given by:

$$q_\psi = \frac{1}{2\pi}\oint d\varphi = \frac{1}{2\pi}\oint \frac{d\varphi}{dl_p} dl_p = \frac{1}{2\pi}\oint \frac{B_t}{RB_p} dl_p$$

$$= \frac{1}{2\pi d\psi}\oint B_t \frac{d\psi}{RB_p} dl_p = \frac{1}{2\pi d\psi}\oint B_t ds dl_p = \frac{1}{2\pi}\frac{d\Phi}{d\psi}$$

where $d\psi = RB_p ds$, ds being the thickness of $d\psi$ and $d\Phi$ is the toroidal flux through the thickness ds corresponding to $d\psi$ (ψ is poloidal flux function (magnetic surface function)). It must be noted that q_I is different from q_ψ with the finite aspect ratio. As an approximate fitting formula of the safety factor at the magnetic surface including 95% of the total toroidal magnetic flux through the total plasma cross-section, the following equation is used

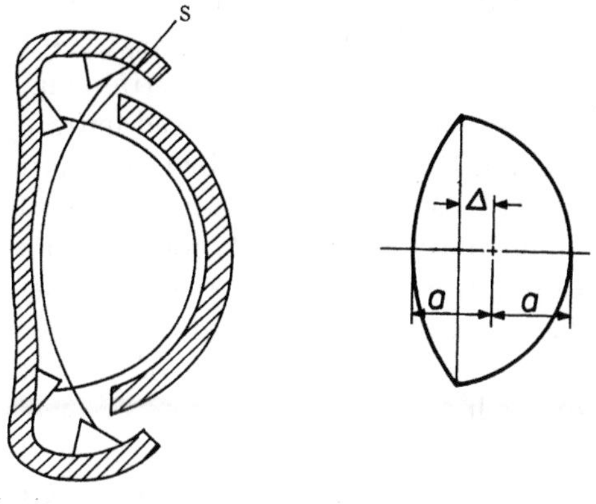

Figure 4.16 Divertor configuration using separatrix S of the magnetic surface (left-hand side). Definition of the triangularity $\delta = \Delta/a$ (right-hand side) (refer to (4.102)).

(including the divertor configuration (refer to Section 4.4)) (p.2160 of [4.23]):

$$q_{95} = q_{\rm I} f_\delta f_A = \frac{a^2 B}{(\mu_0/2\pi)RI}\frac{1+\kappa_{\rm s}^2}{2} f_\delta f_A, \qquad (4.52)$$

where

$$f_\delta = \frac{1+\kappa_{\rm s}^2(1+2\delta^2-1.2\delta^3)}{1+\kappa_{\rm s}^2}, \qquad f_A = \frac{1.17-0.65/A}{(1-1/A^2)^2}.$$

δ is triangularity of the plasma shape defined by (4.102) or $\delta \approx \Delta/a$ (refer to Figure 4.16). f_δ is the factor due to the triangularity of the plasma cross-section and f_A is the factor due to the finite aspect ratio.

4.4 Impurity Control, Scrape Off Layer and Divertor

Radiation loss power $P_{\rm brems}$ by bremsstrahlung due to electron collision with ion per unit volume is

$$P_{\rm brems} = 1.5 \times 10^{-38} Z_{\rm eff} n_{\rm e}^2 (T_{\rm e}/e)^{1/2}. \qquad ({\rm W/m^3})$$

The loss time due to bremsstrahlung defined by $\tau_{\rm brems} = (3/2)n_{\rm e}T_{\rm e}\,/P_{\rm brems}$ is

$$\tau_{\rm brems} = 0.16 \frac{1}{Z_{\rm eff} n_{20}} \left(\frac{T_{\rm e}}{e}\right)^{1/2} \qquad ({\rm sec})$$

where n_{20} is in units of $10^{20}\,\mathrm{m}^{-3}$, T_e/e is in unit of eV. When $n_e \sim 10^{20}\,\mathrm{m}^{-3}$, and $T_e \sim 10\,\mathrm{keV}$, then we have $\tau_{\mathrm{brems}} \sim 16/Z_{\mathrm{eff}}(\mathrm{s})$. Therefore, if the radiation losses such as bremsstrahlung, recombination radiation, and line specter emission are enhanced much by impurity ions, fusion core plasma cannot be realized even by the radiation losses only. When the temperature of the plasma increases, the ions from the plasma hit the walls of the vacuum vessel and impurity ions are sputtered. When the sputtered impurities penetrate the plasma, the impurities are highly ionized and yield a large amount of radiation loss, which causes radiation cooling of the plasma. Therefore, impurity control is one of the most important subjects in fusion research.

The light impurities, such as C and O, can be removed by baking and discharge-cleaning of the vacuum vessel. The sputtering of heavy atoms (Fe, etc.) of the wall material itself can be avoided by covering the metal wall by carbon tiles. Furthermore a divertor, as shown in Figure 4.16, is very effective to reduce the plasma-wall interaction. Plasmas in *Scrape Off Layer* (SOL) flow at the velocity of sound along the lines of magnetic force just outside the *separatrix* S into the neutralized plates, where the plasmas are neutralized. Even if the material of the neutralized plates is sputtered, the atoms are ionized within the divertor regions near the neutralized plates. Since the thermal velocity of the heavy ions is much smaller than the flow velocity of the plasma (which is the same as the thermal velocity of hydrogen ions), they are unlikely to flow back into the main plasma. In the divertor region the electron temperature of the plasma becomes low because of impurity radiation cooling. Because of pressure equilibrium along the lines of magnetic force, the density in the divertor region near the neutralized plates becomes high. Therefore, the velocity of ions from the plasma into the neutralized plates is collisionally damped and sputtering is suppressed. A decrease in the impurity radiation in the main plasma can be observed by using a divertor configuration.

However, the scrape off layer of the divertor is not broad and most of the total energy loss is concentrated to the narrow region of the target divertor plate. The severe heat load to the divertor plate is one of the most critical issues for a reactor design. Physical processes in scrape off layer and divertor region are actively investigated experimentally and theoretically [4.24].

Let us consider the thermal transport in the scrape off layer. It is assumed that the thermal transport parallel to the magnetic line of force is dominated by classical electron thermal conduction and the thermal transport perpendicular to the magnetic field is anomalous thermal diffusion. We use a slab model as is shown in Figure 4.17 and omit the Boltzmann constant in front of temperature. Then we have

$$\nabla q_{\parallel} + \nabla q_{\perp} + Q_{\mathrm{rad}} = 0 \tag{4.53}$$

$$q_{\parallel} = -\kappa_c \frac{\partial T_e}{\partial s} = -\kappa_0 T_e^{5/2} \frac{\partial T_e}{\partial s} = -\frac{2}{7}\kappa_0 \frac{\partial T_e^{7/2}}{\partial s} \tag{4.54}$$

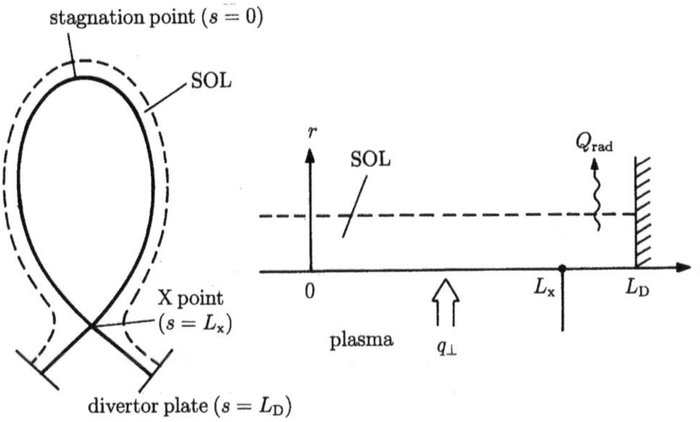

Figure 4.17 Configuration of scrape off layer (SOL) and divertor. The coordinate of the slab model (right-hand side)

$$q_\perp = -n_e\left(\chi_\perp^e \frac{\partial T_e}{\partial r} + \chi_\perp^i \frac{\partial T_i}{\partial r}\right) - \frac{3}{2}D(T_e + T_i)\frac{\partial n}{\partial r}, \tag{4.55}$$

$$\kappa_c \approx n_e \lambda_{ei}^2 \nu_{ei} = \frac{3^{1/2} 36\pi\epsilon^2 T_e^{5/2}}{m_e^{1/2} Z e^4 \ln \Lambda} \approx 1.25 \times 10^{22}\left(\frac{T_e}{e}\right)^{5/2} \mathrm{m^{-1}s^{-1}},$$

where T_e/e is in units of eV ($Z = 1,\ \ln \Lambda = 20$). Here $q_\parallel$ and $q_\perp$ are heat fluxes in the directions of parallel and perpendicular to the magnetic field and Q_{rad} is radiation loss. κ_c is heat conductivity and $\chi_\perp^e, \chi_\perp^i$ are thermal difusion coefficients and D is diffusion coefficient of particles. The stagnation point of heat flow is set as $s = 0$ and the X point of separatrix and divertor plate are set as $s = L_x$ and $s = L_D$, respectively. Then the boundary conditions at $s = 0$ and $s = L_D$ are

$$q_{\parallel 0} = 0, \tag{4.56}$$

$$q_{\parallel D} = \gamma T_D n_D u_D + \frac{1}{2} m_i u_D^2 n_D u_D + \xi n_D u_D$$

$$= n_D M_D c_s((\gamma + M_D^2)T_D + \xi), \tag{4.57}$$

where u_D is flow velocity of plasma at the divertor plate and M_D is Mach number $M_D = u_D/c_s$. $\gamma \approx 7$ is sheath energy transfer coefficient and $\xi \approx 20 \sim 27$eV is ionization energy. The sound velocity is $c_s = \tilde{c}_s T_D^{1/2}$, $\tilde{c}_s = 0.98(2/A_i)^{1/2} 10^4 \mathrm{ms^{-1}(eV)^{-1/2}}$, A_i being ion atomic mass. The first and second terms of (4.57) are the power flux into the sheath and the third term is power consumed within the recycling process. The equations of particles and momentum along the magnetic lines of force are

$$\frac{\partial(nu)}{\partial s} = S_i - S_{cx,r} - \nabla_\perp(nu_\perp) \approx S_i - S_{cx,r}, \tag{4.58}$$

$$mnu\frac{\partial u}{\partial s} = -\frac{\partial p}{\partial s} - muS_{\rm m}, \tag{4.59}$$

where $S_{\rm m} = nn_0\langle\sigma v\rangle_{\rm m}$ is the loss of momentum of plasma flow by collision with neutrals, $S_{\rm i} = nn_0\langle\sigma v\rangle_{\rm i}$ is the ionization term, and $S_{\rm cx,r} = nn_0\langle\sigma v\rangle_{\rm cx,r}$ is ion loss by charge exchange and radiation recombination. Equations (4.58) and (4.59) reduce to

$$\frac{\partial(nmu^2 + p)}{\partial s} = -mu(S_{\rm m} + S_{\rm cx,r}) + muS_{\rm i}. \tag{4.60}$$

The flow velocities at $s = 0$ and $s = L_{\rm D}$ are $u_0 = 0$ and $u_{\rm D} = M_{\rm D}c_{\rm s}, M_{\rm D} \approx 1$, respectively. Equations (4.53),(4.54) and the boundary conditions (4.56),(4.57) reduce to

$$\frac{2\kappa_0}{7}\frac{\partial^2}{\partial s^2}T_{\rm e}^{7/2} = \nabla_\perp q_\perp + Q_{\rm rad}, \tag{4.61}$$

$$\frac{2\kappa_0}{7}(T_{\rm e}^{7/2}(s) - T_{\rm eD}^{7/2}) = \int_{L_{\rm D}}^{s} {\rm d}s' \int_0^{s'} (\nabla_\perp q_\perp + Q_{\rm rad}){\rm d}s''. \tag{4.62}$$

When $\nabla_\perp q_\perp$ =const. $Q_{\rm rad} = 0$ in $0 < s < L_{\rm x}$ and $\nabla_\perp q = 0$, $Q_{\rm rad}$ =const. in $L_{\rm x} < s < L_{\rm D}$, we have

$$\frac{2\kappa_0}{7}(T_{\rm e}^{7/2}(s) - T_{\rm eD}^{7/2}) = 0.5(-\nabla_\perp q_\perp)(2L_{\rm x}L_{\rm D} - L_{\rm x}^2 - s^2) + 0.5Q_{\rm rad}(L_{\rm D} - L_{\rm x})^2,$$

for $(0 < s < L_{\rm x})$. When the radiation term is negligible, $T_{\rm e0} \equiv T_{\rm e}(0)$ becomes

$$T_{\rm e0}^{7/2} = T_{\rm eD}^{7/2} + \frac{7}{4\kappa_0}\left(\frac{2L_{\rm D}}{L_{\rm x}} - 1\right)(-\nabla_\perp q_\perp)L_{\rm x}^2.$$

If $T_{\rm eD} < 0.5T_{\rm e0}$ and $L_{\rm D} - L_{\rm x} \ll L_{\rm x}$, we have

$$T_{\rm e0} \approx 1.17\left(\frac{(-\nabla_\perp q_\perp)L_{\rm x}^2}{\kappa_0}\right)^{2/7} = 1.17\left(\frac{q_\perp L_{\rm x}^2}{\kappa_0\lambda_q}\right)^{2/7}, \tag{4.63}$$

where $1/\lambda_q \equiv -\nabla_\perp q_\perp/q_\perp$. When the scale lengths of gradients of temperature and density are λ_T and λ_n, respectively, $(T(r) = T\exp(-r/\lambda_T),\ n(r) = n\exp(-r/\lambda_n))$ and $\chi_\perp^{\rm i} \ll \chi_\perp^{\rm e}$ and$D \sim \chi_\perp^{\rm e}$ are assumed, (4.55) becomes

$$q_\perp = n\chi_\perp^{\rm e}\frac{T_{\rm e}}{\lambda_T}\left(1 + \frac{3}{2}(1 + \frac{T_{\rm i}}{T_{\rm e}})\frac{\lambda_T}{\lambda_n}\right). \tag{4.64}$$

Therefore, if $\chi^{\rm e}$ is known as a function $\chi^{\rm e}(T_{\rm e}, n, B)$, λ_T is given as $\lambda_T(T_{\rm e}, n, B, q_\perp)$.

Let us consider the relations between $n_{\rm s}, T_{\rm es}, T_{\rm is}$ at stagnation point $s = 0$ and $n_{\rm D}, T_{\rm D}$ at divertor plate $s = L_{\rm D}$. The momentum flux at divertor region

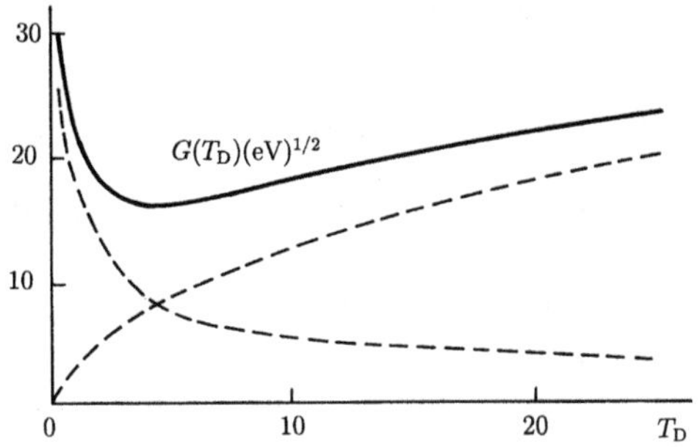

Figure 4.18 Dependence of $G(T_\mathrm{D})(\mathrm{eV})^{1/2}$ on $T_\mathrm{D}(\mathrm{eV})$

decreases due to collision with neutrals, charge exchange and ionization and becomes smaller than that at the stagnation point.

$$f_p \equiv \frac{2(1+M_\mathrm{D}^2)n_\mathrm{D}T_\mathrm{D}}{n_\mathrm{s}(T_\mathrm{es}+T_\mathrm{is})} < 1. \tag{4.65}$$

The power flux to divertor plate is reduced by radiation loss from the power flux $q_\perp L_\mathrm{x}$ into SOL through the separatrix with length of L_x

$$\int_0^\infty q_\parallel \mathrm{d}r = (1-f_\mathrm{rad})q_\perp L_\mathrm{x}, \tag{4.66}$$

where f_rad is the fraction of radiation loss. Equations (4.66) and (4.57) reduce

$$M_\mathrm{D}n_\mathrm{D}\tilde{c}_\mathrm{s}T_\mathrm{D}^{1/2}\left(\frac{(\gamma+M_\mathrm{D}^2)T_\mathrm{D}}{3/2\lambda_T+1/\lambda_n}+\frac{\xi}{1/(2\lambda_T)+1/\lambda_n}\right) = (1-f_\mathrm{rad})q_\perp L_\mathrm{x},$$

that is,

$$(1-f_\mathrm{rad})q_\perp L_\mathrm{x} = \frac{\tilde{c}_\mathrm{s}f_p\lambda_T}{1.5+\lambda_T/\lambda_n}n_\mathrm{s}\frac{T_\mathrm{es}+T_\mathrm{is}}{2}G(T_\mathrm{D}), \tag{4.67}$$

$$G(T_\mathrm{D}) \equiv \frac{M_\mathrm{D}}{1+M_\mathrm{D}^2}(\gamma+M_\mathrm{D})T_\mathrm{D}^{1/2}\left(1+\frac{1}{\gamma+M_\mathrm{D}}\frac{\bar{\xi}}{T_\mathrm{D}}\right). \tag{4.68}$$

The curve of $G(T_\mathrm{D})$ as the function of T_D is shown in Figure 4.18 and $G(T_\mathrm{D})$ has a minimum at $T_\mathrm{D} = \bar{\xi}/(\gamma+M_\mathrm{D}^2)$. In the case of $M_\mathrm{D} \approx 1, \gamma \approx 7, \xi = 24\mathrm{eV}$, $G(T_\mathrm{D})$ is

$$G_\mathrm{D} = 4T_\mathrm{D}^{1/2}\left(1+\frac{4.5}{T_\mathrm{D}}\right).$$

$G(T_\mathrm{D})$ is roughly proportional to $T_\mathrm{D}^{1/2}$ when $T_\mathrm{D} > 15\mathrm{eV}$ in this case. Since T_es depends on n_s through $\lambda_q^{-2/7}$ as is seen in (4.63), the dependence of T_es

on $n_{\rm s}$ is very weak. From (4.67) and (4.65), we have roughly the following relations:

$$T_{\rm D} \propto n_{\rm s}^{-2}, \qquad n_{\rm D} \propto n_{\rm s}^{3}, \tag{4.69}$$

and the density $n_{\rm D}$ at the divertor increases nonlinearly with the density $n_{\rm s}$ of upstream SOL.

When the upstream density $n_{\rm s}$ increases while keeping the left-hand side of (4.67) constant, the solution $T_{\rm D}$ of (4.67) cannot exist beyond a threshold density, since $G(T_{\rm D})$ has the minimum value (Figure 4.18). This is related to the phenomenon of detached plasma above a threshold of upstream density [4.24].

The heat load $\phi_{\rm D}$ of the divertor normal to the magnetic flux surface is given by

$$\phi_{\rm D} \approx \frac{(1-f_{\rm rad})P_{\rm sep}}{2\pi R 2\lambda_{\phi D}} = (1-f_{\rm rad})\pi K \frac{a}{\lambda_T} q_\perp \left(1.5 + \frac{\lambda_T}{\lambda_n}\right) \frac{B_{\theta{\rm D}}}{B_\theta} \tag{4.70}$$

where $P_{\rm sep}$ is the total power flux across the separatrix surface and $\lambda_{\phi D}$ is the radial width of heat flux at divertor plate; that is,

$$P_{\rm sep} = 2\pi a K 2\pi R q_\perp \qquad \lambda_{\phi D} = \lambda_T \frac{1}{1.5 + \lambda_T/\lambda_n} \frac{B_\theta}{B_{\theta{\rm D}}}.$$

The term $B_\theta/B_{\theta{\rm D}} = 2 \sim 3$ is the ratio of separations of magnetic flux surfaces at stagnation point and divertor plate. If the divertor plate is inclined with angle α against the magnetic flux surface, the heat load of the inclined divertor plate becomes $\sin\alpha$ times as small as that of the divertor normal to magnetic flux surface.

4.5 Classical, Neoclassical Transports and Bootstrap Current

Transport of plasmas is among the most important subjects in fusion research, with theoretical and experimental investigations being carried out concurrently. Although a general discussion of transport or confinement requires the consideration of the various instabilities (which will be discussed in subsequent chapters), it is also important to consider simple but fundamental diffusion for the ideal stable cases. A typical example is classical diffusion, in which collisions between electrons and ions are dominant effect. The neoclassical diffusion of toroidal plasmas confined in tokamak, for both the rare-collisional and collisional region is also an important example.

The transport equation of particles is (refer to (3.22))

$$\frac{\partial}{\partial t}n(\boldsymbol{r},t) + \nabla\cdot(n(\boldsymbol{r},t)\boldsymbol{V}(\boldsymbol{r},t)) = 0, \tag{4.71}$$

provided processes of the ionization of neutrals and the recombination of ions are negligible. The particle flux $\boldsymbol{\Gamma} = n\boldsymbol{V}$ is given by

$$n(\boldsymbol{r},t)\boldsymbol{V}(\boldsymbol{r},t) = -D(\boldsymbol{r},t)\nabla n(\boldsymbol{r},t),$$

in many cases, where D is the diffusion coefficient. (Additional terms may be necessary in more general cases.)

Diffusion coefficient D and *particle confinement time* $\tau_{\rm p}$ are related by the diffusion equation of the plasma density n as follows:

$$\nabla\cdot(D\nabla n(\boldsymbol{r},t)) = \frac{\partial}{\partial t}n(\boldsymbol{r},t).$$

Substitution of $n(\boldsymbol{r},t) = n(\boldsymbol{r})\exp(-t/\tau_{\rm p})$ in diffusion equation yields

$$\nabla\cdot(D\nabla n(\boldsymbol{r})) = -\frac{1}{\tau_{\rm p}}n(\boldsymbol{r}).$$

When D is constant and the plasma column is a cylinder of radius a, the diffusion equation is reduced to

$$\frac{1}{r}\frac{\partial}{\partial r}\left(r\frac{\partial n}{\partial r}\right) + \frac{1}{D\tau_{\rm p}}n = 0.$$

The solution satisfying the boundary condition $n(a) = 0$ is

$$n = n_0 J_0\left(\frac{2.4r}{a}\right)\exp\left(-\frac{t}{\tau_{\rm p}}\right),$$

and the particle confinement time is

$$\tau_{\rm p} = \frac{a^2}{2.4^2 D} = \frac{a^2}{5.8D}, \tag{4.72}$$

where J_0 is the zeroth-order Bessel function. The relationship (4.72) between the particle confinement time $\tau_{\rm p}$ and D holds generally, with only a slight modification of the numerical factor. This formula is frequently used to obtain the diffusion coefficient from the observed values of the plasma radius and particle confinement time.

The equation of energy balance is given by (Chapter 6 of [4.25])

$$\frac{\partial}{\partial t}\left(\frac{3}{2}nT\right) + \nabla\cdot\left(\frac{3}{2}Tn\boldsymbol{V}\right) + \nabla\cdot\boldsymbol{q} = Q - p\nabla\cdot\boldsymbol{V} - \sum_{ij}\Pi_{ij}\frac{\partial V_i}{\partial x_j}. \tag{4.73}$$

The first term in the right-hand side is the heat generation due to particle collisions per unit volume per unit time, the second term is the work done by pressure, and the third term is viscous heating. The first term in the left-hand side is the time derivative of the thermal energy per unit volume, the second term is convective energy loss, and the third term is conductive energy loss. Denoting the *thermal conductivity* by κ_{T}, the *thermal flux* due to heat conduction may be expressed by

$$\boldsymbol{q} = -\kappa_{\mathrm{T}}\nabla T.$$

If the *convective loss*, in the second term of the left-hand side of (4.73), is neglected and the terms in the right-hand side are also negligible, we find that

$$\frac{3}{2}\frac{\partial}{\partial t}(nT) - \nabla\cdot\kappa_{\mathrm{T}}\nabla T = 0.$$

In the case of $n = \text{const.}$, this equation reduces to

$$\frac{\partial T}{\partial t} = \nabla\cdot\left(\frac{2}{3}\frac{\kappa_{\mathrm{T}}}{n}\nabla T\right).$$

When the thermal diffusion coefficient χ_{T} is defined by

$$\chi_{\mathrm{T}} = \frac{\kappa_{\mathrm{T}}}{n},$$

the same equation on T is obtained as (4.71). In the case of $\chi_{\mathrm{T}} = \text{const.}$, the solution is

$$T = T_0 J_0\left(\frac{2.4}{a}r\right)\exp\left(-\frac{t}{\tau_{\mathrm{E}}}\right), \qquad \tau_{\mathrm{E}} = \frac{a^2}{5.8(2/3)\chi_{\mathrm{T}}}. \tag{4.74}$$

The term τ_{E} is called *energy confinement time.*

4.5.1 Classical Diffusion (Collisional Diffusion)

Magnetohydrodynamic Treatment

A magnetohydrodynamic (MHD) treatment is applicable to diffusion phenomena when the electron-to-ion collision frequency is large and the mean free path is shorter than the *connection length* of the inside regions of good curvature and the outside region of bad curvature of the torus; i.e.,

$$\frac{v_{\mathrm{Te}}}{\nu_{\mathrm{ei}}} < \frac{2\pi R}{\iota} = q_{\mathrm{s}}R,$$

$$\nu_{\mathrm{ei}} > \nu_{\mathrm{p}} \equiv \frac{1}{q_{\mathrm{s}}R}v_{\mathrm{Te}} = \frac{1}{q_{\mathrm{s}}R}\left(\frac{T_{\mathrm{e}}}{m_{\mathrm{e}}}\right)^{1/2}.$$

MHD treatment can be applied on the plasma diffusion. v_{Te} is electron thermal velocity and ν_{ei} is electron to ion collisional frequency. From Ohm's law (3.28)

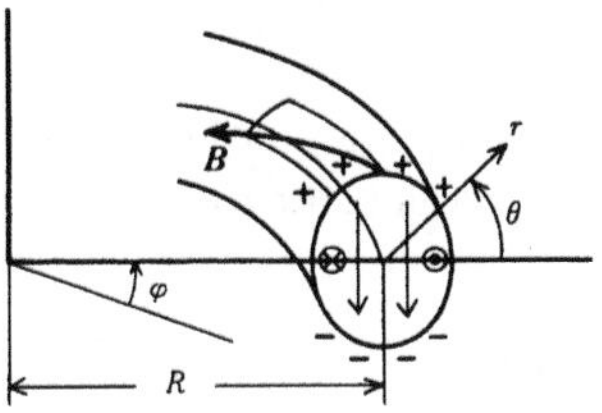

Figure 4.19 Electric field in a plasma confined in a toroidal field. The symbols ⊗ and ⊙ here show the direction of the Pfirsch-Schlüter current.

$$\boldsymbol{E} + \boldsymbol{V} \times \boldsymbol{B} - \frac{1}{en}\nabla p_{\rm i} = \eta \boldsymbol{j},$$

the motion of plasma across the lines of magnetic force is expressed by

$$n\boldsymbol{v}_\perp = \frac{1}{B}\left(\left(n\boldsymbol{E} - \frac{T_{\rm i}}{e}\nabla n\right) \times \boldsymbol{b}\right) - \frac{m_{\rm e}\nu_{\rm ei}}{e^2}\frac{\nabla p}{B^2}$$

$$= \frac{1}{B}\left(\left(n\boldsymbol{E} - \frac{T_{\rm i}}{e}\nabla n\right) \times \boldsymbol{b}\right) - (\rho_{\Omega \rm e})^2 \nu_{\rm ei}\left(1 + \frac{T_{\rm i}}{T_{\rm e}}\right)\nabla n, \qquad (4.75)$$

where $\rho_{\Omega \rm e} = v_{\rm Te}/\Omega_{\rm e}$, $v_{\rm Te} = (T_{\rm e}/m_{\rm e})^{1/2}$ and $\eta = m_{\rm e}\nu_{\rm ei}/e^2 n_{\rm e}$ (see (2.34)).

If the first term in the right-hand side can be neglected, the particle diffusion coefficient D is given by

$$D = (\rho_{\Omega \rm e})^2 \nu_{\rm ei}\left(1 + \frac{T_{\rm i}}{T_{\rm e}}\right). \qquad (4.76)$$

The *classical diffusion coefficient* $D_{\rm ei}$ is defined by

$$D_{\rm ei} \equiv (\rho_{\Omega \rm e})^2 \nu_{\rm ei} = \frac{nT_{\rm e}}{\sigma_\perp B^2} = \frac{\beta_{\rm e}\eta_\parallel}{\mu_0}, \qquad (4.77)$$

where $\sigma_\perp = n_{\rm e}e^2/(m_{\rm e}\nu_{\rm ei})$, $\eta_\parallel = 1/2\sigma_\perp$.

However, the first term of the right-hand side of (4.75) is not always negligible. In toroidal configuration, the charge separation due to the toroidal drift is not completely cancelled along the magnetic field lines due to the finite resistivity and an electric field $\boldsymbol{E}$ arises (see Figure 4.19). Therefore, the $\boldsymbol{E} \times \boldsymbol{b}$ term in (4.75) contributes to the diffusion. Let us consider this term. From the equilibrium equation, the diamagnetic current

$$\boldsymbol{j}_\perp = \frac{\boldsymbol{b}}{B} \times \nabla p, \qquad j_\perp = \left|\frac{1}{B}\frac{\partial p}{\partial r}\right|$$

flows in the plasma. From $\nabla \cdot \boldsymbol{j} = 0$, we find $\nabla \cdot \boldsymbol{j}_\parallel = -\nabla \cdot \boldsymbol{j}_\perp$. By means of the equation $B = B_0(1 - (r/R)\cos\theta)$, $j_\parallel$ may be written as (see (4.44))

$$j_\parallel = 2q_s \frac{1}{B_0}\frac{\partial p}{\partial r}\cos\theta.$$

If the electric conductivity along the lines of magnetic force is $\sigma_\parallel$, the parallel electric field is $E_\parallel = j_\parallel/\sigma_\parallel$. As is clear from Figure 4.19, the relation

$$\frac{E_\theta}{E_\parallel} \approx \frac{B_0}{B_\theta}$$

holds. From $B_\theta/B_0 \approx (r/qR)$, the θ component of the electric field is given by

$$E_\theta = \frac{B_0}{B_\theta}E_\parallel = \frac{qR}{r}\frac{1}{\sigma_\parallel}j_\parallel = \frac{2}{\sigma_\parallel}\frac{R}{r}q_s^2\frac{1}{B_0}\frac{\partial p}{\partial r}\cos\theta.$$

Accordingly (4.75) is reduced to

$$\begin{aligned} nV_r &= -n\frac{E_\theta}{B} - (\rho_{\Omega e})^2\nu_{ei}\left(1+\frac{T_i}{T_e}\right)\frac{\partial n}{\partial r} \\ &= -\left(\frac{R}{r}\cdot 2q_s^2\frac{nT_e}{\sigma_\parallel B_0^2}\cos\theta\left(1+\frac{r}{R}\cos\theta\right)\right. \\ &\quad \left. +\frac{nT_e}{\sigma_\perp B_0^2}\left(1+\frac{r}{R}\cos\theta\right)^2\right)\times\left(1+\frac{T_i}{T_e}\right)\frac{\partial n}{\partial r}. \end{aligned}$$

Noting that the area of a surface element is dependent on θ, and taking the average of nV_r over θ, we find that

$$\begin{aligned} \langle nV_r\rangle &= \frac{1}{2\pi}\int_0^{2\pi} nV_r\left(1+\frac{r}{R}\cos\theta\right)\mathrm{d}\theta \\ &= -\frac{nT_e}{\sigma_\perp B_0^2}\left(1+\frac{T_i}{T_e}\right)\left(1+\frac{2\sigma_\perp}{\sigma_\parallel}q_s^2\right)\frac{\partial n}{\partial r}. \end{aligned}$$

The diffusion coefficient $D_{\mathrm{P.S.}}$ of toroidal plasma is $(1+q_s^2)$ times as large as the diffusion coefficient of (4.76). This value $(1+q_s^2)$ is called the *Pfirsch-Schlüter factor* [4.11]. When the safety factor q_s is about 3, the Pfirsch-Schlüter factor is 10.

A Particle Model

The classical diffusion coefficient of electrons

$$D_{ei} = (\rho_{\Omega e})^2\nu_{ei}$$

is that for electrons which move in a random walk with a step length equal to the Larmor radius. Let us consider a toroidal plasma. For rotational transform angle ι, the displacement Δ of the electron drift surface from the magnetic surface is (see Figure 4.20)

$$\Delta \approx \pm q_s\rho_{\Omega e}.$$

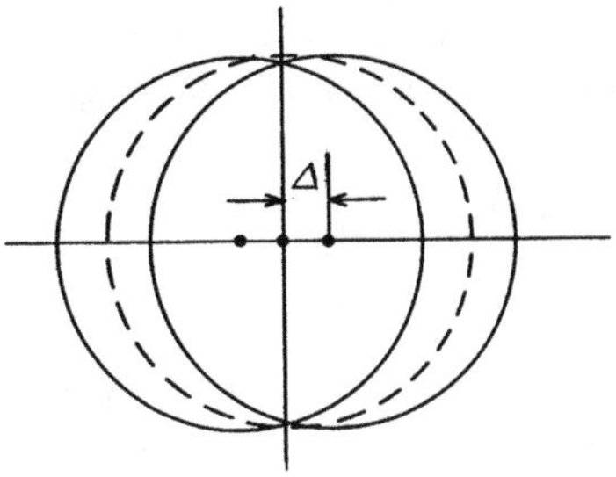

Figure 4.20 Magnetic surface (dotted line) and drift surfaces (solid lines).

The ± signs depend on the direction of electron motion being parallel or antiparallel to the magnetic field (see (2.57)). As an electron can be transferred from one drift surface to the other by collision, the step length across the magnetic field is $\Delta = q_s \rho_{\Omega e}$, the diffusion coefficient of the electron is given by

$$D_{P.S.} = \Delta^2 \nu_{ei} = q_s^2 (\rho_{\Omega e})^2 \nu_{ei}; \tag{4.78}$$

thus, the Pfirsch-Schlüter factor has been reduced ($q_s^2 \gg 1$ is assumed). The diffusion coefficient of (4.78) is called Pfirsch-Schlüter diffusion coefficient [4.11].

4.5.2 Neoclassical Diffusion of Electrons in Tokamak

The magnitude B of the magnetic field of a tokamak is given by

$$B = \frac{RB_0}{R(1 + \epsilon_t \cos\theta)} = B_0(1 - \epsilon_t \cos\theta), \tag{4.79}$$

where

$$\epsilon_t = \frac{r}{R}.$$

Consequently, when the perpendicular component $v_\perp$ of an electron velocity is much larger than the parallel component $v_\parallel$, i.e., when

$$\left(\frac{v_\perp}{v}\right)^2 > \frac{R}{R+r},$$

that is

$$\frac{v_\perp}{v_\parallel} > \frac{1}{\epsilon_t^{1/2}}, \tag{4.80}$$

the electron is trapped outside of the torus, where the magnetic field is weak. Such an electron drifts in a banana orbit (see Figure 2.15). In order to complete a circuit of the banana orbit, the effective collision time $\tau_{\mathrm{eff}} = 1/\nu_{\mathrm{eff}}$ of the trapped electron must be longer than one period τ_{b} of banana orbit

$$\tau_{\mathrm{b}} \approx \frac{q_{\mathrm{s}}R}{v_{\parallel}} = \frac{q_{\mathrm{s}}R}{v_{\perp}\epsilon_{\mathrm{t}}^{1/2}}.$$

The effective collision frequency ν_{eff} of the trapped electron is the frequency in which the condition (4.80) of trapped electron is violated by collision. As the collision frequency ν_{ei} is the inverse of diffusion time required to change the directon of velocity by 1 radian, the effective collisional frequency ν_{eff} is given by

$$\nu_{\mathrm{eff}} = \frac{1}{\epsilon_{\mathrm{t}}}\nu_{\mathrm{ei}}. \tag{4.81}$$

Accordingly, if $\nu_{\mathrm{eff}} < 1/\tau_{\mathrm{b}}$, i.e.,

$$\nu_{\mathrm{ei}} < \nu_{\mathrm{b}} \equiv \frac{v_{\perp}\epsilon_{\mathrm{t}}^{3/2}}{q_{\mathrm{s}}R} = \epsilon_{\mathrm{t}}^{3/2}\frac{1}{q_{\mathrm{s}}R}\left(\frac{T_{\mathrm{e}}}{m_{\mathrm{e}}}\right)^{1/2}, \tag{4.82}$$

the trapped electron can travel the entire banana orbit. When the trapped electron collides, it can shift its position by an amount of the banana width (see (2.60))

$$\Delta_{\mathrm{b}} = \frac{mv_{\parallel}}{eB_{\mathrm{p}}} \approx \frac{mv_{\perp}}{eB}\frac{v_{\parallel}}{v_{\perp}}\frac{B}{B_{\mathrm{p}}} \approx \rho_{\Omega\mathrm{e}}\epsilon_{\mathrm{t}}^{1/2}\frac{q_{\mathrm{s}}R}{r} = q_{\mathrm{s}}\epsilon_{\mathrm{t}}^{-1/2}\rho_{\Omega\mathrm{e}}. \tag{4.83}$$

As the number of trapped electrons is $\epsilon_{\mathrm{t}}^{1/2}$ times the total number of electrons, the trapped-electron contribution to diffusion is

$$D_{\mathrm{G.S.}} = \epsilon_{\mathrm{t}}^{1/2}\Delta_{\mathrm{b}}^{2}\nu_{\mathrm{eff}} = \epsilon_{\mathrm{t}}^{1/2}q_{\mathrm{s}}^{2}\epsilon_{\mathrm{t}}^{-1}(\rho_{\Omega\mathrm{e}})^{2}\frac{1}{\epsilon_{\mathrm{t}}}\nu_{\mathrm{ei}}$$

$$= \epsilon_{\mathrm{t}}^{-3/2}q_{\mathrm{s}}^{2}(\rho_{\Omega\mathrm{e}})^{2}\nu_{\mathrm{ei}}. \tag{4.84}$$

This diffusion coefficient, introduced by Galeev and Sagdeev [4.26], is $\epsilon_{\mathrm{t}}^{-3/2} = (R/r)^{3/2}$ times as large as the diffusion coefficient for collisional case. This derivation is a semi-quantitative discussion. The more rigorous discussion is given in [4.26].

MHD treatment is applicable if the electron to ion collision frequency is larger than the frequency ν_{p} given by

$$\nu_{\mathrm{p}} = \frac{1}{q_{\mathrm{s}}R}v_{\mathrm{Te}} = \frac{1}{q_{\mathrm{s}}R}\left(\frac{T_{\mathrm{e}}}{m_{\mathrm{e}}}\right)^{1/2}. \tag{4.85}$$

When the electron to ion collision frequency is smaller than the frequency

$$\nu_{\mathrm{b}} = \epsilon_{\mathrm{t}}^{3/2}\nu_{\mathrm{p}}, \tag{4.86}$$

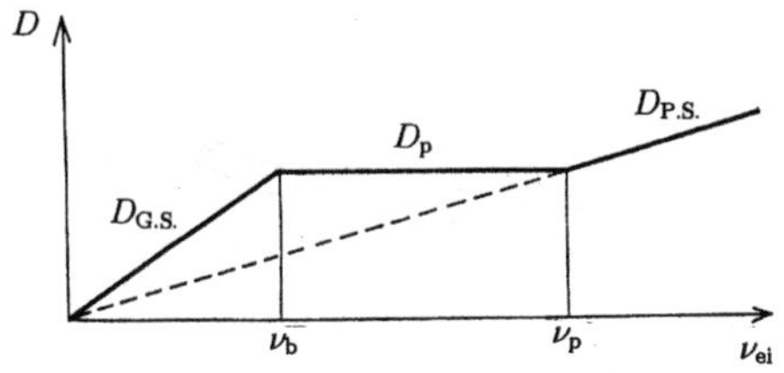

Figure 4.21 Dependence of the diffusion coefficient on collision frequency in a tokamak. $\nu_p = v_{Te}/q_s R$, $\nu_b = \epsilon_t^{3/2}\nu_p$.

the electron can complete a banana orbit. The diffusion coefficients are written by

$$D_{P.S.} = q_s^2(\rho_{\Omega e})^2\nu_{ei}, \quad \nu_{ei} > \nu_p, \tag{4.87}$$

$$D_{G.S.} = \epsilon_t^{-3/2} q_s^2(\rho_{\Omega e})^2\nu_{ei}, \quad \nu_{ei} < \nu_b = \epsilon_t^{3/2}\nu_p. \tag{4.88}$$

If ν_{ei} is in the region $\nu_b < \nu_{ei} < \nu_p$, it is not possible to treat the diffusion phenomena of electrons in this region by means of a simple model. In this region we must resort to the drift approximation of Vlasov's equation. The result is that the diffusion coefficient is not sensitive to the collision frequency in this region and is given by [4.26],[4.27]

$$D_p = q_s^2(\rho_{\Omega e})^2\nu_p, \quad \nu_p > \nu_{ei} > \nu_b = \epsilon_t^{3/2}\nu_p. \tag{4.89}$$

The dependence of the diffusion coefficient on the collision frequency is shown in Figure 4.21. The region $\nu_{ei} > \nu_p$ is called the *MHD region* or *collisional region.* The region $\nu_p > \nu_{ei} > \nu_b$ is the *plateau region* or *intermediate region*; and the region $\nu_{ei} < \nu_b$ is called the *banana region* or *rare collisional region.* These diffusion processes are called neoclassical diffusion. There is an excellent review [4.27] on neoclassical diffusion.

The reason that the electron-electron collison frequency does not affect the electron's particle diffusion coefficient is that the center-of-mass velocity does not change by the Coulomb collision.

The neoclassical thermal diffusion coefficient χ_{Te} is the same order as the particle diffusion coefficient ($\chi_{Te} \sim D_e$). Although ion collision with the same ion species does not affect the ion's particle diffusion coefficient, it does contribute thermal diffusion processes, if temperature gradient exists. Even if the ions are the same species, it is possible to distinguish hot ion (with larger thermal velocity) and cold ion. Accordingly, the ion's thermal diffusion coefficient in banana region is given by $\chi_{Ti} \sim \epsilon_t^{-3/2} q_s^2 \rho_{\Omega i}^2 \nu_{ii}$,

$$\chi_{Ti} \sim (m_i/m_e)^{1/2} D_{ie} \qquad (D_{ie} \sim D_{ei}).$$

Therefore, the ion's thermal diffusion coefficient is about $(m_i/m_e)^{1/2}$ times as large as the ion's particle diffusion coefficient.

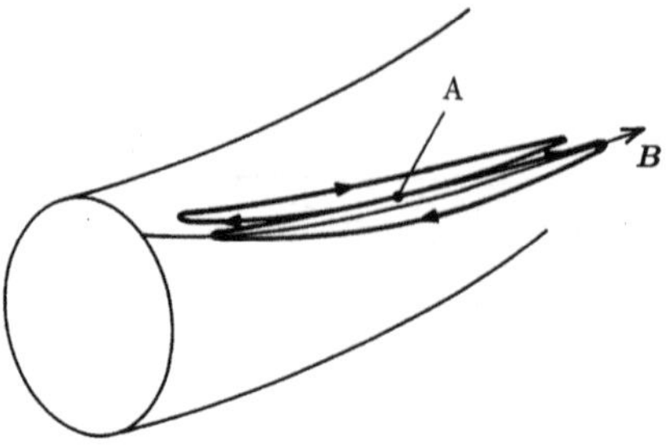

Figure 4.22 Banana orbits of trapped electrons which induce the bootstrap current.

4.5.3 Bootstrap Current

It was predicted theoretically that radial diffusion induces a current in the toroidal direction and the current can be large in the banana region [4.28]~[4.31]. Later this current called *bootstrap current* had been well confirmed experimentally. This is an important process which can provide the means to sustain the plasma current in tokamak in steady state.
As was described in Section 4.5.2, electrons in collisionless region $\nu_{\rm ei} < \nu_{\rm b}$ make complete circuit of the banana orbit. When density gradient exists, there is a difference in particle number on neighboring orbit passing through a point A, as is shown in Figure 4.22. The difference is $({\rm d}n_{\rm t}/{\rm d}r)\Delta_{\rm b}$, $\Delta_{\rm b}$ being the width of the banana orbit. As the component of velocity parallel to the magnetic field is $v_\parallel = \epsilon^{1/2} v_{\rm T}$, the current density due to the trapped electrons with the density $n_{\rm t}$ is

$$j_{\rm banana} = -(ev_\parallel)\left(\frac{{\rm d}n_{\rm t}}{{\rm d}r}\Delta_{\rm b}\right) = -\epsilon^{3/2}\frac{1}{B_{\rm p}}\frac{{\rm d}p}{{\rm d}r}.$$

The untrapped electrons start to drift in the same direction as the trapped electrons due to the collisions with the trapped electrons and the drift becomes steady state due to the collisions with the ions. The drift velocity $V_{\rm untrap}$ of untrapped electrons in steady state is given by

$$m_{\rm e} V_{\rm untrap} \nu_{\rm ei} = \frac{\nu_{\rm ee}}{\epsilon} m_{\rm e} \left(\frac{j_{\rm banana}}{-en_{\rm e}}\right),$$

where $\nu_{\rm ee}/\epsilon$ is effective collision frequency between trapped and untrapped electrons. The current density due to the drift velocity $V_{\rm untrap}$ is

$$j_{\rm boot} \approx -\epsilon^{1/2}\frac{1}{B_{\rm p}}\frac{{\rm d}p}{{\rm d}r}. \tag{4.90}$$

This current is called *bootstrap current.* When the average poloidal beta $\beta_{\rm p} = \langle p\rangle/(B_{\rm p}^2/2\mu_0)$ is used, the ratio of the total bootstrap current $I_{\rm b}$ and the plasma current $I_{\rm p}$ to form $B_{\rm p}$ is given by

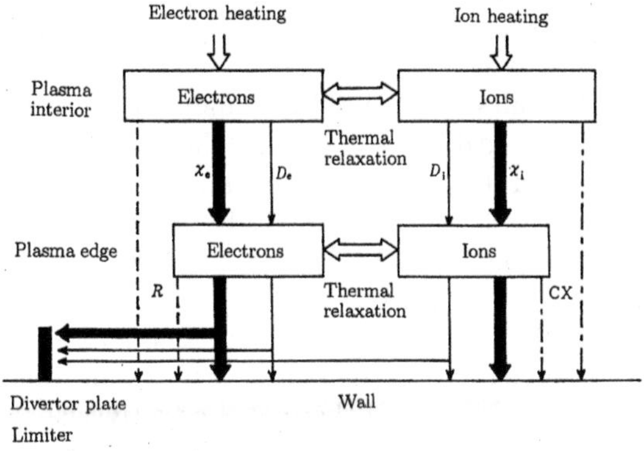

Figure 4.23 Energy flow of ions and electrons in a plasma. Bold arrows, thermal conduction (χ). Light arrows, convective loss (D). Dashed arrows, radiation loss (R). Dot-dashed arrows, charge exchange loss (CX).

$$\frac{I_b}{I_p} \sim c \left(\frac{a}{R}\right)^{1/2} \beta_p, \tag{4.91}$$

where $c \sim 0.3$ is constant. This value can be near 1 if β_p is high ($\beta_p \sim R/a$) and the pressure profile is peaked. Experiments on bootstrap current were carried out in TFTR, JT60U, and JET. 70 %$\sim$ 80 % of I_p=1 MA was bootstrap driven in high β_p operation.

As the bootstrap current profile is hollow, it can produce negative magnetic shear q profile, which is stable against ballooning (refer to Section 8.6). MHD stability of hollow current profile is analyzed in detail in [4.32].

4.6 Confinement Scalings of L Mode and H Mode

The energy flow of ions and electrons inside the plasma is in Figure 4.23. Denote the heating power into the electrons per unit volume by P_{he} and the radiation loss and the energy relaxation of electrons with ions by R and P_{ei}, respectively; then the time derivative of the electron thermal energy per unit volume is given by

$$\frac{d}{dt}\left(\frac{3}{2}n_e T_e\right) = P_{he} - R - P_{ei} + \frac{1}{r}\frac{\partial}{\partial r} r \left(\chi_e \frac{\partial T_e}{\partial r} + D_e \frac{3}{2} T_e \frac{\partial n_e}{\partial r}\right),$$

where χ_e is the electron thermal conductivity and D_e is the electron diffusion coefficient. Concerning the ions, the same relation is derived, but instead of the radiation loss the charge exchange loss L_{ex} of ions with neutrals must be taken into account, and then

$$\frac{\mathrm{d}}{\mathrm{d}t}\left(\frac{3}{2}n_{\mathrm{i}}T_{\mathrm{i}}\right) = P_{\mathrm{hi}} - L_{\mathrm{cx}} + P_{\mathrm{ei}} + \frac{1}{r}\frac{\partial}{\partial r}r\left(\chi_{\mathrm{i}}\frac{\partial T_{\mathrm{i}}}{\partial r} + D_{\mathrm{i}}\frac{3}{2}T_{\mathrm{i}}\frac{\partial n_{\mathrm{i}}}{\partial r}\right).$$

The experimental results of heating by ohmic and neutral beam injection can be explained by classical processes. The efficiency of wave heating can be estimated fairly accurately by theoretical analysis. The radiation and the charge exchange loss are classical processes. In order to evaluate the energy balance of the plasma experimentally, it is necessary to measure the fundamental quantities $n_{\mathrm{e}}(r,t), T_{\mathrm{i}}(r,t), T_{\mathrm{e}}(r,t)$, etc. According to the many experimental results, the energy relaxation between ions and electrons is classical, and the observed ion thermal conductivities in some cases are around $2 \sim 3$ times the neoclassical thermal conductivity;

$$\chi_{\mathrm{i,nc}} = n_{\mathrm{i}} f(q_{\mathrm{s}}, \varepsilon) q_{\mathrm{s}}^2 (\rho_{\Omega\mathrm{i}})^2 \nu_{\mathrm{ii}},$$

($f = 1$ in the Pfirsch-Schlüter region and $f = \epsilon_{\mathrm{t}}^{-3/2}$ in the banana region) and the observed ion thermal conductivities in some other cases are anomalous. The electron thermal conduction estimated by the experimental results is always anomalous and is much larger than the neoclassical one (more than one order of magnitude larger). In most cases the energy confinement time of the plasma is determined mostly by electron thermal conduction loss. The total energy confinement times τ_{E} is defined by

$$\tau_{\mathrm{E}} \equiv \frac{\int (3/2)(n_{\mathrm{e}}T_{\mathrm{e}} + n_{\mathrm{i}}T_{\mathrm{i}})\mathrm{d}V}{P_{\mathrm{in}}}.$$

The energy confinement time τ_{OH} of an ohmically heated plasma is well described by Alcator (neo-Alcator) scaling as follows (units are $10^{20}\mathrm{m}^{-3}$, m):

$$\tau_{\mathrm{OH}}(\mathrm{s}) = 0.103 q_{\mathrm{s}}^{0.5} \bar{n}_{\mathrm{e}20} a^{1.04} R^{2.04}.$$

However, the linearity of τ_{OH} on the average electron density $\bar{n}_{\mathrm{e}}$ deviates in the high-density region $\bar{n}_{\mathrm{e}} > 2.5 \times 10^{20}\,\mathrm{m}^{-3}$ and τ_{OH} tends to saturate. When the plasma is heated by high-power NBI or wave heating, the energy confinement time degrades as the heating power increases. Kaye and Goldston examined many experimental results of NBI heated plasma and derived the so-called Kaye-Goldston scaling on the energy confinement time [4.33], that is,

$$\tau_{\mathrm{E}} = (1/\tau_{\mathrm{OH}}^2 + 1/\tau_{\mathrm{AUX}}^2)^{-1/2},$$

$$\tau_{\mathrm{AUX}}(\mathrm{s}) = 0.037 \kappa_{\mathrm{s}}^{0.5} I_{\mathrm{p}} P_{\mathrm{tot}}^{-0.5} a^{-0.37} R^{1.75}, \tag{4.92}$$

where units are MA, MW, m, κ_{s} is the elongation ratio of noncircularity and P_{tot} is the total heating power in MW.

The ITER team assembled data from larger and more recent experiments. Analysis of the database of L mode experiments (see next section) led to the proposal of the following ITER-P scaling [4.34]:

$$\tau_{\mathrm{E}}^{\mathrm{ITER-P}}(\mathrm{s}) = 0.048 I_{\mathrm{p}}^{0.85} R^{1.2} a^{0.3} \bar{n}_{20}^{0.1} B^{0.2} \left(A_{\mathrm{i}}\kappa_{\mathrm{s}}/P\right)^{1/2}. \tag{4.93}$$

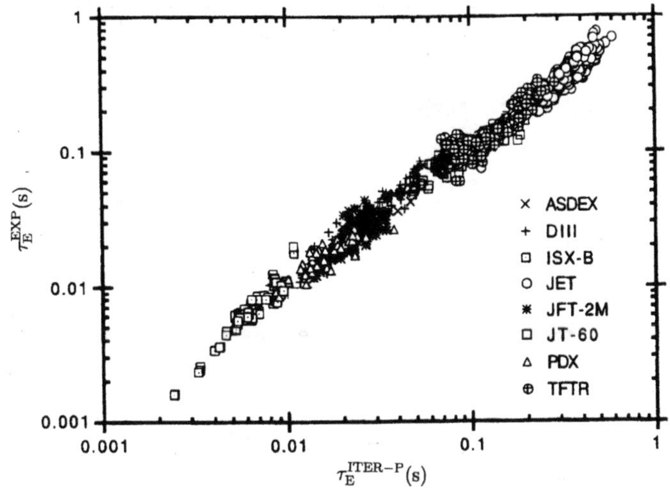

Figure 4.24 Comparison of confinement scaling τ_E^{ITER-P} with experimental data of energy confinement time τ_E^{EXP} of L mode. After [4.34].

where units are MA, m, T, MW and the unit of $\bar{n}_{20}$ is $10^{20}m^{-3}$. P is the heating power corrected for radiation P_R ($P = P_{tot} - P_R$). A comparison of confinement scaling τ_E^{ITER-P} with the experimental data of L mode is presented in Figure 4.24. For burning plasmas, heating power is roughly equal to α particle fusion output power $P_\alpha \approx 0.04 n_{DT20}^2 T^2 A a^3 \kappa_s$ (MW, $10^{20}m^{-3}$, keV, m) at around $T \sim 10$keV (refer to Section 4.8). It is interesting to note that the required condition for sustaining burning plasma by α particle heating depends mainly only on the product of AI_p in the case of Kaye-Goldston scaling ($A = R/a$ is the aspect ratio. Refer to Section 4.8).

An improved confinement state "*H mode*" was found in the ASDEX [4.35, 4.36] experiments with divertor configuration. When the NBI heating power is larger than a threshold value in the divertor configuration, the D_α line of deuterium (atom flux) in the edge region of the deuterium plasma decreases suddenly (time scale of 100 μs) during discharge, and recycling of deuterium atoms near the boundary decreases. At the same time there is a marked change in the edge radial electric field E_r (toward negative). Furthermore, the electron density and the thermal energy density increase and the energy confinement time of NBI heated plasma is improved by a factor of about 2. H mode was observed in PDX, JFT-2, DIII-D, JET, JT60U, and so on. The confinement state following Kaye-Goldston scaling is called the "*L mode*". In the H mode, the gradients of electron temperature and the electron density become steep just at the inside of the plasma boundary determined by the separatrix. In the spontaneous H mode, E_r becomes more negative (inward) (see Figure 4.25) [4.37, 4.38]. Theoretical

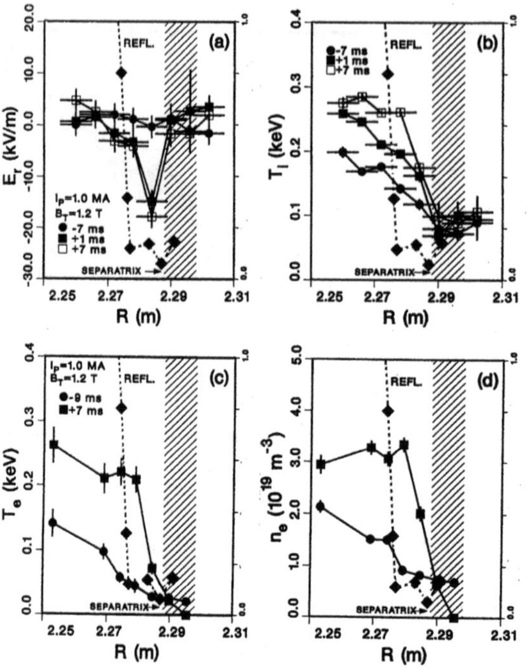

Figure 4.25 Plots of various edge plasma profiles at times spanning the L-H transition in DIIID. (a) E_r profile, (b) profiles of the ion temperature measured by CVII charge exchange recombination spectroscopy, (c),(d) profiles of electron temperature and electron density measured by Thomson scattering. After [4.38].

and experimental studies on L-H transion or bifurcation have been actively curried out [4.39, 4.40]. The radial electric field near plasma boundary is driven by several mechanisms such as momentum injection due to NBI or ion orbit loss near the plasma boundary or non-ambipolar flux.

The radial electric field causes plasma rotation with the velocity of $v_\theta = -E_r/B$ in the poloidal direction and with the velocity $v_\phi = -(E_r/B)(B_\theta/B)$ in the toroidal direction. If the gradient of E_r exists, sheared poloidal rotation and sheared toroidal rotation are generated. The importance of sheared flow for supression of edge turbulence and for improved confinement was pointed out in [4.41].

Let us consider the following fluid model:

$$\left(\frac{\partial}{\partial t} + (\boldsymbol{v}_0 + \tilde{\boldsymbol{v}}) \cdot \nabla + L_{\rm d}\right) \tilde{\xi} = \tilde{s},$$

where $\tilde{\xi}$ is the fluctuating field. $\boldsymbol{v}_0$ is taken to be the equilibrium $\boldsymbol{E} \times \boldsymbol{B}$ flow. $\tilde{s}$ represents a driving source of the turbulence and $L_{\rm d}$ is an operator responsible for dissipation of turbulence. The mutual correlation function $\langle \tilde{\xi}(1)\tilde{\xi}(2)\rangle$ of the fluctuating field $\tilde{\xi}(1)$ at a point 1 and $\tilde{\xi}(2)$ at a point 2 is given by [4.42]

$$\left(\frac{\partial}{\partial t} + (v'_\theta - v_\theta/r_+)r_+\frac{\partial}{\partial y_-} - \frac{\partial}{\partial r_+}D(r_+,y_-)\frac{\partial}{\partial r_+} + L_{\rm d}\right)\langle\tilde{\xi}(1)\tilde{\xi}(2)\rangle = T, \tag{4.94}$$

where D is radial diffusion coefficient of turbulence and T is the driving term and $r_+ = (r_1 + r_2)/2$, $\theta_- = \theta_1 - \theta_2$, $y_- = r_+\theta_-$. The decorrelation time $\tau_{\rm d}$ in the poloidal direction is the time in which the relative poloidal displacement between point 1 and point 2 due to sheared flow becomes the space correlation length of the turbulence k_{0k}^{-1}; that is,

$$k_{0k}\delta y \sim 1,$$

$$\delta y = v'_\theta(\Delta r)\tau_{\rm d},$$

$$\tau_{\rm d} = \frac{1}{v'_\theta \Delta r k_{0k}}.$$

The decorrelation rate $\omega_{\rm s}$ in the poloidal direction is

$$\omega_{\rm s} = \frac{1}{\tau_{\rm d}} = (\Delta r k_{0k})v'_\theta.$$

When Δr is the radial correlation length of the turbulence, the radial decorrelation rate $\Delta\omega_{\rm t}$ is given by

$$\Delta\omega_{\rm t} = \frac{D}{(\Delta r)^2}.$$

Since there is strong mutual interaction between radial and poloidal decorrelation processes, the decorrelation rate $1/\tau_{\rm corr}$ becomes a hybrid of two decorrelation rates; that is,

$$\frac{1}{\tau_{\rm corr}} = (\omega_{\rm s}^2\Delta\omega_{\rm t})^{1/3} = \left(\frac{\omega_{\rm s}}{\Delta\omega_{\rm t}}\right)^{2/3}\Delta\omega_{\rm t}. \tag{4.95}$$

The decorrelation rate $1/\tau_{\rm corr}$ becomes $(\omega_{\rm s}/\Delta\omega_{\rm t})^{2/3}$ times as large as $\Delta\omega_{\rm t}$; $\Delta\omega_{\rm t}$ is the decorrelation rate of the turbulence in the case of shearless flow. Since the saturation level of the fluctuating field $\tilde{\xi}$ is

$$|\tilde{\xi}|^2 \sim T \times \tau_{\rm corr},$$

the saturation level of the fluctuating field is reduce to

$$\frac{|\tilde{\xi}|^2}{|\tilde{\xi}_0|^2} \sim \left(\frac{\Delta\omega_{\rm t}}{\omega_{\rm s}}\right)^{2/3} \sim \left(\frac{1}{({\rm d}v_\theta/{\rm d}r)t_0}\right)^{2/3}\frac{1}{(k_{0y}\Delta r)^2},$$

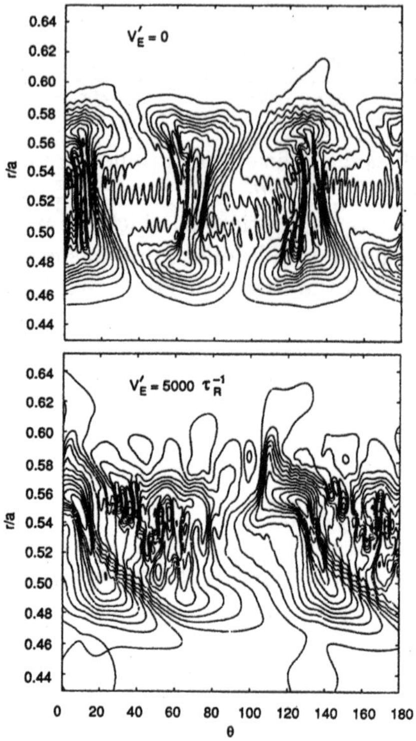

Figure 4.26 Snapshot of equidensity contour for shearless (top) and strongly sheared (bottom) flows. After [4.41]. Vertical axis is radial coordinate r/a and the horizontal axis is poloidal angle θ (degree).

$$t_0^{-1} \equiv \langle k_{0y}^2 \rangle D$$

where $|\tilde{\xi}_0|$ is the level in the case of shearless flow. The effect of sheared flow on the saturated resistive pressure gradient driven turbulence is shown in Figure 4.26. The coupling between poloidal and radial decorrelation in shearing fluctuation is evident in this figure. Since the thermal diffusion coefficient is proportional to $|\tilde{\xi}|^2$, the thermal diffusion is reduced; that is, a thermal barrier near the plasma edge is formed.

Active theoretical studies on H mode physics are being carried out. In addition to the standard H mode as observed in ASDEX and others, the other types of improved confinement modes have been observed. In the TFTR experiment [4.43], outgassing of deuterium from the wall and the carbon limiter located on the inner (high-field) side of the vacuum torus was extensively carried out before the experiments. Then balanced neutral beam injections of co-injection (beam direction parallel to the plasma current) and counterinjection (beam direction opposite to that of co-injection) were applied to the deuterium plasma, and an improved confinement "supershot" was observed. In supershot, the electron density profile is strongly peaked ($n_e(0)/\langle n_e \rangle = 2.5 \sim 3$).

In DIII-D experiment, VH mode [4.44] was observed, in which the region of strong radial electric field was expanded from the plasma edge to the plasma interior ($r/a \sim 0.6$) and $\tau_E/\tau_E^{\rm ITER-P}$ becomes 3.6.

In JT60U experiment, high beta-poloidal H mode [4.45] was observed, in which β_p was high (1.2~1.6) and the density profile was peaked ($n_e(0)/\langle n_e \rangle = 2.1 \sim 2.4$). Furthermore, the edge thermal barrier of H mode was formed.

Hinton et al. [4.46] pointed out the peaked pressure and density profiles induce the gradient of the radial electric field. From the radial component of the equation of motion (3.7) of ion fluid or (3.28), we have

$$E_r \simeq B_p u_t - B_t u_p + \frac{1}{en_i}\frac{dp_i}{dr}.$$

Differentiation of E_r by r is

$$\frac{dE_r}{dr} \sim -\frac{1}{en_i^2}\frac{dn_i}{dr}\frac{dp_i}{dr},$$

since the contribution from the other terms is small in the usual experimental condition of H mode.

Recently a high performance mode of negative magnetic shear configuration was demonstrated in DIII-D, TFTR, JT60U, JET, and Tore Supra [4.47]. As described in Section 9.6, ballooning mode is stable in the negative shear region

$$S = \frac{r}{q_s}\frac{dq_s}{dr} < 0. \tag{4.96}$$

An example of radial profiles of temperature, density, and q profile of JT60U is shown in Figure 4.27. By combination of the central heating and the magnetic negative shear, the steep gradients in temperature and density appear at around the q minimum point. This internal transport barrier is formed by the effects of the negative magnetic shear (refer to Section 14.4) and $\boldsymbol{E} \times \boldsymbol{B}$ flow shear.

As a measure of high performance of improved confinement mode, the ratio, H_L factor, of observed energy confinement time $\tau_E^{\rm EXP}$ to ITER-P scaling $\tau_E^{\rm ITER-P}$ is widely used.

$$H_L \equiv \frac{\tau_E^{\rm EXP}}{\tau_E^{\rm ITER-P}}. \tag{4.97}$$

Observed H_L factors are in the range of 2~3.

ITER H mode database working group assembled standard experimental data of H mode from ASDEX, ASDEX-U, DIII-D, JET, JFT-2M, PDX,

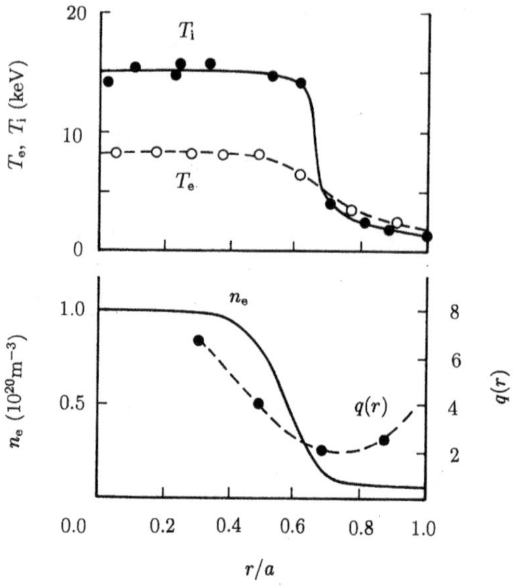

Figure 4.27 Radial profiles of ion and electron temperatures and density and q profiles in the negative magnetic shear configuration of JT60U. After [4.47].

PBX, Alcator C-Mod, and so on. Results of regression analysis of H mode experiments led to the following thermal energy confinement time (chap. 2 in [4.23]):

$$\tau_{\rm E,th}^{\rm IPB98y2} = 0.0562 I_{\rm p}^{0.93} B_{\rm t}^{0.15} P^{-0.69} M_{\rm i}^{0.19} R^{1.97} \bar{n}_{\rm e19}^{0.41} \epsilon^{0.58} \kappa^{0.78}, \qquad (4.98)$$

where units of sec, MA, T, MW, amu, m, 10^{19}m^{-3} are used and the total heating power corrected for shine-through of NBI heating, orbit loss, and charge exchange loss, less the time derivative of stored energy. This scaling is used when edge-localized-modes (ELM) exist [4.48]. A comparison of thermal energy confinement scaling with experimental data of ELMy H mode is presented in Figure 4.28.

In most experiments of hot plasmas, neutral beam injections are used to heat the plasma. With improved confinement mode operations, such as H mode, supershot and high $\beta_{\rm p}$ mode in large tokamaks, fusion grade plasmas are produced by neutral beam injection. The plasma parameters of typical shots of JET [4.49], JT60U [4.50], and TFTR [4.51] are listed in Table 4.2.

Deuterium-tritium experiments were carried out on TFTR [4.51]. Fusion power of 9.3 MW ($Q \sim 0.27$) was obtained in supershot (refer to Table 4.2). JET set records of DT fusion output of 16.1 MW ($Q \sim 0.62$) with the heating power of 25.7 MW [4.52].

In the present neutral beam source, the positive hydrogen ions are accelerated and then passed through the cell filled with neutral hydrogen

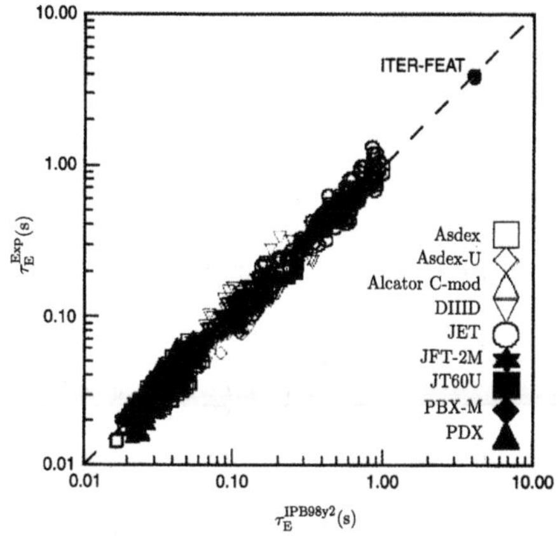

Figure 4.28 Comparison of IPB98y2 confinement scaling $\tau_{E\,th,\,scaling}$ with experimental data of energy confinement time $\tau_{E\,th}$ of H mode. After [4.23].

gas, where ions are converted to a fast neutral beam by charge exchange (attachment of electron). However, the conversion ratio of positive hydrogen ions to neutral becomes small when the ion energy is larger than 100 keV (2.5% at 200 keV of H^+). On the other hand, the conversion ratio of negative hydrogen ions (H^-) to neutral (stripping of electron) does not decrease in the high energy range (∼60%); a neutral beam source with a *negative ion source* is being developed as a high-efficiency source.

Wave heating is another method of plasma heating and will be described in detail in Chapter 13. The similar heating efficiency of wave heating in ICRF (ion cyclotron range of frequency) to that of NBI was observed in PLT. In the ICRF experiments of JET, the parameters $T_i(0) = 5.4\,\text{keV}, T_e(0) = 5.6\,\text{keV}$, $n_e(0) = 3.7 \times 10^{13}\,\text{cm}^{-3}$, $\tau_E \sim 0.3\,\text{s}$ were obtained by $P_{ICRF} = 7\,\text{MW}$.

4.7 Steady-State Operation

Non-Inductive Current Drives and Bootstrap Current

As long as the plasma current is driven by electromagnetic induction of a current transformer in a tokamak device, the discharge is a pulsed operation with finite duration. In such cases, heat cycles of heavy thermal load to the first wall (plasma faced wall) may cause serious technological difficulty. Furthermore, pulsed reactors are not competitive as commercial power plants. If the plasma current is driven by the non-inductive way, a

Table 4.2 Plasma parameters of large tokamaks JET [4.49], JT60U [4.50], and TFTR [4.51].

	JET	JT60U	TFTR
	ELM free No.26087	ELMy No.E21140	supershot
I_p(MA)	3.1	2.2	2.5
B_t(T)	2.8	4.4	5.1
R/a(m/m)	3.15/1.05	3.05/0.72	~2.48/0.82
κ_s	1.6	1.7	1
q's	q_{95}=3.8	$q_{\rm eff}$=4.6	q^*=3.2
q_I	2.8	3.0	2.8
$n_e(0)(10^{19}\mathrm{m}^{-3})$	5.1	7.5	8.5
$n_e(0)/\langle n_e\rangle$	1.45	2.4	-
$n_i(0)(10^{19}\mathrm{m}^{-3})$	4.1	5.5	6.3
$T_e(0)$(keV)	10.5	10	11.5
$T_e(0)/\langle T_e\rangle$	1.87	-	-
T_i(keV)	18.6	30	44
$W_{\rm dia}$ (MJ)	11.6	7.5	6.5
$\mathrm{d}W_{\rm dia}/\mathrm{d}t$(MJ/s)	6.0	-	7.5
$Z_{\rm eff}$	1.8	2.2	2.2
β_p	0.83	1.2	~1.1
β_t(%)	2.2	~1.3	~1.2
β_N(Troyon factor)	2.1	~1.9	2
$P_{\rm NB}$(MW)	14.9	24.8	33.7
$E_{\rm NB}$(keV)	135, 78	95	110
$\tau_E^{\rm tot} = W/P_{\rm tot}$(s)	0.78	0.3	0.2
$H = \tau_E^{\rm tot}/\tau_E^{\rm ITER-P}$	~ 3.0	~ 2.1	~ 2.0
$n_i(0)\tau_E^{\rm tot}T_i(0)(10^{20}\mathrm{keVm}^{-3}\mathrm{s})$	5.9	5	5.5
$n_T(0)/(n_T(0)+n_D(0))$	0	0	0.5
$P_{\rm fusion}$(MW)	-	-	9.3

$n_i(0)\tau_E^{\rm tot}T_i(0)$ is fusion triple product. κ_s is the ratio of vertical radius to horizontal radius. qs are the effective safety factors near plasma boundary. q_I is the factor defined in (4.50). $E_{\rm NB}$ is a particle energy of neutral beam injection.

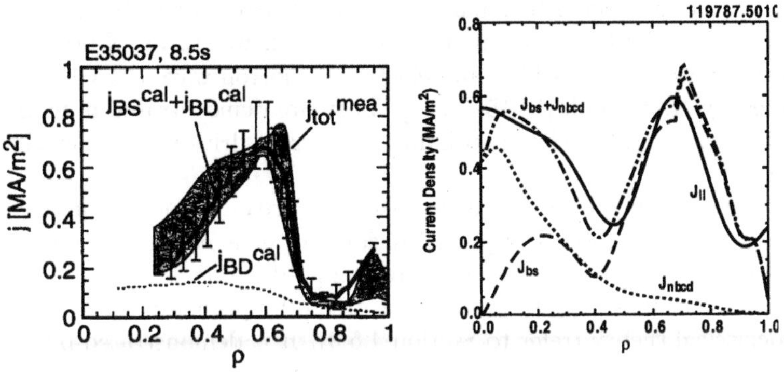

Figure 4.29 Left-hand side: Radial profiles of measured total current density $j_{\rm tot}^{\rm mea}$, calculated beam driven current density $j_{\rm BD}^{\rm cal}$, and sum of calculated bootstrap current density $j_{\rm BS}^{\rm cal}$ and $j_{\rm BD}^{\rm cal}$ of JT60U discharge E35037. After [4.53].
Right-hand side: Current profiles for discharge 119787 of DIIID: total ($J_{\|}$), bootstrap ($J_{\rm bs}$), beam driven ($J_{\rm NBCD}$) current densities, and the sum of the bootstrap and beam-driven current densities. After [4.54].

Table 4.3 Experimental data of full non-inductive current drive with large fraction of bootstrap current in JT60U and DIIID.

	JT60U (E35037) Reversed shear config.	DIIID (119787) ELMy H, standard q profile
R(m)	3.34	(1.66)
a(m)	0.8	(0.67)
$I_{\rm p}$(MA)	0.8	0.6
$B_{\rm t}$(T)	3.4	(< 2.1)
κ	1.5	1.85
δ	0.42	0.7
q_{95}	~9	~10
$q_{\min}$	3.6	2.94
$P_{\rm NB}$(MW)	5	–
$E_{\rm b}$(keV)	85	–
$\beta_{\rm N}$	1.9~2.2	3.08
$\beta_{\rm p}$	2.7~3.0	3.18
$\beta_{\rm t}$(%)	0.65	1.5
$\tau_{\rm E}$(s)	0.4~0.5	0.098
HH_{98y2}	2.1~2.3	2.03
$n_{\rm e19}(0)/\langle n_{\rm e19}\rangle$	3.5/–	6.7/4.37
$T_{\rm i}(0))/\langle T_{\rm i}\rangle$(keV)	7.5/–	3.8/2.08
$T_{\rm e}(0))/\langle T_{\rm e}\rangle$(keV)	5.0/–	2.5/1.49
$f_{\rm BS}$(%)	$(78 \sim 84) \pm 11$	80.8
$f_{\rm BD}$(%)	~25	19.2
$\tau_{\rm duration}$(s)	2.7	0.7

HH_{98y2} is energy confinement enhance factor over IPB98y2 scaling (9.48) and $\tau_{\rm duration}$ is the duration time of steady-state, high bootstrap current fraction and full current drive. $f_{\rm BS}$ and $f_{\rm BD}$ are fractions of bootstrap current and beam driven current, respectively. δ is triangularity of plasma shape. The parameters of R, a, $B_{\rm t}$ in the DIIID column are not given in [4.54], so the parameters are refered from [4.55].

steady-state operation of tokamak is possible. Current drive by neutral beam injection (NBCD) was predicted by T. Ohkawa [12.13]. Current drive by lower hybrid wave (LHCD) and electron cyclotron wave (ECCD) were predicted by N. J. Fisch [12.15, 12.16]. They are demonstrated in many experiments. The physics in non-inductive current drive will be described in Chapter 12.

However, even if the total plasma current is driven by non-inductive current drive, the substantial part of fusion output is necessary by the present experimental results and theoretical predictions.

On the other hand, bootstrap current driven by plasma itself is predicted by neoclassical theory (refer to Section 4.5.3). It is demonstrated by experiments that the fraction of bootstrap current can reach 70∼80% of the total plasma current in the case of high poloidal beta $\beta_{\rm p}$ operation. The experimental results of full non-inductive current drive operation with large bootstrap current fraction in JT60U [4.53] and DIIID [4.54] are shown in Figure 4.29. Radial profiles of measured total current density and the sum of calculated bootstrap current density and beam driven current density are compared and they are in good agreement within the experimental error. The parameters of experimental conditions and measured data are listed in Table 4.3.

Although the plasma currents in these experiments are relatively low in order to obtain high poloidal beta plasma, steady-state plasmas with high bootstrap current fraction and full non-inductive current drive are realized with good energy confinement HH_{98y2} in the cases of $\beta_{\rm p} \sim 3$ and these are very encouraging results.

Neoclassical Tearing Mode

When the duration of sustaining high toroidal beta plasma is longer than the resistive diffusion time scale $\tau_{\rm R} = \mu_0 a_{\rm s}^2/\eta$ (η is specific resistivity of plasma at the radius $a_{\rm s}$ of rational surface), neoclassical tearing mode (NTM) appears in the magnetic island, when the poloidal beta exceeds critical values (refer to (9.46)). Physics of neoclassical tearing mode will be described in Section 9.2. It is possible to control neoclassical tearing mode by local current drive in rational surface [9.7].

Resistive Wall Mode

When the discharge duration exceeds the resistive skin time $\tau_{\rm w} = \mu_0 d\delta w/\eta_{\rm w}$ ($\eta_{\rm w}$ is specific resistivity of the wall with radius d and thickness δw), the resistive wall mode (RWM) appears as is described in Section 9.4. In such case, high beta plasma in the configuration of perfect conducting wall is no longer stable and the beta ratio of confined plasma decreases to the beta ratio in the configuration without conductive wall.

It is demonstrated in DIIID as well as the other devices that RWM with long wavelength external kink mode is stabilized by the feedback control of internal coil currents, which produces non-symmetric magnetic field [4.56] (Ex/3-1Ra). Beta ratio well above the no-wall limit is sustained beyond the

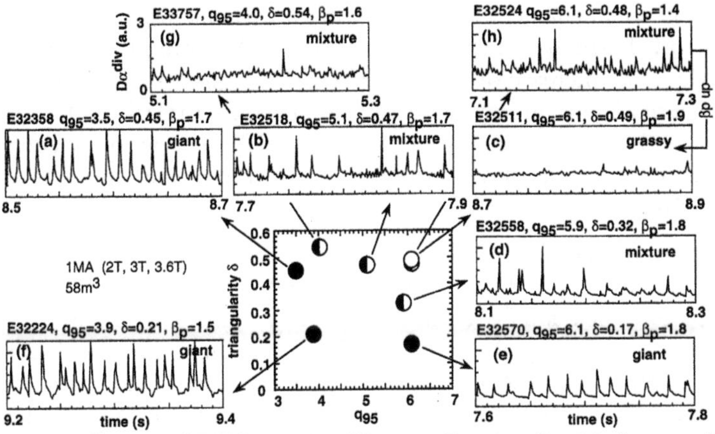

Figure 4.30 Change of ELM activity (divertor D_α signal) with δ, q_{95} and β_p at $I_p = 1$MA. Giant ELMs are replaced by grassy ELMs as q_{95} increases with fixed δ and β_p ((a) to (b) to (c)); as δ increases with fixed q_{95} and β_p ((e) to (d) to (c)); and as β_p increases with fixed δ and q_{95} ((h) to (c)). After [4.57].

resistive time of resistive wall.

Furthermore, rapid toroidal plasma rotation past a resistive wall affects the stability of high beta plasma [4.56] (Ex/3-1Rb). When the plasma rotation frequency is the order of a few percent of the inverse Alfv́en transit time, RWM is stabilized.

Edge Localized Mode

The improved confinement regime H mode is often perturbed by the onset of a quasi-periodic series of relaxation oscillations involving bursts of MHD activity and D_α emission, known as edge localized mode (ELM). These transit bursts of energy and particles into SOL produce high peak heat load and erosion on the divertor plates. On the other hand, the ELMs are effective in removing density and impurities accumlation. ELMy H mode is a standard operation of ITER (refer to Section 4.8). There are reviews on ELM [4.48].

ELM activities are classified as three types:

Type I ELMs: ELM repetition frequency $\nu_{\rm ELM}$ decreases with heating power. In this case, the plasma edge is always close to the stability limit of normalized pressure gradient $\alpha \sim \alpha_{\rm cr}$. They appear as isolated sharp bursts on D_α

Type III ELMs: ELM repetition frequency decreases with heating power, contrary to the type I. The plasma edge pressure gradient is below the ideal ballooning limit $0.3 < \alpha/\alpha_{\rm cr} < 0.5$.

Type II ELMs: They become more frequent and the magnitude of D_α burst decreases (grassy ELM, small ELM).

H mode operation with grassy or small ELM is one of the candidates to satisfy good confinement with low impurities without high peak heat load to the divertor. There are trials to control ELMs. An example is the control of plasma shape by increase of triangularity. At a higher triangularity of $\delta = 0.54$, almost pure grassy phase appears even at a relatively low $q_{95} = 4.5$ in JT60U as is seen in Figure 4.31 [4.57]. ASDEX-U experiment confirms that periodic injections of deuterium ice-pellet into plasma mitigate the type I ELM [4.58].

Disruption Control

Loss of the plasma equilibrium control at the burning phase of tokamak fusion reactors such as ITER will cause serious damage to the first wall materials and divertor plate due to the high heat flux on the wall and due to the generation of high energy runaway electrons. Furthermore, toroidally asymmetric halo current generated at the vertical displacement during the plasma current quench causes intense electromagnetic force locally which may break in-vessel materials. Halo current, which flows helically along the wall-intersecting plasma flux surfaces, makes a complete circuit by flowing from the strike points at one end of the open SOL field lines, through the conducting first wall structures and out onto the other end of the SOL field lines.

It is demonstrated in the experiments [4.59],[4.60] that the injection of a solid impurity pellet (commonly termed *killer pellet*) into a tokamak plasma is capable of effecting a non-disruptive fast plasma energy and current shutdown, provided that the pellet size and velocity are chosen such that the pellet can penetrate deeply into the plasma core.

Neural-net predictor for beta limit disruption is developed in JT60U [4.61].

4.8 International Tokamak Experimental Reactor (ITER)

Although there are many parameters to specify a tokamak device, there are also many relations and constraints between them [4.62]. If a plasma radius a, aspect ratio A, and toroidal field B_{t} are specified, the rato Q of fusion output power to auxiliary heating power and the other parameters of tokamak are determined by use of scaling laws of electron density, beta, energy confinement time, and burning condition, when cylindrical safety factor q_{I} (or the safety factor q_{95} at 95% toroidal flux surface), the elongation ratio κ_{s} and triangularity δ of plasma cross-section are given. By the definition of q_{I}, we have

$$q_{\mathrm{I}} \equiv \frac{Ka}{R}\frac{B_{\mathrm{t}}}{B_{\mathrm{p}}} = \frac{5K^2 a B_{\mathrm{t}}}{AI_{\mathrm{p}}}, \qquad B_{\mathrm{p}} = \frac{\mu_0 I_{\mathrm{p}}}{2\pi Ka} = \frac{I_{\mathrm{p}}}{5Ka},$$

the plasma current is

$$I_{\rm p} = \frac{5K^2 a B_{\rm t}}{A q_{\rm I}},$$

where $K^2 = (1+\kappa_{\rm s}^2)/2$ ($I_{\rm p}$ in MA, $B_{\rm t}$ in T, and a in m). The safety factor q_{95} is approximately given by (see (4.52), [4.23])

$$q_{95} \approx q_{\rm I} f_\delta f_{\rm A}$$

$$f_\delta = \frac{1+\kappa^2(1+2\delta^2-1.2\delta^3)}{1+\kappa^2} \qquad f_{\rm A} = \frac{1.17-0.65/A}{(1-1/A^2)^2}.$$

The volume average electron density n_{20} in unit of $10^{20}{\rm m}^{-3}$ is

$$n_{20} = N_{\rm G}\frac{I_{\rm p}}{\pi a^2}, \tag{4.99}$$

where $N_{\rm G}$ is Greenward normalized density. The beta ratio of thermal plasma

$$\beta_{\rm th} \equiv \frac{\langle p\rangle}{B_{\rm t}^2/2\mu_0} = 0.0403(1+f_{\rm DT}+f_{\rm He}+f_{\rm I})\frac{\langle n_{20}T_{\rm keV}\rangle}{B_{\rm t}^2} \tag{4.100}$$

is expressed by

$$\beta_{\rm th} = f_{\rm th}\beta_{\rm total}, \quad \beta_{\rm total} = 0.01\beta_{\rm N}\frac{I_{\rm p}}{aB_{\rm t}},$$

where $\beta_{\rm N}$ is normalized beta. $\beta_{\rm total}$ is the sum of $\beta_{\rm th}$ (thermal plasma) and $\beta_{\rm fast}$ (fast (α) particle component) and $f_{\rm th} = \beta_{\rm th}/\beta_{\rm total}$. The notations $f_{\rm DT}$, $f_{\rm He}$, and $f_{\rm I}$ are the ratios of fuel DT, He, and impurity density to electron density, respectively, and the unit of T_{keV} is keV. $\langle X\rangle$ means volume average of X. Thermal energy of plasma $W_{\rm th}$ is

$$W_{\rm th} = \frac{3}{2}\beta_{\rm th}\frac{B_{\rm t}^2}{2\mu_0}V = 0.5968\beta_{\rm th}B_{\rm t}^2 V, \tag{4.101}$$

where $W_{\rm th}$ is in the unit of MJ and plasma volume V is in the unit of ${\rm m}^{-3}$. Plasma shape with elongation ratio $\kappa_{\rm s}$ and triangularity δ is given by

$$R = R_0 + a\cos(\theta+\delta\sin\theta), \tag{4.102}$$

$$z = a\kappa_{\rm s}\sin\theta.$$

Plasma volume V is given by

$$V \approx 2\pi^2 a^2 R\kappa_{\rm s} f_{\rm shape},$$

where $f_{\rm shape}$ is a correction factor due to triangularity. When the plasma cross-section is given by (4.102), $f_{\rm shape}$ is given by

$$f_{\rm shape} \approx 1 - \frac{\delta^2}{8} - \frac{3a}{4R}\delta.$$

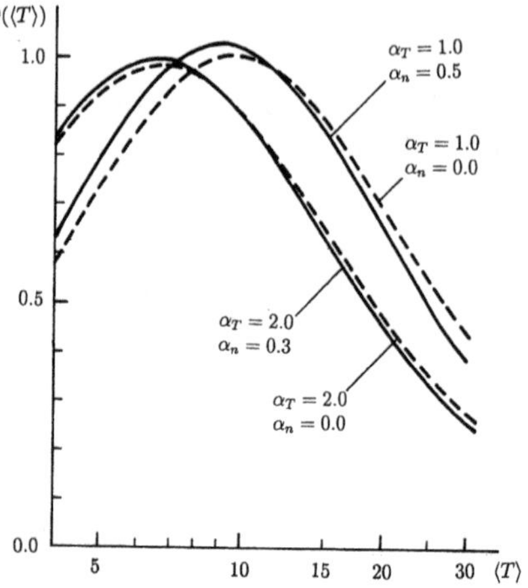

Figure 4.31 Θ is a function of average temperature $\langle T \rangle$(keV) in cases with profile parameters ($\alpha_T = 1.0, \alpha_n = 0.0$), ($\alpha_T = 2.0, \alpha_n = 0.0$), ($\alpha_T = 1.0, \alpha_n = 0.5$), and ($\alpha_T = 2.0, \alpha_n = 0.3$).

f_{shape} changes in the case of divertor configuration. We utilize the thermal energy confinement scaling of IPB98y2

$$\tau_{\text{E}} = 0.0562 \times 10^{0.41} H_{\text{y2}} I^{0.93} B_{\text{t}}^{0.15} M^{0.19} n_{20}^{0.41} a^{1.97} A^{1.39} \kappa_{\text{s}}^{0.78} P^{-0.69}, \tag{4.103}$$

where $M(= 2.5)$ is the average ion mass unit and P is the loss power in MW by transport and is equal to necessary absorbed heating power subtracted by radiation loss power P_{rad}. The total α particle fusion output power P_α is

$$P_\alpha = \frac{Q_\alpha}{4} \langle n_{\text{DT}}^2 \langle \sigma v \rangle_v \rangle V,$$

where $Q_\alpha = 3.515$MeV. $\langle \sigma v \rangle_v$ is a function of T and a fitting equation for $\langle \sigma v \rangle_v$ is given in (1.5). Since the fusion rate σv near T =10 keV is approximated by

$$\langle \sigma v \rangle_v \approx 1.1 \times 10^{-24} T_{\text{keV}}^2 (\text{m}^3/s),$$

the folowing Θ ratio is introduced:

$$\Theta(\langle T \rangle) \equiv \frac{\langle n_{\text{DT}}^2 \langle \sigma v \rangle_v \rangle}{1.1 \times 10^{-24} \langle n_{\text{DT}}^2 T^2 \rangle}.$$

Θ is a function of average temperature $\langle T \rangle$ in keV and the profiles of density and temperature and has a peak of around 1 near $\langle T \rangle \approx 8 \sim 10$ keV. The curves of Θ versus $\langle T \rangle$ in cases of $n(\rho) = (1 + \alpha_n)\langle n \rangle (1 - \rho^2)^{\alpha_n}$, $T(\rho) = (1 + \alpha_T)\langle T \rangle (1 - \rho^2)^{\alpha_T}$ are shown in Figure 4.31. Then P_α is reduced to

$$P_\alpha = 0.9551 \frac{f_{\rm prof} f_{\rm DT}^2}{(1 + f_{\rm DT} + f_{\rm He} + f_{\rm I})^2} \beta_{\rm th}^2 B_{\rm t}^4 \Theta V, \tag{4.104}$$

where $f_{\rm prof} \equiv \langle n^2T^2\rangle/\langle nT\rangle^2 \approx (\alpha_n + \alpha_T + 1)^2/(2\alpha_n + 2\alpha_T + 1)$ is the profile effect of temperature and density. When absorbed auxiliary heating power is denoted by $P_{\rm aux}$ and heating efficiency of α heating is f_α, the total heating power is $f_\alpha P_\alpha + P_{\rm aux}$. When the fraction of radiation loss power from the plasma to the total heating power is $f_{\rm rad}$, the heating power to sustain burning plasma is given by

$$P = (1 - f_{\rm rad})(f_\alpha P_\alpha + P_{\rm aux}).$$

When Q ratio is defined by the ratio of total fusion output power $P_{\rm n} + P_\alpha = 5P_\alpha$ ($P_{\rm n}$ is neutron output power) to absorbed auxiliary heating power $P_{\rm aux}$, Q is

$$Q = \frac{5P_\alpha}{P_{\rm aux}}.$$

Then the heating power to the plasma P is reduced to

$$P = (1 - f_{\rm rad}) \left(f_\alpha + \frac{5}{Q} \right) P_\alpha.$$

Therefore, burning condition is

$$\frac{W_{\rm th}}{\tau_{\rm E}} = (1 - f_{\rm rad}) \left(f_\alpha + \frac{5}{Q} \right) P_\alpha. \tag{4.105}$$

From (4.101), (4.103), and (4.105), we have

$$\beta_{\rm th} B_{\rm t}^2 \tau_{\rm E} = 0.625 \frac{(1 + f_{\rm DT} + f_{\rm He} + f_{\rm I})^2/f_{\rm DT}^2}{(1 - f_{\rm rad})(f_\alpha + 5/Q) f_{\rm prof} \Theta(T)}. \tag{4.106}$$

The term in the left-hand side is proportional to the triple product $\langle n_{20} T_{\rm keV}\rangle \tau_{\rm E}$ (refer to (4.100)). Equations (4.103) to (4.105) reduce to the following burning condition [4.63]:

$$\frac{B_{\rm t}^{0.73} a^{0.42}}{A^{0.26}} (f_\alpha + \frac{5}{Q})^{0.31} = 1.95 \left[\frac{(1 + f_{\rm DT} + f_{\rm He} + f_{\rm I})^2}{(1 - f_{\rm rad}) f_{\rm prof} f_{\rm DT}^2 \Theta} \right]^{0.31} \times$$

$$\frac{q_{\rm I}^{0.96} (f_{\rm th}\beta_{\rm N})^{0.38} f_{\rm shape}^{0.69}}{H_{\rm y2} M^{0.19} N_{\rm G}^{0.41} K^{1.92} \kappa_{\rm s}^{0.09}}. \tag{4.107}$$

When parameters a, $B_{\rm t}$, A are specified, Q value and other parameters can be evaluated and are shown in Table 4.4. The results of this simple analysis are relatively consistent with ITER design parameters given by Table 4.5 [4.64].

Table 4.4 Specified design parameters.

a	B_t	A	q_{95}	κ_s	δ	f_{rad}	f_α	β_N/f_{th}	N_G	f_{DT}	f_{shape}	H_{y2}
2.0	5.3	3.1	3.0	1.7	0.35	0.27	0.95	1.77/0.928	0.85	0.82	0.995	1.05

Specified value of β_N is the normalized beta of plasma which includes the contribution of energetic ion components. ($\alpha_T = 1.0, \alpha_n = 0.1$ and $f_{He} = 0.04,\ f_I = 0.02$ are assumed.)

Reduced parameters.

Q	R	I_p	τ_E	n_{20}	$\langle T\rangle$	W_{th}	P_n	P_α	P_{aux}	P_{rad}	β_{total}	q_I	Θ
11	6.2	15.0	3.7	1.01	8.7	325	425	85	41	32.6	0.025	2.22	0.95

$P_n, P_\alpha, P_{aux}, P_{rad}$ are in the unit of MW and W_{th} is in the unit of MJ.

Table 4.5 Parameters of ITER.

	inductive operation	non-inductive operation
I_p(MA)	15	9
B_t(T)	5.3	5.17
R/a(m)	6.2/2.0	6.35/1.84
A	3.1	3.45
$\kappa_{s\,95}/\delta_{95}$	1.7/0.33	1.84/0.41
$\langle n_e\rangle(10^{20}\mathrm{m}^{-3})$	1.01	0.67
N_G	0.85	0.83
$\langle 0.5\times(T_e+T_i)\rangle$(keV)	8.5	~11
$W_{thermal}/W_{fast}$(MJ)	325/25	273/60~255/50
τ_E^{tr}(s)	3.7	2.54~2.32
$H_{y2}=\tau_E^{tr}/\tau_E^{IPB98y2}$	1.0	1.41~1.3
P_{fus}(MW)	410	361~338
P_{aux}(MW)	41	P_{NB}=34/P_{LH}=29~33.7
P_{rad}(MW)	48	
Z_{eff}	1.65	2.2~2.17
β_t(%)	2.5	2.64~2.42
β_p	0.67	3.55~3.26
β_N@@	1.77	2.8~2.56
q_{95}	3.0	5.16~5.13
q_I	2.22	3.44
l_i	0.86	2.2~2.17
Q	10	5
f_R	0.39	
f_{DT}/f_{He}(%)	82/4.1	/
f_{Be}/f_{Ar}(%)	2/0.12	/

Inductive operation scenario in outline design in 2000 and non-inductive operation scenario in 2002 [4.64]. In non-inductive operation scenario, the bootstrap current and driven current are 5 MA and 4 MA, respectively.

When Kaye-Goldston scaling (4.92) is used instead of IPB98y2 scaling, the burning condition becomes

$$\frac{I_p A^{1.25}}{a^{0.12}}\left(f_\alpha + \frac{5}{Q}\right)^{0.5} = \frac{75.1 f_{\rm shape}(1+f_{\rm DT}+f_{\rm He}+f_{\rm I})}{H_{\rm KG}[(1-f_{\rm rad})f_{\rm prof}\Theta]^{0.5} f_{\rm DT}}. \tag{4.108}$$

H_{KG} is the improved factor of energy confinement time over the Kaye-Goldston scaling. The value of H_{KG} necessary for the case of ITER parameters is 2.57. It is interesting to note that the burning condition mainly depends on AI_p in the case of Kaye-Goldston scaling.

When the distance of plasma separatrix and the conductor of toroidal field coil is Δ and the maximum field of toroidal field coil is B_{max} (see Figure 4.32), there is a constraint of

$$\frac{B_t}{B_{max}} = \frac{R-a-\Delta}{R} = 1 - \left(1 + \frac{\Delta}{a}\right)\frac{1}{A},$$

and

$$1 - \frac{2}{A} < \frac{B_t}{B_{max}} < 1 - \frac{1}{A},$$

under the assumption $a > \Delta > 0$. By specification of Δ and B_{max}, B_t is a function of a.

The ratio ξ of the flux swing $\Delta\Phi$ of ohmic heating coil and the flux of plasma ring $L_p I_p$ is given by

$$\xi \equiv \frac{\Delta\Phi}{L_p I_p} = \frac{5B_{max}({\rm T})((R_{\rm OH}+d_{\rm OH})^2 + 0.5 d_{\rm OH}^2)}{(\ln(8A/\kappa_s^{1/2}) + l_i - 2)RI_p({\rm MA})},$$

where $R_{OH} = R - (a + \Delta + d_{TF} + d_s + d_{OH})$, d_{TF} and d_{OH} being the thickness of TF and OH coil conductors and d_s being the separation of TF and OH coil conductors in meter unit, respectively (refer to Figure 4.32). The average current densities j_{TF}, j_{OH} of TF and OH coil conductors in MA/m^2=A/(mm)^2 are

$$j_{\rm TF}({\rm MA/m^2}) = \frac{2.5}{\pi}\frac{B_{max}({\rm T})}{d_{\rm TF}}\frac{1}{1 - 0.5 d_{\rm TF}/(R-a-\Delta)},$$

$$j_{\rm OH}({\rm MA/m^2}) = \frac{2.5}{\pi}\frac{B_{max}({\rm T})}{d_{\rm OH}}.$$

The conceptual design of tokamak reactors has been actively pursued according to the development of tokamak experimental research. INTOR (International Tokamak Reactor) [4.65] and ITER (International Thermonuclear Experimental Reactor)[4.64] are representative of international activity in this field. ITER aims achievement of extended burn

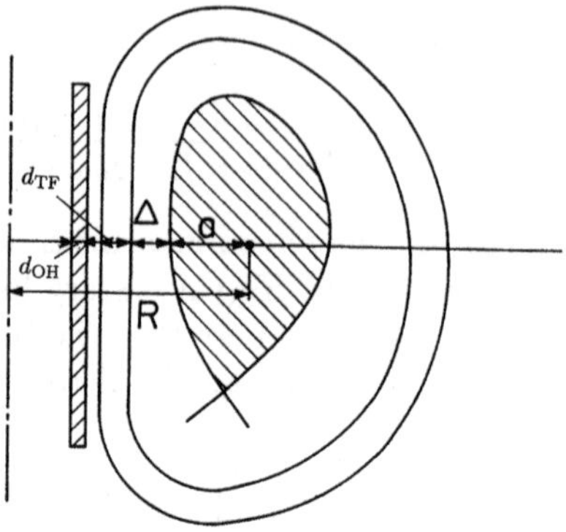

Figure 4.32 Geometry of plasma, toroidal field coil, and central solenoid of current transformer in tokamak.

in inductively driven plasmas with $Q \sim 10$ and aims at demonstrating steady-state operation using non-inductive drive with $Q \sim 5$.

The main parameters of ITER in 2000 are given in Table 4.5. τ_E^{tr} is the energy confinement time corrected for radiation loss. κ_s is the ratio of vertical radius to horizontal radius. q_{95} is the safety factor at 95% flux surface. The maximum field of toroidal field coils is B_{max}=11.8 T. The number of toroidal field coils is 18. Single null divertor configuration. One turn loop voltage is $V_{loop} = 89$mV. Inductive pulse flat-top under $Q = 10$ condition is several hundred seconds. P_{fus} is the total fusion output power. N_G is defined in (4.99). f_R is the fraction of radiation loss and f_{DT}, f_{Be}, f_{He}, f_{He} are fractions of DT, Be, He, Ar densities to the electron density.

A cross-section of ITER outline design in 2000 is shown in Figure 4.6.

4.9 Trials to Innovative Tokamaks

4.9.1 Spherical Tokamak

Potential theoretical advantages of spherical tokamak (ST) have been outlined by Peng and Strickler [4.66], in which aspect ratio A/a of the standard tokamak is substantially reduced toward unity. Predicted advantages include a natural high elongation ($\kappa_s > 2$), high toroidal beta and tokamak-like confinement. These predictions have been confirmed experimentally, in particular by START (Small Tight Aspect Ratio Tokamak) [4.67] at Culham ($R/a = 0.3/0.28 = 1.31$), $I_p \approx 0.25$MA, $B_t \approx 0.15$T. The toroidal beta reached 40% $\beta_N = 3.5 \sim 5.9$ (see Figure 4.33)

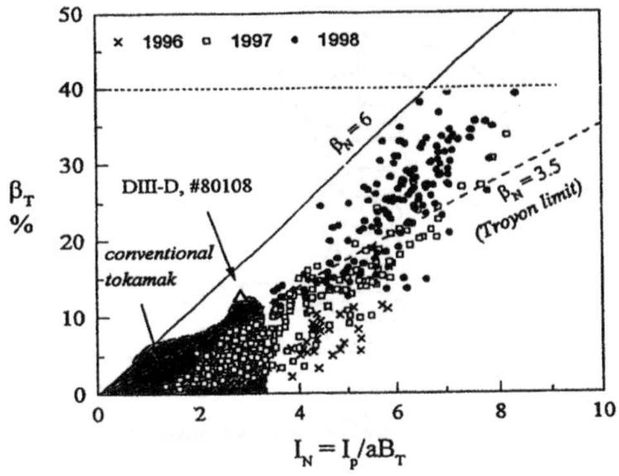

Figure 4.33 Average β_t measured in START versus normalized current $I_N = I_p/aB_t$. After [4.68].

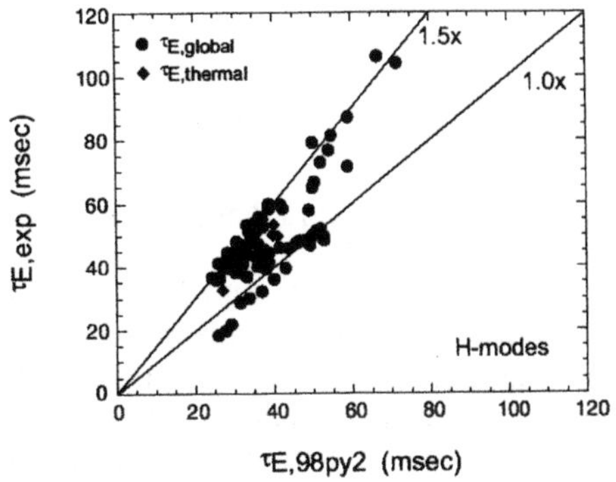

Figure 4.34 NSTX H-mode experimental confinement data points are shown compared to ITER98y2 H-mode scaling. After [4.69].

[4.68]. Observed energy confinement times follow similar scaling to standard tokamak and it was demonstrated experimentally that the energy confinement times follow ITER98y2 scaling up to $\tau_E \sim 0.1$ s in the experment NSTX (National Spherical Torus Experiment) in Princeton [4.69] ($R/a = 0.85/0.65 = 1.3$, $L_p = 1.5$MA, $B_t = 0.3 \sim 0.6$T) and MAST (Mega Ampere ST) in Culham [4.70] ($R/a = 0.85/0.65 = 1.3$, $I_p = 1.35$MA, $B_t = 0.52$T (see Figure 4.34). Density scaling is $N_G \sim 1$. Therefore, the fundamental scalings are almost the same as those of standard tokamaks (ITER) except a slight modification of q_{95} dependence on the aspect ratio; that is,

$$q_I \equiv \frac{aB_t}{RB_p} = \frac{5K^2 aB_t}{AI_p} = \frac{5}{AI_N}\frac{1+\kappa_s^2}{2}, \qquad I_N \equiv \frac{I_p}{aB_t},$$

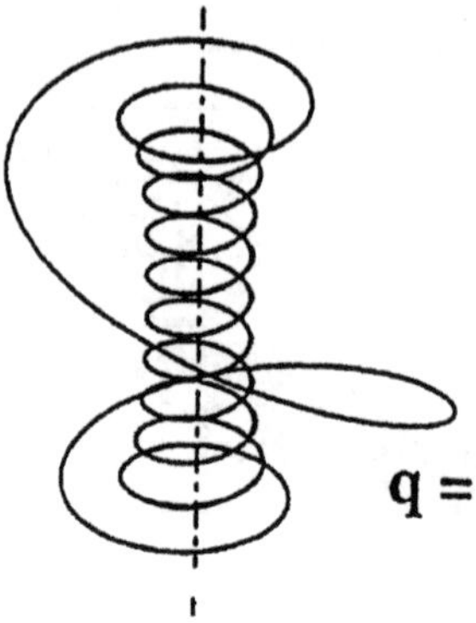

Figure 4.35 Magnetic field line in the plasma edge of ST. After [4.68].

$$q_{95} = \left(\frac{5}{AI_{\mathrm{N}}}\frac{1+\kappa_{\mathrm{s}}^2}{2}\right)\frac{1.22-0.68/A}{(1-1/A^2)2} \qquad (A<3), \tag{4.109}$$

$$\beta_{\mathrm{c}}(\%) = \beta_{\mathrm{N}} I_{\mathrm{N}}, \qquad \beta_{\mathrm{c}}(\%)\beta_{\mathrm{p}} = 0.25\beta_{\mathrm{N}}^2 K^2,$$

$$n_{\mathrm{e}\,20} = N_{\mathrm{G}}\frac{I_{\mathrm{p}}}{\pi a^2},$$

$$\tau_{\mathrm{E}} = 0.0562\times 10^{0.41} H_{\mathrm{y2}} I^{0.93} B_{\mathrm{t}}^{0.15} M^{0.19} n_{20}^{0.41} a^{1.97} A^{1.39} \kappa_{\mathrm{s}}^{0.78} P^{-0.69},$$

where units of R, a, B_{t}, and I_{p} are m, T, and MA, respectively, and I_{N} is normalized current. In addition, there are advantageous characteristics of ST. Refering to Figure 4.35, in the center of plasma, the flux surfaces appear as in a standard tokamak but near the edge of plasma (if very low aspect ratio), the field lines dwell in the high field inboard side, so that q in edge is greatly increased. As A approaches toward 1, q_{95} becomes large as is seen in (4.109), even if the cylindrical safety factor $q_{\mathrm{I}} \approx 2$. Therefore, it is possible to take I_{N} large value regardless of MHD stability limit (say $q_{95} > 3$) (refer to the horizontal axis of Figure 4.33). Accordingly Troyon limit of β_{c} can be large and the diamagnetic effect of high beta plasma counters the paramagnetic effect of the high normalized current I_{N}. This should improve MHD stability [4.68].

The charcteristic MHD activity is internal reconnection event (IRE). IRE occurs commonly in ST. It appears in low m mode activity. IRE can be modeled using an adaption of MHD relaxation theory (conservation of helicity) that has been used successfully to model the equilibrium profiles of reversed field pinch (RFP) (refer to Section 5.2). IRE event in START is driven by the current profiles (high internal inductance l_{i}, that is, peaked current profile). There is an interesting non-linear MHD simulation of IRE process [4.71] by solving the following MHD differential equations.

$$\frac{\partial(\rho_{\mathrm{m}}\boldsymbol{v})}{\partial t} = -\nabla\cdot(\rho_{\mathrm{m}}\boldsymbol{v}\boldsymbol{v}) - \nabla p + \rho\boldsymbol{E} + \boldsymbol{j}\times\boldsymbol{B} + \mu\left(\nabla^2\boldsymbol{v} + \frac{1}{3}\nabla(\nabla\cdot\boldsymbol{v})\right),$$

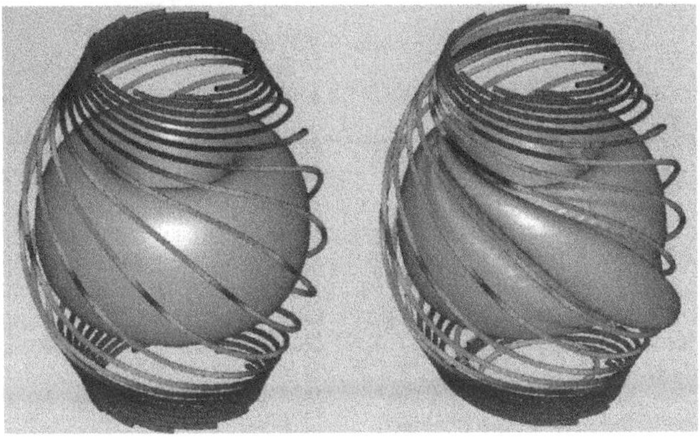

Figure 4.36 Time developement of the 3D profiles of plasma pressure and magnetic field lines in the simulation of IRE of ST at the initial time (left) and $t = 197\tau_{\mathrm{A}}$. After [4.71].

$$\frac{\partial p}{\partial t} = \nabla \cdot (p\boldsymbol{v}) - (\gamma - 1)p\nabla \cdot \boldsymbol{v} + (\gamma - 1)(\eta j^2 + \Phi),$$

$$\frac{\partial \rho_{\mathrm{m}}}{\partial t} = -\nabla(\rho_{\mathrm{m}}\boldsymbol{v}),$$

$$\frac{\partial \boldsymbol{B}}{\partial t} = -\nabla \times \boldsymbol{E},$$

where

$$\boldsymbol{j} = \frac{1}{\mu_0}\nabla \times \boldsymbol{B}, \qquad \boldsymbol{E} = -\boldsymbol{v} \times \boldsymbol{b} + \eta\boldsymbol{j}, \qquad \Phi \equiv \sum_{ij} \mu e_{ij} e_{ji} - \frac{1}{3}(\nabla \cdot \boldsymbol{v}).$$

The μ term represents the viscosity effect and the other notations are the same as those of Chapter 3. The result of numerical simultion is shown in Figure 4.36, which reproduces the experimental results quite well. According to this simulation, magnetic reconnection between the internal and external field occurs and the dynamics of magnetic reconnection is studied by 3D structure of pressure, flow, and the magnetic fields lines.

4.9.2 Trials to Innovative Tokamak Reactors

There are several trials of ST reactors with use of a solid copper center rod for TF (toroidal field) coils, and tight aspect tokamak with super-conducting TF coils. The parameters of STPP (ST Power Plant) [4.72], ARIES-ST [4.73] and Vector (Very Compact Tokamak Reactor) [4.74], SlimCS [4.75] are given in Table 4.6.

Cross-section of STPP is shown in Figure 4.38. TF coils of STPP and

Table 4.6 Parameters of STPP, ARIES-ST and Vector, SlimCS.

	STPP	ARIES-ST	Vector('02)	SlimCS
R/a(m)	3.42 / 2.44	3.20 / 2.00	3.75 / 1.9	5.5 / 2.1
A	1.4	1.6	2	2.6
κ_s/δ	3.2 / 0.55	3.42 / 0.64	3.75 /$\sim$ 0.1	2.0 /$\sim$0.4
I_p(MA)	31	30.8	18.3	16.7
B_t/B_{max}(T)	1.77 / -	2.14 / 7.6	4.7 / 19.6	6.0 / 16.4
q_0/q_{edge}	3 / 15	- / -	- / 5.2	- / 5.4
$\beta_t(\%)/\beta_N$	59 / 8.2	54 / 8	7.7 / 3.75	5.7 / 4.3
$\langle n_{e\,20}\rangle/\bar{n}_{e\,20}$	1.08 /	- / 2.74	1.6 /	1.18 /
n_G	1.66	2.45	1.6	1.2
N_G	0.65	1.12	1	0.98
$\langle T_e\rangle$(keV)	22	-	-	17.0
$H_{98y2}/H_{ITER93H}$	1.6 /	- / 1.83	1.8 /	1.3 /
P_{fusion}(GW)	3.1	2.86	1.8	3
CD power (MW)	50	31	$\sim$40	59
BS current fraction (%)	92.6	95	$\sim$0.8	0.77
neutron wall load (MW/m^2)	3.5	4.1	3.5	3.2
TF coil ohmic loss (MW)	-	288	$\sim$0	$\sim$0
recirculating power fraction		0.32	$Q \sim 45$	$Q \sim 52$

Notations are the same as Table 4.5. TF coils of STPP and ARIES-ST are normal conductor and TF coils of Vector and SlimCS are super-conductor.

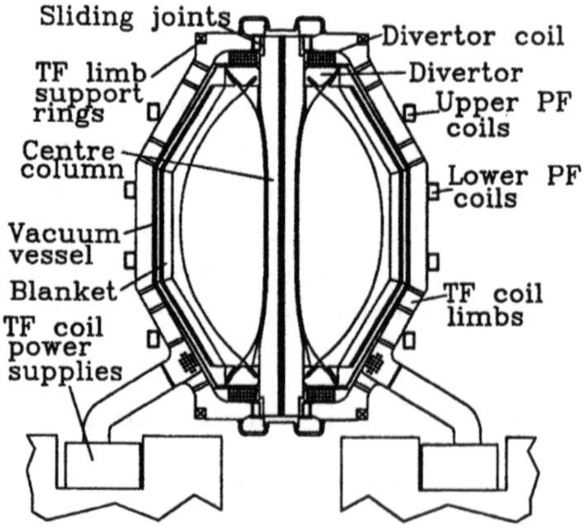

Figure 4.37 Cross section of STPP base line tokamak engineering design, highlighting the flared center column. After [4.72].

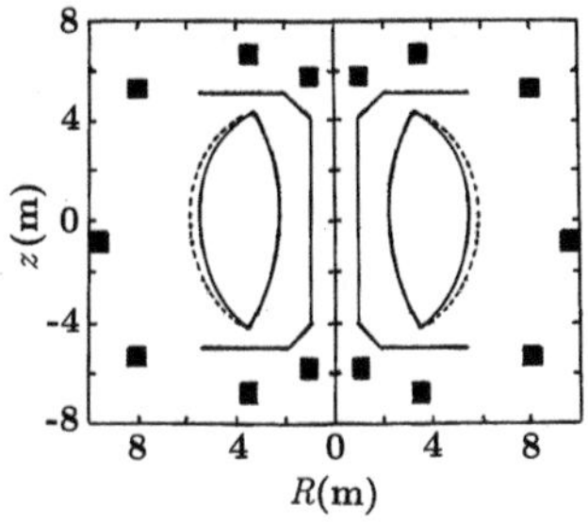

Figure 4.38 Cross-section of CS (central solenoid)-less tokamak Vector ('04). After [4.74].

ARIES-ST use a solid copper center rod, so that B_t is small and requirements of $\beta_N \sim 8$ and $H_{98y2} \sim 1.6$ or 1.8 are demanding.

Vector is a tokamak reactor with super-conducting TF coils (Bi2212/Ag/AgMgSb (20 K)) and low aspect ratio $A \sim 2$ by removing the center solenoid (CS) coil system from standard tokamak ($A \sim 3$). This brings relatively high beta plasma due to low aspect ratio, low recirculating power fraction. However, research and development of superconducting TF coils with very high magnetic field are required. Issue of start-up feasibility and current ramp up are being studied experimentally [4.76]. Cross-section of Vector is shown in Figure 4.38.

Motivated by the Vector concept, the more realistic demo reactor SlimCS is proposed. The parameters are $A = 2.6$, $B_{max} = 16.4$T (Nb_3Al), $\beta_N = 4.3$ and $H_{98y2} = 1.3$ (refer to Table 4.6). Slim CS is capable of raising the plasma current up to 3.8MA, while the full plasma current in steady-state is 16.7 MA.

Equilibrium and stability of ST including the effect of a large fraction of bootstrap current is analyzed in [4.77].

Design of ST reactor NSST (Next Step ST) with the other name CTF (compact Component Test Facility) is being carried out as a volume neutron source [4.78].

Problems

1. Magnetic Pressure and Plasma Pressure Calculate the magnetic pressure of $B = 5$T and plasma pressure $p = 2n_eT$ with $n_e = 10^{20}\text{m}^{-3}$ and $T_i = T_e = 10\,\text{keV}$.

2. Internal Inductance Show that the magnetic energy of poloidal field B_p inside the plasma region is $\mu_0(Rl_i/2)I_p^2/2$ by use of l_i defined by $l_i \equiv \int_0^a B_p(r)^2 2\pi r \mathrm{d}r/(B_p(a)^2\pi a^2)$. Calculate the normalized internal inductance l_i in the case of flat current profile and parabolic current profile $j(r) \propto (1 - r^2/a^2)$. Refer to (4.29).

3. Toroidal Coordinates Toroidal coordinates (ψ, ω, φ) are defined by

$$r = \frac{R_0(1-\psi^2)}{1-\psi\cos\omega}, \qquad z = -\frac{R_0\psi\sin\omega}{1-\psi\cos\omega}, \qquad \varphi = \varphi,$$

where (r, φ, z) are cylindrical coordinates. Show that the surfaces $\psi = \text{const.}$

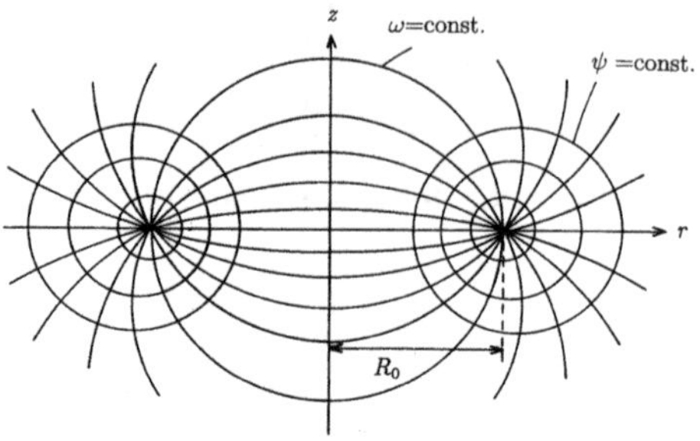

Figure 4.39 Toroidal coordinates.

are circles with the center at $(r, z) = (R_0(1-\psi^2)^{-1/2}, 0)$ and $\omega = \text{const.}$ are also circles with the center at $(r, z) = (0, R_0\cos\omega/\sin\omega)$ (refer Figure 4.39). Calculate the radii of circles $\psi = \text{const.}$ and $\omega = \text{const.}$ Confirm

$$\mathrm{d}s^2 \equiv \mathrm{d}r^2 + r^2\mathrm{d}\varphi^2 + \mathrm{d}z^2 = \frac{R_0^2}{(1-\psi\cos\omega)^2}\left(\frac{\mathrm{d}\psi^2}{1-\psi^2} + \psi^2\mathrm{d}\omega^2 + (1-\psi^2)\mathrm{d}\varphi^2\right).$$

4. Positional Instability The decay index of the vertical field n of tokamaks with vertically elongated cross-section is negative and the plasma is positionally unstable for vertical direction. Estimate the growth rate of this instability from (4.33). Note that the total plasma mass M that appeared in (4.33) is equal to $M \sim 2\pi R\pi^2 a^2 \rho_m$, where R and a are the major and minor radius of plasma, respectively, and the growth rate is proportional to the inverse poloidal Alfvën transit time a/v_{Ap}, $v_{Ap}^2 = B_p^2/\mu_0\rho_m = \mu_0 I_p^2/4\pi^2 a^2 \rho_m$. We assume that the elongation ratio κ_s is not much larger than 1.

5. Shell Stabilization of Positional Instability A conductive shell with finite resistivity of the minor radius a_s is settled around the plasma and the resistive decay time of the induced shell current is τ_s. Let the growth rate of vertical positional instability without shell be $\alpha v_{Ap}/a$. When the shell is introduced, the time variation of vertical position of plasma z is given by $\mathrm{d}^2z/\mathrm{d}t^2 = (\alpha v_{Ap}/a)^2 z - 2\pi R I_p B_s/M$, $B_s = \mu_0 I_s/(4a_s)$, and $\alpha^2 = (a/R)^2(-n)[\ln(8R/a) + \Lambda - 1/2]$. B_s is the radially horizontal field produced by the current I_s in the shell induced by the movement of plasma-current-ring in z direction. B_s yields the stabilizing effect against fast movement of plasma-current-ring; that is, $\mathrm{d}^2z/\mathrm{d}t^2 = (\alpha v_{Ap}/a)^2 z - (v_{Ap}/a)^2\pi(a/a_s)aI_s/I_p$. The induced current in shell I_s is given by $\mathrm{d}I_s/\mathrm{d}t + I_s/\tau_s = 2(I_p/\pi a_s)(\mathrm{d}z/\mathrm{d}t)$. Estimate the growth rate γ_s of the vertical movement ($\propto \exp\gamma_s t$) in the case with shell using approximation $v_{Ap}/a \gg \gamma_s$ and $1 \gg a/R$. Refer to [4.79].

6. Bootstrap Current Confirm the ratio of the bootstrap current I_b to the plasma current I_p is given by (4.91).

7. Non-Inductive Operation of ITER Parameters of ITER in non-inductive operation scenario are given in Table 4.5. When the parameters a, B_t, A etc. are specified as listed in Table 4.4, estimate the reduced parameters Q, R, I_p, τ_E in non-inductive operation scenario by the same way as that of inductive operation scenario. Refer to Section 4.8.

5

Reversed Field Pinch (RFP)

5.1 RFP Configuration

Reversed field pinch (RFP) is an axisymmetric toroidal field used as a tokamak. The magnetic field configuration is composed of the poloidal field B_{p} produced by the toroidal component of the plasma current and the toroidal field B_{t} produced by the external toroidal field coil and the poloidal component of the plasma current. The particle orbit loss is as small as that of tokamak. However, RFP and tokamaks have quite different characteristics. In RFP, the magnitudes of the poloidal field B_{p} and the toroidal field B_{t} are comparable and the safety factor

$$q_{\mathrm{s}}(r) = \frac{r}{R}\frac{B_z(r)}{B_\theta(r)}$$

is much less than 1 ($q_{\mathrm{s}}(0) \sim a/(R\Theta)$, $\Theta \sim 1.6$). The radial profile of the toroidal field is shown in Figure 5.1. The direction of the boundary toroidal field is reversed with respect to the direction of the on-axis field, and the magnetic shear is strong. Therefore, high-beta ($\langle\beta\rangle = 10 \sim 20\%$) plasmas can be confined in an MHD stable way. The plasma current can be much larger than the Kruskal-Shafranov limit ($q < 1$).

RFP started in an early phase of nuclear fusion research. A stable quiescent phase of discharge was found in Zeta in Harwell in 1968 [5.1]. The configuration of the magnetic field in the quiescent phase was the reversed field pinch configuration, as shown in Figure 5.1. The electron temperature, the energy confinement time, and the average beta of Zeta were $T_{\mathrm{e}} = 100 \sim 150\mathrm{eV}$, $\tau_{\mathrm{E}} = 2\mathrm{ms}$, $\langle\beta\rangle \sim 10\%$ at the time of the IAEA conference at Novosibirsk. However, the epoch-making result of tokamak T-3 with high electron temperature ($T_{\mathrm{e}} = 1\mathrm{keV}$, $\tau_{\mathrm{E}} =$ several ms, $\beta \sim 0.2\,\%$) was also presented in the same conference, and Zeta was shut down because of the better confinement characteristics in tokamaks. On the other hand, RFP can confine higher beta plasma and has been actively investigated to improve the confinement characteristics (ZT-40 M, OHTE, HBTX1-B, TPE-1RM 20, MST and RFX, TPE-RX) [5.2]~[5.5]. The important issues of RFP are confinement scaling and impurity control in the high-temperature region.

5.2 MHD Relaxation

5.2.1 Taylor's Theory on Relaxation

Even if the plasma is initially MHD unstable in the formation phase, it has been observed in RFP experiments that the plasma turns out to be a stable RFP configuration irrespective of the initial condition. J. B. Taylor pointed out in 1974 that RFP configuration is a minimum energy state by relaxation processes under certain constraints [5.6].

Let us introduce a physical quantity "*magnetic helicity*" for the study of this subject. By use of scalar and vector potentials ϕ, $\boldsymbol{A}$ of electric and magnetic field $\boldsymbol{E}$, $\boldsymbol{B}$, the magnetic helicity K is defined by the integral of the scalar product $\boldsymbol{A}\cdot\boldsymbol{B}$ over the volume V surrounded by a magnetic surface

$$K = \int_V \boldsymbol{A}\cdot\boldsymbol{B}\,\mathrm{d}\boldsymbol{r}, \tag{5.1}$$

where $\mathrm{d}\boldsymbol{r} \equiv \mathrm{d}x\,\mathrm{d}y\,\mathrm{d}z$. Since

$$\boldsymbol{E} = -\nabla\phi - \frac{\partial \boldsymbol{A}}{\partial t}, \qquad \boldsymbol{B} = \nabla\times\boldsymbol{A},$$

we find from Maxwell equations [5.7]

$$\begin{aligned}\frac{\partial}{\partial t}(\boldsymbol{A}\cdot\boldsymbol{B}) &= \frac{\partial \boldsymbol{A}}{\partial t}\cdot\boldsymbol{B} + \boldsymbol{A}\cdot\frac{\partial \boldsymbol{B}}{\partial t} = (-\boldsymbol{E}-\nabla\phi)\cdot\boldsymbol{B} - \boldsymbol{A}\cdot(\nabla\times\boldsymbol{E})\\ &= -\boldsymbol{E}\cdot\boldsymbol{B} - \nabla\cdot(\phi\boldsymbol{B}) + \nabla\cdot(\boldsymbol{A}\times\boldsymbol{E}) - \boldsymbol{E}\cdot(\nabla\times\boldsymbol{A})\\ &= -\nabla\cdot(\phi\boldsymbol{B} + \boldsymbol{E}\times\boldsymbol{A}) - 2(\boldsymbol{E}\cdot\boldsymbol{B}).\end{aligned}$$

When the plasma is surrounded by a perfect conductive wall, then the conditions $(\boldsymbol{B}\cdot\boldsymbol{n}) = 0$, $\boldsymbol{E}\times\boldsymbol{n} = 0$ hold ($\boldsymbol{n}$ is unit outward vector normal to the wall), so that we find

$$\frac{\partial K}{\partial t} = \frac{\partial}{\partial t}\int_V \boldsymbol{A}\cdot\boldsymbol{B}\mathrm{d}\boldsymbol{r} = -2\int_V \boldsymbol{E}\cdot\boldsymbol{B}\mathrm{d}\boldsymbol{r}. \tag{5.2}$$

The right-hand side in (5.2) is the loss term of the magnetic helicity. When Ohm's law

$$\boldsymbol{E} + \boldsymbol{v}\times\boldsymbol{B} = \eta\boldsymbol{j}$$

is applicable, the loss term is reduced to

$$\frac{\partial K}{\partial t} = -2\int_V \eta\boldsymbol{j}\cdot\boldsymbol{B}\mathrm{d}\boldsymbol{r}. \tag{5.3}$$

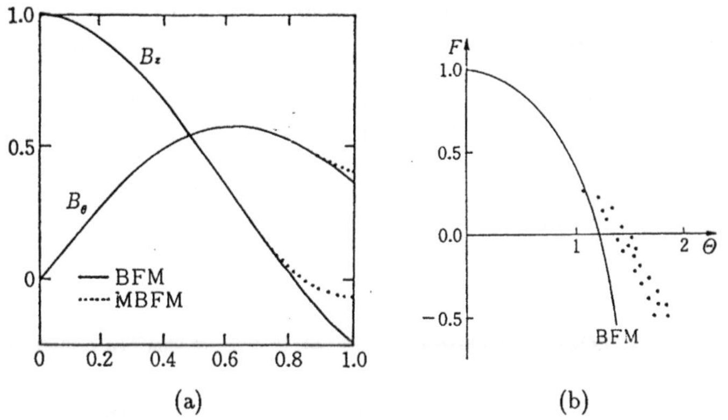

Figure 5.1 (a) Toroidal field $B_z(r)$ and poloidal field $B_\theta(r)$ of RFP. The radial profiles of the Bessel function model (BFM) and the modified Bessel function model (MBFM) are shown. (b) F-Θ curve.

When $\eta = 0$, the magnetic helicity is conserved; in other words, if a plasma is perfectly conductive, K integral over the volume surrounded by arbitrary closed magnetic surfaces is constant. However, if there is small resistivity in the plasma, the local reconnections of the lines of magnetic force are possible and the plasma can relax to a more stable state and the magnetic helicity may change locally. But J. B. Taylor postulates that the global magnetic helicity $K_{\rm T}$ integrated over the whole region of the plasma changes much more slowly. It is assumed that $K_{\rm T}$ is constant within the time scale of relaxation processes. Under the constraint of $K_{\rm T}$ invariant

$$\delta K_{\rm T} = \int \boldsymbol{B} \cdot \delta \boldsymbol{A} {\rm d}\boldsymbol{r} + \int \delta \boldsymbol{B} \cdot \boldsymbol{A} {\rm d}\boldsymbol{r} = 2 \int \boldsymbol{B} \cdot \delta \boldsymbol{A} {\rm d}\boldsymbol{r} = 0,$$

the condition of minimum energy of magnetic field

$$(2\mu_0)^{-1} \delta \int (\boldsymbol{B} \cdot \boldsymbol{B})\, {\rm d}\boldsymbol{r} = \mu_0^{-1} \int \boldsymbol{B} \cdot \nabla \times \delta \boldsymbol{A} {\rm d}\boldsymbol{r} = \mu_0^{-1} \int (\nabla \times \boldsymbol{B}) \cdot \delta \boldsymbol{A} {\rm d}\boldsymbol{r}$$

can be obtained by the method of undetermined multipliers, and we have

$$\nabla \times \boldsymbol{B} - \lambda \boldsymbol{B} = 0. \tag{5.4}$$

This solution is the minimum energy state in the force-free or pressureless plasma ($\boldsymbol{j} \times \boldsymbol{B} = \nabla p = 0,\ \boldsymbol{j} \| \boldsymbol{B}$). The axisymmetric solution in cylindrical coordinates is

$$B_r = 0, \qquad B_\theta = B_0 J_1(\lambda r), \qquad B_z = B_0 J_0(\lambda r), \tag{5.5}$$

and is called a *Bessel function model* (BFM). The profiles of $B_\theta(r)$ and $B_z(r)$ are shown in Figure 5.1(a). In the region $\lambda r > 2.405$, the toroidal field

B_z is reversed. The pinch parameter Θ and the field reversal ratio F are used commonly to characterize the RFP magnetic field as follows:

$$\Theta = \frac{B_\theta(a)}{\langle B_z \rangle} = \frac{(\mu_0/2)I_{\rm p}a}{\int B_z 2\pi r\, {\rm d}r}, \qquad F = \frac{B_z(a)}{\langle B_z \rangle}, \tag{5.6}$$

where $\langle B_z \rangle$ is the volume average of the toroidal field. For given Θ and F, average toroidal field $\langle B_z \rangle$ and toroidal field at the plasma edge (field of TF coil) $B_z(a)$ are

$$\langle B_z \rangle = \frac{1}{\Theta} B_\theta(a), \qquad B_z(a) = \frac{F}{\Theta} B_\theta(a),$$

where the range of Θ and F are 1.3∼3.5 and $-0.2 \sim -2$, respectively. The safety factor at the plasma edge is $q_{\rm s}(a) = (a/R)(F/\Theta)$.

The values of F and Θ for the Bessel function model are

$$\Theta = \frac{\lambda a}{2}, \qquad F = \frac{\Theta J_0(2\Theta)}{J_1(2\Theta)} \tag{5.7}$$

and the F-Θ curve is plotted in Figure 5.1(b). The quantity

$$\lambda = \frac{\mu_0 \boldsymbol{j} \cdot \boldsymbol{B}}{B^2} = \frac{(\nabla \times \boldsymbol{B}) \cdot \boldsymbol{B}}{B^2} = {\rm const.}$$

is constant in Taylor model. The safety factor $q_{\rm s}(r)$ of Bessel function model is

$$q_{\rm s}^{\rm BFM}(r) = \frac{rB_z}{RB_\theta} = \frac{a}{R\Theta}\frac{J_0(x)}{2J_1(x)/x}, \qquad q_{\rm s}^{\rm BFM}(0) = \frac{a}{R\Theta},$$

where $x = 2\Theta r/a$.

The observed RFP fields in experiments deviate from the Bessel function model due to the finite beta effect and the imperfect relaxation state. The λ value is no longer constant in the outer region of plasma and tends to 0 in the boundary. The solution of

$$\nabla \times \boldsymbol{B} - \lambda(r)\boldsymbol{B} = 0$$

with $\lambda(r)$, is called modified Bessel function model (MBFM).

The stability condition of the local MHD mode [5.8] is

$$\frac{1}{4}\left(\frac{q_{\rm s}'}{q_{\rm s}}\right)^2 + \frac{2\mu_0 p'}{rB_z^2}(1 - q_{\rm s}^2) > 0. \tag{5.8}$$

This formula indicates that the strong shear can stabilize the RFP plasma in the $p'(r) < 0$ region but that the flat pressure profile, $p'(r) \sim 0$, is preferable in the central region of weak shear. When $q_s < 1$, the local MHD mode is unstable near $q_s' = 0$ (*pitch minimum*).

When the effect of finite resistivity of a plasma is taken into account, it is expected by the classical process of magnetic dissipation that the RFP configuration can be sustained only during the period of $\tau_{\rm cl} = \mu_0 \sigma a^2$, where σ is the specific conductivity. However, ZT-40M experiments [5.9] demonstrated that RFP discharge was sustained more than three times ($\sim 20\,$ms) as long as $\tau_{\rm cl}$. This is clear evidence that the regeneration process of the toroidal flux exists during the relaxation process, which is consumed by classical magnetic dissipation, so that that RFP configuration can be sustained as long as the plasma current is sustained.

5.2.2 Relaxation Process

As the electron temperature in the core of RFP plasma is higher than in the outer region and the parallel component of inductive toroidal electric field to the magnetic field line is large in the core region compared with the outer region (even reversed in the edge region due to RFP configuration), the current profile tends to be a peaked one. The $m = 1$ modes become unstable and relaxation phenomena occurs like the internal disruption of tokamak (refer to Section 4.3). But the physical process is quite different from the Kadomtsev type one.

When the current profile is peaked and $q(0)$ becomes below $1/n$, global non-linear deformation of the flux surface builds up the antiparallel radial components of the magnetic field near the separatrix on the rear side of the displacement. Then the non-linear driven reconnection starts to develop.

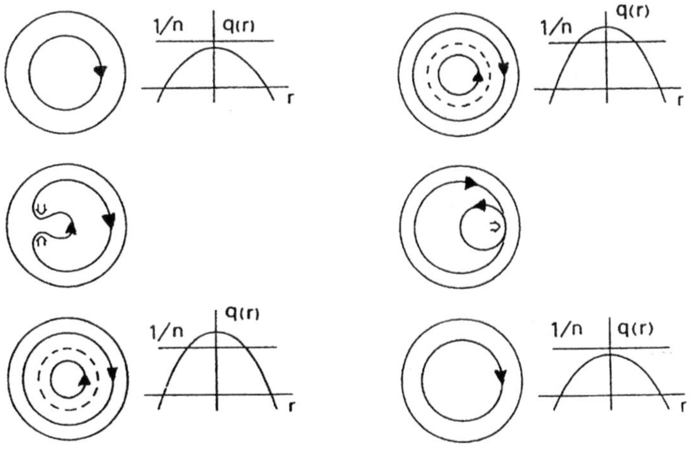

Figure 5.2 Driven reconnection by $m = 1$ global mode (left-hand side) and Kadomtsev type reconnction by tearing mode at the rational surface (right-hand side).

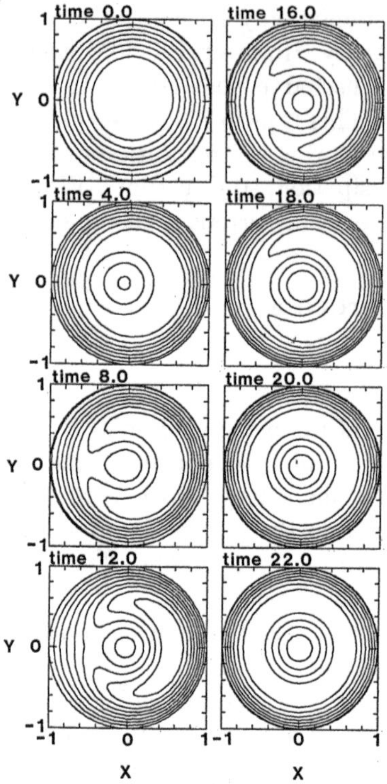

Figure 5.3 Contour plots of the helical flux of helicity $m = 1, n = 5$ mode which are extracted from mutiple helicity simulation. τ_A is Alfvén transit time. After [5.11].

According to the topology of magnetic flux, the rational surface appears and the $q(0)$ value increases larger than $1/n$ after reconnection (refer to Figure 5.2). The current profile becomes more flat and the RFP plasma is relaxed to a more stable state. The time variation of $q(0)$, which was measured in REPUTE RFP plasma [5.10], is consistent with this scenario. If Kadomtsev type reconnection happens in RFP $q(r)$ profile (reversed shear) configuration, the $q(0)$ value decreases after the reconnection and current profile is more peaked according to the topology of the magnetic flux. Then $m = 1$ global mode becomes unstable.

Studies of full MHD non-linear simulations have been done extensively [5.11, 5.12]. Contour plots of the helical flux for $m/n = 1/5$ mode which are extracted from mutiple helicity simulation are shown in Figure 5.3 [5.11]. As is evident in this figure, the occurrence of helical driven reconnection is observed in the period of 12$\sim$ 20τ_A (Alfvén transit time). In the multiple helicity relaxation, unstable $m = 1$ modes with different n feed most of their energy to (0,1) mode by non-linear coupling and axisymmetric non-linear

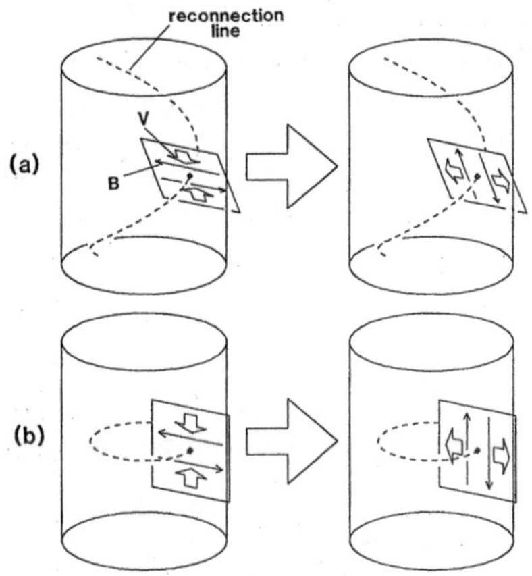

Figure 5.4 Schematic diagram of (a) helical non-linear reconnection and (b) axisymmetric non-linear reconnection. The left-hand side represents the state before reconnection and the right-hand side represents the state after reconnection. After [5.11].

reconnection of the $m = 0$ islands play a more dominant role in the relaxation process than helical non-linear reconnection of $m = 1$ islands. The geometrical difference between helical non-linear reconnection and axisymmetric non-linear reconnection is shown in Figure 5.4. In the helical non-linear reconnection (a), the reconnection (X)-line is the helical line and the converging flow leads reconnection (open arrows). After the reconnection, magnetic fields (arrows in the right-hand side) are created. These magnetic fields have toroidal and poloidal components. In the axisymmetric non-linear driven reconnection, the reconnection line is a poloidal ring and toroidal field only is created by axisymmetric reconnection. The latter process is more efficient for the generation of axisymmetric reversed field (positive field in the inner region and reversed field in the outer region of rational surface).

When there are fluctuations in plasmas, the magnetic field $\boldsymbol{B}$ in the plasma is expressed by the sum $\boldsymbol{B} = \langle \boldsymbol{B} \rangle_{\mathrm{t}} + \tilde{\boldsymbol{B}}$ of the time average $\langle \boldsymbol{B} \rangle_{\mathrm{t}}$ and the fluctuation term $\tilde{\boldsymbol{B}}$. The time average of Ohm's law (3.21)

$$\boldsymbol{E} + \left(\boldsymbol{v} - \frac{\boldsymbol{j}}{en_{\mathrm{e}}} \right) \times \boldsymbol{B} + \frac{1}{en_{\mathrm{e}}} \nabla p_{\mathrm{e}} = \eta \boldsymbol{j}$$

is reduced to

$$\langle \boldsymbol{E} \rangle_{\mathrm{t}} + \langle \boldsymbol{v} \rangle_{\mathrm{t}} \times \langle \boldsymbol{B} \rangle_{\mathrm{t}} + \langle \tilde{\boldsymbol{v}} \times \tilde{\boldsymbol{B}} \rangle_{\mathrm{t}} - \frac{1}{en_{\mathrm{e}}} \langle \tilde{\boldsymbol{j}} \times \tilde{\boldsymbol{B}} \rangle = \langle \eta \boldsymbol{j} \rangle_{\mathrm{t}} \tag{5.9}$$

where $\langle \ \rangle_{\mathrm{t}}$ denotes the time average. New terms appear due to fluctuations. The third term in the left-hand side of (5.9) is called MHD dynamo and the forth term is called Hall dynamo [5.17]. Since the time average of the toroidal flux $\Phi_z = \int B_z dS$ within the plasma cross-section is constant during quasi-stationary state, the time average of the electric field in θ direction is 0 ($\oint E_\theta dl = -d\Phi_z/dt = 0$) and $\langle \boldsymbol{v}_r \rangle_t = 0$. Steady-state RFP plasmas require the following condition:

$$\langle \eta j_\theta \rangle_t = \langle (\tilde{\boldsymbol{v}} \times \tilde{\boldsymbol{B}})_\theta \rangle_t - \frac{1}{en_e} \langle (\tilde{\boldsymbol{j}} \times \tilde{\boldsymbol{B}})_\theta \rangle_t. \quad (5.10)$$

In other words, resistive dissipation is compensated for by the effective electric field due to the fluctuations. This process is called *dynamo mechanism*.

When electron mean free path is very long, local relations such as Ohm's law may not be applicable. Instead of MHD dynamo theory, kinetic dynamo theory was proposed [5.13], in which anomalous transport of electron momentum across magnetic surfaces plays an essential role in sustaining RFP configuration.

Magnetic fluctuations of dynamo sustaining the poloidal plasma current, on the other hand, enhance the electron diffusion, since the electron diffusion coefficient is given by $D_e \sim v_{Te} a \langle (\delta B_r / B)^2 \rangle$ (see Section 13.2) and the energy confinement of RFP is deteriorated.

5.3 Confinement of RFP

The energy confinement time τ_E in an ohmically heated plasma can be obtained by the energy balance equation

$$\frac{(3/2)\langle n(T_e + T_i) \rangle_v 2\pi R \pi a^2}{\tau_E} = V_z I_p,$$

where V_z is the loop voltage and I_p is the plasma current. The notation $\langle\ \rangle_v$ means the volume average. Using the definition of the poloidal beta

$$\beta_\theta \equiv \frac{\langle n(T_e + T_i) \rangle_v}{B_\theta^2 / 2\mu_0} = \frac{8\pi^2 a^2 \langle n(T_e + T_i) \rangle_v}{\mu_0 I_p^2},$$

the energy confinement time is given by

$$\tau_E = \frac{3\mu_0}{8} R \beta_\theta \frac{I_p}{V_z}. \quad (5.11)$$

Therefore, the scalings of β_θ and V_z are necessary for the scaling of τ_E. In order to apply a loop voltage on RFP plasma, a cut in the toroidal direction is necessary in the shell conductor surrounding the plasma. In this case, the contribution of the surface integral must be added in (5.1) of magnetic helicity as follows:

$$\frac{\partial K}{\partial t} = -2\int \boldsymbol{E}\cdot\boldsymbol{B}\,\mathrm{d}\boldsymbol{r} - \int (\phi\boldsymbol{B} + \boldsymbol{E}\times\boldsymbol{A})\cdot\boldsymbol{n}\,\mathrm{d}S.$$

The induced electric field in the (conductive) shell surface is zero and is concentrated between both edges of the shell cut. The surface integral consists of the contribution $2V_z\Phi_z$ from the shell cut and the contribution from the other part of surface S_-; that is,

$$\frac{\partial K}{\partial t} = -2\int \eta\boldsymbol{j}\cdot\boldsymbol{B}\,\mathrm{d}\boldsymbol{r} + 2V_z\Phi_z - \int_{\mathrm{S}_-} (\phi\boldsymbol{B} + \boldsymbol{E}\times\boldsymbol{A})\cdot\boldsymbol{n}\,\mathrm{d}S, \qquad (5.12)$$

where Φ_z is the volume average of toroidal magnetic flux $\Phi_z = \pi a^2\langle B_z\rangle_{\mathrm{v}}$. In quasi-steady-state, the time average $\langle\partial K/\partial t\rangle_{\mathrm{t}}$ is zero. Then the time average of (5.12) yields

$$V_z = \frac{\int\langle\eta\boldsymbol{j}\cdot\boldsymbol{B}\rangle_{\mathrm{t}}\mathrm{d}\boldsymbol{r} + (1/2)\int_{\mathrm{S}_-}\langle\phi\boldsymbol{B} + \boldsymbol{E}\times\boldsymbol{A}\rangle_{\mathrm{t}}\cdot\boldsymbol{n}\,\mathrm{d}S}{\langle\Phi_z\rangle_{\mathrm{t}}}$$

$$= \frac{2\pi R}{\pi a^2}\eta_0 I_{\mathrm{p}}\zeta + V_{\mathrm{B}},$$

$$V_{\mathrm{B}} = \frac{2\pi R}{a}\frac{\langle\langle(\phi\boldsymbol{B} + \boldsymbol{E}\times\boldsymbol{A})\cdot\boldsymbol{n}\rangle_{\mathrm{t}}\rangle_{\mathrm{S}_-}}{\langle\langle B_z\rangle_{\mathrm{t}}\rangle_{\mathrm{v}}},$$

where $\langle\quad\rangle_{\mathrm{S}_-}$ is the average in the surface region S_-. The notation ζ is a non-dimensional factor determined by the radial profiles of specific resistivity and magnetic field as follows:

$$\zeta \equiv \frac{\langle\langle\eta\boldsymbol{j}\cdot\boldsymbol{B}\rangle_{\mathrm{t}}\rangle_{\mathrm{v}}}{\eta_0\langle\langle j_z\rangle_{\mathrm{t}}\rangle_{\mathrm{v}}\langle\langle B_z\rangle_{\mathrm{t}}\rangle_{\mathrm{v}}} = \frac{\langle\langle\eta\boldsymbol{j}\rangle_{\mathrm{t}}\cdot\langle\boldsymbol{B}\rangle_{\mathrm{t}}\rangle_{\mathrm{v}} + \langle\langle\widetilde{(\eta\boldsymbol{j})}\cdot\widetilde{\boldsymbol{B}}\rangle_{\mathrm{t}}\rangle_{\mathrm{v}}}{\eta_0\langle\langle j_z\rangle_{\mathrm{t}}\rangle_{\mathrm{v}}\langle\langle B_z\rangle_{\mathrm{t}}\rangle_{\mathrm{v}}}.$$

Here η_0 is the specific resistivity at the plasma center. When the term of fluctuation is negligible, the value ζ of modified Bessel function model is $\zeta \sim 10$, but the value is generally $\zeta > 10$ due to fluctuation. The value of V_{B} is 0 when the whole plasma boundary is a conductive shell. In reality, the plasma boundary is liner or protecting material for the liner. Lines of magnetic force can cross the wall by the magnetic fluctuation or shift of the plasma position ($\boldsymbol{B}\cdot\boldsymbol{n}\neq 0$, $\boldsymbol{E}\neq 0$). Then the term V_{B} has a finite value. The substitution of V_z into the equation of energy confinement time τ_{E} gives

$$\tau_{\mathrm{E}} = \frac{3}{8}\beta_\theta\left(\frac{\mu_0 a^2}{\eta_0}\frac{1}{2\zeta}\right)\left(1 + \frac{V_{\mathrm{B}}/2\pi R}{aB_\theta(a)}\frac{\mu_0 a^2}{\eta_0}\frac{1}{2\zeta}\right)^{-1}. \qquad (5.12)$$

When plasmas become hot, the resistive term becomes small and the contribution of V_{B} is no longer negligible.

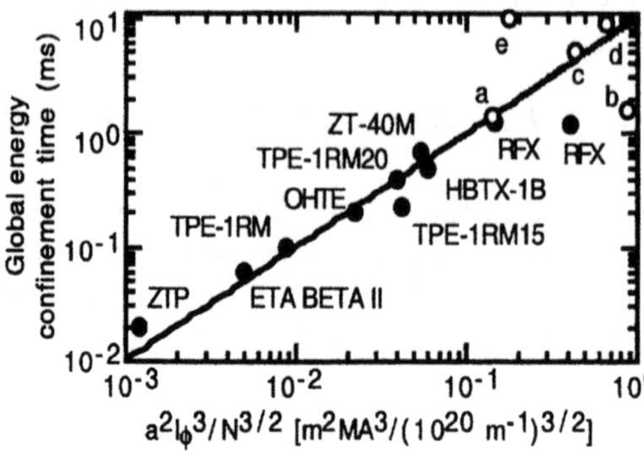

Figure 5.5 Scaling of energy confinement time of RFP in standard operation. I_ϕ in the figure is equivalent to I_p. After [5.19].

Energy confinement scaling of RFPs (ohmic heating only) was presented by ZT-40M group from the data of ZT-40M, RFX, TPE-1RM and the others as follows [5.14]:

$$\tau_{\mathrm{E\,stand.}} = 10.2a^2 I_p^{1.5}\left(\frac{I_p}{N}\right)^{1.5}, \tag{5.13}$$

where units are τ_E(ms), I_p(MA) and line density $N = \pi a^2 n_e (10^{20}\mathrm{m}^{-1})$. I_p/N in (5.13) has been used as an important parameter to specify PFP plasmas, since the experiments of RFP started and is the same as the inverse of the normalized Greenwald density (4.46), that is, $I_p/N = n_G^{-1}$. I_p/N is usually larger than 2 and the lowest limit is 1 as tokamaks [5.15]. Experimental results of energy confinement time from various RFP devices are compared with the scaling (5.13) and are shown in Figure 5.5 [5.14, 5.19]. Open circles in the figure are results of MST including improved confinement plasmas, which will described in the next paragraph.

Pulsed Parallel Current Drive (PPCD)

In recent years confinement has improved greatly in RFP through current density profile control. In standard operation of RFP, the parallel electric field is strongly peaked in the center due to RFP configuration. The resulting peaked current density profile is unstable to tearing instability as well as global kink mode. The instabilities grow and force the current density profile to one that is less peaked through the robust reconnections. The magnetic fluctuations in the saturated state cause substantial anomalous transport. To improve confinement, the applied electric field is adjusted so that the plasma current induced by the electric field is less peaked, yielding a more stable plasma with smaller fluctuation. During a plasma discharge, an additional ohmic electric field is applied that has a strong parallel (mostly poloidal) component in the outer region of plasma, a

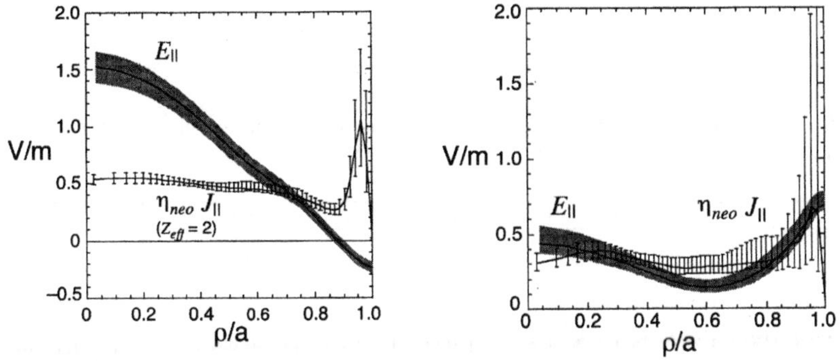

Figure 5.6 Radial profile of mean electric field and current density terms in Ohm's law for standard operation (left-hand side) and PPCD operation (right-hand side). The difference between the two curves indicates the dynamo effect. After [5.17].

technique that is often referred to as pulsed parallel current drive (PPCD). By PPCD, the energy confinement time of MST was doubled (2.2 ms) in 1994 [5.16] and energy confinement time in PPCD improved up to 10 ms in 2001 [5.19] and the dynamo strongly reduced.

The presence of the dynamo is most clearly seen through measurement of the terms in the parallel mean-field Ohm's law [5.17]. Figure 5.6 shows the measured radial profiles of the parallel electric field and current density terms in Ohm's law. In standard plasma (Figure 5.6, left-hand side), the current density profile is much different than the electric field profile. In the core, current density is smaller than would be expected from the electric field, while in the edge the current density is directed opposite the electric field. A strong dynamo exists which drives current opposite to the electric field over much of the cross-section. In PCCD plasma (Figure 5.6, right-hand side), the electric field profile changes substantially and, within experimental uncertainty, the current is fully accounted for by the applied electric field over most of the plasma cross-section.

Parameters of operating conditions and confinement properties in PCCD experiments of MST, RFX, and TPE-RX are listed in Table 5.1. Note for the plasma confined in MST with $\beta = 18\%$, $\tau_{\rm E} = 10\,{\rm ms}$ listed in the second line of Table 5.1, the toroidal field at the plasma edge (field of TF coil) is only 0.024T.

Empirical scaling of energy confinement time for RFP PCCD operation was presented by TPE group [5.23, 5.24] as follows:

$$\tau_{\rm E\_PCCD} = 0.84\tau_{\rm E\_scaling}, \qquad \tau_{\rm E\_scaling} = 8.07\times 10^{-3} a^{1.63} I_{\rm p}^{0.78} \left(\frac{I_{\rm p}}{N}\right)^{0.33} \Theta^{2.97},$$

Table 5.1 Parameters of operating conditions and confinement properties in PPCD experiments of MST, RFX, and TPE-RX.

	R/a	I_p	I_p/N	$\bar{n}_{e\,20}$	$T_{e\,0}/T_{i\,0}$	F/Θ	β	τ_E	Ref.
MST	1.5/0.5	0.34	4.0	0.1	0.39/0.2	-0.65/2.2	9	5	[5.18]
MST	1.5/0.5	0.21	3.5	0.07	0.6/0.18	-2.1/3.5	18	10	[5.19]
MST	1.5/0.5	0.4	5.1	0.1	0.8/0.3			5	[5.20]
MST^s	1.5/0.5	0.4	5.1	0.1	0.32/0.3			1	[5.20]
RFX	2/0.46	0.79	2.5	0.47	0.33/0.33	-0.20/1.5	6	1.9	[5.21]
TPE-RX	1.72/0.45	0.34	8.0	0.067	0.84/0.37	-0.69/2.0	9	3.5	[5.22]

Values are evaluated around the time when the energy confinement time has a peak value in PPCD period. Line density is evaluated by $N \equiv \bar{n}_e \pi a^2$ where $\bar{n}_e$ is line average electron density, as the profile of n_e is not available from the references. MST^s in the third line means standard operating conditions in MST. Units are R, a(m), I_p(MA), $I_p/N(10^{-20}$MA m), $\bar{n}_{e\,20}(10^{20}\text{m}^{-3})$, $T_{e\,0}, T_{i\,0}$(keV), β(%), τ_E(ms). Referred from [5.24],[5.20].

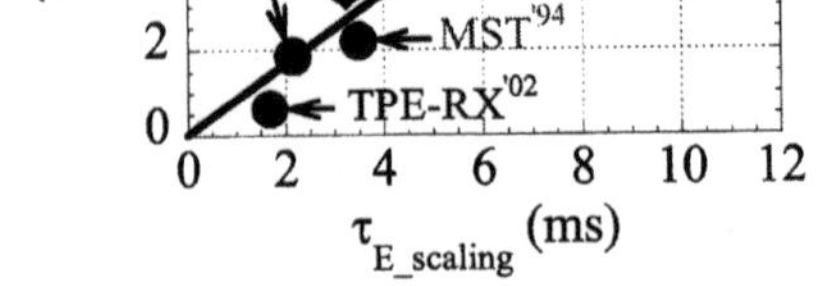

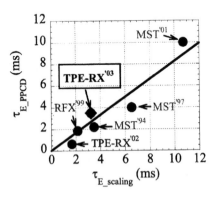

Figure 5.7 Comparison of experimental values of τ_{E_PCCD} in PPCD versus scaled values of $\tau_{E\_scaling}$. After [5.23].

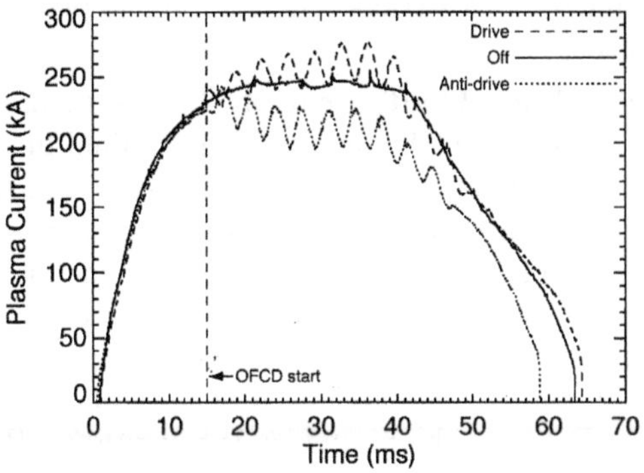

Figure 5.8 Plasma current vs. time for three cases. Oscillating field current drive (OFCD) with toloidal and poloidal loop voltages relative phases set for maximum helicity injection (dashed curve), OFCD with voltage phases set for maximum helicity removal, and OFCD off. After Ref.[5.17].

where units are $\tau_{\rm E}$(ms), a(m), $I_{\rm p}$(kA), $I_{\rm p}/N(10^{-20}$MAm). Comparison for experimental values of energy confinement time and $\tau_{\rm E\_scaling}$ is shown in Figure 5.8.

It is desirable that pulsed parallel current drive is replaced by stationary current drive, and experiments of additional heating are accelerated in RFP researches.

Oscillating Field Current Drive

RFP plasmas tend to be MBFM due to non-linear phenomena of MHD relaxaton. Oscillating field current drive (OFCD) was proposed [5.25] for sustaining the plasma current and preliminary experiments have been done [5.26]. More recently, still preliminary experiment of OFCD is tested in MST [5.17]. The result is shown in Figure 5.8. If terms V_z and Φ_z of the second term in the right-hand side of the magnetic helicity balance equation (5.12) are modulated as $V_z(t) = \tilde{V}_z \cos\omega t$, $\Phi_z(t) = \Phi_{z0} + \tilde{\Phi}_z \cos\omega t$, a direct current component $\tilde{V}_z\tilde{\Phi}_z$ in the product of $2V_z \cdot \Phi_z$ appears and compensates the resistive loss of the magnetic helicity. The period of the oscillating field must be longer than the characteristic time of relaxation and must be shorter than magnetic diffusion time. The disturbing effect of the oscillating field to RFP plasma must be evaluated further.

Problems

1. Magnetic Helicity When $\boldsymbol{A}$ and ϕ are a set of vector potential and scalor potential, any other set of $\boldsymbol{A}'$ and ϕ' given by $\boldsymbol{A}' = \boldsymbol{A} - \nabla\psi$ and $\phi' = \phi + (\partial\psi/\partial t)$ are also a set of vector potential and scalor potential which give the same $\boldsymbol{E}$ and $\boldsymbol{B}$. Prove that the magnetic helicity $\int \boldsymbol{A}\cdot\boldsymbol{B}{\rm d}\boldsymbol{r}$ defined by (5.1) is not changed by this transformation, if the plasma is surrounded by an ideal conductor.

2. Reversed Field Pinch Configuration Let us assume that the uniform plasma is produced in the uniform longitudinal field B_{z0} surrounded by ideal conductive cylinder with the radius of a at the initial time. When the longitudinal plasma current is driven, the plasma starts to z pinch and $B_z(r)$ in the internal region is increased, while total magnetic flux is conserved due to ideal shell. It is assumed that plasma changes according to Talor relaxation and the profiles of magetic field components follow Bessel function model (5.5), (5.6). Estimate the threshold plasma current to realize RFP configuration ($B_z(a)$ at the boundary is reversed to $B_z(0)$).

3. Oscillating Field Current Drive Estimate the necessary condition on the magnitudes of oscillating loop votage $\tilde{V}_z$ and oscillating component $\tilde{\Phi}_z$ of the toroidal flux to compensate the resistive diffusion for sustaining RFP configuration. Assume that plasma boundary is an ideal conductor. Refer to (5.12).

6

Stellarator

A stellarator field can provide a steady-state magneto-hydrodynamic equilibrium configuration of plasma only by the external field produced by the coils outside the plasma. The rotational transform, which is necessary to confine the toroidal plasma, is formed by the external coils so that the stellarator has the merit of steady-state confinement. Although Stellarator C [6.1] was rebuilt as the ST tokamak in 1969 at the Princeton Plasma Physics Laboratory, confinement experiments by Wendelstein 7A, 7AS, Heliotoron-E, and ATF are being carried out, because there is a merit of steady-state confinement without current-driven instabilities. Large helical device LHD started experiments in 1998 and advanced stellarator WVII-X is under construction.

6.1 Helical Field

Let us consider a magnetic field of helical symmetry. By means of cylindrical coordinates $(r,\ \theta,\ z)$, we can express the field in terms of $(r,\ \varphi \equiv \theta - \delta\alpha z)$, where $\alpha > 0$, $\delta = \pm 1$. A magnetic field in a current-free region ($\boldsymbol{j} = 0$) can be expressed by a scalar potential ϕ_B, satisfying $\Delta\phi_B = 0$, and we can write

$$\phi_B = B_0 z + \frac{1}{\alpha}\sum_{l=1}^{\infty} b_l I_l(l\alpha r)\sin(l\varphi), \tag{6.1}$$

$$\varphi \equiv \theta - \delta\alpha z.$$

The field components $(B_r,\ B_\theta,\ B_z)$ of $\boldsymbol{B} = \nabla\phi_B$ are given by

$$B_r = \sum_{l=1}^{\infty} l b_l I_l'(l\alpha r)\sin(l\varphi), \tag{6.2}$$

$$B_\theta = \sum_{l=1}^{\infty}\left(\frac{1}{\alpha r}\right) l b_l I_l(l\alpha r)\cos(l\varphi), \tag{6.3}$$

$$B_z = B_0 - \delta\sum_{l=1}^{\infty} l b_l I_l(l\alpha r)\cos(l\varphi). \tag{6.4}$$

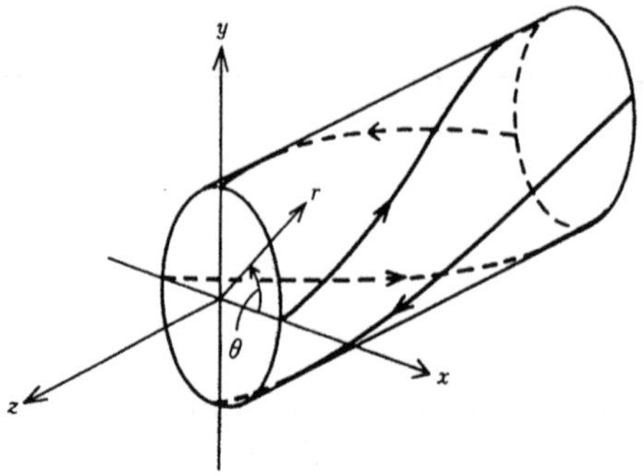

Figure 6.1 Current of helical coils.

The vector potential corresponding to this field has components

$$A_r = -\frac{\delta}{\alpha^2 r}\sum_{l=1}^{\infty} b_l I_l(l\alpha r)\sin(l\varphi),$$

$$A_\theta = \frac{B_0}{2}r - \frac{\delta}{\alpha}\sum_{l=1}^{\infty} b_l I_l'(l\alpha r)\cos(l\varphi),$$

$$A_z = 0.$$

Using these, we can write

$$B_r = -\frac{\partial A_\theta}{\partial z}, \qquad B_\theta = \frac{\partial A_r}{\partial z}, \qquad B_z = \frac{1}{r}\frac{\partial (rA_\theta)}{\partial r} - \frac{1}{r}\frac{\partial A_r}{\partial \theta}.$$

The magnetic surface $\psi = A_z + \delta\alpha r A_\theta = \delta\alpha r A_\theta = \text{const.}$ is given by

$$\psi(r,\varphi) = B_0\frac{\delta\alpha r^2}{2} - r\sum_{l=1}^{\infty} b_l I_l'(l\alpha r)\cos(l\varphi) = \text{const.} \tag{6.5}$$

Such a helically symmetric field can be produced by a helical current distribution as is shown in Figure 6.1. Let the magnetic fluxes in z and θ directions inside the magnetic surface be denoted by Φ and X (X is the integral over the pitch along z, i.e., over $2\pi/\alpha$); then these may be expressed by

$$\Phi = \int_0^{2\pi}\int_0^{r(\varphi)} B_z(r,\ \varphi) r\, \mathrm{d}r\, \mathrm{d}\theta,$$

$$X = \int_0^{2\pi/\alpha}\int_0^{r(\varphi)} B_\theta(r,\ \varphi)\, \mathrm{d}r\, \mathrm{d}z = \frac{1}{\alpha}\int_0^{2\pi}\int_0^{r(\varphi)} B_\theta(r,\ \varphi)\, \mathrm{d}r\, \mathrm{d}\theta.$$

Since $\alpha r B_z - \delta B_\theta = \alpha\partial(rA_\theta)/\partial r = \delta\partial\psi/\partial r$, we find that

$$\Phi - \delta X = 2\pi\psi/\delta\alpha.$$

Let us consider only one harmonic component of the field. The scalar potential and the magnetic surface are expressed by

$$\phi_B = B_0 z + \frac{b}{\alpha} I_l(l\alpha r)\sin(l\theta - \delta l\alpha z),$$

$$\psi = \frac{B_0}{2\delta\alpha}\left((\alpha r)^2 - \frac{2\delta(\alpha r)b}{B_0} I_l'(l\alpha r)\cos(l\theta - \delta l\alpha z)\right) = \frac{B_0}{2\delta\alpha}(\alpha r_0)^2.$$

The singular points $(r_{\rm s}, \theta_{\rm s})$ in the $z = 0$ plane are given by

$$\frac{\partial\psi}{\partial r} = 0, \qquad \frac{\partial\psi}{\partial\theta} = 0.$$

Since the modified Bessel function $I_l(x)$ satisfies $I_l''(x) + I_l'(x)/x - (1 + l^2/x^2)I_l = 0$, the singular points are given by

$$\sin(l\theta_{\rm s}) = 0,$$

$$\alpha r\left(1 - \frac{\delta bl}{B_0}\left(1 + \frac{1}{(\alpha r_{\rm s})^2}\right) l_l(l\alpha r_{\rm s})\cos(l\theta_{\rm s})\right) = 0$$

or

$$\theta_{\rm s} = 2\pi(j-1)/l, \qquad \delta b/B_0 > 0,$$

$$= 2\pi\left(j - \frac{1}{2}\right)/l, \qquad \delta b/B_0 < 0, \qquad j = 1, \cdots, l,$$

$$\left|\frac{\delta bl}{B_0}\right| = \frac{1}{(1 + (\alpha r_{\rm s})^{-2})I_l(l\alpha r_{\rm s})}.$$

The magnetic surfaces for $l = 1Cl = 2Cl = 3$ are shown in Figure 6.2. The magnetic surface which passes through the hyperbolic singular point or X point is called *separatrix*. When $x \ll 1$, the modified Bessel function is

$$I_l(x) \approx \frac{1}{l!}\left(\frac{x}{2}\right)^l.$$

The magnetic surfaces in the region $\alpha r \ll 1$ are expressed by

$$(\alpha r)^2 - \frac{\delta b(l/2)^{l-1}}{B_0(l-1)!}(\alpha r)^l \sin l(\theta - \delta\alpha z) = \text{const.}$$

The magnitude B is

$$\left(\frac{B}{B_0}\right)^2 = 1 - 2\frac{\delta lb}{B_0} I_l \cos(l\varphi) + \left(\frac{lb}{B_0}\right)^2 \left(I_l^2\left(1 + \frac{1}{(\alpha r)^2}\right)\cos^2(l\varphi) + (I_l')^2\sin^2(l\varphi)\right).$$

The magnitude B at X point $(r_{\rm s}, \theta_{\rm s})$ is

$$\left(\frac{B}{B_0}\right)^2 = 1 - \frac{(\alpha r)^2}{1 + (\alpha r)^2},$$

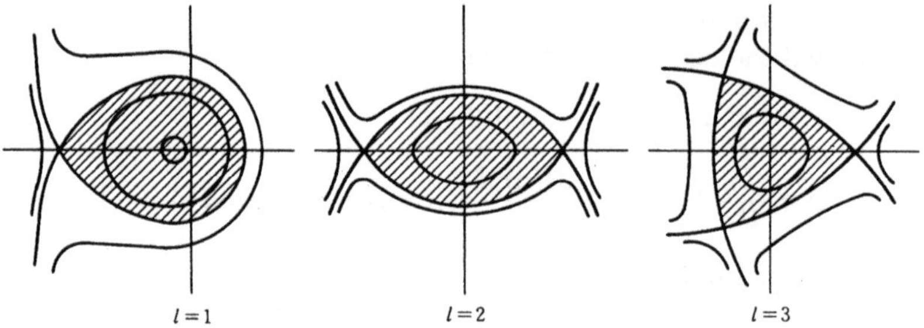

Figure 6.2 Magnetic surfaces, showing X points and separatrices, of the helical field.

and B at the point $(r_s, \theta_s + \pi/l)$ is

$$\left(\frac{B}{B_0}\right)^2 = 1 + \frac{(\alpha r)^2}{1+(\alpha r)^2}.$$

Therefore, the magnitude B is small at X points.

Let us estimate the rotational transform angle ι. As the line of magnetic force is expressed by

$$\frac{\mathrm{d}r}{B_r} = \frac{r\mathrm{d}\theta}{B_\theta} = \frac{\mathrm{d}z}{B_z},$$

the rotational transform angle is given by

$$\frac{r\iota}{2\pi R} = \left\langle \frac{r\mathrm{d}\theta}{\mathrm{d}z} \right\rangle = \left\langle \frac{B_\theta}{B_z} \right\rangle = \left\langle \frac{(1/\alpha r) l b I_l(l\alpha r)\cos l(\theta - \delta z)}{B_0 - l b I_l(l\alpha r)\cos l(\theta - \delta z)} \right\rangle.$$

Here r and θ are the values on the line of magnetic force and are functions of z and $\langle \quad \rangle$ denotes the average over z. In a vaccum field, $\oint B_\theta \mathrm{d}l = \int (\nabla \times \boldsymbol{B}) \cdot \mathrm{d}\boldsymbol{S} = 0$ holds, so that the rotational transform angle is 0 in the first order of b/B_0. However, the first order components of B_θ and B_z resonate to yield the resultant second order rotational transform angle. The average method gives the formula of the *rotational transform angle* [6.2], ch.2 of [2.2]

$$\frac{\iota}{2\pi} = \delta \left(\frac{b}{B}\right)^2 \frac{l^3}{2} \left(\frac{\mathrm{d}}{\mathrm{d}x}\left(\frac{I_l I_l'}{x}\right)\right)_{x=l\alpha r} \frac{R}{r}. \tag{6.6}$$

By use of the expansion

$$I_l(x) = \left(\frac{x}{2}\right)^l \left(\frac{1}{l!} + \frac{1}{(l+1)!}x^2 + \frac{1}{2!(l+2)!}x^4 + \cdots\right),$$

we find

$$\frac{\iota}{2\pi} = \delta \left(\frac{b}{B}\right)^2 \left(\frac{1}{2^l l!}\right)^2 l^5 (l-1)\alpha R \left((l\alpha r)^{2(l-2)} + \ldots\right). \quad (l \geq 2)$$

An example of the analysis of toroidal helical field is given in [6.3].

Equilibrium Equation of a Helically Symmetric System
In a helically symmetric system, there exists an equilibrium equation similar to the Grad-Shafranov equation in an axi-symmetric system. Since physical quantities depend only on r and $\phi = \theta - hz$, the magnetic surface ψ ($\nabla\psi \cdot \boldsymbol{B} = 0, \nabla \cdot \boldsymbol{B} = 0$) satisfies

$$B_r \frac{\partial \psi}{\partial r} + \left(\frac{1}{r} B_\theta - h B_z\right) \frac{\partial \psi}{\partial \phi} = 0,$$

$$\frac{\partial (r B_r)}{\partial r} + \frac{\partial}{\partial r}(B_\theta - hr B_z) = 0.$$

Accordingly, we have

$$r B_r = \frac{\partial \psi}{\partial \phi}, \qquad B_\theta - hr B_z = -\frac{\partial \psi}{\partial r}.$$

From $\nabla p \cdot \boldsymbol{B} = 0$, it follows $p = p(\psi)$. Writing

$$B^*(r, \phi) \equiv B_z + hr B_\theta$$

and using $\nabla p \cdot \boldsymbol{j} = 0$, we find

$$B^*(r, \phi) = B^*(\psi),$$

and

$$B_r = \frac{1}{r}\frac{\partial \psi}{\partial \phi}, \qquad B_\theta = \frac{-\frac{\partial \psi}{\partial r} + hr B^*}{1 + (hr)^2}, \qquad B_z = \frac{hr \frac{\partial \psi}{\partial r} + B^*}{1 + (hr)^2}.$$

The current density is calculated from $\boldsymbol{B}$ as follows:

$$\mu_0 \boldsymbol{j}_r = \frac{B^{*\prime}}{r} \frac{\partial \psi}{\partial \phi},$$

$$\mu_0 \boldsymbol{j}_\theta = -hr L_h(\psi) - \frac{B^{*\prime} \frac{\partial \psi}{\partial r}}{1 + (hr)^2} + \frac{2h^2 r B^*}{(1 + (hr)^2)^2},$$

$$\mu_0 \boldsymbol{j}_z = -hr L_h(\psi) + \frac{hr B^{*\prime} \frac{\partial \psi}{\partial r}}{1 + (hr)^2} + \frac{2h B^*}{(1 + (hr)^2)^2}.$$

The equilibrium equation for helically symmetric system is [6.4]

$$L_h(\psi) + \frac{B^* B^{*\prime}}{1 + (hr)^2} - \frac{2h B^*}{(1 + (hr)^2)^2} = -\mu_0 p', \tag{6.7}$$

$$L_h(\psi) \equiv \left(\frac{1}{r}\frac{\partial}{\partial r}\frac{r}{1 + (hr)^2}\frac{\partial}{\partial r} + \frac{1}{r^2}\frac{\partial^2}{\partial \phi^2}\right)\psi.$$

There is an analytical solution, when $B^*(\psi)$ and $p(\psi)$ are linear functions of ψ [6.5].

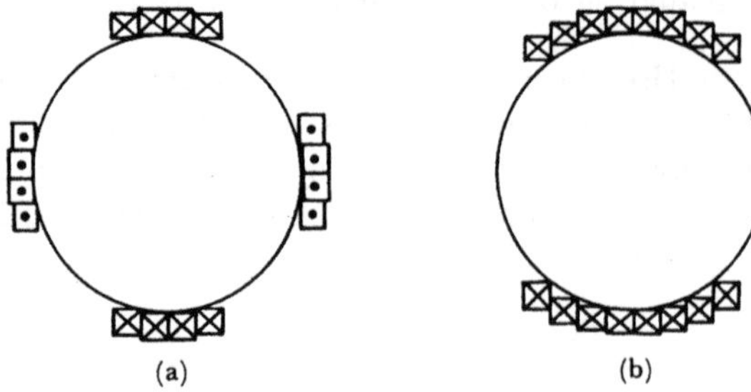

Figure 6.3 Cross-sectional views of helical coils in the $l = 2$ case. (a) Standard stellarator. (b) Heliotron/Torsatron.

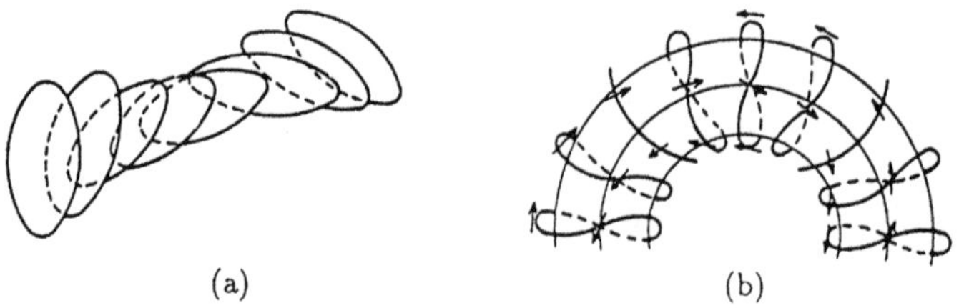

Figure 6.4 (a) Arrangement of elliptical coils used to produce an $l = 2$ linear helical field. (b) Twisted toroidal coils that produce the $l = 2$ toroidal helical field.

6.2 Stellarator Devices

Familiar helical fields are of pole number $l = 2$ or $l = 3$. The three-dimensional magnetic axis system of Heliac has $l = 1$ component. When the ratio of the minor radius a_h of a helical coil to the helical pitch length R/m (R is the major radius and m is the number of field periods) is much less than 1, that is, $ma_h/R \ll 1$, the rotational transform angle is $\iota_2(r) = \text{const.}$ for $l = 2$ and $\iota_3(r) = \iota(r/a)^2$ for $l = 3$. In this case, the shear is small for the $l = 2$ configuration, and $\iota_3(r)$ is very small in the central region for the $l = 3$ configuration. However, if $ma_h/R \sim 1$, then $\iota_2(r) = \iota_0 + \iota_2(r/z)^2 + \cdots$, so that the shear can be large even when $l = 2$.

The arrangement of coils in the $l = 2$ case is shown in Figure 6.3. Figure 6.3(a) is the standard type of stellarator [6.6, 6.7] and Figure 6.3(b) is a heliotron/torsatron type [6.8, 6.9]. Usually helical fields are produced by the toroidal field coils and the helical coils. In the heliotron/torsatron configuration, the current directions of the helical coils are the same so that the toroidal field and the helical field can be produced by the helical coils alone [6.10, 6.11], if the pitch is properly chosen. Closed magnetic surfaces can be formed even without toroidal field coils [6.12, 6.13]. The typical devices of this type are Heliotron E, ATF and LHD. The device of LHD is

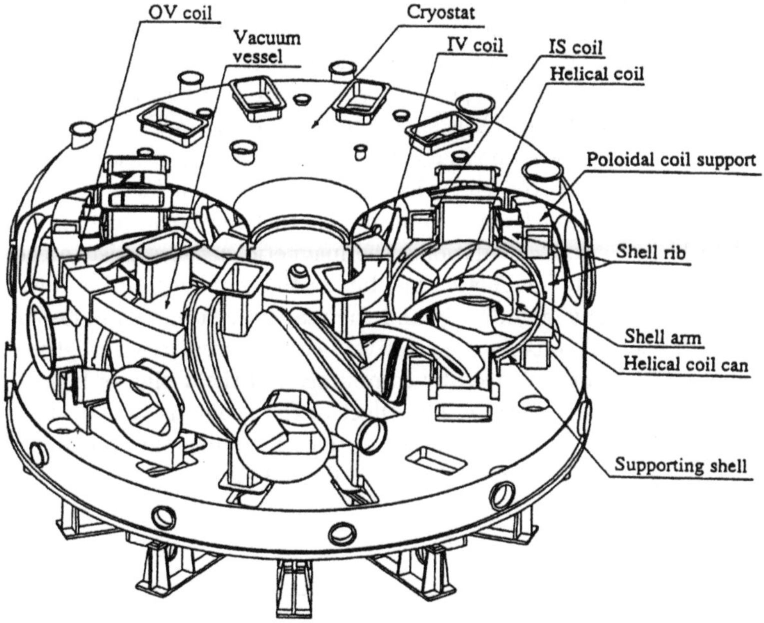

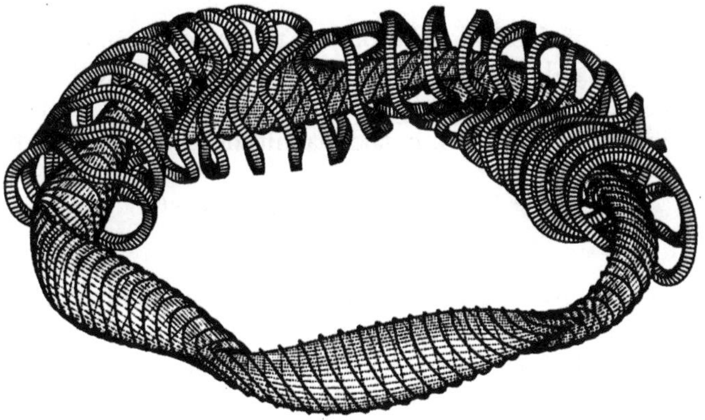

Figure 6.5 Upper figure is schematic view of the LHD device in Toki ($R = 3.9\,\mathrm{m}$, $a \sim 0.6\,\mathrm{m}$, $B = 3\mathrm{T}$). After [6.16]. Lower figure is a modular coil system and a magnetic surface of the optimized stellerator Wendelstein 7-X under the construction in Greifswald ($R = 5.5\mathrm{m}$, $a = 0.55\mathrm{m}$, $B = 2.5\mathrm{T}$). After [6.17].

shown in the top of Figure 6.5 [6.16].

When elliptical coils are arranged as shown in Figure 6.4(a), an $l = 2$ helical field can be obtained [6.14]. The currents produced by the twisted toroidal coil system shown in Figure 6.4(b) can simulate the currents of toroidal field coils and the helical coils taken together [6.15]. The typical devices of this modular coil type are Wendelstein 7AS and 7X. The modular coil system of Wendelstein 7X is shown in the bottom of Figure 6.5 [6.17].

For linear helical fields, the magnetic surface $\Psi = rA_\theta$ exists due to its helical symmetry. However, the existence of magnetic surfaces in toroidal helical fields has not yet been proven in the strict mathematical sense. According to numerical calculations, the magnetic surfaces exist in the central region near the magnetic axis, but in the outer region the lines of magnetic force behave ergodically and the magnetic surfaces are destroyed. Although the helical coils have a relatively complicated structure, the lines of magnetic force can be traced by computer, and the design of helical field devices becomes less elaborate. The effect of the geometrical error to the helical field can be estimated, and accurate coil windings are possible with numerically controlled devices ($\Delta l/R < 0.05 \sim 0.1\%$).

6.3 Neoclassical Diffusion in Helical Field

For the analysis of classical diffusion due to Coulomb collision, the study of the orbit of charged particles is necessary. In a helical field or even in a tokamak toroidal field produced by a finite number of coils, there is an asymmetric inhomogenous term in the magnitude B of magnetic field

$$\frac{B}{B_0} \approx 1 - \epsilon_{\rm h} \cos(l\theta - m\varphi) - \epsilon_t \cos\theta, \tag{6.8}$$

in addition to the toroidal term $-\epsilon_{\rm t}\cos\theta$. The variation of B along the lines of magnetic force is shown in Figure 6.6. Particles trapped by the inhomogeneous field of helical ripples drift across the magnetic surfaces and contribute to the particle diffusion in addition to the banana particles as was discussed in tokamak. The curvature of line of magnetic force near the helically trapped region is convex outward and is denoted by $R_{\rm h}$. Helically trapped particles drift in poloidal direction (θ direction) due to ∇B drift with the velocity of $v_{\rm h} \approx mv_\perp^2/(qBR_{\rm h})$ (see Figure 6.7). The angular velocity of poloidal rotation is

$$w_{\rm h} = v_{\rm h}/r \approx (r/R_{\rm h})(T/qBr^2) \approx \epsilon_{\rm h}(T/qBr^2). \tag{6.9}$$

In the case of a linear helical field ($\epsilon_{\rm t} = 0$), helically trapped particles rotate along the magnetic surface. However, in the case of a toroidal helical field,

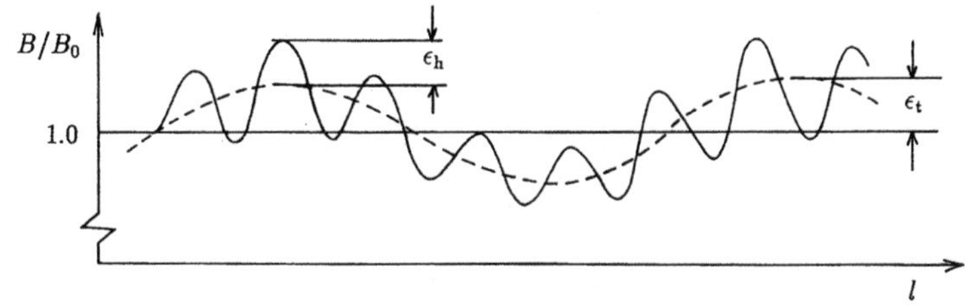

Figure 6.6 Variation of the magnitude B along the length l of the line of magnetic force.

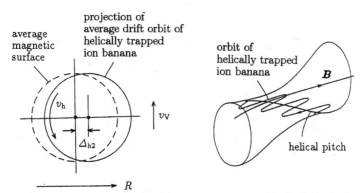

Figure 6.7 Orbit of helical banana ion trapped in helical ripple.

the toroidal drift is superposed and the toroidal drift velocity is $v_{\rm v} = T/(qBR)$ in the vertical direction (refer Section 2.5.4). When the effective collision time $(\nu_{\rm eff})^{-1} = (\nu/\epsilon_{\rm h})^{-1}$ is shorter than one period $(\omega_{\rm h})^{-1}$ of poloidal rotation, the deviation of orbit of helical banana from the magnetic surface is

$$\Delta_{\rm h1} = v_{\rm v}\frac{\epsilon_{\rm h}}{\nu} = \epsilon_{\rm h}\frac{T}{qBR}\frac{1}{\nu}.$$

Then the coefficient of particle diffusion becomes [6.18]

$$D_{\rm h1} \sim \epsilon_{\rm h}^{1/2}\Delta_{\rm h1}^2\nu_{\rm eff} = \epsilon_{\rm h}^{3/2}\left(\frac{T}{qBR}\right)^2\frac{1}{\nu} = \epsilon_t^2\epsilon_{\rm h}^{3/2}\left(\frac{T}{qBr^2}\frac{1}{\nu}\right)\left(\frac{T}{qB}\right).$$

Since $R_{\rm h} \sim r/\epsilon_{\rm h}$, the other expression is

$$D_{\rm h1} \sim \gamma_{\rm h}\epsilon_{\rm h}^{1/2}\epsilon_{\rm t}^2\left(\frac{\omega_{\rm h}}{\nu}\right)\left(\frac{T}{qB}\right), \quad (\nu/\epsilon > \omega_{\rm h}), \tag{6.10}$$

where $\gamma_{\rm h}$ is a coefficient with the order of $O(1)$ (Figure 6.8).

When the effective collision time $(\nu_{\rm eff})^{-1}$ is longer than $(\omega_{\rm h})^{-1}$, the deviation $\Delta_{\rm h2}$ of the orbit and the magnetic surface is

$$\Delta_{\rm h2} \approx v_{\rm v}/\omega_{\rm h} \approx \frac{R_{\rm h}}{R}r \sim \frac{\epsilon_{\rm t}}{\epsilon_{\rm h}}r,$$

and the particle diffusion coefficient $D_{\rm h2}$ in this region becomes (Figure 6.8)

$$D_{\rm h2} \approx \epsilon_{\rm h}^{1/2}\Delta_{\rm h2}^2\nu_{\rm eff} = \left(\frac{\epsilon_{\rm t}}{\epsilon_{\rm h}}\right)^2\frac{1}{\epsilon_{\rm h}^{1/2}}r^2\nu. \qquad (\nu/\epsilon_{\rm h} < \omega_{\rm h})$$

When a particle is barely trapped in a local helical mirror, the particle moves very slowly near the reflection point where the magnetic field is locally maximum and the field line is concave to outward. The effective curvature, which the particle feels in time average, becomes negative (concave). The orbit of the trapped particle in this case becomes *superbanana* [6.18]. However, this theoretical treatment is based on the assumption of the longitudinal adiabatic invariant $J_\parallel$ =const. along the orbit of the helically trapped particle. The adiabatic invariance is applicable when the poloidal rotation angle, during the one period of back and forth motion in the helical local mirror, is small. As the one period of back and forth motion of barely trapped particles becomes long, the adiabatic invariance may not be applicable. The orbit trace by numerical calculations shows that the superbanana does not appear [6.19] in the realistic case of $\epsilon_{\rm h} \sim \epsilon_{\rm t}$. If a particle orbit crosses the wall, the particle is lost. This is called *orbit loss*. A loss region in velocity space appears due to orbit loss in some cases [6.20]. When a radial electric field appears, the angular frequency of the poloidal drift rotation becomes $\omega_{\rm h} + \omega_{\rm E}$ ($\omega_{\rm E} = E_r/B_0$), and the orbit is affected by the radial electric field.

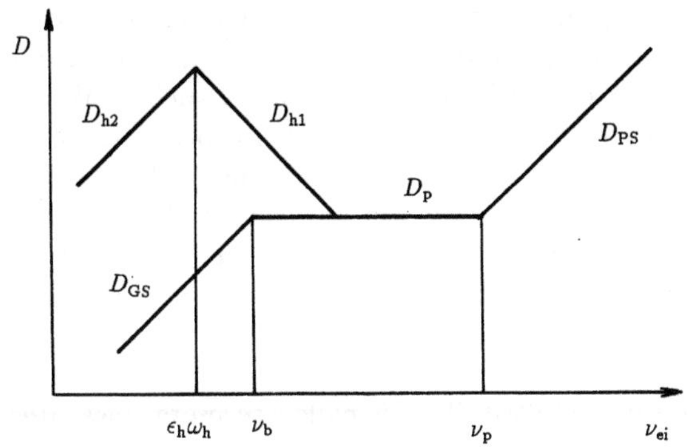

Figure 6.8 Dependence of the neoclassical diffusion coefficient of helical field on collision frequency. $\nu_p = (\iota/2\pi)v_{Te}/R$, $\nu_b = \epsilon_t^{3/2}\nu_p$, $\omega_h = \epsilon_h\kappa T_e/(qBr^2)$.

The thermal diffusion coefficient χ_{h1} due to helically trapped particles in the region of $\nu/\epsilon_h > \omega_h$ is given by

$$\chi_{h1} \sim \gamma_T\epsilon_t^2\epsilon_h^{3/2}\left(\frac{T}{qBr}\right)^2\frac{1}{\nu} \qquad (\gamma_T \sim 50). \tag{6.11}$$

Since $\nu \propto T^{-1.5}$, it means $\chi_{h1} \propto T^{3.5}$. This may suggest that the thermal conduction loss becomes large in hot plasma and the suppression of helical ripple loss is very important [6.21]~[6.23].

Since toroidal helical systems lose helical symmetry as well as axisymmetry, the generalized momentum correponding cyclic coordinate is not conserved (angular momentum $mr^2\dot{\theta} + qrA_\theta = \text{const.}$ for axisymmetric system $\boldsymbol{A}(r,z)$ and $m(\dot{z} + \alpha r^2\dot{\theta}) + q(A_z + \alpha rA_\theta) = \text{const.}$ for helically symmetric system $\boldsymbol{A}(r, \theta - \alpha z)$). Therefore, the orbit loss of energetic ions produced by heatings or fusion produced alpha particles with 3.5 MeV becomes large and the heating efficiencies may be deteriorated.

If quasi-symmetric stellarators are designed, particle orbits are quasi-closed due to the constant of motion. The neoclassical diffusion in classical stellarator field must be modified. There are active efforts to design quasi-axisymmetric, quasi-helically symmetric, and quasi-isodynamic or quasi-omnigenous stellerators, which will be introduced in Section 6.5.

6.4 Confinement of Stellarator

After Stellarator C [6.24], the basic experiments were carried out in small but accurate stellarator devices (Clasp, Proto Cleo, Wendelstein IIb, JIPP I, Heliotron D, L1, Uragan 1). Alkali plasmas, or afterglow plasmas produced by wave heating or gun injection, were confined quiescently. The effect of shear on the stability and confinement scaling was investigated.

The $l = 2$ stellarators with long helical pitch, such as Wendelstein IIa or JIPP I-b, have nearly constant rotational transform angles and the shears are small. When the transform angle is rational, $\iota/2\pi = n/m$, a line of magnetic force comes back to the initial position after m turns of the torus and is closed. If electric charges are localized in some place, they cannot be dispersed uniformly within the magnetic surface in the case of rational surfaces. A resistive drift wave or resistive MHD instabilities are likely to be excited, and convective loss is also possible [6.25]. The enhanced loss is observed in the rational case. This is called *resonant loss.* Resonant loss can be reduced by the introduction of shear.

Medium-scale stellarator devices (Wendelstein 7A, Cleo, JIPP T-II, Heliotron-E, L2, Uragan 2, Uragan 3) have been constructed. NBI heating or wave heatings, which were developed in tokamaks, have been applied to plasma production in helical devices. In Wendelstein 7A, a target plasma was produced by ohmic heating; then the target plasma was sustained by NBI heating while the plasma current was gradually decreased, and finally a high-temperature plasma with $T_{\rm i}\sim$ several hundred eV, $n_{\rm e}\sim$ several 10^{13} $\rm cm^{-3}$ was confined without plasma current. In Heliotron-E, a target plasma was produced by electron cyclotron resonance heating with $T_{\rm e}\sim 800\,{\rm eV}, n_{\rm e}\sim 0.5\times 10^{13}\,{\rm cm}^{-3}$, and the target plasma was heated by NBI heating with 1.8 MW to the plasma with $T_{\rm i}\sim 1{\rm keV}, n_{\rm e}=2\times 10^{13}{\rm cm}^{-3}$. These experimental results demonstrate the possibility of steady-state confinement by stellarator configurations [6.26].

Experimental scaling laws of energy confinement time are presented from Heliotron-E group [6.27] as follows:

$$\tau_{\rm E}^{\rm LHD} = 0.17a^{2.0}R^{0.75}n_{20}^{0.69}B^{0.84}P^{-0.58}, \tag{6.12}$$

where the unit of n_{20} is $10^{20}{\rm m}^{-3}$. W7AS group presented W7AS confinement scaling [6.28] of

$$\tau_{\rm E}^{\rm W7AS} = 0.115a^{2.21}R^{0.74}n_{19}^{0.5}B^{0.73}P^{-0.54}(\iota/2\pi)^{0.43}. \tag{6.13}$$

The scaling law of the international stellarator database [6.28] is

$$\tau_{\rm E}^{\rm ISS95} = 0.079a^{2.21}R^{0.65}n_{19}^{0.51}B^{0.83}P^{-0.59}(\iota/2\pi)^{0.4}, \tag{6.14}$$

where the units of n_{19} and n_{20} are $10^{19}{\rm m}^{-3}$ and $10^{20}{\rm m}^{-3}$, respectively. $\iota/2\pi$ is the value at $r=(2/3)a$. Units are s, m, T, and MW. The density scaling proposed by [6.27] is

$$n_{20}^{\rm s} = 0.25N_{\rm s}\left(\frac{PB_{\rm t}}{a^2R}\right)^{0.5}. \tag{6.15}$$

Electron thermal transport barrier was observed in ECH heated plasma of Compact Helical System (CHS) (R=1m, a=0.2m, $B_{\rm t}$=1T) in 1999

Table 6.1 The maximal plasma parameters achieved in W7-AS [6.35].

	B_t	$\hat{\iota}$	P_{NB}/P_{EC}	n_e	T_{e0}	T_{i0}	β	τ_E	Remarks
T_{e0}	2.5	0.34	/2	0.2	6.8				e-root
T_{i0}	2.5	0.52	1.3/0.5	0.5		1.7			H-NBI
n_e	2.5	0.55	2.4/0	4	0.35			10*	HDH
β	0.9	0.5	2.8/0	2	$\sim 0.37^*$		3.4		HDH
τ_E	2.5	0.345	0.33/0	1.1				60	H-NBI
$\star$	2.5	0.345	0.85/0.35	0.6			$\star = 50$		HDH

Units are T, MW, 10^{20}m^{-3}, keV, %, and ms. $\hat{\iota} \equiv \iota/2\pi$ and $\star$ mark in the bottom line means the triple product $n_e T_{i0} \tau_E$ in units of 10^{20}m^{-3}eV·s. * marks in the third and fourth lines are the values referred from [6.36].

[6.29]. When the ECH power is increased from 150 kW to 200 kW, L state to H state transition occurs. The electrostatic potential profile measured by HIBP (heavy ion beam probe) in H state exhibits a prominent peak around the core that is not seen in the L state. The electron temperature increases from 1.4 keV to 2.0 keV. The development of the electron root with positive (outward) electric field gives rise to low core transport in H state. Electron thermal transport barrier is also observed in W7AS [6.30] and LHD [6.31, 6.32]. Furthermore zonal flow, which will be discussed in Section 13.5, is identified by use of two sets of HIBP in CHS [13.35].

W7-AS is a stellarator with modular field coils ($l = 2, 3/n = 5$) which were designed to optimize the magnetic configuration with respect to low neoclassical transport and a reduction of Pfirsch-Schlüter current ($R = 2$ m, $a = 0.17$ m, $B_t = 2.5$ T). Confinement of W7-AS L mode discharge follows W7AS scaling (6.13) which is about 25% enhanced compared to ISS95 scaling (6.14).

High confinement NBI discharge (H-NBI) [6.33] of W7-AS was found in 1999. $\boldsymbol{E} \times \boldsymbol{B}$ shear flow and transport barrier are formed at $r/a \sim 0.7$ and confinement time is enhanced by a factor of 2 of W7AS scaling, that is, up to factor of 2.5 above ISS95 scaling.

High Density H mode (HDH) [6.34] was observed in 2000. The density reaches $2 \sim 4 \times 10^{20}\text{m}^{-3}$. Observed confinement time is around $2\tau_E^{\text{W7AS}}$ and the maximum density is about 1.6 times the density n_{20}^{s} (6.15).

The maximal plasma parameters achieved in W7-AS are listed in Table 6.1 [6.35].

Large Helical Device (LHD) is a superconducting device with $R = 3.5 \sim 3.9$ m, $a \sim 0.6$ m, $B_t = 0.45 \sim 3$ T, $l = 2$, $m = 10$, $P_{NBI} = 13$ MW with the beam energy of 180 keV, $P_{ECH} = 2$MW with 168 GHz and 84 GHz and $P_{ICRF} = 2.7$ MW with 38.5MHz (see Figure 6.5). The energy confinement time in the LHD inward shifted discharge ($R_{axis} = 3.6$ m) is consistent to ISS95 scaling with an enhancement factor of up to 1.5. The maximal

Table 6.2 The maximal plasma parameters achieved in LHD.

	B_t/R	P_{NB}/P_{EC}	n_e	T_{e0}	T_{i0}	β	τ_E	Ref.
T_{e0}	~3/~3.75	1.3/0.88	0.02	9.5				[6.32]
T_{i0}	2.75/3.6	1.8/	0.13	3.3	3.5		90	[6.36]
T_{i0}			0.07		5			[6.31]
n_e	~2.8/3.6	11/0	1.6					[6.37]
β	0.45/3.6	1.2/0	0.3			4.1		[6.37]
τ_E	2.75/3.6	1.8/0	0.65	1.1			300	[6.36]
$\star$		1.5(P_{abs})	0.48		1.3	$\star = 220$	360	[6.31]

Units are T, m, MW, 10^{20}m^{-3}, keV, %, and ms. $\star$ mark in the bottom line means the triple product $n_e T_{i0} \tau_E$ in units of 10^{20}m^{-3}eV·s.

parameters achieved in LHD are listed in Table 6.2.

6.5 Quasi-Symmetric Stellarators

6.5.1 Magnetic Coordinates (Boozer Coordinates) and Natural Coordinates (Hamada Coordinates)

A suitable choice of the coordinates system is very important to analyze particle orbits and MHD stabilities, especially in non-axisymmetric toroidal sytems. The natural (Hamada) coordinates [6.38] and magnetic (Boozer) coordinates [6.39] are often used on particle orbits analysis and MHD stabilities study. In this subsection, the relation between them is described according to [6.40]. The equiliblium equations derived in Section 4.2.1 are as follows:

$$\nabla p = \boldsymbol{j} \times \boldsymbol{B}, \tag{6.16}$$

$$\mu_0 \boldsymbol{j} = \nabla \times \boldsymbol{B}, \tag{6.17}$$

$$\nabla \cdot \boldsymbol{B} = 0, \tag{6.18}$$

$$\nabla \cdot \boldsymbol{j} = 0, \tag{6.19}$$

$$\boldsymbol{B} \cdot \nabla p = 0, \tag{6.20}$$

$$\boldsymbol{j} \cdot \nabla p = 0. \tag{6.21}$$

We assume that the toroidal MHD equilibrium given by (6.16)~(6.21) has nested magnetic flux surface having a single magnetic axis, each of which is specified by ρ.

$$p = p(\rho). \tag{6.22}$$

From (6.18), (6.20), and (6.22), we have

$$\boldsymbol{B} = \nabla\rho \times \nabla v. \tag{6.23}$$

The vector potential $\boldsymbol{A}$ is given by

$$\boldsymbol{A} = \rho\nabla v.$$

From (6.19), (6.21), and (6.22), we have

$$\mu_0\boldsymbol{j} = \nabla w \times \nabla\rho. \tag{6.24}$$

Since there is the relation of

$$\nabla \times (\boldsymbol{B} - w\nabla\rho) = 0$$

due to (6.17) and (6.24), we have

$$\boldsymbol{B} = \nabla u + w\nabla\rho. \tag{6.25}$$

Equations (6.16), (6.22), (6.23), and (6.24) reduce to

$$\mu_0\frac{\mathrm{d}p}{\mathrm{d}\rho}\nabla\rho = \mu_0\nabla p = (\nabla w \times \nabla\rho) \times (\nabla\rho \times \nabla v)$$

$$= [(\nabla w \times \nabla\rho)\cdot\nabla v]\nabla\rho = (\boldsymbol{B}\cdot\nabla w)\nabla\rho = \mu_0(\boldsymbol{j}\cdot\nabla v)\nabla\rho;$$

that is,

$$\mu_0\frac{\mathrm{d}p}{\mathrm{d}\rho} = \boldsymbol{B}\cdot\nabla w = \mu_0\boldsymbol{j}\cdot\nabla v. \tag{6.26}$$

Equations (6.23),(6.25) and (6.24),(6.25) reduce to

$$\boldsymbol{B}\cdot\boldsymbol{B} = \boldsymbol{B}\cdot\nabla u, \tag{6.27}$$

$$\boldsymbol{j}\cdot\boldsymbol{B} = \boldsymbol{j}\cdot\nabla u. \tag{6.28}$$

Let θ and ζ be the poloidal and toroidal angle variables in cordinates system ρ, θ, ζ, respectively, and they have a period of 2π. The functions u, v, w must give single-valued $\boldsymbol{B}$ and $\boldsymbol{j}$ for the angle variables θ and ζ. Therefore, we can put them as follows:

$$v = \frac{\mathrm{d}\psi_{\mathrm{t}}}{\mathrm{d}\rho}\theta - \frac{\mathrm{d}\psi_{\mathrm{p}}}{\mathrm{d}\rho}\zeta + \tilde{v}(\rho,\theta,\zeta), \tag{6.29}$$

$$w = -\mu_0\frac{\mathrm{d}I_{\mathrm{t}}}{\mathrm{d}\rho}\theta - \mu_0\frac{\mathrm{d}I_{\mathrm{p}}}{\mathrm{d}\rho}\zeta + \tilde{w}(\rho,\theta,\zeta), \tag{6.30}$$

$$u = \mu_0 I_{\mathrm{t}}\theta + \mu_0 I_{\mathrm{p}}\zeta + \tilde{u}(\rho,\theta,\zeta). \tag{6.31}$$

$2\pi\psi_{\rm t}$ and $2\pi\psi_{\rm p}$ are toroidal and poloidal flux inside a flux surface ρ and $2\pi I_{\rm t}$ is toroidal current inside a flux surface ρ and $2\pi I_{\rm p}$ is poloidal current outside a flux surface ρ.

Equations (6.26) and (6.30) reduce to

$$\boldsymbol{B}\cdot\nabla\tilde{w} = \mu_0\frac{{\rm d}p}{{\rm d}\rho} + \mu_0\frac{{\rm d}I_{\rm t}}{{\rm d}\rho}\boldsymbol{B}\cdot\nabla\theta + \mu_0\frac{{\rm d}I_{\rm p}}{{\rm d}\rho}\boldsymbol{B}\cdot\nabla\zeta, \tag{6.32}$$

and Equations (6.27) and (6.31) reduce to

$$\boldsymbol{B}\cdot\nabla\tilde{u} = |\boldsymbol{B}|^2 - \mu_0 I_{\rm t}\boldsymbol{B}\cdot\nabla\theta - \mu_0 I_{\rm p}\boldsymbol{B}\cdot\nabla\zeta. \tag{6.33}$$

These equations are magnetic differential equations. Solvability condtion of

$$\boldsymbol{B}\cdot\nabla F = S, \qquad \boldsymbol{b}\cdot\nabla F = \frac{S}{B}, \tag{6.34}$$

are

$$\oint S\frac{{\rm d}l}{B} = 0, \tag{6.35}$$

for a closed field line and

$$\int S{\rm d}V = 0 \tag{6.36}$$

for the integral region inside a magnetic surface $\rho = $ const. (6.32) and (6.36) reduce to

$$\frac{{\rm d}p}{{\rm d}V} = -(2\pi)^2\left(\frac{{\rm d}I_{\rm p}}{{\rm d}V}\frac{{\rm d}\psi_{\rm t}}{{\rm d}V} + \frac{{\rm d}I_{\rm t}}{{\rm d}V}\frac{{\rm d}\psi_{\rm p}}{{\rm d}V}\right). \tag{6.37}$$

In the reduction process of (6.37), we used the following relation:

$$\boldsymbol{B}\cdot\nabla\theta = (\nabla\theta\times\nabla\rho)\cdot\nabla v = (\nabla\theta\times\nabla\rho)\cdot\nabla\zeta\frac{\partial v}{\partial\zeta} = g^{-1/2}\left(\frac{{\rm d}\psi_{\rm p}}{{\rm d}\rho} - \frac{\partial\tilde{v}}{\partial\zeta}\right).$$

As is shown in Chapteter 8, (8.18) and (8.19) yield

$$\frac{1}{N}\oint\frac{{\rm d}l}{B} = \frac{1}{2\pi}\frac{{\rm d}V}{{\rm d}\psi_{\rm t}}. \qquad (6.38).$$

Equations (6.32) and (6.36) reduce to

$$\langle|\boldsymbol{B}|^2\rangle = (2\pi)^2\frac{{\rm d}\psi_{\rm t}}{{\rm d}V}(I_{\rm p} + q^{-1}I_{\rm t}), \tag{6.39}$$

where $q(\psi_{\rm t})$ is safety factor $q \equiv {\rm d}\psi_{\rm t}/{\rm d}\psi_{\rm p}$ and $\langle A\rangle$ is volume average of A. (6.33) reduces to

$$\frac{1}{N}\oint\boldsymbol{B}\cdot{\rm d}\boldsymbol{l} = 2\pi\mu_0(I_{\rm p} + q^{-1}I_{\rm t}). \tag{6.40}$$

Equation (6.23) indicates that a magnetic field line is determined as the intersection of a flux surface $\rho = $ const. with a surface $v = $ const. Then as is clear from (6.29) in a coordinates system where $\tilde{v} = 0$, magnetic field lines are expressed as straight lines. An adequate coordinates transformation with respect to periodic coordinates makes two of $\tilde{u}, \tilde{v},$ and $\tilde{w}$ to zero.

Natural (Hamada) Coordinates System is the system with $\tilde{u} = 0,\ \tilde{w} = 0$. Both magnetic field line and current line are straight in Hamada coordinates and we have

$$v = \frac{\mathrm{d}\psi_{\mathrm{t}}}{\mathrm{d}\rho}\theta_{\mathrm{H}} - \frac{\mathrm{d}\psi_{\mathrm{p}}}{\mathrm{d}\rho}\zeta_{\mathrm{H}}, \tag{6.41}$$

$$w = -\mu_0\frac{\mathrm{d}I_{\mathrm{t}}}{\mathrm{d}\rho}\theta_{\mathrm{H}} - \mu_0\frac{\mathrm{d}I_{\mathrm{p}}}{\mathrm{d}\rho}\zeta_{\mathrm{H}}, \tag{6.42}$$

$$u = \mu_0 I_{\mathrm{t}}\theta_{\mathrm{H}} + \mu_0 I_{\mathrm{p}}\zeta_{\mathrm{H}} + \tilde{u}_{\mathrm{H}}. \tag{6.43}$$

The magnetic field is expressed by

$$\boldsymbol{B} = \frac{\mathrm{d}\psi_{\mathrm{t}}}{\mathrm{d}\rho}\nabla\rho\times\nabla\theta_{\mathrm{H}} - \frac{\mathrm{d}\psi_{\mathrm{p}}}{\mathrm{d}\rho}\nabla\rho\times\nabla\zeta_{\mathrm{H}}, \tag{6.44}$$

$$\boldsymbol{B} = \mu_0 I_{\mathrm{t}}\nabla\theta_{\mathrm{H}} + \mu_0 I_{\mathrm{p}}\nabla\zeta + \nabla\tilde{u}_{\mathrm{H}}. \tag{6.45}$$

The current density is expressed by

$$\boldsymbol{j} = \mu_0\frac{\mathrm{d}I_{\mathrm{t}}}{\mathrm{d}\rho}\nabla\rho\times\nabla\theta_{\mathrm{H}} + \mu_0\frac{\mathrm{d}I_{\mathrm{p}}}{\mathrm{d}\rho}\nabla\rho\times\nabla\zeta_{\mathrm{H}}. \tag{6.46}$$

The Jacobian $g^{1/2}$ ($\mathrm{d}x\mathrm{d}y\mathrm{d}z = g^{1/2}\mathrm{d}\rho\mathrm{d}\theta\mathrm{d}\zeta$) is given by the substitution of (6.41) and (6.46) into (6.26) with use of (6.37) (refer to Table 6.3)

$$g_{\mathrm{H}}^{1/2} = [\nabla\rho\cdot(\nabla\theta\times\nabla\zeta)]^{-1} = \frac{1}{(2\pi)^2}\frac{\mathrm{d}V}{\mathrm{d}\rho}. \tag{6.47}$$

Magnetic (Boozer) Coordinates System is the system with $\tilde{u} = 0,\ \tilde{v} = 0$. The magnetic field line is straight in Boozer coordinates and we have

$$v = \frac{\mathrm{d}\psi_{\mathrm{t}}}{\mathrm{d}\rho}\theta_{\mathrm{B}} - \frac{\mathrm{d}\psi_{\mathrm{p}}}{\mathrm{d}\rho}\zeta_{\mathrm{B}}, \tag{6.48}$$

$$w = -\mu_0\frac{\mathrm{d}I_{\mathrm{t}}}{\mathrm{d}\rho}\theta_{\mathrm{B}} - \mu_0\frac{\mathrm{d}I_{\mathrm{p}}}{\mathrm{d}\rho}\zeta_{\mathrm{B}} + \tilde{w}_{\mathrm{B}}, \tag{6.49}$$

$$u = \mu_0 I_{\mathrm{t}}\theta_{\mathrm{B}} + \mu_0 I_{\mathrm{p}}\zeta_{\mathrm{B}}. \tag{6.50}$$

The magnetic field is expressed by

$$\boldsymbol{B} = \frac{\mathrm{d}\psi_{\mathrm{t}}}{\mathrm{d}\rho}\nabla\rho\times\nabla\theta_{\mathrm{B}} - \frac{\mathrm{d}\psi_{\mathrm{p}}}{\mathrm{d}\rho}\nabla\rho\times\nabla\zeta_{\mathrm{B}}, \tag{6.51}$$

$$\boldsymbol{B} = \mu_0 I_{\mathrm{t}}\nabla\theta_{\mathrm{B}} + \mu_0 I_{\mathrm{p}}\nabla\zeta + \tilde{w}_{\mathrm{B}}\nabla\rho. \tag{6.52}$$

Table 6.3 Vector calculus in general coordinates.

$$\boldsymbol{a}_j \equiv \frac{\partial \boldsymbol{r}}{\partial u^j}, \qquad \boldsymbol{a}^i \equiv \nabla u^i, \qquad V \equiv \boldsymbol{a}_1 \cdot (\boldsymbol{a}_2 \times \boldsymbol{a}_3), \tag{1}$$

$$\mathrm{d}\boldsymbol{r} = \sum_j \frac{\partial \boldsymbol{r}}{\partial u^j} \mathrm{d}u^j, \qquad \boldsymbol{a}^i \cdot \boldsymbol{a}_j = \delta^i_j, \tag{2}$$

$$\boldsymbol{a}^1 = V^{-1}(\boldsymbol{a}_2 \times \boldsymbol{a}_3), \qquad \boldsymbol{a}^2 = V^{-1}(\boldsymbol{a}_3 \times \boldsymbol{a}_1), \qquad \boldsymbol{a}^3 = V^{-1}(\boldsymbol{a}_1 \times \boldsymbol{a}_2), \tag{3}$$

$$\boldsymbol{a}_1 = V(\boldsymbol{a}^2 \times \boldsymbol{a}^3), \qquad \boldsymbol{a}_2 = V(\boldsymbol{a}^3 \times \boldsymbol{a}^1), \qquad \boldsymbol{a}_3 = V(\boldsymbol{a}^1 \times \boldsymbol{a}^2), \tag{4}$$

$$\boldsymbol{a}^1 \cdot (\boldsymbol{a}^2 \times \boldsymbol{a}^3) = V^{-1}, \tag{5}$$

$$g_{i\,j} \equiv \boldsymbol{a}_i \cdot \boldsymbol{a}_j = g_{j\,i}, \qquad g^{i\,j} \equiv \boldsymbol{a}^i \cdot \boldsymbol{a}^j = g^{j\,i}, \tag{6}$$

$$\boldsymbol{F} = \sum_i f^i \boldsymbol{a}_i, \qquad f^i \equiv \boldsymbol{F} \cdot \boldsymbol{a}^i \quad \text{(contravariant)}, \tag{7}$$

$$\boldsymbol{F} = \sum_i f_i \boldsymbol{a}^i, \qquad f_i \equiv \boldsymbol{F} \cdot \boldsymbol{a}_i \quad \text{(covariant)}, \tag{8}$$

$$f_j = \sum_i g_{j\,i} f^i, \qquad f^i = \sum_j g^{i\,j} f_j, \tag{9}$$

$$g \equiv |g_{i\,j}| = V^2, \quad \mathrm{d}x\mathrm{d}y\mathrm{d}z = g^{1/2}\mathrm{d}u^1\mathrm{d}u^2\mathrm{d}u^3, \quad g^{1/2} = [\nabla u^1 \cdot (\nabla u^2 \times \nabla u^3)]^{-1}, \tag{10}$$

$$(\mathrm{d}s)^2 = (\mathrm{d}\boldsymbol{r})^2 = \sum_{i\,j} g_{i\,j}\mathrm{d}u^i\mathrm{d}u^j = \sum_{i\,j} g^{i\,j}\mathrm{d}u_i\mathrm{d}u_j, \tag{11}$$

$$(\boldsymbol{a} \times \boldsymbol{b})^1 = g^{-1/2}(a_2 b_3 - a_3 b_2), \qquad (\boldsymbol{a} \times \boldsymbol{b})_1 = g^{1/2}(a^2 b^3 - a^3 b^2), \tag{12}$$

$$\nabla \phi = \sum_i \frac{\partial \phi}{\partial u^i} \boldsymbol{a}^i, \qquad \nabla \cdot \boldsymbol{F} = \frac{1}{g^{1/2}} \sum_i \frac{\partial}{\partial u^i}(g^{1/2} f^i), \tag{13}$$

$$\nabla \times \boldsymbol{F} = \frac{1}{g^{1/2}} \left(\left(\frac{\partial f_3}{\partial u^2} - \frac{\partial f_2}{\partial u^3} \right) \boldsymbol{a}_1 + \left(\frac{\partial f_1}{\partial u^3} - \frac{\partial f_3}{\partial u^1} \right) \boldsymbol{a}_2 + \left(\frac{\partial f_2}{\partial u^1} - \frac{\partial f_1}{\partial u^2} \right) \boldsymbol{a}_3 \right), \tag{14}$$

$$\nabla^2 \phi = \nabla \cdot (\nabla \phi) = \frac{1}{g^{1/2}} \sum_{i\,j} \frac{\partial}{\partial u^i} \left(g^{1/2} g^{i\,j} \frac{\partial \phi}{\partial u^j} \right), \tag{15}$$

The current density is expressed by

$$\mu_0 \boldsymbol{j} = \mu_0 \frac{\mathrm{d}I_\mathrm{t}}{\mathrm{d}\rho} \nabla\rho \times \nabla\theta_\mathrm{B} + \mu_0 \frac{\mathrm{d}I_\mathrm{p}}{\mathrm{d}\rho} \nabla\rho \times \nabla\zeta_\mathrm{B} + \nabla\tilde{w}_\mathrm{B} \times \nabla\rho. \tag{6.53}$$

Jacobian is given by substitution of (6.50) and (6.51) into (6.27) with use of (6.39) as follows:

$$g_\mathrm{B} = \mu_0 \frac{I_\mathrm{p} + q^{-1} I_\mathrm{t}}{|B|^2} \frac{\mathrm{d}\psi_\mathrm{t}}{\mathrm{d}\rho}. \tag{6.54}$$

6.5.2 Boozer Equation of Drift Motion

The velocity of drift motion of guiding center was given in (2.64) as follows:

$$\boldsymbol{v} = \frac{v_\parallel}{B} \frac{1}{1 + \rho_\parallel \boldsymbol{b} \cdot \nabla \times \boldsymbol{b}} (\boldsymbol{B} + \nabla \times (\rho_\parallel \boldsymbol{B})), \tag{6.55}$$

where

$$\rho_\parallel = \frac{m v_\parallel}{eB}. \tag{6.56}$$

The magnetic field in Boozer coordinates (ρ, θ, ζ) is given by (6.51) and (6.52) as follows $(v \leftrightarrow \theta_0,\ u \leftrightarrow \chi)$:

$$\boldsymbol{B} = \nabla\psi \times \nabla\theta_0, \qquad \theta_0 = \theta - q^{-1}\zeta, \qquad q^{-1} = \frac{\mathrm{d}\psi_\mathrm{p}}{\mathrm{d}\psi}, \tag{6.57}$$

$$\boldsymbol{B} = \nabla\chi + \beta\nabla\psi, \quad \psi \equiv \psi_\mathrm{t}, \quad \chi = \mu_0 I_\mathrm{t}\theta + \mu_0 I_\mathrm{p}\zeta, \quad \beta \equiv \mu_0 I_\mathrm{t}'\theta + \mu_0 I_\mathrm{p}' + \tilde{w}. \tag{6.58}$$

The term $\boldsymbol{b} \cdot \nabla \times \boldsymbol{b}$ in (6.55) is expressed by

$$\boldsymbol{b} \cdot \nabla \times \boldsymbol{b} = \frac{1}{B^2} \boldsymbol{B} \cdot \nabla \times \boldsymbol{B} = \frac{1}{B^2} \nabla\chi \cdot (\nabla\beta \times \nabla\psi) = \frac{1}{B^2} (\nabla\psi \times \nabla\chi) \cdot \nabla\beta.$$

(6.57), (6.58) reduce to $(\nabla\psi \times \nabla\theta_0) \cdot \nabla\chi = B^2$. The substitution of $u^1 = \psi$, $u^2 = \theta_0$, and $u^3 = \chi$ into Equations (1), (4), and (5) in Table 6.3 reduces $(\nabla\psi \times \nabla\chi) = -B^2 (\mathrm{d}\boldsymbol{r}/\mathrm{d}\theta_0)$. Therefore, we obtain

$$\nabla \times (\rho_\parallel \boldsymbol{B}) \cdot \nabla\chi = \rho_\parallel (\nabla\beta \times \nabla\psi) \cdot \nabla\chi + (\nabla\rho_\parallel \times \boldsymbol{B}) \cdot \nabla\chi$$

$$= \rho_\parallel (\nabla\psi \times \nabla\chi) \cdot \nabla\beta + \beta(\nabla\psi \times \nabla\chi) \cdot \nabla\rho_\parallel = -B^2 \rho_\parallel \frac{\partial\beta}{\partial\theta_0} - B^2 \beta \frac{\partial\rho_\parallel}{\partial\theta_0}.$$

and

$$\dot{\chi} = \boldsymbol{v} \cdot \nabla\chi = \frac{v_\parallel}{B} \left(\frac{1}{1 - \rho_\parallel (\partial\beta/\partial\theta_0)} \right) (\boldsymbol{B} + \nabla \times (\rho_\parallel \boldsymbol{B})) \cdot \nabla\chi$$

$$= v_\parallel B \left(1 - \frac{\beta(\partial\rho_\parallel/\partial\theta_0)}{1 - \beta(\partial\rho_\parallel/\partial\theta_0)} \right). \tag{6.59}$$

$(\partial \rho_\parallel/\partial \theta_0)$ means differentiation by θ_0 while keeping ψ, χ, and Hamiltonian H_0 constant. Hamiltonian is

$$H_0 \equiv \frac{e}{2m}B^2\rho_\parallel^2 + \frac{\mu}{e}B + \Phi, \tag{6.60}$$

where $B = B(\psi, \theta_0, \chi)$ and $\Phi = \Phi(\psi, \theta_0, \chi)$ and $\rho_\parallel = (m/e)(v_\parallel/B)$. We define

$$\theta_{0c} \equiv \theta_0 - \beta(\psi, \theta_0, \chi)\rho_\parallel. \tag{6.61}$$

There are the following relations:

$$\left.\frac{\partial \theta_0}{\partial \rho_\parallel}\right|_{\theta 0c} = \beta + \rho_\parallel \frac{\partial \beta}{\partial \theta_0}\left.\frac{\partial \theta_0}{\partial \rho_\parallel}\right|_{\theta 0c}, \qquad \left.\frac{\partial \theta_0}{\partial \rho_\parallel}\right|_{\theta 0c} = \frac{\beta}{1-\rho_\parallel(\partial\beta/\partial\theta_0)},$$

$$-\frac{e}{m}B^2\rho_\parallel \left.\frac{\partial \rho_\parallel}{\partial \theta_0}\right|_{H0} = \frac{e}{m}\rho_\parallel^2\frac{\partial B^2}{\partial \theta_0} + \frac{\mu}{e}\frac{\partial B}{\partial \theta_0} + \frac{\partial \Phi}{\partial \theta_0},$$

$$\left.\frac{\partial \theta_0}{\partial \chi}\right|_{\theta 0c} = \frac{\rho_\parallel \partial\beta/\partial\chi}{1-\rho_\parallel(\partial\beta/\partial\theta_0)}, \qquad \frac{\partial \theta_0}{\partial \theta_{0c}} = \frac{1}{1-\rho_\parallel(\partial\beta/\partial\theta_0)},$$

$$\left.\frac{\partial \theta_0}{\partial \psi}\right|_{\theta 0c} = \frac{\rho_\parallel \partial\beta/\partial\psi}{1-\rho_\parallel(\partial\beta/\partial\theta_0)}.$$

Let us change the independent variable θ_0 to $\theta_{0c} = \theta_0 - \beta\rho_\parallel$ and $H(\rho_\parallel, \psi, \theta_{0c}, \chi) \equiv H_0(\rho_\parallel, \psi, \theta_{0c} + \beta\rho_\parallel, \chi)$, that is,

$$H(\rho_\parallel, \psi, \theta_{0c}, \chi) = \frac{1}{2}\frac{e}{m}B^2\rho_\parallel^2 + \frac{\mu}{e}B + \Phi, \tag{6.62}$$

$$B = B(\psi, \theta_{0c} + \beta\rho_\parallel, \chi), \qquad \Phi = \Phi(\psi, \theta_{0c} + \beta\rho_\parallel, \chi). \tag{6.62}$$

Then we have

$$\left.\frac{\partial H}{\partial \rho_\parallel}\right|_{\theta 0c, \psi, \chi} = \frac{\partial H_0}{\partial \rho_\parallel} + \frac{\partial H_0}{\partial \theta_0}\left.\frac{\partial \theta_0}{\partial \rho_\parallel}\right|_{\theta 0c} = \frac{e}{m}B^2\rho_\parallel + \frac{\partial H_0}{\partial \theta_0}\left.\frac{\partial \theta_0}{\partial \rho_\parallel}\right|_{\theta 0c}$$

$$= \frac{e}{m}B^2\rho_\parallel + \left(\frac{e}{2m}\rho_\parallel^2\frac{\partial B^2}{\partial \theta_0} + \frac{\mu}{e}\frac{\partial B}{\partial \theta_0} + \frac{\partial \Phi}{\partial \theta_0}\right) \times \left(\frac{\beta}{1-\rho_\parallel(\partial\beta/\partial\theta_0)}\right)$$

$$= v_\parallel B - \left(\frac{\partial \rho_\parallel}{\partial \theta_0} v_\parallel B\right)\frac{\beta}{1-\rho_\parallel(\partial\beta/\partial\theta_0)} = v_\parallel B\left(1 - \frac{\beta(\partial\rho_\parallel/\theta_0)}{1-\rho_\parallel(\partial\beta/\partial\theta_0)}\right).$$

Finally we reduce desirous result $\dot{\chi} = \partial H/\partial \rho_\parallel|_{\theta 0\,\mathrm{c}}$. Then the equation of drift motion is expressed in Hamilton's canonical form similar to $\dot{\chi}$ [6.41]

$$\dot{\chi} = \frac{\partial H}{\partial \rho_\parallel}, \qquad \dot{\rho}_\parallel = -\frac{\partial H}{\partial \chi}, \tag{6.63}$$

$$\dot{\theta}_{0c} = \frac{\partial H}{\partial \psi}, \qquad \dot{\psi} = -\frac{\partial H}{\partial \theta_{0c}}. \tag{6.64}$$

The canonical transformation to new coordinates

$$(\theta_{0c}, \chi, \psi, \rho_\|) \to (\theta_c, \zeta, P_\theta, P_\zeta)$$

is given by the following generating function:

$$F(\psi, \rho_\|, \theta_c, \zeta) = \mu_0 \rho_\| (g\zeta + I\theta_c) + \psi\theta_c - \zeta\psi/q, \tag{6.65}$$

that is,

$$\theta_{0c} = \frac{\partial F}{\partial \psi} = \left(\theta_c - \frac{\zeta}{q}\right) + \mu_0 \rho_\| (\zeta g' + I'), \tag{6.66}$$

$$\chi = \frac{\partial F}{\partial \rho_\|} = \mu_0 (g\zeta + I\theta_c), \tag{6.67}$$

$$P_\theta = \frac{\partial F}{\partial \theta_c} = \mu_0 \rho_\| I + \psi, \tag{6.68}$$

$$P_\zeta = \frac{\partial F}{\partial \zeta} = \mu_0 \rho_\| g + \psi/q. \tag{6.69}$$

New Hamilton is [6.42]

$$H(\theta_c, \zeta, P_\theta, P_\zeta) = \frac{1}{2}\frac{e}{m} B^2 \rho_\|^2 (P_\theta, P_\zeta, \psi) + \frac{\mu}{e} B + \Phi, \tag{6.70}$$

where

$$\rho_\| = \frac{P_\theta/q - P_\zeta}{\mu_0 (I/q - g)}, \qquad \psi = \frac{P_\theta g - P_\zeta I}{I/q - g}, \tag{6.71}$$

$$\theta = \theta_c + \rho_\| \delta, \qquad \delta = \zeta g' + \theta I' + \beta. \tag{6.72}$$

$$B,\ \Phi = B,\ \Phi(\psi, \chi, \theta) = B,\ \Phi(\psi, g\zeta + I\theta_c, \theta). \tag{6.73}$$

Then Hamilton's equation in new coordinates is

$$\dot{\theta}_c = \frac{\partial H}{\partial P_\theta}, \qquad \dot{P}_\theta = -\frac{\partial H}{\partial \theta_c}, \tag{6.74}$$

$$\dot{\zeta} = \frac{\partial H}{\partial P_\zeta}, \qquad \dot{P}_\zeta = -\frac{\partial H}{\partial \zeta}. \tag{6.75}$$

The substitution of θ_c by θ gives formally

$$H(\theta, \zeta, P_\theta, P_\zeta) = \frac{1}{2}\frac{e}{m} \rho_\| (P_\theta, P_\zeta, \psi) + \frac{\mu}{e} B + \Phi \qquad (B, \Phi) = (B, \Phi)(\psi, \theta, \zeta) \tag{6.76}$$

$$\dot{\theta} = \frac{\partial H}{\partial P_\theta}, \qquad \dot{P}_\theta = -\frac{\partial H}{\partial \theta}. \tag{6.77}$$

$$\dot{\zeta} = \frac{\partial H}{\partial P_\zeta}, \qquad \dot{P}_\zeta = -\frac{\partial H}{\partial \zeta}. \tag{6.78}$$

The distinction between the solutions of (6.74), (6.75) and (6.77), (6.78) is the order of gyroradius ($\theta = \theta_c + \rho_\| \delta$). Therefore, the nonresonant difference between them is negligible [6.43]. In the currentless case ($\beta = \delta = 0$), they are identical.

6.5.3 Quasi-Symmetric Stellarator System

The particle orbits in Boozer coordinates depend on the magnetic field strength $|B|$ and do not depend on the components of $\boldsymbol{B}$ as is seen in (6.70). If $|B|$ can be made symmetric, the particle orbit is closed, subject to a constant of motion. There are three types of symmetry: (1) axisymmetry $B(\rho, \zeta)$, (2) helical symmetry $B(\rho, \theta - \alpha\zeta)$, and (3) poloidal symmetry $B(\rho, \theta)$ in Boozer coordinates.

Quasi-Axisymmetric Stellarator (QAS)
NCSX (National Compact Stellarator Experiment) is repesentative QAS under construction in Princeton. Optimized modular coils of NCSX and radial profile of non-axisymmetric components of magnetic field strength are shown in Figure 6.9. NCSX has $R = 1.4\,\mathrm{m}$, $B_\mathrm{t} = 1.2 \sim 2.0\,\mathrm{T}$, $A = 4 \sim 4.4$. It is expected theoretically that, even with bootstrap current consistent profiles, they are passively stable to the ballooning, kink, mercier, and neoclassical tearing mode for $\beta > 4\%$ [6.44].

Design activity of CHS-qa is also carried out in NIFS ($R = 1.5\,\mathrm{m}, A = 3.2$, $B_\mathrm{t} = 1.5\,\mathrm{T}$) [6.45].

Quasi-Helically Symmetric Stellarator (QHS)
HSX is QHS with $B/B_0 \approx (1 + \varepsilon_\mathrm{H} \cos(\theta - 4\zeta))$ ($R = 1.2\,\mathrm{m},\ a = 0.15\,\mathrm{m}$, $B_\mathrm{t} = 1\,\mathrm{T}$). Experiments are carried out at University of Wisconsin [6.46].

Quasi-Poloidally Symmetric Stellarator (QPS)
QPS device with $R = 0.9\,\mathrm{m},\ A = 2.7, B_\mathrm{t} = 1\,\mathrm{T}$ is under construction in ORNL. QPS coil set has two field periods and will operate in an external vacuum tank [6.47]. Contours of magnetic field strength at $\psi/\psi_\mathrm{edge} = 0.25$ is shown in Figure 6.10.

Quasi-Isodynamic or Quasi-Omnigenous Stellarator
Quasi-isodynamicity addresses the property that trapped particle gyrocenters precess poloidally with the confinement of thermal and energetic particles ensured by poloidally closed contours of the longitudinal invariant $J_\parallel$ (refer to (2.37)). Wendelstein-7X is fully optimized following the equasi-isodynamicity principle. The optimization of W7-X leads to good and nested flux surfaces, low Shafranov shift due to $\langle j_\parallel^2/j_\perp^2 \rangle \sim 0.5$, equilibrium and stability of $\langle\beta\rangle \sim 5\%$, and low neoclassical flux with a confinement determined by turbulent transport [6.48]. Device parameters are $R = 5.5\,\mathrm{m}$, $a = 0.55\,\mathrm{m}$, $B_\mathrm{t} = 2.5\,\mathrm{T}$ (superconductor coils), volume of plasma $V = 30\,\mathrm{m}^3$, pulse duration 30 min., $P_\mathrm{ECH} = 60\,\mathrm{MW}$, $P_\mathrm{NBI} = 5\,\mathrm{MW}(20^*)$ and $P_\mathrm{ICRF} = 3\,\mathrm{MW}(9^*)$ (*: stage II heating). Components of W7-X device are shown in Figure 6.11. Modular coils and magnetic flux surface were shown in Figure 6.5.

Quasi-omnigenous approach targets to align second (longitudinal) invariant $J_\parallel$ contours for the trapped particles with magnetic flux surfaces [6.49]. An example of $J_\parallel$ contours is shown in Figure 6.12. Concentric circles

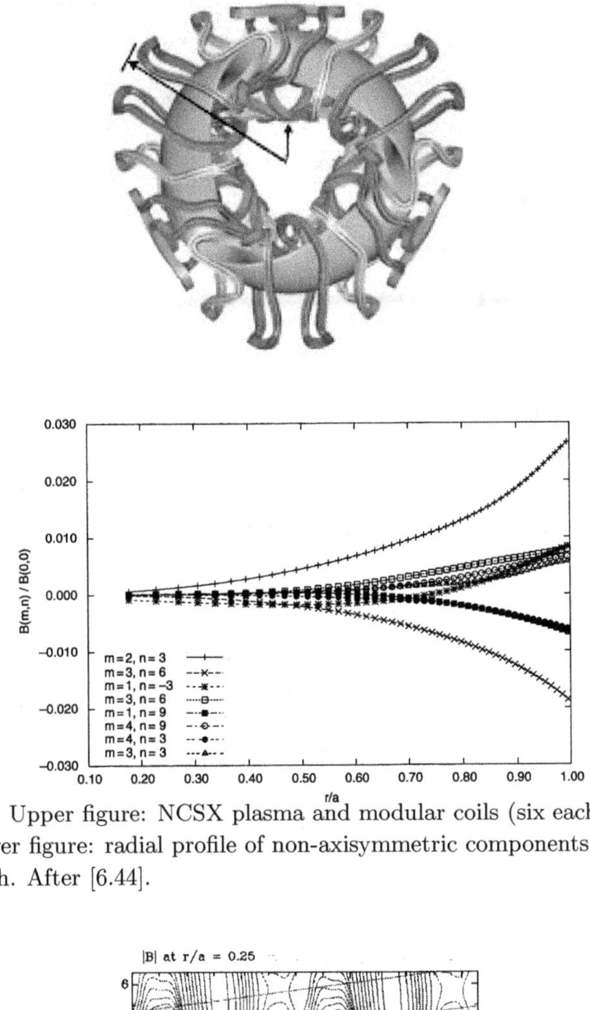

Figure 6.9 Upper figure: NCSX plasma and modular coils (six each of three coil types). Lower figure: radial profile of non-axisymmetric components of magnetic field strength. After [6.44].

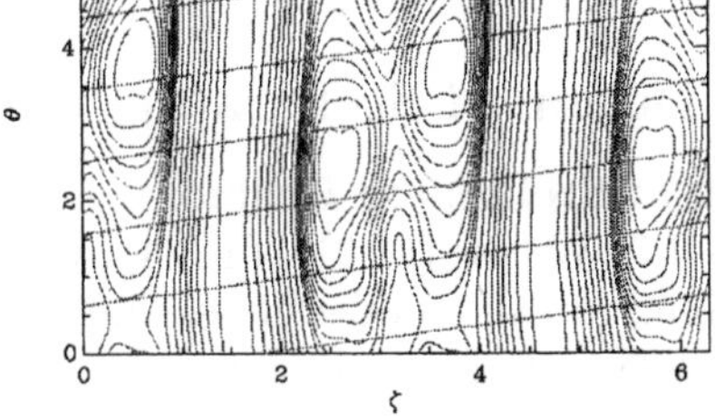

Figure 6.10 Contours of magnetic field strength and magnetic field lines of QPS (straight lines in Boozer coordinates on the flux surface $(\psi/\psi_{\mathrm{edge}})^{1/2} = 0.25$. After [6.47]. $B > 1$T in the regions that contain closed contours plotted with thin lines and $B < 1$T in the regions without closed contours. θ is the poloidal angle variable and ζ is the toroidal angle variable.

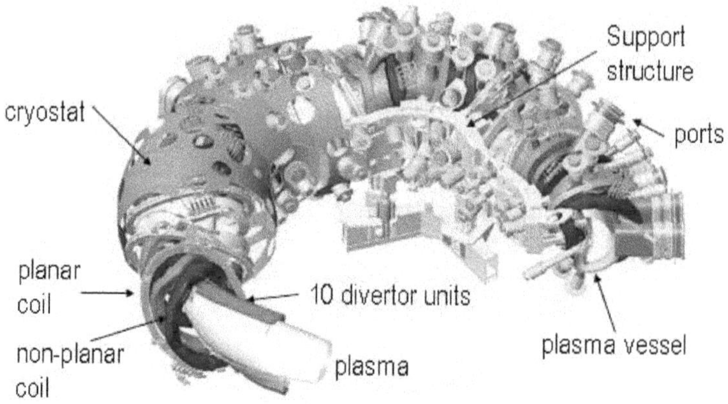

Figure 6.11 Components of W7-X device. After [6.48].

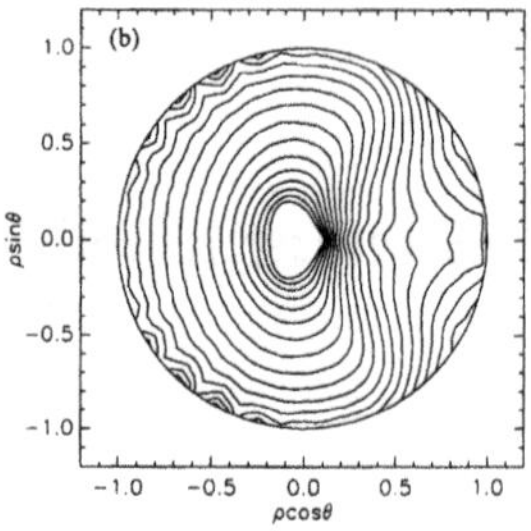

Figure 6.12 $J_{\parallel}$ contours for trapped particles. Concentric circles are the flux surfaces. After [6.49].

are the flux surfaces.

These new concepts of stellarator configurations are reviewed in [6.50].

6.6 Conceptual Design of Stellarator Reactor

Burning Condition of Stellarator Reactor can be reduced in the same way as in Section 4.8. We use (4.101), (4.104), and (4.105), ISS95 energy confinement scaling (6.14), and density scaling (6.15). Then burning condition with use of ISS95 scaling is given by

$$a^{1.09}B^{1.745} = \frac{13.26(1+f_{\rm DT}+f_{\rm He}+f_{\rm I})^{1.33}}{H^{\rm ISS95}N_{\rm s}^{0.51}\beta_{\rm th}[(1-f_{\rm rad\,c})(f_\alpha+5/Q)f_{\rm prof}f_{\rm DT}^2\Theta]^{0.665}(\iota/2\pi)^{0.4}A^{0.06}}, \quad (6.79)$$

Table 6.4 The parameters of HSR5/22 and FFHR2m2 reactors.

	HSR5/22 modular coils	FFHR2m2 LHD type coils
R / a(m)	22 / 1.8	16 / 2.8
$V(\mathrm{m}^3)$	1400	-
No. of field period	5	$l = 2$ $m = 10$
$B_\mathrm{t}(\mathrm{T})$	4.75	4.43
$B_\mathrm{max}(\mathrm{T})$	10	13
coil radius (m)	5.4	4.33
$W_\mathrm{mag}(\mathrm{GJ})$	100	142
fusion power (GW)	3	3
$\bar{n}_\mathrm{e}$ $(10^{20}\mathrm{m}^{-3})$	2.12	
$n_\mathrm{e}(0)$ $(10^{20}\mathrm{m}^{-3})$	3.0	1.9
$\langle T_\mathrm{e}\rangle(\mathrm{keV})$	4.96	-
$T_\mathrm{e0}/T_\mathrm{i0}(\mathrm{keV})$	15 / -	/ 16.1
$\beta(\%)$	4.24	4.1
$\tau_\mathrm{E}(\mathrm{s})$ (required)	1.62	-
$\tau_\mathrm{E}^\mathrm{ISS95}(\mathrm{s})$	0.96	-
H_ISS95	1.69	1.76

V: volume of plasma, W_mag: stored energy of magnetic field.
The number of the module coils in HSR5/22 is 50.

where H^ISS95 is confinement enhancement factor over ISS95 scaling.

The Helias stellerator reactor is an upgraded version of the Wendelstein 7-X device taking into account the design criteria of a power reactor [6.51]. FFHR2 is LHD-type D-T demo-reactor [6.52]. The parameters of two reactors are listed in Table 6.4.

Problems

1. Equilibrium Equation of a Helically Symmetric System Confirm the equilibrium equation of a helically symmetric system (6.7).

2. Boozer Equation of Particle Orbit Prove that the Boozer equation of particle orbit is given by (6.74) and (6.75). Use the following relations:

$$\nabla\rho_\parallel = (\partial\rho_\parallel/\partial\psi)\nabla\psi + (\partial\rho_\parallel/\partial\theta_0)\nabla\theta_0 + (\partial\rho_\parallel/\partial\chi)\nabla\chi,$$

$$\nabla\times(\rho_\parallel \boldsymbol{B})\cdot\nabla\rho_\parallel = \rho_\parallel(\nabla\times\nabla\boldsymbol{B})\cdot\nabla\rho_\parallel = \rho_\parallel(\nabla\beta\times\nabla\psi)\cdot\nabla\rho_\parallel = \rho_\parallel(\nabla\rho_\parallel\times\nabla\beta)\cdot\nabla\psi$$

$$= \rho_\parallel B^2\left(\frac{\partial\rho_\parallel}{\partial\theta_0}\frac{\partial\beta}{\partial\chi} - \frac{\partial\rho_\parallel}{\partial\chi}\frac{\partial\beta}{\partial\theta_0}\right).$$

3. Burning Condition Derive a burning condition (6.79) of stellarator reactor with use of ISS95 energy confinement scaling (6.14) and density scaling (6.15).

4. Precession of Helical Banana in Stellarator The magnitude of the magnetic field of stellarator with large aspect ratio is given by $B = B_0(1 - \epsilon_\mathrm{t}(r/R)\cos\theta - \epsilon_\mathrm{h}(r)\cos(l\theta - m\varphi))$, so that the longitudinal adiabatic invariant $J_\parallel$ of banana is identical to (7.9) of trapped particles in

mirror, if κ^2 is taken to be
$\kappa^2 = (2\mu_{\rm m} B_0 \epsilon_{\rm h})^{-1}\big(W - \mu_{\rm m} B(1 - \epsilon_{\rm t}(r/R)\cos\theta - \epsilon_{\rm h})\ - q\phi(r)\big)$. Generalized coordinates are $u^1 \approx r$, $u^2 \approx \theta - (\iota/2\pi)\varphi$ and $u^3 \approx R\varphi$ in stellarator. Furthermore, there is constraint $(l\theta - m\varphi) = 0$ on helical banana center trapped by helical ripples in stellarator field where the magnitue of B is locally minimum. Discuss the orbit of precession of helical banana by use of longitudinal adiabatic invariant $J_{\|}$ as described in Section 7.1. Refer to Section 3.5 of [4.25].

7

Mirror and Tandem Mirror

7.1 Trapped Particles in Mirror and Confinement Time

Mirror is of a simpler configuration than toroidal systems. The attainment of absolute minimum-B configurations is possible with mirror systems, whereas only average minimum-B configurations can be realized in toroidal systems. Although absolute minimum-B configurations are MHD stable, the velocity distribution of the plasma becomes non-Maxwellian due to end losses, and the plasma will be prone to velocity-space instabilities.

The most critical issue of mirror with the open-ends is the suppression of end loss. The end plug of the mirror due to electrostatic potential has been studied by tandem mirrors.

Particles are trapped in a mirror field when the velocity components $v_\perp$ and $v_\parallel$, perpendicular and parallel to the magnetic field, satisfy the condition (refer to Section 4.2)

$$\frac{v_\perp^2}{v^2} > \frac{B_0}{B_\mathrm{M}}, \qquad (v^2 = v_\perp^2 + v_\parallel^2)$$

where B_0 and B_M are the magnitudes of the magnetic field at the center and at the end, respectively. Denoting the mirror ratio B_M/B_0 by R_M, so that

$$\frac{B_0}{B_\mathrm{M}} = \frac{1}{R_\mathrm{M}} \equiv \sin^2\theta_\mathrm{M},$$

the trapping condition is reduced to (refer to Figure 7.1)

$$\frac{v_\perp}{v} > \sin\theta_\mathrm{M}. \tag{7.1}$$

When ions (as well as electrons) are trapped and bounce between both ends of the mirror field, the bounce orbit drifts by the bounce averaged acceleration of centrifugal force and ∇B force due to the inhomogenous mirror field; that is (refer to (2.6)),

$$\boldsymbol{v} = \frac{m}{qB}\frac{v_\parallel^2 + v_\perp^2/2}{R}(\boldsymbol{n}\times\boldsymbol{b}), \tag{7.2}$$

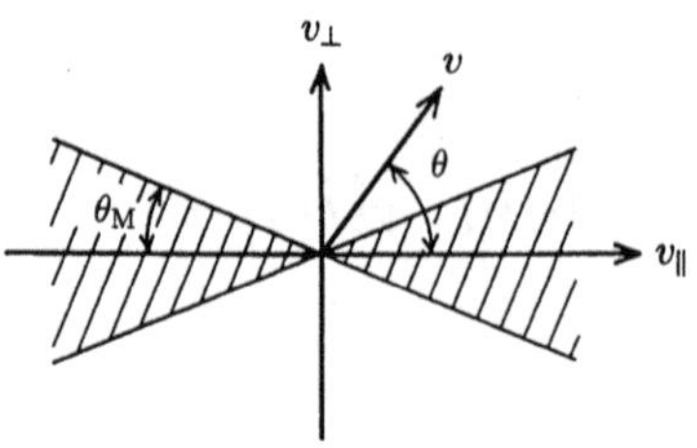

Figure 7.1 Mirror field and loss cone in $v_\parallel - v_\perp$ space.

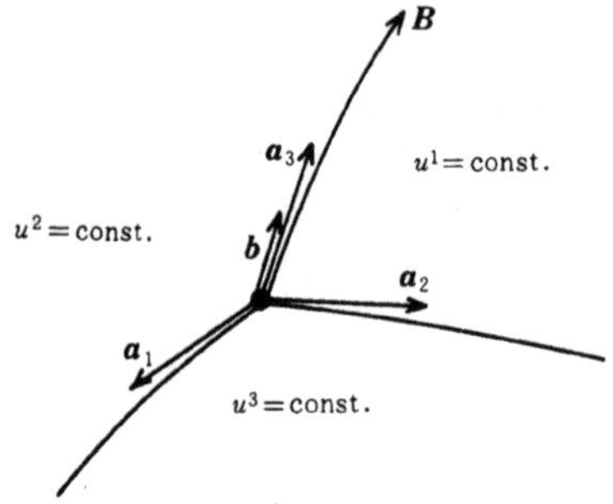

Figure 7.2 General coordinates (u^1, u^2, u^3) used for the analysis of the orbit of a trapped particle.

where R^{-1} is the curvature of field line and $\boldsymbol{n}$ is the unit vector from the center of curvature toward the field line. The orbit of the crossing point (r, θ) of trapped ions at the center plane of the mirror is given by (2.37) as follows:

$$J_\parallel(r, \theta, \mu_m, W) = m \oint v_\parallel \mathrm{d}l = \text{const.} \tag{7.3}$$

For the analysis of motion of a trapped particle, we use (2.64):

$$\boldsymbol{v} = \left(v_\parallel - \frac{m v_\parallel^2}{qB} (\boldsymbol{b} \cdot \nabla \times \boldsymbol{b}) \right) \boldsymbol{b} + \frac{v_\parallel}{qB} \nabla \times (m v_\parallel \boldsymbol{b}), \tag{7.4}$$

$$v_\parallel = \pm \big((2/m)(W - \mu_m B - q\phi) \big)^{1/2}, \tag{7.5}$$

where W is the total energy of the trapped particle. The general coordinates (u^1, u^2, u^3) used here are shown in Figure 7.2. u^3 is the coordinate along the field line. A vector $\boldsymbol{F}$ is expressed by $\boldsymbol{F} = \sum f^i \boldsymbol{a}_i$, $f^i = \boldsymbol{F} \cdot \boldsymbol{a}^i$ (contravariant), or $\boldsymbol{F} = \sum f_i \boldsymbol{a}^i$, $f_i = \boldsymbol{F} \cdot \boldsymbol{a}_i$ (covariant), where $\boldsymbol{a}_j \equiv \partial \boldsymbol{r} / \partial u^j$ and $\boldsymbol{a}^i \equiv \nabla u^i$ (refer to Table 6.1 in Section 6.5). The unit vector $\boldsymbol{b}$ of magnetic field is expressed by $\boldsymbol{b} = b^3 \boldsymbol{a}_3 = (\boldsymbol{a}_3 \cdot \boldsymbol{a}_3)^{-1/2} \boldsymbol{a}_3$. Equation (7.4), by use of the coordinates (u^1, u^2, u^3), is given by (refer to Table 6.1)

$$\frac{\mathrm{d}u^1}{\mathrm{d}t} = \frac{v_\parallel}{qBg^{1/2}}\left(\frac{\partial}{\partial u^2}(mv_\parallel b_3) - \frac{\partial}{\partial u^3}(mv_\parallel b_2)\right),$$

$$\frac{\mathrm{d}u^2}{\mathrm{d}t} = \frac{v_\parallel}{qBg^{1/2}}\left(\frac{\partial}{\partial u^3}(mv_\parallel b_1) - \frac{\partial}{\partial u^1}(mv_\parallel b_3)\right).$$

The variations of u^1 and u^2 during one period are

$$\Delta u^1 = \oint \frac{1}{qB}\left(\frac{g_{33}}{g}\right)^{1/2}\left(\frac{\partial}{\partial u^2}(mv_\parallel b_3) - \frac{\partial}{\partial u^3}(mv_\parallel b_2)\right)\mathrm{d}u_3$$

$$= \left(\frac{1}{qB}\left(\frac{g_{33}}{g}\right)^{1/2}\right)_{\mathrm{m}}\frac{\partial J_\parallel}{\partial u^2},$$

$$\Delta u^2 = -\left(\frac{1}{qB}\left(\frac{g_{33}}{g}\right)^{1/2}\right)_{\mathrm{m}}\frac{\partial J_\parallel}{\partial u^1},$$

where $J_\parallel = \oint(mv_\parallel b_3)\mathrm{d}u^3 = \oint(mv_\parallel)\mathrm{d}l$ $\quad(\mathrm{d}l = (\boldsymbol{v}\cdot\boldsymbol{b})\mathrm{d}t = \mathrm{d}u^3(\boldsymbol{a}_3\cdot\boldsymbol{b})$ $= \mathrm{d}u^3 g_{33}^{1/2}$, the definitions of g and g_{ij} are given in Table 6.1). The notation of $(f)_{\mathrm{m}}$ means the value of f at a point within the orbit (mean value theorem). The period τ of one cycle is

$$\tau = \oint\frac{\mathrm{d}l}{v_\parallel} = m\frac{\partial}{\partial W}\oint v_\parallel \mathrm{d}l = \frac{\partial J_\parallel}{\partial W}. \tag{7.6}$$

Consequently, the drift velocity of the trapped particle orbit is given by

$$\frac{\mathrm{d}u^1}{\mathrm{d}t} = \left(\frac{1}{qB}\left(\frac{g_{33}}{g}\right)^{1/2}\right)_{\mathrm{m}}\frac{\partial J_\parallel/\partial u^2}{\partial J_\parallel/\partial W}, \tag{7.7}$$

$$\frac{\mathrm{d}u^2}{\mathrm{d}t} = -\left(\frac{1}{qB}\left(\frac{g_{33}}{g}\right)^{1/2}\right)_{\mathrm{m}}\frac{\partial J_\parallel/\partial u^1}{\partial J_\parallel/\partial W}, \tag{7.8}$$

provided the period τ is much smaller than the time scale of the drift motion. An analytical example of mirror field strength is

$$B = B_0\frac{R_{\mathrm{M}}+1}{2}\left(1 - \frac{R_{\mathrm{M}}-1}{R_{\mathrm{M}}+1}\cos\frac{2\pi z}{L}\right) \equiv \bar{B}\left(1 - \epsilon\cos\frac{2\pi z}{L}\right),$$

where R_{M} is the mirror ratio and the maximum field strength is $R_{\mathrm{M}}B_0$ at $z = \pm L/2$, $\bar{B} = B_0(R_{\mathrm{M}} + 1/2)$, $\epsilon = (R_{\mathrm{M}} - 1)/(R_{\mathrm{M}} + 1)$. The coordinates u^1, u^2, u^3 are approximately given by $u^1 \approx r(B/B_0)^{1/2}$, $u^2 = \theta$, and $u^3 \approx z$. Then longitudinal adiabatic invariant $J_\parallel$ is

$$J_\parallel = m\oint\left((2/m)(W - \mu_{\mathrm{m}}\bar{B}(1 - \epsilon\cos(2\pi z/L))\right)^{1/2}\mathrm{d}l$$

$$\approx (2m\mu_m \bar{B}\epsilon)^{1/2} \oint (2\kappa^2 + \cos(2\pi z/L) - 1)^{1/2} \mathrm{d}z$$

$$= 4 \cdot 4(m\mu_m \bar{B}\epsilon)^{1/2}(L/2\pi) \int_0^{\varphi_0/2} (\kappa^2 - \sin^2(\varphi/2))\mathrm{d}(\varphi/2)$$

$$= 16(m\mu_m \bar{B}\epsilon)^{1/2}(L/2\pi)\big(E(\kappa) - (1-\kappa^2)K(\kappa)\big), \tag{7.9}$$

where

$$\kappa^2 \equiv \frac{W - \mu_\mathrm{m}(1-\epsilon)\bar{B}}{2\epsilon\mu_\mathrm{m}\bar{B}} < 1, \qquad \sin^2(\varphi_0/2) = \kappa^2, \qquad \varphi = \frac{2\pi z}{L}.$$

$K(\kappa)$ and $E(\kappa)$ are complete elliptic integral of the first kind and the second kind, respectively, as follows:

$$K(\kappa) = \int_0^{\pi/2} \frac{\mathrm{d}\phi}{(1-\kappa^2 \sin^2\phi)^{1/2}} = \int_0^1 \frac{\mathrm{d}x}{((1-x^2)(1-\kappa^2 x^2))^{-1/2}},$$

$$E(\kappa) = \int_0^{\pi/2} (1-\kappa^2\sin^2\phi)^{1/2}\mathrm{d}\phi = \int_0^1 \left(\frac{1-\kappa^2x^2}{1-x^2}\right)^{1/2} \mathrm{d}x,$$

$$\frac{\partial K}{\partial \kappa} = \frac{1}{\kappa}\left(\frac{E}{1-\kappa^2} - K\right), \quad \frac{\partial E}{\partial \kappa} = \frac{1}{\kappa}(E-K).$$

$K(\kappa) = \pi/2$, 1.854, 2.257, 2.578, ∞ and $E(\kappa) = \pi/2$, 1.351, 1.178, 1.105, 1 for $\kappa = 0$, 0.5, 0.8, 0.9, 1, respectively.

The period of one cycle is

$$\tau = \frac{8L}{\pi}(m\mu_m\bar{B}\epsilon)^{1/2}\frac{1}{4\epsilon\mu_m\bar{B}}K(\kappa) = \frac{2L}{\pi}\left(\frac{m}{\epsilon\mu_\mathrm{m}\bar{B}}\right)^{1/2} K(\kappa).$$

Since $\partial J_\parallel/\partial u^2 = \partial J_\parallel/\partial\theta = 0$,

$$\frac{\mathrm{d}r}{\mathrm{d}t} = 0,$$

and we have

$$\frac{\partial J_\parallel}{\partial u^1} \approx \frac{8L}{\pi}(m\epsilon\mu_\mathrm{m}\bar{B})^{1/2}K(\kappa)\frac{W}{4\epsilon\mu_\mathrm{m}B}\frac{-\partial B/\partial r}{B} = \frac{2L}{\pi}\left(\frac{m}{\epsilon\mu_m\bar{B}}\right)^{1/2} K(\kappa)W\frac{1}{R},$$

where R^{-1} is the curvature of field line and $1/R \approx -B'/B$. The drift velocity in θ direction is roughly

$$r\frac{\mathrm{d}\theta}{\mathrm{d}t} \sim -\frac{mv^2}{2qBR}.$$

This value is consistent with (7.2). This analytical method can be applied to the toroidal precession of banana in tokamak ($u^1 \approx r,\ u^2 \approx \theta - \varphi/q_\mathrm{s}$) and

the precession of helically trapped banana in stellarator
$(u^1 \approx r,\ u^2 \approx \theta - (\iota/2\pi)\varphi)$.

When the particles enter the loss cone, they escape immediately, so that the confinement time is determined by the velocity space diffusion to the loss cone. The *particle confinement time* $\tau_{\rm p}$ of a mirror field is essentially determined by the ion-ion collision time $\tau_{\rm ii}$ as follows [7.1]:

$$\tau_{\rm p} \approx \tau_{\rm ii} \ln R_{\rm M}. \tag{7.10}$$

Even if the mirror ratio $R_{\rm M}$ is increased, the confinement time is increased only as fast as $\ln R_{\rm M}$. The confinement time is independent of the magnitude of the magnetic field and the plasma size. If the density is $n \approx 10^{20}\,{\rm m}^{-3}$, the ion temperature $T_{\rm i} \approx 100\,{\rm keV}$, the atomic mass number $A = 2$, and ion charge $Z = 1$, the ion-ion collision time is $\tau_{\rm ii} = 0.3\,{\rm s}$, so that $n\tau_{\rm p} = 0.3 \times 10^{20}{\rm m}^{-3}\,{\rm s}$. This value is not enough for a fusion reactor even if $\tau_{\rm p} \sim \tau_{\rm E}$ is assumed. It is necessary to increase the efficiency of recovery of the kinetic energy of charged particles escaping through the ends or to suppress the end loss by innovative concept.

7.2 Mirror Experiments

One of the most important mirror confinement experiments was done by Ioffe and his colleagues, who demonstrated that the minimum-B mirror configuration is quite stable [7.2]. Addition of stabilizing coils as shown in Figure 7.3(a) results in a mirror field with the minimum-B property (Figure 7.3(c)).

Define the *wall mirror ratio* as the ratio of the magnitude of the magnetic field at the radial boundary to its magnitude at the center. When the wall mirror ratio is increased, flute instability disappears and the confinement time becomes long. In the PR-5 device, a plasma of density $10^9 \sim 10^{10}\,{\rm cm}^{-3}$ and ion temperature $3 \sim 4$ keV has been confined for 0.1 s [7.3]. However, when the density is larger than $10^{10}\,{\rm cm}^{-3}$ $(\Pi_{\rm i} > |\Omega_{\rm i}|)$, the plasma suffers from loss-cone instability.

One of the most promising mirror confinement results was brought about in the 2X experiments [7.4]. Here a plasmoid of mean energy of 2.5 keV produced by a plasma gun is injected into a quadrupole mirror and then compressed. The initial density of the trapped plasma is $n \approx 3 \times 10^{13}\,{\rm cm}^{-3}$, and $n\tau_{\rm p} \approx 10^{10}\,{\rm cm}^{-3}{\rm s}$. The magnitude of the magnetic field is 1.3 T and $\beta \approx 5\%$. The average energy of the ions is 6-8 keV, and the electron temperature is about 200 eV. Since $n\tau_{\rm p}$ for an ideal case (classical Coulomb collision time) is $n\tau_{\rm p} \approx 3 \times 10^{10}\,{\rm cm}^{-3}{\rm s}$, the results obtained are $1/3 \sim 1/15$ that of the ideal. In the experiments of 2XIIB with yin-yang coil (refer to

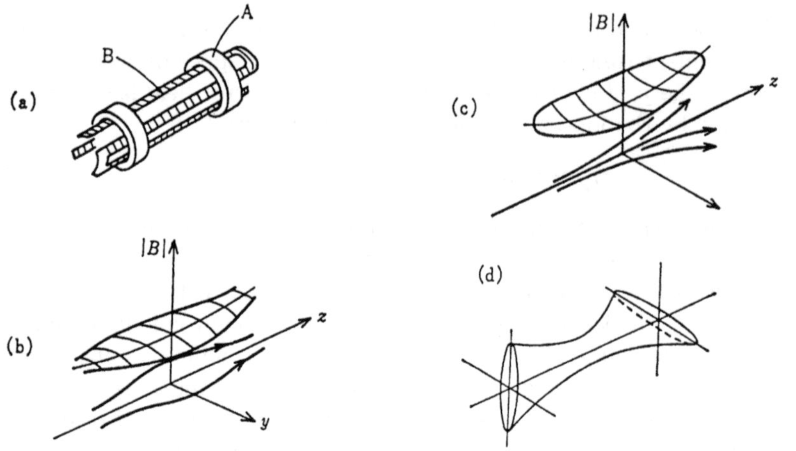

Figure 7.3 Minimum-B mirror field. (a) Coil system with mirror coil (A) and stabilizing coils (Ioffe bars) (B). (b) Magnitude of the magnetic field in a simple mirror. (c) Magnitude of the magnetic field in the minimum-B mirror field. (d) The shape of a quadrupole minimum-B (fishtail).

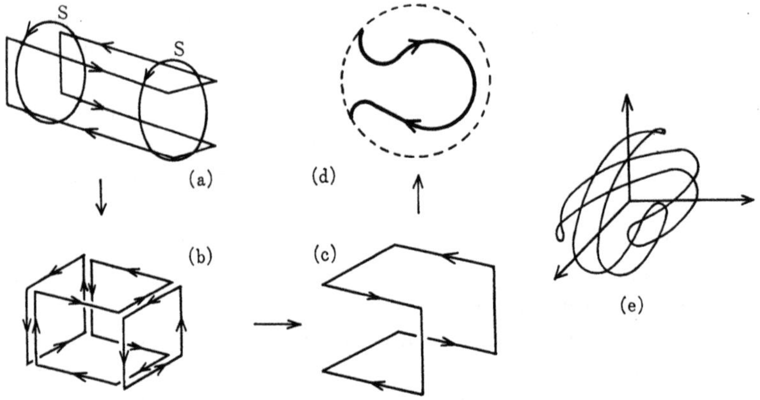

Figure 7.4 Ioffe-field coils. (a) Ring coils produce a simple mirror and four straight bars produce the quadrapole field, (b) and (c) modification of this array, (d) baseball coil, (e) yin-yang coil.

Figure 7.4) [7.5], microscopic instability is suppressed by adding a small stream of warm plasma to smooth out the loss cone. A 15-19 keV, 260 A neutral beam is injected, and the resulting plasma, of 13 keV ion temperature, is confined to $n\tau_{\rm p} \approx 10^{11}\,{\rm cm}^{-3}{\rm s}$ ($n \approx 4 \times 10^{13}\,{\rm cm}^{-3}$).

7.3 Instabilities in Mirror System

Instabilities of mirror systems are reviewed in [7.6]~[7.8]. MHD instability can be suppressed by the minimum-B configuration. However, the particles in the loss-cone region are not confined, and the non-Maxwellian distribution gives rise to electrostatic perturbations at the ion cyclotron frequency and its harmonics, which scatter the particles into the loss cone region and so enhance the end loss. It can be shown that instabilities are induced when the cyclotron wave couples with other modes, such as the electron plasma wave or the drift wave.

(i) Instability in the Low-Density Region ($\Pi_{\rm e} \approx l|\Omega_{\rm i}|$)
Let us consider the low-density case. When the plasma electron frequency $\Pi_{\rm e}$ reaches the ion cyclotron frequency $|\Omega_{\rm i}|$, there is an interaction between the ion Larmor motion and the electron Langmuir oscillation, and *Harris instability* occurs [7.9] (refer to Section 11.6.2).

When the density increases further, the oblique Langmuir wave with $\omega = (k_\parallel/k_\perp)\Pi_{\rm e}$ couples with the harmonics $l|\Omega_{\rm i}|$ of the ion cyclotron wave. The condition $\omega \approx k_\perp v_{\perp{\rm i}}$ is necessary for the ions to excite the instability effectively; and, if $\omega > 3k_\parallel v_{\parallel{\rm e}}$, then the Landau damping due to electrons is ineffective. Thus, excitation will occur when

$$\omega \approx l|\Omega_{\rm i}| \approx \Pi_{\rm e} k_\parallel/k \approx k_\perp v_{\perp{\rm i}} > 3k_\parallel v_{\parallel{\rm e}}$$

where $k_\parallel$, $k_\perp$ are the parallel and perpendicular components of the propagation vector and $v_\parallel$, $v_\perp$ are the parallel and perpendicular components of the velocity. Therefore, the instability condition is

$$\frac{\Pi_{\rm e}}{l|\Omega_{\rm i}|} > \left(1 + \frac{9v_{\parallel{\rm e}}^2}{v_{\perp{\rm i}}^2}\right)^{1/2} \approx \left(1 + \frac{9MT_{\rm e}}{mT_{\rm i}}\right)^{1/2}. \qquad (7.11)$$

Let L be the length of the device. As the relation $k_\parallel > 2\pi/L$ holds, we have [7.10, 7.11]

$$\frac{L}{\rho_{\Omega{\rm i}}} > \frac{6\pi}{l}\frac{v_{\parallel{\rm e}}}{v_{\perp{\rm i}}}.$$

Harris instability has been studied in detail experimentally [7.12].

(ii) Instability in the High Density ($\Pi_i > |\Omega_i|$)
When the density increases further, so that Π_i is larger than $|\Omega_i|$ (while $\Omega_e > \Pi_i$ still holds), *loss-cone instability* with $\omega_r \approx \omega_i \approx \Pi_i$ will occur [7.13]. This is a convective mode and the length of the device must be less than a critical length, given by

$$L_{crit} = 20A\rho_{\Omega i}\left(\frac{\Omega_e^2}{\Pi_e^2}+1\right)^{1/2} \tag{7.12}$$

for stability. Here A is of the order of 5 ($A \approx 1$ for complete reflection and $A \approx 10$ for no reflection at the open end). Therefore, the instability condition is $L > 100\rho_{\Omega i}$. Loss-cone instability can occur for a homogeneous plasma. When there is a density gradient, this type of instability couples with the drift wave, and drift cyclotron loss-cone instability may occur [7.13]. When the characteristic length of the density gradient is comparable to the radial dimension R_p of the plasma, the instability condition of this mode is

$$R_p < \rho_{\Omega i}\left(\frac{\Pi_i}{|\Omega_i|}\left(1+\frac{\Pi_e^2}{\Omega_e^2}\right)^{-1/2}\right)^{4/3}. \tag{7.13}$$

(iii) Mirror Instability
When the beta ratio becomes large, the anisotropy of plasma pressure induces electromagnetic-mode *mirror instability*. (Note that the Harris and loss-cone instabilities are electrostatic.) The instability condition is

$$\left(\frac{T_{\perp i}}{T_{\parallel i}}-1\right)2\beta > 1. \tag{7.14}$$

To avoid the instabilities described in (i)-(iii), the following stability conditions must be met:

$$L < 100 \sim 200\rho_{\Omega i}, \tag{7.15}$$

$$R_p > 200\rho_{\Omega i}, \tag{7.16}$$

$$\beta < \frac{1}{2}\left(\frac{T_{\perp i}}{T_{\parallel i}}-1\right)^{-1} \approx 0.3 \sim 0.5. \tag{7.17}$$

(iv) Negative Mass Instability
Let us assume that charged particles in Larmor motion are uniformly distributed at first and then a small perturbation occurs, so that positive charge (for example) accumulates at the region A shown in Figure 7.5. The electric field decelerates ions in the region to the right of A, and their kinetic energy ϵ corresponding to the velocity component perpendicular to the

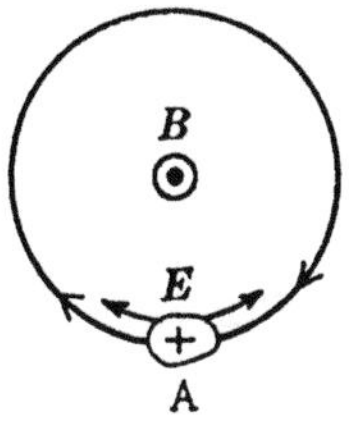

Figure 7.5 Negative mass instability.

magnetic field is decreased. Ions to the left of A are accelerated, and their ϵ is increased. When the rotational frequency ω depends on ϵ through the relation $d\omega/d\epsilon < 0$, the frequency ω of the ions in the region to the right of A is increased and the ions approach A despite the deceleration. These ions behave as if they had negative mass. This situation is exactly the same for the ions in the left-hand region. Therefore, the charge accumulates at A, and the perturbation is unstable. This type of instability is called *negative mass instability* [7.14]. The condition $d\omega/d\epsilon < 0$ is satisfied when the magnitude of the magnetic field decreases radially. Thus, as expected, simple mirrors, where B decreases radially and the Larmor radius is large, have been reported to exhibit negative mass instability. The PR-5 device is of minimum-B configuration, so that the magnitude of the magnetic field increases radially. However, here another type of negative mass instability is observed [7.15]. When the perpendicular energy ϵ decreases, ions can enter the mirror region more deeply, so that the ion cyclotron frequency is increased. Thus, the condition $d\omega/d\epsilon < 0$ is satisfied even in PR-5.

(v) Instability in Hot Electron Plasmas

A hot electron plasma can be produced by electron cyclotron resonant heating in mirror fields. The temperature of the hot component is raised up to the range of several keV to several hundred keV, and the density range is $10^{10} \sim 10^{11}\,\mathrm{cm}^{-3}$. The electromagnetic *whistler instability* [7.16, 7.17] is excited by anistropy of the velocity distribution function ($T_\perp > T_\parallel$). This whistler instability of hot electron plasmas has been observed experimentally [7.18].

7.4 Tandem Mirrors

7.4.1 Theory of Tandem Mirror

The input and output energy balance of a classical mirror reactor is quite marginal even in the ideal confinement condition. Therefore, the suppression of the end loss is the critical issue for realistic mirror reactors.

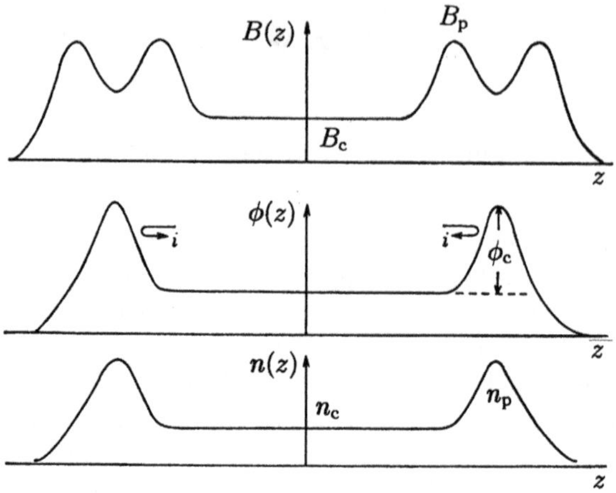

Figure 7.6 The magnitudes of magnetic field $B(z)$, electrostatic potential $\phi(z)$, and density $n(z)$ along the z axis (mirror axis) of a tandem mirror.

This section describes the research on the end plug by the use of electrostatic potential. In a simple mirror case, the ion confinement time is of the order of ion-ion collision time, and the electron confinement time is of the order of electron-electron or electron-ion collision times,($\tau_{\text{ee}} \sim \tau_{\text{ei}}$). Since $\tau_{\text{ee}} \ll \tau_{\text{ii}}$, the plasma is likely to be ion rich, and the plasma potential in the mirror tends to be positive. When the two mirrors are arranged in the ends of the central mirror in tandem, it is expected that the plasma potentials in the plug mirrors (plug cell) at both ends become positive with respect to the potential of the central mirror (central cell). This configuration is called a *tandem mirror* [7.19, 7.20]. The loss-cone region in velocity space of the tandem mirror, as shown in Figure 7.6, is given by

$$\left(\frac{v_\perp}{v}\right)^2 < \frac{1}{R_{\text{m}}}\left(1 - \frac{q\phi_{\text{c}}}{mv^2/2}\right), \tag{7.18}$$

where q is the charge of the particle and R_{m} is the mirror ratio. The loss-cone regions of electrons and ions are shown in Figure 7.7 in the positive case of the potential ϕ_{c}. Solving the Fokker-Planck equation, Pastukhov [7.21] derived the *ion confinement time of a tandem mirror* with positive potential as

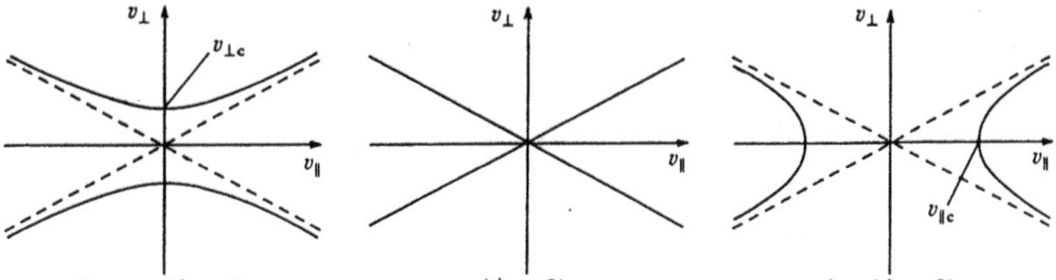

Figure 7.7 Loss-cone regions of an ion ($q = Ze$) and electron ($-e$) for positive electrostatic potential $\phi_c > 0$ and the mirror ratio R_m. $(v_{\perp c}/v)^2 = (e\phi_c/(R_m - 1))/(m_e v^2/2)$, in the electron case. $(v_{\parallel c}/v)^2 = Ze\phi_c/(m_i v^2/2)$ in the ion case.

$$\tau_{\mathrm{PAST}} = \tau_{\mathrm{ii}} g(R_m) \left(\frac{e\phi_c}{T_{\mathrm{ic}}}\right) \exp\left(\frac{e\phi_c}{T_{\mathrm{ic}}}\right), \tag{7.19}$$

$$g(R_m) = \pi^{1/2}(2R_m + 1)(4R_m)^{-1} \ln(4R_m + 2),$$

where T_{ic} is the ion temperature of the central cell and ϕ_c is the potential difference of the plug cells in both ends with respect to the central cell. Denote the electron densities of the central cell and the plug cell by n_c and n_p, respectively; then the Boltzmann relation $n_p = n_c \exp(e\phi_c/T_e)$ gives

$$\phi_c = \frac{T_e}{e} \ln\left(\frac{n_p}{n_c}\right). \tag{7.20}$$

By application of neutral beam injection into the plug cell, it is possible to increase the density in the plug cell to be larger than that of the central cell. When $R_m \sim 10$ and $e\phi_c/T_{\mathrm{ic}} \sim 2.5$, $\tau_{\mathrm{PAST}} \sim 100\tau_{\mathrm{ii}}$, so that the theoretical confinement time of the tandem mirror is much longer than that of the simple mirror.

In this type of tandem mirror it is necessary to increase the density n_p in the plug cell in order to increase the plug potential ϕ_c, and the necessary power of the neutral beam injection becomes large. Since the plug potential ϕ_c is proportional to the electron temperature T_e, an increase in T_e also increases ϕ_c. If the electrons in the central cell and the electrons in the plug cells can be thermally isolated, the electrons in the plug cells only can be heated, so that efficient potential plugging may be expected. For this purpose a *thermal barrier* [7.22] is introduced between the central cell and the plug cell, as shown in Figure 7.8. When a potential dip is formed in the thermal barrier in an appropriate way, the electrons in the plug cell and the central cell are thermally isolated.

Since the electrons in the central cell are considered to be Maxwellian

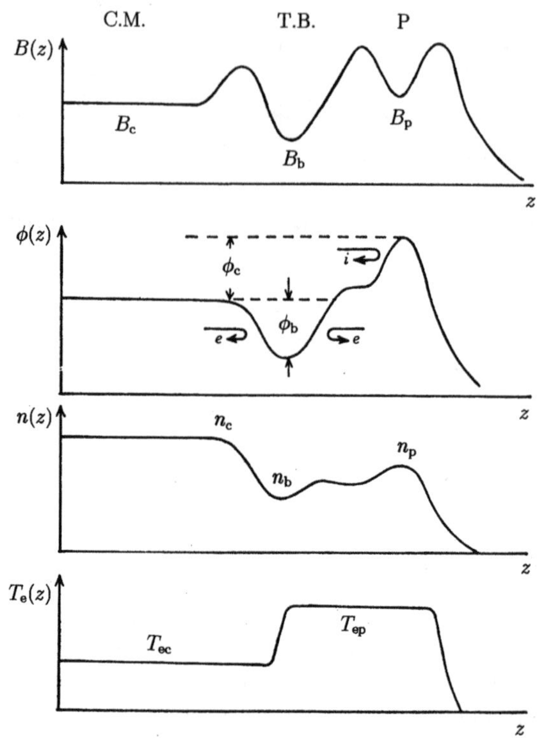

Figure 7.8 The magnitudes of magnetic field $B(z)$, electrostatic potential $\phi(z)$, density $n(z)$, and electron temperature $T_e(z)$ at the thermal barrier. C.M., central mirror. T.B., thermal barrier. P, plug mirror.

with temperature T_{ec}, we have the relation

$$n_c = n_b \exp\left(\frac{e\phi_b}{T_{ec}}\right). \tag{7.21}$$

The electrons in the plug cell are modified Maxwellian [7.23] and the following relation holds:

$$n_p = n_b \exp\left(\frac{e(\phi_b + \phi_c)}{T_{ep}}\right) \times \left(\frac{T_{ep}}{T_{ec}}\right)^{\nu}, \tag{7.22}$$

where $\nu \sim 0.5$. These relations reduce to

$$\phi_c = \frac{T_{ep}}{e} \ln\left(\frac{n_p}{n_b}\left(\frac{T_{ec}}{T_{ep}}\right)^{\nu}\right) - \frac{T_{ec}}{e} \ln\left(\frac{n_c}{n_b}\right), \tag{7.23}$$

$$\phi_b = \frac{T_{ec}}{e} \ln\left(\frac{n_c}{n_b}\right). \tag{7.24}$$

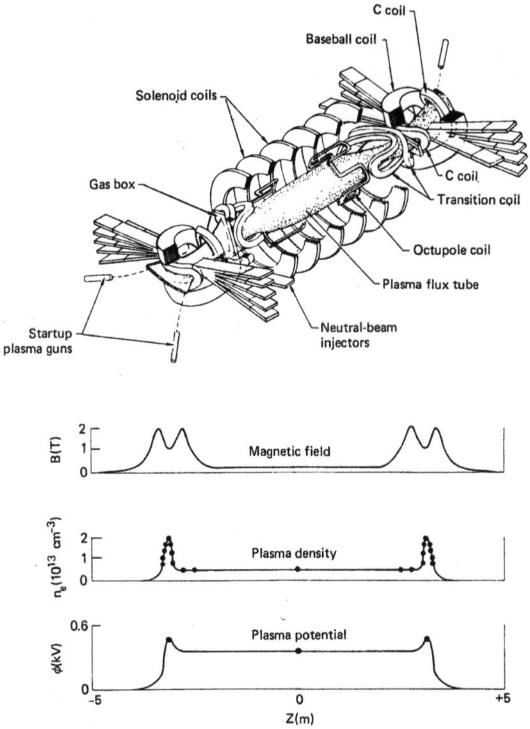

Figure 7.9 Schematic diagram of TMX magnet geometry and measured axial profiles of magnetic field, electron density, and plasma potential. After [7.24].

Therefore, if the electron temperature T_{ep} in the plug cell is increased, ϕ_{c} can be increased without the condition $n_{\mathrm{p}} > n_{\mathrm{c}}$; thus, efficient potential plugging may be expected.

7.4.2 Tandem Mirror Experiments

Schematic diagram of **TMX** (tandem mirror experiment) at Lawrence Livermore Laboratory is shown in Figure 7.9. At both sides of central solenoid coils confining the center-cell plasma, baseball coil type minimum B plugs provide an average minimum B configuration (refer to Section 8.1.2) against interchange instability. Neutral beams with $13\,\mathrm{keV}\times 200\,\mathrm{A}$ (equivalent) per plug are injected and about $400\,\mathrm{kW}$ is trapped by the end plug plasmas provided by titanium-washer gun of the type used in 2XIIB. Axial profile data of magnetic field, electron density, and plasma potential are plotted in Figure 7.9 [7.24]. In plug cells, the densities are higher than that of center-cell and ion confining potentials are formed. The mean temperature of center-cell plasma is $T_{\mathrm{i}} \sim T_{\mathrm{e}} \sim 70\,\mathrm{eV}$ and center-cell β value as high as 5∼10% is observed. Ion confining potential is $\phi_c \sim 120\,\mathrm{V}$ and center-cell confinement parameter is $n_e\tau_p \sim 3 \times 10^{10}\,\mathrm{cm}^{-3}\cdot\mathrm{s}$,($n_e \sim 0.5 \times 10^{13}$ cm^{-3}). Electrostatic enhancement in confinement is ∼4.

TMX-U is the tandem mirror with thermal barrier and demonstrates the formation of microstable sloshing-ion distribution. By sloshing-ion produced by neutral beam injectors and electron cyclotron resonance heating of electrons, both ion confining potential ϕ_c and thermal barrier potential ϕ_b are formed in single plug cell [7.25]. When the neutral beam is injected with the appropriate angle to the magnetic field, fast neutrals are ionized and are trapped and bounce back and forth in the plug cell. The density profile of sloshing-ions is higher in the reflecting regions than in the middle region (double peaks). Furthermore, when the electrons at the (outside) peak region are heated by ECRH with fundamental cyclotron frequency (ω_{ce}), hot electrons are produced. Hot electrons are pushed out by the force $-\mu\nabla B$ and ion confining potential ϕ_c is built up to keep the charge neutrality. Furthermore, when hot electrons are produced by 2nd harmonic ECRH ($2\omega_{ce}$) in the middle region of the plug cell, Equation (7.24) of the thermal barrier potential ϕ_b is modified to

$$e\phi_b = T_{ec} \ln\left(\frac{n_c}{n_b - n_{hot}}\right), \tag{7.25}$$

where n_{hot} is the density of hot electrons, which are trapped in the plug cell and cannot transit to the center-cell. The thermal barrier is then formed.

Axial profile of the magnetic field and axial measured potential profile are shown in Figure 7.10. When both sloshing-ions and ECRH powers are present, Figure 7.10 shows remarkable reduction of the end-loss currents out one end when both ends are plugged. The pulsed nature of the end-loss current seen in Figure 7.10 results from sweeping the analyzer ion-repeller voltage to measure the energy distribution. Ion confining potential ϕ_c is larger than 1.5 kV and barrier potential ϕ_b is 0.45 kV. The center-cell beta resulting mainly from the neutral beam injected ions reaches 6% average over the radius to the limiter. During strong end plugging at low density ($7 \times 10^{11}\,\mathrm{cm}^{-3}$), the axial ion confinement time $\tau_\parallel$ reaches 50∼100 ms, while the non-ambipolar radial ion confinement time $\tau_\perp$ is 14 ms. At high densitiy and lower center-cell potential, $\tau_\parallel$ is 6∼12 ms and $\tau_\perp$ exceeds 100 ms. Ions have been heated up to $T_{\perp i} \sim 2\,\mathrm{keV}$.

GAMMA 10 at University of Tukuba is the tandem mirror with axisymmetrized end plugging with thermal barrier with use of neutral beam injection and ECRH. Schematics of GAMMA 10 device and magnetic flux tube, together with heating systems, are shown in Figure 7.11. The device length is 27 m. It consists of a center-cell ($B \sim 0.5$T), anchor cells and axisymmetric end mirror cells with thermal barrier and ion confining potential. The anchor cell is a minimum B mirror for suppressing interchange instability. Ramp coils set at both ends of the center cell

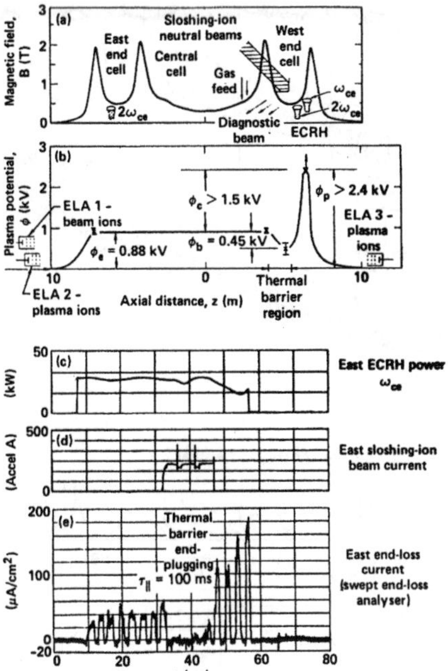

Figure 7.10 Thermal barrier measurements in TMX-U. (a) Axial profile of the magnetic field with illustration of heating systems used for single-end plugging operation, (b) axial measured potential profile, (c) time history of ECRH power, (d) time history of sloshing-beam current, and (e) time history of resulting end-loss current measured with a sweep end-loss analyzer showing that both ECRH and sloshing beam are required to reduce end losses. After [7.25].

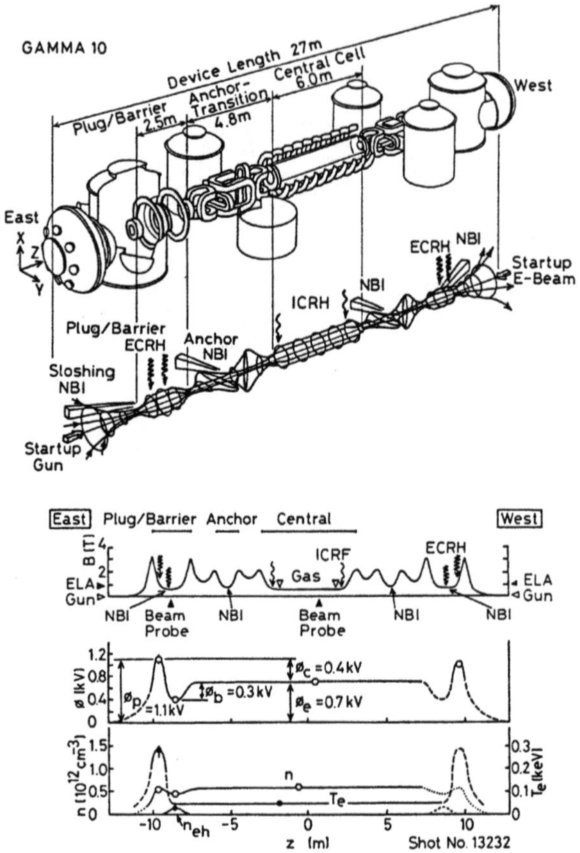

Figure 7.11 Schematics of GAMMA 10 device and magnetic flux tube with heating systems. Axial distribution of magnetic field strength and positions of heating, potential distribution measured by center-cell and barrier beam probes and end-loss analyzer, and axial distribution of density, electron temperature. After [7.26].

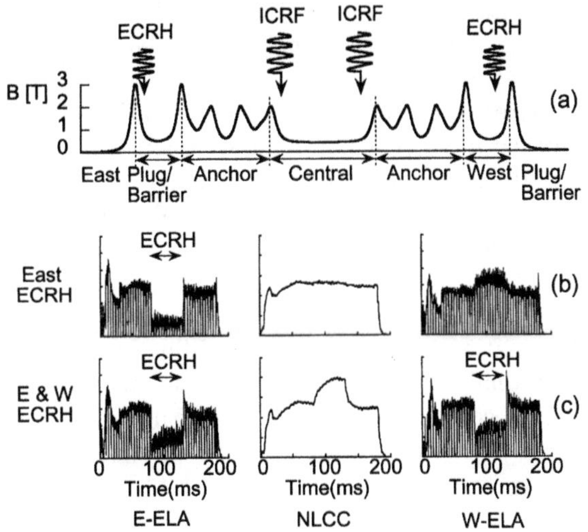

Figure 7.12 (a) Axial profile of magnetic field and location of heating systems in GAMMA 10. (b) and (c) Center-cell line density (NLCC) and end-loss currents to the east end-loss ion energy analyzer (E-ELA) and west (W-ELA), where one division of NLCC is $10^{13}\,\mathrm{cm}^{-2}$ and that of ELA is $10^{-4}\,\mathrm{A/cm}^2$. ECRH is applied from 80 ms to 130 ms. After [7.27].

throttle the fraction of passing particles flowing into the minimum B anchors and plug/barrier cells. When the axial potential hill is kept low in the anchor transition region, the center-cell ions are either reflected by the ramp coils or pass through the anchors and are reflected by the axisymmetric plugs with thermal barrier by use of NBI for sloshing-ions and ECRH. Axisymmetric plugs could reduce the resonant diffusion of ions trapped by non-axisymmetric plugs. Axial profile of magnetic field, potential, and density are shown in Figure 7.11 [7.26].

The improvement of plasma confinement in the hot ion mode was obtained in 1998 [7.27]. RF heating in the ion cyclotron range of frequency (ICRF) is mainly used for plasma production and heating in center-cell. ECRHs are used for the formation of ion confining potential without sloshing-ions. The optimizations of the axisymmetrized heating patterns of ECRH and ICRH heating power are made. Furthermore, conducting plates in anchor cells are installed near the surface of plasma with the fanning magnetic flux tube in order to fix the potential at the boundary for reducing irregular electric field. Center-cell line density and end-loss currents to the end-loss ion energy analyzer (ELA) located in the east and west open ends are shown in Figure 7.12. The effects of potential confinement are clearly seen in Figure 7.12. The center-cell plasma with average ion temperature of 2.8 keV ($T_{\mathrm{i}0} \sim 4.5$ keV) and the density on axis of $2.7 \times 10^{12}\,\mathrm{cm}^{-3}$ are obtained. An ion confining potential is ~ 0.6 kV. The particle confinement time is $\tau_\parallel \approx 40\,\mathrm{ms}$, which is one order of magnitude improvement due to potential confinement. The observed energy confinement time is $\tau_\mathrm{E} \approx 10\,\mathrm{ms}$. The electron temperature is $\sim$80 eV.

A factor of 3 progresses up to 2.1 kV in the formation of ion confining potential was achieved in the hot ion mode with ion temperature of several keV and $n_\mathrm{e} \sim 2 \times 10^{12}\,\mathrm{cm}^{-3}$ in 2004 [7.28]. The advance in the potential formation gives the basis for finding the remarkable effects of sheared flow due to radial electric field on the suppression of turbulent fluctuations in GAMMA 10. From the viewpoint of both higher potential confinement in the axial direction and the suppressed turbulence due to a strong shear flow, a high potential formation plays a role in providing improved ion confinement both radially and axially simultaneously. In the case of $\phi_\mathrm{c} = 0.84\,\mathrm{kV}$, particle confinement time is $\tau = 75\,\mathrm{ms}$ ($\tau_\parallel = 95\,\mathrm{ms}$, $\tau_\perp > 4\tau_\parallel$) and Pastukhov's ion energy confinement time is $\tau_{\mathrm{E\,Pastu}} = 50\,\mathrm{ms}$.

Problems

1. Why does the plasma potential tend to be positive in a single mirror?

2. Let the radius of curvature of field lines near the center region of a simple mirror be R and positive (convex toward outward). Then interchange instability may occur. Estimate the growth rate, assuming the propagation constant $k_\perp \sim m/a$ in perpendicular to the magnetic field, a being the plasma radius.

3. An example of a simple mirror field is given by a magnetic scalar potential ϕ_M (in the currentless region) as follows: $\boldsymbol{B}_\mathrm{M} = \nabla\phi_\mathrm{M}$, $\nabla^2\phi_\mathrm{M} = 0$, $\phi_\mathrm{M} = B_0(z - (b/k)I_0(kr)\sin(kz))$, where $I_0(z) = 1 + (z/2)^2 + \cdots$ is 0th modified Bessel funtion. Quadrupole field produced by Ioffe bar (refer to Figure 7.3(a)) is given by the magnetic potential $\phi_\mathrm{I} = (aB_\mathrm{I}/2)(r/a)^2 \sin 2\theta$. Calculate the square magnitude of the mirror field with Ioffe bar, $\boldsymbol{B} = \boldsymbol{B}_\mathrm{M} + \boldsymbol{B}_\mathrm{I}$, and expand it by $(kr)^2 \ll 1$ and $(kz)^2 \ll 1$ up to the second order. Derive the necessary condition for the absolute minimum B.

8

Magnetohydrodynamic (MHD) Instabilities

The stability of plasmas in magnetic fields is one of the primary research subjects in the area of controlled thermonuclear fusion and both theoretical and experimental investigations have been actively pursued. If a plasma is free from all possible instabilities and if the confinement is dominated by neoclassical diffusion in the banana region, then the energy confinement time $\tau_{\rm E}$ is given by

$$\tau_{\rm E} \approx \frac{(3/2)a^2}{5.8\chi_{\rm G.S.}} \approx \frac{(3/2)}{5.8} q^2 \epsilon^{3/2} \left(\frac{a}{\rho_{\Omega \rm i}} \right)^2 \frac{1}{\nu_{\rm ii}},$$

where a is the plasma radius, $\rho_{\Omega \rm i}$ is the ion Larmor radius, and $\nu_{\rm ii}$ is the ion-ion collision frequency. For such an ideal case, a device of a reasonable size satisfies ignition condition. (For example, with $B = 5\,\rm T, a = 1\,\rm m$, $T_{\rm i} = 20\,\rm keV$, $q = 3$, and inverse aspect ratio $\epsilon = 0.2$, the value of $n\tau_{\rm E} \sim 3.5 \times 10^{20}\,\rm cm^{-3} \cdot sec.$)

A plasma consists of many moving charged particles and has many magnetohydrodynamic degrees of freedom. When a certain mode of perturbation grows and induces fluctuations in the electric and magnetic fields, it enhances anomalous diffusion. These instabilities are called the *magnetohydrodynamic* (MHD) *instabilities* or *macroscopic instabilities.*

Therefore, it is very important to study whether any particular perturbed mode is stable (damping mode) or unstable (growing mode). In the stability analysis, it is assumed that the deviation from the equilibrium state is small so that a linearized approximation can be used. In this chapter we will consider instabilities that can be described by linearized MHD equations.

A small perturbation $\boldsymbol{F}(\boldsymbol{r}, t)$ of the first order is expanded in terms of its Fourier components

$$\boldsymbol{F}(\boldsymbol{r}, t) = \boldsymbol{F}(\boldsymbol{r}) \exp(-i\omega t), \qquad \omega = \omega_{\rm r} + i\omega_{\rm i},$$

and each term can be treated independently in the linearized approximation. The dispersion equation is solved for ω and the stability of the perturbation depends on the sign of the imaginary part $\omega_{\rm i}$ (unstable for $\omega_{\rm i} > 0$ and stable for $\omega_{\rm i} < 0$). When $\omega_{\rm r} \neq 0$, the perturbation is oscillatory. When $\omega_{\rm r} = 0$, it grows or damps monotonously.

In the following sections, typical MHD instabilities are introduced. In Section 8.1, interchange instability is explained in an intuitive manner. In

Section 8.2 the MHD equations are linearized and the boundary conditions are implemented. Alfvén wave and interchange instability are analyzed by the linearized MHD equation in this section. A cylindrical plasma with sharp boundary is studied as a simple but important example in Section 8.3. Kruskal-Sharanov stability criterion is described in this section. To solve the linearized MHD equation is equivalent to solving the calculus of variation of energy integral. This energy principle is described in Section 8.4. In Section 8.5 energy integral of more realistic cylindrical plasma with diffuse boundary is introduced and stabilities of tokamak with large aspect ratio are examined. Hain-Lüst MHD equation of cylindrical diffused configuration is described in Section 8.6. Ballooning instability (Section 8.7), $\eta_{\rm i}$ mode due to density and temperature gradients (Section 8.8) and Alfv̈en Eigenmode (Section 8.9) are described in this chapter. It should be understood that there are many other instabilities. General reviews of MHD instabilities may be found in [8.1].

8.1 Interchange Instabilities

8.1.1 Interchange Instability

Let $x = 0$ be the boundary between plasma and vacuum and let the z axis be taken in the direction of the magnetic field $\boldsymbol{B}$. The plasma region is $x < 0$ and the vacuum region is $x > 0$. It is assumed that the acceleration $\boldsymbol{g}$ is applied in the x direction (see Figure 8.1). Ions and electrons drift in opposite directions, due to the acceleration, with drift velocities of

$$\boldsymbol{v}_{\rm G,i} = \frac{M}{e}\frac{\boldsymbol{g}\times\boldsymbol{B}}{B^2},$$

$$\boldsymbol{v}_{\rm G,e} = -\frac{m}{e}\frac{\boldsymbol{g}\times\boldsymbol{B}}{B^2},$$

where M and m are the masses of ion and electron, respectively. Let us assume that, due to a perturbation, the boundary of the plasma is displaced from the surface $x = 0$ by the amount

$$\delta x = a(t)\sin(k_y y).$$

The charge separation due to the opposite ion and electron drifts yields an electric field. The resultant $(\boldsymbol{E}\times\boldsymbol{B})$ drift enhances the original perturbation if the direction of the acceleration $\boldsymbol{g}$ is outward from the plasma. We see that this is the same as saying that the magnetic flux originally inside but near the plasma boundary is displaced so that it is outside the boundary, while the flux outside moves in to fill the depression thus left in the boundary; because of this geometrical picture of the process, this type of instability has

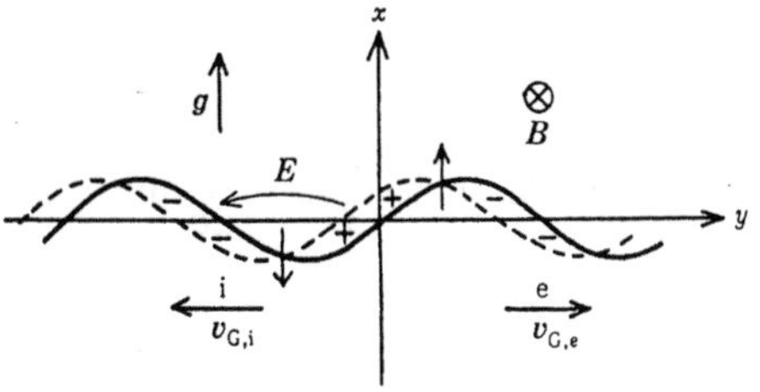

Figure 8.1 Ion and electron drifts and the resultant electric field for interchange instability.

come to be called *interchange instability*. As the perturbed plasma boundary is in the form of flutes along the lines of magnetic force, this instability is also called *flute instability*. Similar phenomena occur in hydrodynamics when a dense fluid is supported against gravity by a fluid with less density. Therefore,interchange instability is also called *Rayleigh-Taylor instability*.

The drift due to the acceleration yields a surface charge on the plasma, of charge density

$$\sigma_s = \sigma(t)\cos(k_y y)\delta(x) \tag{8.1}$$

(see Figure 8.1). The electrostatic potential ϕ of the induced electric field $\boldsymbol{E} = -\nabla\phi$ is given by

$$\epsilon_\perp \frac{\partial^2 \phi}{\partial y^2} + \frac{\partial}{\partial x}\left(\epsilon_\perp \frac{\partial \phi}{\partial x}\right) = -\sigma_s. \tag{8.2}$$

The boundary condition is

$$\epsilon_0 \left(\frac{\partial \phi}{\partial x}\right)_{+0} - \left(\epsilon_\perp \frac{\partial \phi}{\partial x}\right)_{-0} = -\sigma_s,$$

$$\phi_{+0} = \phi_{-0}.$$

Under the assumption $k_y > 0$, the solution ϕ is

$$\phi = \frac{\sigma(t)}{k_y(\epsilon_0 + \epsilon_\perp)}\cos(k_y y)\exp(-k_y|x|). \tag{8.3}$$

The velocity of the boundary $\mathrm{d}(\delta x)/\mathrm{d}t$ is equal to $\boldsymbol{E} \times \boldsymbol{B}/B^2$ at $x = 0$, with $\boldsymbol{E}$ found from the potential (8.3). The velocity is

$$\frac{\mathrm{d}a(t)}{\mathrm{d}t}\sin(k_y y) = \frac{\sigma(t)}{(\epsilon_0 + \epsilon_\perp)B}\sin(k_y y). \tag{8.4}$$

The charge flux in the y direction is

$$ne|\boldsymbol{v}_{\mathrm{G,i}}| = \frac{\rho_{\mathrm{m}} g}{B},$$

where $\rho_{\mathrm{m}} = nM$. Accordingly the changing rate of charge density is

$$\frac{\mathrm{d}\sigma(t)}{\mathrm{d}t}\cos(k_y y) = \frac{\rho_{\mathrm{m}} g}{B} a(t) \frac{\mathrm{d}}{\mathrm{d}y}\sin(k_y y), \tag{8.5}$$

and

$$\frac{\mathrm{d}^2 a}{\mathrm{d}t^2} = \frac{\rho_{\mathrm{m}} g k_y}{(\epsilon_0 + \epsilon_\perp)B^2} a. \tag{8.6}$$

The solution is in the form $a \propto \exp \gamma t$; the growth rate γ is given by

$$\gamma = \left(\frac{\rho_{\mathrm{m}}}{(\epsilon_0 + \epsilon_\perp)B^2}\right)^{1/2} (g k_y)^{1/2}. \tag{8.7}$$

In the low-frequency case (compared with the ion cyclotron frequency), the dielectric constant is given by

$$\epsilon_\perp = \epsilon_0 (1 + \frac{\rho_{\mathrm{m}}}{B^2 \epsilon_0}) \gg \epsilon_0, \tag{8.8}$$

as will be explained in Section 11.4.1. Accordingly the growth rate γ is [8.2]

$$\gamma = (g k_y)^{1/2}. \tag{8.9}$$

When the acceleration is outward, a perturbation with the propagation vector $\boldsymbol{k}$ normal to the magnetic field $\boldsymbol{B}$ is unstable as is shown in Figure 8.2(a); i.e.,

$$(\boldsymbol{k} \cdot \boldsymbol{B}) = 0. \tag{8.10}$$

However, if the acceleration is inward ($g < 0$), γ of (8.9) is imaginary and the perturbation is oscillatory and stable.

The origin of interchange instability is charge separation due to the acceleration. When lines of magnetic force are curved, as is shown in Figure 8.2, the charged particles are subjected to a centrifugal force. If the magnetic lines of force are convex outward (Figure 8.2a), this centrifugal acceleration induces interchange instability. If the lines are concave outward, the plasma is stable. Accordingly, the plasma is stable when the magnitude B of the magnetic field increases outward. In other words, if B is a minimum at the plasma region, the plasma is stable. This is the *minimum-B condition* for stability. More general treatment of the interchange instability is discussed in Section 8.2.3.

The drift motion of charged particles is expressed by (refer to Section 2.1.2)

$$\boldsymbol{v}_{\mathrm{G}} = \frac{\boldsymbol{E} \times \boldsymbol{b}}{B} + \frac{\boldsymbol{b}}{\Omega} \times \left(\boldsymbol{g} + \frac{(v_\perp^2/2) + v_\parallel^2}{R}\boldsymbol{n}\right) + v_\parallel \boldsymbol{b},$$

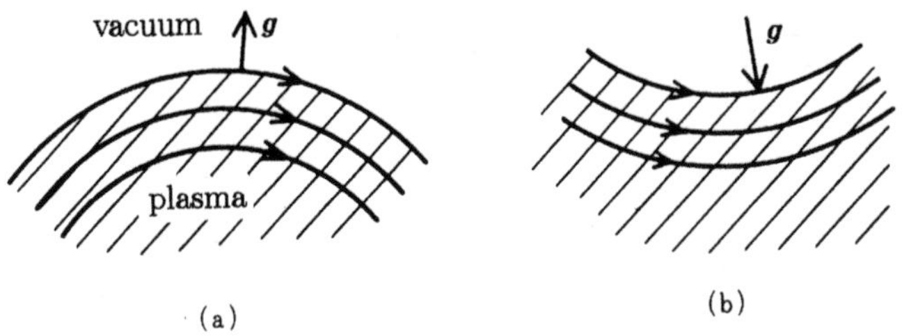

Figure 8.2 Centrifugal force due to the curvature of line of magnetic force.

where $\boldsymbol{n}$ is the normal unit vector from the center of curvature to a point on a line of magnetic force. R is the radius of curvature of line of magnetic force. The equivalent acceleration is

$$\boldsymbol{g} = \frac{(v_\perp^2/2) + v_\parallel^2}{R}\boldsymbol{n}. \tag{8.11}$$

The growth rate becomes $\gamma \approx (a/R)^{1/2}(v_{\rm T}/a)$ in this case.

When the growth rate $\gamma \sim (gk_y)^{1/2}$ is not very large and the ion Larmor radius $\rho_\Omega^{\rm i}$ is large enough to satisfy

$$(k_y\rho_\Omega^{\rm i})^2 > \frac{\gamma}{|\Omega_{\rm i}|},$$

the perturbation is stabilized [8.3]. When the ion Larmor radius becomes large, the average perturbed electric field felt by the ions is different from that felt by the electrons, and the $\boldsymbol{E}\times\boldsymbol{B}/B^2$ drift velocities of the ion and the electrons are different. The charge separation thus induced has opposite phase from the charge separation due to acceleration and stabilizes the instabiltity.

8.1.2 Stability Criterion for Interchange Instability, Magnetic Well

Let us assume that a magnetic line of force has "good" curvature at one place B and "bad" curvature at another place A (Figure 8.3(b)). Then the directions of the centrifugal force at A and B are opposite, as is the charge separation. The charge can be short-circuited easily along the magnetic lines of the force, so that the problem of stability has a different aspect. Let us consider perturbations in which the magnetic flux of region 1 is interchanged with that of region 2 and the plasma in region 2 is interchanged with the plasma in region 1 (interchange perturbations, Figure 8.3b). It is assumed that the plasma is low-beta so that the magnetic field is nearly identical to the vacuum field. Any deviation from the vacuum magnetic field is accompanied by an increase in the magnetic energy of the disturbed field.

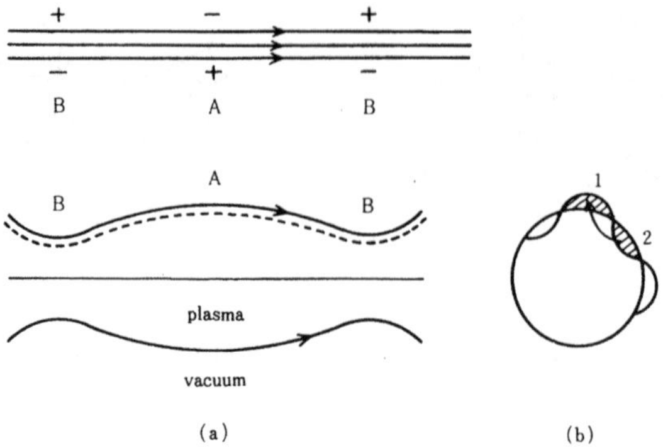

Figure 8.3 Charge separation in interchange instability. (a) The lower figure shows the unstable part A and the stable part B along a magnetic line of force. The upper figure shows the charge separation due to the acceleration along a flute. (b) Cross-section of the perturbed plasma.

This is the consequence of Maxwell equation. It can be shown that the most dangerous perturbations are those that exchange equal magnetic fluxes, as follows.

The energy $Q_{\rm M}$ of the magnetic field inside a magnetic tube is

$$Q_{\rm M} = \int {\rm d}\boldsymbol{r}\frac{B^2}{2\mu_0} = \int {\rm d}lS\frac{B^2}{2\mu_0}, \tag{8.12}$$

where l is the length taken along a line of magnetic force and S is the cross-section of the magnetic tube. As the magnetic flux $\Phi = B \cdot S$ is constant, the energy is

$$Q_{\rm M} = \frac{\Phi^2}{2\mu_0}\int\frac{{\rm d}l}{S}.$$

The change $\delta Q_{\rm M}$ in the magnetic energy due to the interchange of the fluxes of regions 1 and 2 is

$$\delta Q_{\rm M} = \frac{1}{2\mu_0}\left(\left(\Phi_1^2\int_2\frac{{\rm d}l}{S} + \Phi_2^2\int_1\frac{{\rm d}l}{S}\right) - \left(\Phi_1^2\int_1\frac{{\rm d}l}{S} + \Phi_2^2\int_2\frac{{\rm d}l}{S}\right)\right). \tag{8.13}$$

If the exchanged fluxes Φ_1 and Φ_2 are the same, the energy change $\delta Q_{\rm M}$ is zero, so that perturbations resulting in $\Phi_1 = \Phi_2$ are the most dangerous.

The kinetic energy $Q_{\rm p}$ of a plasma of volume $\mathcal{V}$ is

$$Q_{\rm p} = \frac{nT\mathcal{V}}{\gamma - 1} = \frac{p\mathcal{V}}{\gamma - 1}, \tag{8.14}$$

where γ is the specific-heat ratio. As the perturbation is adiabatic,

$$p\mathcal{V}^\gamma = {\rm const}.$$

is conserved during the interchange process. The change in the plasma energy is

$$\delta Q_{\mathrm{p}} = \frac{1}{\gamma - 1}\left(p_2'\mathcal{V}_2 - p_1\mathcal{V}_1 + p_1'\mathcal{V}_1 - p_2\mathcal{V}_2\right),$$

where p_2' is the pressure after interchange from the region $\mathcal{V}_1$ to $\mathcal{V}_2$ and p_1' is the pressure after interchange from the region $\mathcal{V}_2$ to $\mathcal{V}_1$. Because of adiabaticity, we have $p_2' = p_1\left(\frac{\mathcal{V}_1}{\mathcal{V}_2}\right)^{\gamma}$, $p_1' = p_2\left(\frac{\mathcal{V}_2}{\mathcal{V}_1}\right)^{\gamma}$ and δQ_{p} becomes

$$\delta Q_{\mathrm{p}} = \frac{1}{\gamma - 1}\left(p_1\left(\frac{\mathcal{V}_1}{\mathcal{V}_2}\right)^{\gamma}\mathcal{V}_2 - p_1\mathcal{V}_1 + p_2\left(\frac{\mathcal{V}_2}{\mathcal{V}_1}\right)^{\gamma}\mathcal{V}_1 - p_2\mathcal{V}_2\right). \tag{8.15}$$

Setting

$$p_2 = p_1 + \delta p,$$

$$\mathcal{V}_2 = \mathcal{V}_1 + \delta\mathcal{V},$$

we can write δQ_{p} as

$$\delta Q_{\mathrm{p}} = \delta p\delta\mathcal{V} + \gamma p\frac{(\delta\mathcal{V})^2}{\mathcal{V}}. \tag{8.16}$$

Since the stability condition is $\delta Q_{\mathrm{p}} > 0$, the sufficient condition is

$$\delta p\delta\mathcal{V} > 0.$$

Since the volume is

$$\mathcal{V} = \int \mathrm{d}l S = \Phi\int\frac{\mathrm{d}l}{B},$$

the stability condition for interchange instability is written as

$$\delta p\delta\int\frac{\mathrm{d}l}{B} > 0.$$

Usually the plasma pressure p decreases ($\delta p < 0$), so that the stability condition is

$$\delta\int\frac{\mathrm{d}l}{B} < 0 \tag{8.17}$$

in the outward direction [8.4]. The integral is to be taken only over the plasma region. Let the volume inside a magnetic surface ψ be V and the magnetic flux in the toroidal direction φ inside the magnetic surface ψ be Φ. We define the *specific volume* U by

$$U = \frac{\mathrm{d}V}{\mathrm{d}\Phi}. \tag{8.18}$$

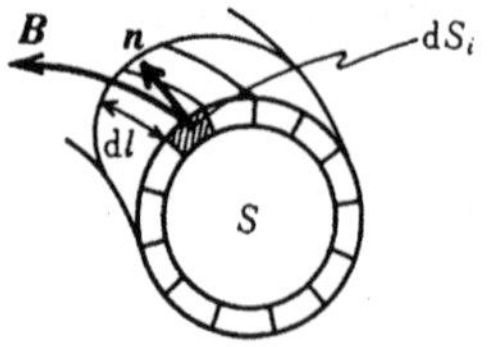

Figure 8.4 Specific volume of a toroidal field.

If the unit vector of the magnetic field $\boldsymbol{B}$ is denoted by $\boldsymbol{b}$ and the normal unit vector of the infinitesimal cross-sectional area $\mathrm{d}S$ is denoted by $\boldsymbol{n}$ (see Figure 8.4), then we have

$$\mathrm{d}V = \int \sum_i (\boldsymbol{b}\cdot\boldsymbol{n})_i \mathrm{d}S_i \mathrm{d}l, \qquad \mathrm{d}\Phi = \sum_i (\boldsymbol{b}\cdot\boldsymbol{n})_i B_i \mathrm{d}S_i.$$

When lines of magnetic force close upon a single circuit of the torus, the specific volume U is

$$U = \frac{\oint \left(\sum_i (\boldsymbol{b}\cdot\boldsymbol{n})_i \mathrm{d}S_i\right) \mathrm{d}l}{\sum_i (\boldsymbol{b}\cdot\boldsymbol{n})_i B_i \mathrm{d}S_i} = \frac{\sum_i (\boldsymbol{b}\cdot\boldsymbol{n})_i B_i \mathrm{d}S_i \oint \frac{\mathrm{d}l}{B_i}}{\sum_i (\boldsymbol{b}\cdot\boldsymbol{n})_i B_i \mathrm{d}S_i}.$$

As the integral over l is carried out along a small tube of the magnetic field, $\sum_i (\boldsymbol{b}\cdot\boldsymbol{n})_i \mathrm{d}S_i B_i$ is independent of l (conservation of magnetic flux). As $\oint \mathrm{d}l/B_i$ on the same magnetic surface is constant, U is reduced to

$$U = \oint \frac{\mathrm{d}l}{B}.$$

When the lines of magnetic force close at N circuits, U is

$$U = \frac{1}{N}\int_N \frac{\mathrm{d}l}{B}. \tag{8.19}$$

When the lines of magnetic force are *not* closed, U is given by

$$U = \lim_{N\to\infty} \frac{1}{N}\int_N \frac{\mathrm{d}l}{B}.$$

Therefore, U may be considered to be an average of $1/B$. When U decreases outward, it means that the magnitude B of the magnetic field increases outward in an average sense, so that the plasma region is the so-called *average minimum-B* region. In other words, the stability condition for interchange instability is reduced to average minimum-B condition;

$$\frac{\mathrm{d}U}{\mathrm{d}\Phi} = \frac{\mathrm{d}^2 V}{\mathrm{d}\Phi^2} < 0. \tag{8.20}$$

When the value of U on the magnetic axis and on the outermost magnetic surface are U_0 and U_a, respectively, we define a *magnetic well depth* $-\Delta U/U$ as

$$-\frac{\Delta U}{U} = \frac{U_0 - U_a}{U_0}. \tag{8.21}$$

8.2 Formulation of MHD Instabilities

8.2.1 Linearization of MHD Equations

The stability problems of plasmas can be studied by analyzing infinitesimal perturbations from the equilibrium state. If the mass density, pressure, flow velocity, and magnetic field are denoted by $\rho_{\mathrm{m}}, p, \boldsymbol{V}$, and $\boldsymbol{B}$, respectively, the equation of motion, conservation of mass, Ohm's law, and the adiabatic relation are

$$\rho_{\mathrm{m}} \frac{\partial \boldsymbol{V}}{\partial t} = -\nabla p + \boldsymbol{j} \times \boldsymbol{B}, \qquad \frac{\partial \rho_{\mathrm{m}}}{\partial t} + \nabla \cdot (\rho_{\mathrm{m}} \boldsymbol{V}) = 0,$$

$$\boldsymbol{E} + \boldsymbol{V} \times \boldsymbol{B} = 0, \qquad \left(\frac{\partial}{\partial t} + \boldsymbol{V} \cdot \nabla\right)(p \rho_{\mathrm{m}}^{-\gamma}) = 0$$

respectively (γ is the ratio of specific heat). Maxwell's equations are then

$$\nabla \times \boldsymbol{E} = -\frac{\partial \boldsymbol{B}}{\partial t}, \qquad \nabla \times \boldsymbol{B} = \mu_0 \boldsymbol{j}, \qquad \nabla \cdot \boldsymbol{B} = 0.$$

These are the MHD equations of a plasma with zero specific resistivity. The values of ρ_{m}, p, $\boldsymbol{V}$, and $\boldsymbol{B}$ in the equilibrium state are $\rho_{\mathrm{m}0}$, p_0, $\boldsymbol{V}_0 = 0$, and $\boldsymbol{B}_0$, respectively. The first-order small quantities are $\rho_{\mathrm{m}1}$, p_1, $\boldsymbol{V}_1 = \boldsymbol{V}$, and $\boldsymbol{B}_1$. The zeroth-order equations are

$$\nabla p_0 = \boldsymbol{j}_0 \times \boldsymbol{B}_0, \qquad \nabla \times \boldsymbol{B}_0 = \mu_0 \boldsymbol{j}_0, \qquad \nabla \cdot \boldsymbol{B}_0 = 0. \tag{8.22}$$

The first-order linearized equations are

$$\frac{\partial \rho_{\mathrm{m}1}}{\partial t} + \nabla \cdot (\rho_{\mathrm{m}0} \boldsymbol{V}) = 0, \tag{8.23}$$

$$\rho_{m0}\frac{\partial \boldsymbol{V}}{\partial t} + \nabla p_1 = \boldsymbol{j}_0 \times \boldsymbol{B}_1 + \boldsymbol{j}_1 \times \boldsymbol{B}_0, \tag{8.24}$$

$$\frac{\partial p_1}{\partial t} + (\boldsymbol{V}\cdot\nabla)p_0 + \gamma p_0 \nabla\cdot\boldsymbol{V} = 0, \tag{8.25}$$

$$\frac{\partial \boldsymbol{B}_1}{\partial t} = \nabla \times (\boldsymbol{V} \times \boldsymbol{B}_0). \tag{8.26}$$

If displacement of the plasma from the equilibrium position $\boldsymbol{r}_0$ is denoted by $\boldsymbol{\xi}(\boldsymbol{r}_0, t)$, it follows that

$$\boldsymbol{\xi}(\boldsymbol{r}_0, t) = \boldsymbol{r} - \boldsymbol{r}_0, \qquad \boldsymbol{V} = \frac{\mathrm{d}\boldsymbol{\xi}}{\mathrm{d}t} \approx \frac{\partial \boldsymbol{\xi}}{\partial t}. \tag{8.27}$$

Equation (8.26) is reduced to

$$\frac{\partial \boldsymbol{B}_1}{\partial t} = \nabla \times \left(\frac{\partial \boldsymbol{\xi}}{\partial t} \times \boldsymbol{B}_0\right), \qquad \text{and} \qquad \boldsymbol{B}_1 = \nabla \times (\boldsymbol{\xi} \times \boldsymbol{B}_0). \tag{8.28}$$

From $\mu_0 \boldsymbol{j} = \nabla \times \boldsymbol{B}$, it follows that

$$\mu_0 \boldsymbol{j}_1 = \nabla \times \boldsymbol{B}_1. \tag{8.29}$$

Equations (8.23) and (8.25) yield

$$\rho_{m1} = -\nabla\cdot(\rho_{m0}\boldsymbol{\xi}), \tag{8.30}$$

$$p_1 = -\boldsymbol{\xi}\cdot\nabla p_0 - \gamma p_0 \nabla\cdot\boldsymbol{\xi}. \tag{8.31}$$

The substitution of these equations into (8.24) gives

$$\rho_{m0}\frac{\partial^2 \boldsymbol{\xi}}{\partial t^2} = \nabla(\boldsymbol{\xi}\cdot\nabla p_0 + \gamma p_0 \nabla\cdot\boldsymbol{\xi}) + \frac{1}{\mu_0}(\nabla \times \boldsymbol{B}_0) \times \boldsymbol{B}_1 + \frac{1}{\mu_0}(\nabla \times \boldsymbol{B}_1) \times \boldsymbol{B}_0$$
$$= -\nabla\left(p_1 + \frac{\boldsymbol{B}_0\cdot\boldsymbol{B}_1}{\mu_0}\right) + \frac{1}{\mu_0}((\boldsymbol{B}_0\cdot\nabla)\boldsymbol{B}_1 + (\boldsymbol{B}_1\cdot\nabla)\boldsymbol{B}_0). \tag{8.32}$$

This is the linearized equation of motion in terms of $\boldsymbol{\xi}$.

Next let us consider the boundary conditions. Where the plasma contacts an ideal conductor, the tangential component of the electric field is zero, i.e., $\boldsymbol{n} \times \boldsymbol{E} = 0$. This is equivalent to $\boldsymbol{n} \times (\boldsymbol{\xi} \times \boldsymbol{B}_0) = 0$, $\boldsymbol{n}$ being taken in the outward direction. The conditions $(\boldsymbol{\xi}\cdot\boldsymbol{n}) = 0$ and $(\boldsymbol{B}_1\cdot\boldsymbol{n}) = 0$ must also be satisfied.

When plasma is in contact with a vacuum, the total pressure must be continuous at the boundary surface between plasma and vacuum and

$$p - p_0 + \frac{B_{\rm in}^2 - B_{0,\rm in}^2}{2\mu_0} = \frac{B_{\rm ex}^2 - B_{0,\rm ex}^2}{2\mu_0},$$

where $\boldsymbol{B}_{\rm in}, \boldsymbol{B}_{0,\rm in}$ give the internal magnetic field of the plasma and $\boldsymbol{B}_{\rm ex}, \boldsymbol{B}_{0,\rm ex}$ give the external field. When $\boldsymbol{B}_{\rm in}(\boldsymbol{r}), \boldsymbol{B}_{\rm ex}(\boldsymbol{r})$, and $p(\boldsymbol{r})$ are expanded in $\boldsymbol{\xi} = \boldsymbol{r} - \boldsymbol{r}_0$ $(f(\boldsymbol{r}) = f_0(\boldsymbol{r}_0) + (\boldsymbol{\xi}\cdot\nabla)f_0(\boldsymbol{r}) + f_1)$, the boundary condition is reduced to

$$-\gamma p_0 \nabla\cdot\boldsymbol{\xi} + \frac{\boldsymbol{B}_{0,\rm in}\cdot(\boldsymbol{B}_{1,\rm in} + (\boldsymbol{\xi}\cdot\nabla)\boldsymbol{B}_{0,\rm in})}{\mu_0}$$
$$= \frac{\boldsymbol{B}_{0,\rm ex}\cdot(\boldsymbol{B}_{1,\rm ex} + (\boldsymbol{\xi}\cdot\nabla)\boldsymbol{B}_{0,\rm ex})}{\mu_0}. \tag{8.33}$$

From Maxwell's equations, there are the following relations:

$$\boldsymbol{n}_0\cdot(\boldsymbol{B}_{0,\rm in} - \boldsymbol{B}_{0,\rm ex}) = 0, \tag{8.34}$$

$$\boldsymbol{n}_0\times(\boldsymbol{B}_{0,\rm in} - \boldsymbol{B}_{0,\rm ex}) = \mu_0\boldsymbol{K}, \tag{8.35}$$

where $\boldsymbol{K}$ is the surface current. Ohm's law yields

$$\boldsymbol{E}_{\rm in} + \boldsymbol{V}\times\boldsymbol{B}_{0,\rm in} = 0 \tag{8.36}$$

in the plasma. As the electric field $\boldsymbol{E}^*$ in coordinates moving with the plasma is $\boldsymbol{E}^* = \boldsymbol{E} + \boldsymbol{V}\times\boldsymbol{B}_0$ and the tangential component of the electric field $\boldsymbol{E}^*$ is continuous across the plasma boundary, the boundary condition can be written as

$$\boldsymbol{E}_{\rm t} + (\boldsymbol{V}\times\boldsymbol{B}_{0,\rm ex})_{\rm t} = 0, \tag{8.37}$$

where the subscript t indicates the tangential component. Since the normal component of $\boldsymbol{B}$ is given by the tangential component of $\boldsymbol{E}$ by the relation $\nabla\times\boldsymbol{E} = -\partial\boldsymbol{B}/\partial t$, (8.37) is reduced to

$$(\boldsymbol{n}_0\cdot\boldsymbol{B}_{1,\rm ex}) = \boldsymbol{n}_0\cdot\nabla\times(\boldsymbol{\xi}\times\boldsymbol{B}_{0,\rm ex}). \tag{8.38}$$

The electric field $\boldsymbol{E}_{\rm ex}$ and the magnetic field $\boldsymbol{B}_{\rm ex}$ in the external (vacuum) region can be expressed in terms of a vector potential:

$$\boldsymbol{E}_{\rm ex} = -\frac{\partial\boldsymbol{A}}{\partial t}, \qquad \boldsymbol{B}_{1,\rm ex} = \nabla\times\boldsymbol{A}, \qquad \nabla\cdot\boldsymbol{A} = 0.$$

If no current flows in the vacuum region, $\boldsymbol{A}$ satisfies

$$\nabla\times\nabla\times\boldsymbol{A} = 0. \tag{8.39}$$

Using the vector potential, we may express (8.37) as

$$\boldsymbol{n}_0\times\left(-\frac{\partial\boldsymbol{A}}{\partial t} + \boldsymbol{V}\times\boldsymbol{B}_{0,\rm ex}\right) = 0.$$

For $\boldsymbol{n}_0 \cdot \boldsymbol{B}_{0,\mathrm{in}} = \boldsymbol{n} \cdot \boldsymbol{B}_{0,\mathrm{ex}} = 0$, the boundary condition is

$$\boldsymbol{n}_0 \times \boldsymbol{A} = -\xi_n \boldsymbol{B}_{0,\mathrm{ex}}. \tag{8.40}$$

The boundary condition at the wall of an ideal conductor is

$$\boldsymbol{n} \times \boldsymbol{A} = 0. \tag{8.41}$$

The stability problem now becomes one of solving (8.32) and (8.39) under the boundary conditions (8.33), (8.38) or (8.40) and (8.41). When a normal mode $\boldsymbol{\xi}(\boldsymbol{r}, t) = \boldsymbol{\xi}(\boldsymbol{r}) \exp(-i\omega t)$ is considered, the problem is reduced to the eigenvalue problem $\rho_0 \omega^2 \boldsymbol{\xi} = -\boldsymbol{F}(\boldsymbol{\xi})$. If any eigenvalue is negative, the plasma is unstable; that is, if all the eigenvalues are positive, the plasma is stable.

8.2.2 Magnetoacoustic Wave

Let us consider the case where $\boldsymbol{B}_0 = \text{const.}$, $p_0 = \text{const.}$ and the displacement is expressed by $\boldsymbol{\xi}(\boldsymbol{r}, t) = \boldsymbol{\xi}_1 \exp i(\boldsymbol{k} \cdot \boldsymbol{r} - \omega t)$, then (8.32) is reduced to

$$-\rho_{\mathrm{m}0} \omega^2 \boldsymbol{\xi}_1 = -\gamma p_0 (\boldsymbol{k} \cdot \boldsymbol{\xi}_1) \boldsymbol{k} - \mu_0^{-1} (\boldsymbol{k} \times (\boldsymbol{k} \times (\boldsymbol{\xi}_1 \times \boldsymbol{B}_0))) \times \boldsymbol{B}_0. \tag{8.42}$$

Using the vector formula $\boldsymbol{a} \times (\boldsymbol{b} \times \boldsymbol{c})$ (refer to Table 3.1), we can write (8.42) as

$$\left((\boldsymbol{k} \cdot \boldsymbol{B}_0)^2 - \mu_0 \omega^2 \rho_{\mathrm{m}0}\right) \boldsymbol{\xi}_1 + \left((B_0^2 + \mu_0 \gamma p_0)\boldsymbol{k} - (\boldsymbol{k} \cdot \boldsymbol{B}_0)\boldsymbol{B}_0\right)(\boldsymbol{k} \cdot \boldsymbol{\xi}_1)$$
$$-(\boldsymbol{k} \cdot \boldsymbol{B}_0)(\boldsymbol{B}_0 \cdot \boldsymbol{\xi}_1)\boldsymbol{k} = 0.$$

If the unit vectors of $\boldsymbol{k}$, $\boldsymbol{B}_0$ are denoted by $\widehat{\boldsymbol{k}} \equiv \boldsymbol{k}/k$, $\boldsymbol{b} \equiv \boldsymbol{B}_0/B_0$, and the notations $V \equiv \omega/k$, $v_{\mathrm{A}}^2 \equiv B_0^2/(\mu_0 \rho_{\mathrm{m}0})$, $\beta \equiv p_0/(B_0^2/2\mu_0)$, $\cos\theta \equiv (\widehat{\boldsymbol{k}} \cdot \widehat{\boldsymbol{b}})$ are introduced, we find

$$\left(\cos^2\theta - \frac{V^2}{v_{\mathrm{A}}^2}\right) \boldsymbol{\xi}_1 + \left(\left(1 + \frac{\gamma\beta}{2}\right)\widehat{\boldsymbol{k}} - \cos\theta \boldsymbol{b}\right)(\widehat{\boldsymbol{k}} \cdot \boldsymbol{\xi}_1) - \cos\theta(\boldsymbol{b} \cdot \boldsymbol{\xi}_1)\widehat{\boldsymbol{k}} = 0. \tag{8.43}$$

The scalar product of (8.43) with $\widehat{\boldsymbol{k}}$ and $\boldsymbol{b}$, and the vector product of $\widehat{\boldsymbol{k}}$ with (8.43), yield

$$\left(1 + \frac{\gamma\beta}{2} - \frac{V^2}{v_{\mathrm{A}}^2}\right)(\widehat{\boldsymbol{k}} \cdot \boldsymbol{\xi}_1) - \cos\theta(\boldsymbol{b} \cdot \boldsymbol{\xi}_1) = 0,$$

$$\frac{\gamma\beta}{2} \cos\theta(\widehat{\boldsymbol{k}} \cdot \boldsymbol{\xi}_1) - \frac{V^2}{v_{\mathrm{A}}^2}(\boldsymbol{b} \cdot \boldsymbol{\xi}_1) = 0,$$

$$\left(\cos^2\theta - \frac{V^2}{v_{\mathrm{A}}^2}\right) \boldsymbol{b} \cdot (\widehat{\boldsymbol{k}} \times \boldsymbol{\xi}_1) = 0.$$

The solutions of these equations are magnetoacoustic waves. One solution is

$$V^2 = v_{\rm A}^2 \cos^2\theta, \qquad (\boldsymbol{\xi}_1 \cdot \boldsymbol{k}) = 0, \qquad (\boldsymbol{\xi}_1 \cdot \boldsymbol{B}_0) = 0.$$

Since $\boldsymbol{\xi}_1$ of this solution is orthogonal to $\boldsymbol{k}$ and $\boldsymbol{B}_0$, this is called torsional Alfvén wave (refer to Section 12.4.1). The other solutions are given by

$$\left(\frac{V}{v_{\rm A}}\right)^4 - (1 + \frac{\gamma\beta}{2})\left(\frac{V}{v_{\rm A}}\right)^2 + \frac{\gamma\beta}{2}\cos^2\theta = 0, \tag{8.44}$$

$$\boldsymbol{B}_0 \cdot (\boldsymbol{k} \times \boldsymbol{\xi}_1) = 0.$$

If the velocity of sound is denoted by $c_{\rm s}^2 = \gamma p_0/\rho_{\rm m0}$, (8.44) becomes

$$V^4 - (v_{\rm A}^2 + c_{\rm s}^2)V^2 + v_{\rm A}^2 c_{\rm s}^2 \cos^2\theta = 0$$

and

$$V_{\rm f}^2 = \frac{1}{2}\left(v_{\rm A}^2 + c_{\rm s}^2) + ((v_{\rm A}^2 + c_{\rm s}^2)^2 - 4v_{\rm A}^2 c_{\rm s}^2 \cos^2\theta)^{1/2}\right), \tag{8.45}$$

$$V_{\rm s}^2 = \frac{1}{2}\left(v_{\rm A}^2 + c_{\rm s}^2) - ((v_{\rm A}^2 + c_{\rm s}^2)^2 - 4v_{\rm A}^2 c_{\rm s}^2 \cos^2\theta)^{1/2}\right). \tag{8.46}$$

The solution of (8.45) is called compressional Alfvén wave (see Section 12.4.1) and the solution of (8.46) is called *magnetoacoustic slow wave.* Characteristic velocity

$$v_{\rm A}^2 = \frac{B^2}{\mu_0 \rho_{\rm m0}} \tag{8.47}$$

is called *Alfvén velocity.* The plasma with zero resistivity is frozen to the magnetic field. There is tension $B^2/2\mu_0$ along the magnetic field line. As the plasma, of mass density $\rho_{\rm m}$, sticks to the field lines, the magnetoacoustic waves can be considered as waves propagating along the strings of magnetic field lines (see Section 12.4.1).

8.2.3 Rayleigh-Taylor (Interchange) Instability

Let the z-axis be the direction of the magnetic field and the x-axis be the (anti-)direction of gravitation; that is, $\boldsymbol{g} = -g\hat{\boldsymbol{x}}$. The zeroth order of the magnetic field $\boldsymbol{B}_0$, current $\boldsymbol{j}_0$, density ρ_0, and pressure p_0 are given as follows:

$$\boldsymbol{B}_0 = (0, 0, B_0(x)), \quad \mu_0 \boldsymbol{j}_0 = -\frac{{\rm d}B_0}{{\rm d}x}\hat{\boldsymbol{y}}, \quad \rho_0(x), \quad p_0(x).$$

The zeroth order equation is

$$\frac{\rm d}{{\rm d}x}\left(p_0 + \frac{B_0^2}{2\mu_0}\right) = 0.$$

We assume that the first order perturbations are in the form of

$$f_1(x, y, z) = f_1(x) \exp(ik_y y + ik_z z + \gamma t). \quad (8.48)$$

The equation of motion (8.32) with gravitation term is

$$\gamma \rho_0 \boldsymbol{v} == -\nabla \left(p_1 + \frac{\boldsymbol{B}_0 \cdot \boldsymbol{B}_1}{\mu_0} \right) + \frac{1}{\mu_0}((\boldsymbol{B}_0 \cdot \nabla)\boldsymbol{B}_1 + (\boldsymbol{B}_1 \cdot \nabla)\boldsymbol{B}_0) - \rho_1 g \hat{\boldsymbol{x}}. \quad (8.49)$$

(8.30) reduces

$$\gamma \rho_1 + v_x \frac{\partial \rho_0}{\partial x} = 0, \quad (8.50)$$

and incompressible condition is

$$\frac{\partial v_x}{\partial x} + ik_y v_y + ik_z v_z = 0. \quad (8.51)$$

The components of $\boldsymbol{B}_1$ are

$$\boldsymbol{B}_1 = (ik_z v_x B_0, \quad ik_z v_y B_0, \quad ik_y v_y B_0 - \frac{\partial v_x B_0}{\partial x} B_0).$$

Then, each componet of (8.49) is

$$\gamma \rho_0 v_x = -\frac{\partial}{\partial x} p_1 + \frac{1}{\gamma \mu_0} \frac{\partial}{\partial x} \left(ik_y v_y B_0^2 + \frac{\partial}{\partial x}(v_x B_0) B_0 \right) - \frac{1}{\gamma \mu_0} k_z^2 B_0^2 v_x - \rho_1 g, \quad (8.52)$$

$$\gamma \rho_0 v_y = -ik_y p_1 - \frac{1}{\gamma \mu_0} k_y^2 B_0^2 v_y + ik_y \frac{\partial}{\partial x}(v_x B_0) B_0 - k_z^2 \frac{B_0^2}{\gamma \mu_0} v_y, \quad (8.53)$$

$$\gamma \rho_0 v_z = -ik_z p_1 + ik_z \frac{1}{\gamma \mu_0} B_0 \frac{\partial}{\partial x} B_0 v_x. \quad (8.54)$$

(8.53) $\times\, ik_y$ + (8.54) $\times\, ik_z$ gives p_1 as follows ($k^2 = k_y^2 + k_z^2$)

$$p_1 = \frac{\gamma \rho_0}{k^2} \frac{\partial v_x}{\partial x} + \frac{k_y^2}{k^2} \frac{B_0^2}{\gamma \mu_0} \frac{\partial v_x}{\partial x} + \frac{1}{\gamma \mu_0} B_0 \frac{\partial B_0}{\partial x} v_x + ik_y \frac{B_0^2}{\gamma \mu_0} v_y. \quad (8.55)$$

Substitution of (8.55) into (8.52) reduces [8.1]

$$\frac{\partial}{\partial x} \left(\left(\rho_0 \gamma^2 + \frac{(k_z B_0)^2}{\mu_0} \right) \frac{\partial v_x}{\partial x} \right) - k^2 \left(\rho_0 \gamma^2 + \frac{(k_z B_0)^2}{\mu_0} - g \frac{\partial \rho_0}{\partial x} \right) v_x = 0. \quad (8.56)$$

Variational form of (8.56) is

$$\gamma^2 = \frac{\int \left(g \frac{\partial \rho_0}{\partial x} v_x^2 - \frac{(\boldsymbol{k} \cdot \boldsymbol{B}_0)^2}{\mu_0} \left(v_x^2 + k^{-2} \left(\frac{\partial v_x}{\partial x} \right)^2 \right) \right) \mathrm{d}x}{\int \rho_0 \left(v_x^2 + k^{-2} \left(\frac{\partial v_x}{\partial x} \right)^2 \right) \mathrm{d}x}. \quad (8.57)$$

This integral has the advantage to estimate γ^2 by use of a proper test function of $v_x(x)$. Perturbation with wave number perpendicular to the magnetic field is most dangerous as is seen from (8.57). Let us consider the case of $\boldsymbol{k} \cdot \boldsymbol{B} = 0$ and study the following example of $\rho_0(x)$; that is,

$$\rho_0(x) = \rho_0 \quad (x > d), \qquad \rho_0(x) = \rho_0 \frac{x}{d} \quad (d > x > 0), \qquad \rho_0(x) = 0 \quad (0 > x).$$

When the width of transition layer is smaller than the wavelength of the perturvation in y direction $kd \ll 1$, a trial function $v_x \propto \exp(-|x|/\lambda)$ $(\lambda \gg d)$ gives

$$\gamma^2 \approx \frac{2gk}{k\lambda + (k\lambda)^{-1}}. \tag{8.58}$$

The maximum value is $\gamma^2 = gk$ when $\lambda = k^{-1}$. This is a natural result since $v_x \propto \exp(-k|x|)\exp(iky)$, $k_z = 0$ can satisfy (8.51) and is the same result as (8.9).

8.3 Cylindrical Plasma with Sharp-Boundary Configuration

Let us consider a sharp-boundary plasma of radius a, with a longitudinal magnetic field B_{0z} inside the boundary and a longitudinal magnetic field B_{ez} and an azimuthal magnetic field $B_\theta = \mu_0 I/(2\pi r)$ outside. B_{0z} and B_{ez} are assumed to be constant (refer to Figure 8.5). We can consider the displacement

$$\boldsymbol{\xi}(r)\exp(im\theta + ikz). \tag{8.59}$$

Any displacement may be expressed by a superposition of such modes. Since the term in $\nabla \cdot \boldsymbol{\xi}$ in the energy integral is positive, incompressible perturbation is the most dangerous. We examine only the worst mode,

$$\nabla \cdot \boldsymbol{\xi} = 0. \tag{8.60}$$

The perturbation of the magnetic field $\boldsymbol{B}_1 = \nabla \times (\boldsymbol{\xi} \times \boldsymbol{B}_0)$ inside the plasma is

$$\boldsymbol{B}_1 = ikB_{0z}\boldsymbol{\xi}. \tag{8.61}$$

The equation of motion (8.32) becomes

$$\left(-\omega^2 \rho_{m0} + \frac{k^2 B_{0z}^2}{\mu_0}\right)\boldsymbol{\xi} = -\nabla\left(p_1 + \frac{\boldsymbol{B}_0 \cdot \boldsymbol{B}_1}{\mu_0}\right) \equiv -\nabla p^*. \tag{8.62}$$

As $\nabla \cdot \boldsymbol{\xi} = 0$, it follows that $\Delta p^* = 0$, i.e.,

$$\left(\frac{\mathrm{d}^2}{\mathrm{d}r^2} + \frac{1}{r}\frac{\mathrm{d}}{\mathrm{d}r} - \left(k^2 + \frac{m^2}{r^2}\right)\right)p^*(r) = 0. \tag{8.63}$$

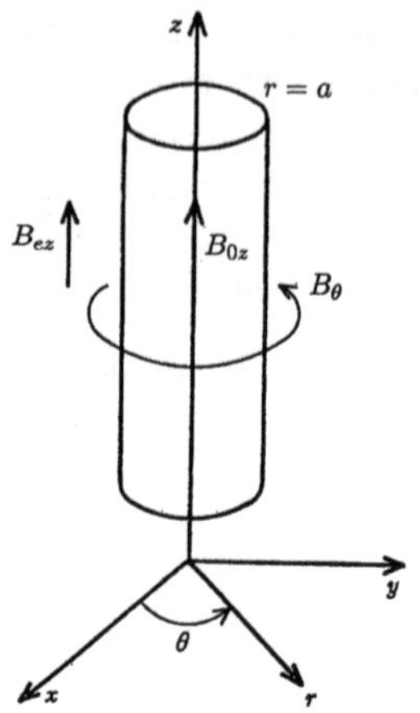

Figure 8.5 Sharp-boundary plasma.

The solution without singularity at $r = 0$ is given by the modified Bessel function $I_m(kr)$ of the first kind, so that $p^*(r)$ is

$$p^*(r) = p^*(a)\frac{I_m(kr)}{I_m(ka)}.$$

Accordingly, we find

$$\xi_r(a) = \frac{\dfrac{kp^*(a)}{I_m(ka)}}{\omega^2\rho_{\mathrm{m}0} - \dfrac{k^2B_0^2}{\mu_0}} I'_m(ka). \tag{8.64}$$

As the perturbation of the vacuum magnetic field $\boldsymbol{B}_{1\mathrm{e}}$ satisfies $\nabla \times \boldsymbol{B} = 0$ and $\nabla \cdot \boldsymbol{B} = 0$, $\boldsymbol{B}_{1\mathrm{e}}$ is expressed by $\boldsymbol{B}_{1\mathrm{e}} = \nabla\psi$. The scalar magnetic potential ψ satisfies $\Delta\psi = 0$ and $\psi \to 0$ as $r \to \infty$. Then

$$\psi = C\frac{K_m(kr)}{K_m(ka)}\exp(im\theta + ikz), \tag{8.65}$$

where $K_{\mathrm{m}}(kr)$ is the modified Bessel function of the second kind. From the boundary condition (8.33), we have

$$p^*(r) = p_1 + \frac{1}{\mu_0}\boldsymbol{B}_0 \cdot \boldsymbol{B}_1 = \frac{1}{\mu_0}\boldsymbol{B}_{\mathrm{e}} \cdot \boldsymbol{B}_{1\mathrm{e}} + (\boldsymbol{\xi}\cdot\nabla)\left(\frac{B_{\mathrm{e}}^2}{2\mu_0} - \frac{B_0^2}{2\mu_0} - p_0\right)$$
$$= \frac{1}{\mu_0}\boldsymbol{B}_{\mathrm{e}} \cdot \boldsymbol{B}_{1\mathrm{e}} + (\boldsymbol{\xi}\cdot\nabla)\left(\frac{B_\theta^2}{2\mu_0}\right).$$

Since $B_\theta \propto 1/r$ and $\boldsymbol{B}_{1\mathrm{e}} = \nabla\psi$ with (8.65), $p^*(a)$ is given by

$$p^*(a) = \frac{i}{\mu_0}(kB_{ez} + \frac{m}{a}B_\theta)C - \frac{B_\theta^2}{\mu_0 a}\xi_r(a). \tag{8.66}$$

Due to $B_{1r} = \partial\psi/\partial r$, the boundary condition (8.38) is reduced to

$$Ck\frac{K_m'(ka)}{K_m(ka)} = i(kB_{ez} + \frac{m}{a}B_\theta)\xi_r(a). \tag{8.67}$$

From Equations (8.56), (8.58), and (8.59), the dispersion equation is

$$\frac{\omega^2}{k^2} = \frac{B_{0z}^2}{\mu_0\rho_{m0}} - \frac{(kB_{ez} + (m/a)B_\theta)^2}{\mu_0\rho_{m0}k^2}\frac{I_m'(ka)}{I_m(ka)}\frac{K_m(ka)}{K_m'(ka)} - \frac{B_\theta^2}{\mu_0\rho_{m0}}\frac{1}{(ka)}\frac{I_m'(ka)}{I_m(ka)}. \tag{8.68}$$

The first and second terms represent the stabilizing effect of B_{0z} and B_{ez} ($K_m/K_m' < 0$). If the propagation vector $\boldsymbol{k}$ is normal to the magnetic field, i.e., if

$$(\boldsymbol{k}\cdot\boldsymbol{B}_e) = kB_{ez} + \frac{m}{a}B_\theta = 0,$$

the second stabilizing term of (8.68) becomes zero, so that a flutelike perturbation is dangerous. The third term is the destabilizing term.

(i) The $m = 0$ mode with $B_{ez} = 0$ @ Let us consider the $m = 0$ mode with $B_{ez} = 0$. This azimuthally symmetric perturbation constricts the plasma like a sausage. This mode is called *sausage instability.* Equation (8.60) reduces to

$$\omega^2 = \frac{B_{0z}^2 k^2}{\mu_0\rho_{m0}}\left(1 - \frac{B_\theta^2}{B_{0z}^2}\frac{I_0'(ka)}{(ka)I_0(ka)}\right). \tag{8.69}$$

Since $I_0'(x)/xI_0(x) < 1/2$, the stability condition is

$$B_{0z}^2 > B_\theta^2/2.$$

(ii) The $m = 1$ mode with $B_{ez} = 0$ The $m = 1$ mode perpurbation kinks the plasma column so that it is called the *kink mode.* For the $m = 1$ mode with $B_{ez} = 0$C(8.60) is

$$\omega^2 = \frac{B_{0z}^2 k^2}{\mu_0\rho_{m0}}\left(1 + \frac{B_\theta^2}{B_{0z}^2}\frac{1}{(ka)}\frac{I_1'(ka)}{I_1(ka)}\frac{K_1(ka)}{K_1'(ka)}\right). \tag{8.70}$$

For perturbations with long characteristic length, (8.62) becomes

$$\omega^2 = \frac{B_{0z}^2 k^2}{\mu_0\rho_{m0}}\left(1 - \left(\frac{B_\theta}{B_{0z}}\right)^2 \ln\frac{1}{ka}\right). \tag{8.71}$$

This dispersion equation shows that the kink mode is unstable for the perturbation with long wavelength.

(iii) Instability in the case of $|B_{ez}| > |B_\theta|$ When $|B_{ez}| \gg |B_\theta|$, the term including $|ka| \ll 1$ predominates. Expanding the modified Bessel function ($m > 0$ is assumed), we find

$$\mu_0 \rho_{m0} \omega^2 = k^2 B_{0z}^2 + \left(k B_{ez} + \frac{m}{a} B_\theta\right)^2 - \frac{m}{a^2} B_\theta^2. \tag{8.72}$$

ω^2 becomes minimum at $\partial\omega/\partial k = 0$, i.e., $k(B_{0z}^2 + B_{ez}^2) + (m/a)B_\theta B_{ez} = 0$. In this case, ω^2 is the minimum

$$\omega_{\min}^2 = \frac{B_\theta^2}{\mu_0 \rho_{m0} a^2}\left(\frac{m^2 B_{0z}^2}{B_{ez}^2 + B_{0z}^2} - m\right) = \frac{B_\theta^2}{\mu_0 \rho_{m0} a^2} m \left(m \frac{1-\beta}{2-\beta} - 1\right), \tag{8.73}$$

where β is the beta ratio. Accordingly, the plasma is unstable when $0 < m < (2-\beta)/(1-\beta)$. For a low-beta plasma only the modes $m = 1$ and $m = 2$ become unstable. However, if

$$\left(\frac{B_\theta}{B_z}\right)^2 < (ka)^2 \tag{8.74}$$

is satisfied, the plasma is stable even for $m = 1$. Usually the length of the plasma is finite so that k cannot be smaller than $2\pi/L$. Accordingly, when

$$\left|\frac{B_\theta}{B_z}\right| < \frac{2\pi a}{L}, \tag{8.75}$$

the plasma is stable. This stability condition is called the *Kruskal-Shafranov condition* [8.5, 8.6].

When a cylindrical conducting wall of radius b surrounds the plasma, the scalar magnetic potential of the external magnetic field is

$$\psi = \left(c_1 \frac{K_m(kr)}{K_m(ka)} + c_2 \frac{I_m(kr)}{I_m(ka)}\right) \exp(im\theta + ikz)$$

instead of (8.65). The boundary condition $B_{1er} = 0$ at $r = b$ yields

$$\frac{c_1}{c_2} = -\frac{I_m'(kb) K_m(ka)}{K_m'(kb) I_m(ka)}.$$

The dispersion equation becomes

$$\frac{\omega^2}{k^2} = \frac{B_{0z}^2}{\mu_0 \rho_{m0}} - \frac{(kB_{ez} + (m/a)B_\theta)^2}{\mu_0 \rho_{m0} k^2} \frac{I_m'(ka)}{I_m(ka)} \left(\frac{K_m(ka) I_m'(kb) - I_m(ka) K_m'(kb)}{K_m'(ka) I_m'(kb) - I_m'(ka) K_m'(kb)}\right)$$
$$- \frac{B_\theta^2}{\mu_0 \rho_{m0}} \frac{1}{(ka)} \frac{I_m'(ka)}{I_m(ka)}.$$

Expanding the modified Bessel functions under the conditions $ka \ll 1, kb \ll 1$, we find

$$\mu_0\rho_{m0}\omega^2 = k^2B_{0z}^2 + \frac{1+(a/b)^{2m}}{1-(a/b)^{2m}}(kB_{ez} + \frac{m}{a}B_\theta)^2 - \frac{m}{a^2}B_\theta^2.$$

The closer the wall is to the plasma boundary, the more effective is the wall stabilization.

In toroidal systems, the propagation constant is $k = n/R$ where n is an integer and R is the major radius of the torus. If the *safety factor* q_a at the plasma boundary $r = a$

$$q_a = \frac{aB_{ez}}{RB_\theta} \tag{8.76}$$

is introduced, $(\boldsymbol{k}\cdot\boldsymbol{B})$ may be written as

$$(\boldsymbol{k}\cdot\boldsymbol{B}) = (kB_{ez} + \frac{m}{a}B_\theta) = \frac{nB_\theta}{a}\left(q_a + \frac{m}{n}\right).$$

The Kruskal-Shafranov condition (8.66) of $m = 1,\ n = -1$ mode can then be expressed in terms of the safety factor as

$$q_a > 1. \tag{8.77}$$

This is the reason why q_a is called the safety factor.

8.4 Energy Principle

8.4.1 Energy Integral in General

The eigenvalue problem is complicated and difficult to solve in general. When we introduce a potential energy associated with the displacement $\boldsymbol{\xi}$, the stability problem can be simplified. The equation of motion has the form

$$\rho_{m0}\frac{\partial^2\boldsymbol{\xi}}{\partial t^2} = \boldsymbol{F}(\boldsymbol{\xi}) = -\widehat{\boldsymbol{K}}\cdot\boldsymbol{\xi}. \tag{8.78}$$

where $\widehat{\boldsymbol{K}}$ is a linear operator. When this equation is integrated, the equation of energy conservation is reduced:

$$\frac{1}{2}\int\rho_{m0}\left(\frac{\partial\boldsymbol{\xi}}{\partial t}\right)^2 \mathrm{d}\boldsymbol{r} + \frac{1}{2}\int\boldsymbol{\xi}\cdot\widehat{\boldsymbol{K}}\boldsymbol{\xi}\,\mathrm{d}\boldsymbol{r} = \mathrm{const.}$$

The kinetic energy T and the potential energy W are

$$T \equiv \frac{1}{2}\int\rho_{m0}\left(\frac{\partial\boldsymbol{\xi}}{\partial t}\right)^2 \mathrm{d}\boldsymbol{r}, \qquad W \equiv \frac{1}{2}\int\boldsymbol{\xi}\cdot\widehat{\boldsymbol{K}}\boldsymbol{\xi}\mathrm{d}\boldsymbol{r} = -\frac{1}{2}\int\boldsymbol{\xi}\cdot\boldsymbol{F}(\boldsymbol{\xi})\mathrm{d}\boldsymbol{r},$$

respectively. Accordingly if

$$W > 0$$

for all possible displacements, the system is stable. This is the stability criterion of the *energy principle* [8.7]. W is called the *energy integral.*

It is possible to prove that the operator $\widehat{\boldsymbol{K}}$ is a Hermite operator (self-adjoint operator) [8.8], Chapter 9 of [2.3]. A displacement $\boldsymbol{\eta}$ and a vector potential $\boldsymbol{Q}$ are introduced which satisfy the same boundary conditions as $\boldsymbol{\xi}$ and $\boldsymbol{A}$, i.e.,

$$\boldsymbol{n}_0 \times \boldsymbol{Q} = -\eta_n \boldsymbol{B}_{0,\mathrm{ex}}$$

at the plasma-vacuum boundary and

$$\boldsymbol{n}_0 \times \boldsymbol{Q} = 0$$

at the conducting wall. By substitution of (8.32), the following integral in the plasma region V_{in} is seen to be $(\nabla \cdot (\boldsymbol{B}_1 \times (\boldsymbol{\eta} \times \boldsymbol{B}_0)) = (\boldsymbol{\eta} \times \boldsymbol{B}_0) \cdot \nabla \times \boldsymbol{B}_1 - \boldsymbol{B}_1 \cdot \nabla \times (\boldsymbol{\eta} \times \boldsymbol{B}_0))$

$$\int_{V_{\mathrm{in}}} \boldsymbol{\eta} \cdot \widehat{\boldsymbol{K}} \boldsymbol{\xi} \, \mathrm{d}\boldsymbol{r}$$

$$= \int_{V_{\mathrm{in}}} \left(\gamma p_0 (\nabla \cdot \boldsymbol{\eta})(\nabla \cdot \boldsymbol{\xi}) + (\nabla \cdot \boldsymbol{\eta})(\boldsymbol{\xi} \cdot \nabla p_0) + \frac{1}{\mu_0} (\nabla \times (\boldsymbol{\eta} \times \boldsymbol{B}_0)) \cdot \nabla \times (\boldsymbol{\xi} \times \boldsymbol{B}_0) \right.$$

$$\left. - \frac{1}{\mu_0} (\boldsymbol{\eta} \times (\nabla \times \boldsymbol{B}_0)) \cdot \nabla \times (\boldsymbol{\xi} \times \boldsymbol{B}_0) \right) \mathrm{d}\boldsymbol{r}$$

$$+ \int_S \boldsymbol{n}_0 \cdot \boldsymbol{\eta} \left(\frac{\boldsymbol{B}_{0,\mathrm{in}} \cdot \nabla \times (\boldsymbol{\xi} \times \boldsymbol{B}_{0,\mathrm{in}})}{\mu_0} - \gamma p_0 (\nabla \cdot \boldsymbol{\xi}) - (\boldsymbol{\xi} \cdot \nabla p_0) \right) \mathrm{d}S. \qquad (8.79)$$

Next let us consider the surface integral in (8.79). Due to the boundary condition $\boldsymbol{n}_0 \times \boldsymbol{Q} = -\eta_n \boldsymbol{B}_{0,\mathrm{ex}}$, we find that

$$\int_S \eta_n \boldsymbol{B}_{0,\mathrm{ex}} \cdot \boldsymbol{B}_{1,\mathrm{ex}} \mathrm{d}S = \int_S \eta_n \boldsymbol{B}_{0,\mathrm{ex}} (\nabla \times \boldsymbol{A}) \, \mathrm{d}S = - \int_S (\boldsymbol{n}_0 \times \boldsymbol{Q}) \cdot (\nabla \times \boldsymbol{A}) \, \mathrm{d}S$$

$$= - \int_S \boldsymbol{n}_0 \cdot (\boldsymbol{Q} \times (\nabla \times \boldsymbol{A})) \, \mathrm{d}S = \int_{V_{\mathrm{ex}}} \nabla \cdot (\boldsymbol{Q} \times (\nabla \times \boldsymbol{A})) \, \mathrm{d}\boldsymbol{r}$$

$$= \int_{V_{\mathrm{ex}}} ((\nabla \times \boldsymbol{Q}) \cdot (\nabla \times \boldsymbol{A}) - \boldsymbol{Q} \cdot \nabla \times (\nabla \times \boldsymbol{A})) \, \mathrm{d}\boldsymbol{r}$$

$$= \int_{V_{\mathrm{ex}}} (\nabla \times \boldsymbol{Q}) \cdot (\nabla \times \boldsymbol{A}) \, \mathrm{d}\boldsymbol{r}.$$

From the boundary condition (8.33), the difference between the foregoing surface integral and the surface integral in (8.43) is reduced to

$$\int \eta_n \left(\frac{\boldsymbol{B}_{0,\mathrm{in}} \cdot \boldsymbol{B}_{1,\mathrm{in}} - \boldsymbol{B}_{0,\mathrm{ex}} \cdot \boldsymbol{B}_{1,\mathrm{ex}}}{\mu_0} - \gamma p_0 (\nabla \cdot \boldsymbol{\xi}) - (\boldsymbol{\xi} \cdot \nabla) p_0 \right) \mathrm{d}S$$

$$= \int_S \eta_n (\boldsymbol{\xi} \cdot \nabla) \left(\frac{B_{0,\mathrm{ex}}^2}{2\mu_0} - \frac{B_{0,\mathrm{in}}^2}{2\mu_0} - p_0 \right) \mathrm{d}S$$

$$= \int_S \eta_n \xi_n \frac{\partial}{\partial n} \left(\frac{B_{0,\mathrm{ex}}^2}{2\mu_0} - \frac{B_{0,\mathrm{in}}^2}{2\mu_0} - p_0 \right) \mathrm{d}S$$

where the relation $\boldsymbol{n}_0 \times \nabla (p_0 + B_{0,\mathrm{in}}^2/2\mu_0 - B_{0,\mathrm{ex}}^2/2\mu_0) = 0$ is used. The integral region V_{ex} is the region outside the plasma. Finally, the energy integral is reduced to

$$\int_{V_{\mathrm{in}}} \boldsymbol{\eta} \cdot \widehat{\boldsymbol{K}} \boldsymbol{\xi} \, \mathrm{d}\boldsymbol{r} = \int_{V_{\mathrm{in}}} \Bigg(\gamma p_0 (\nabla \cdot \boldsymbol{\eta})(\nabla \cdot \boldsymbol{\xi}) + \frac{1}{\mu_0} (\nabla \times (\boldsymbol{\eta} \times \boldsymbol{B}_0)) \cdot (\nabla \times (\boldsymbol{\xi} \times \boldsymbol{B}_0))$$

$$+ (\nabla \cdot \boldsymbol{\eta})(\boldsymbol{\xi} \cdot \nabla p_0) - \frac{1}{\mu_0} (\boldsymbol{\eta} \times (\nabla \times \boldsymbol{B}_0)) \cdot \nabla \times (\boldsymbol{\xi} \times \boldsymbol{B}_0) \Bigg) \mathrm{d}\boldsymbol{r} + \frac{1}{\mu_0} \int_{V_{\mathrm{ex}}} (\nabla \times \boldsymbol{Q}) \cdot (\nabla \times \boldsymbol{A}) \, \mathrm{d}\boldsymbol{r}$$

$$+ \int_S \eta_n \xi_n \frac{\partial}{\partial n} \left(\frac{B_{0,\mathrm{ex}}^2}{2\mu_0} - \frac{B_{0,\mathrm{in}}^2}{2\mu_0} - p_0 \right) \mathrm{d}S. \tag{8.80}$$

The energy integral W is divided into three parts W_{P}, W_{S}, and W_{V}, the contributions of the plasma internal region V_{in}, the boundary region SC and the external vacuum region V_{ex}, i.e.,

$$W = \frac{1}{2} \int_{V_{\mathrm{in}}} \boldsymbol{\xi} \cdot \widehat{\boldsymbol{K}} \boldsymbol{\xi} \, \mathrm{d}\boldsymbol{r} = W_{\mathrm{p}} + W_{\mathrm{S}} + W_{\mathrm{V}}, \tag{8.81}$$

$$W_{\mathrm{p}} = \frac{1}{2} \int_{V_{\mathrm{in}}} \Bigg(\gamma p_0 (\nabla \cdot \boldsymbol{\xi})^2 + \frac{1}{\mu_0} (\nabla \times (\boldsymbol{\xi} \times \boldsymbol{B}_0))^2 + (\nabla \cdot \boldsymbol{\xi})(\boldsymbol{\xi} \cdot \nabla p_0)$$

$$- \frac{1}{\mu_0} (\boldsymbol{\xi} \times (\nabla \times \boldsymbol{B}_0)) \cdot \nabla \times (\boldsymbol{\xi} \times \boldsymbol{B}_0) \Bigg) \mathrm{d}\boldsymbol{r}$$

$$= \frac{1}{2} \int_{V_{\mathrm{in}}} \left(\frac{\boldsymbol{B}_1^2}{\mu_0} - p_1 (\nabla \cdot \boldsymbol{\xi}) - \boldsymbol{\xi} \cdot (\boldsymbol{j}_0 \times \boldsymbol{B}_1) \right) \mathrm{d}\boldsymbol{r}, \tag{8.82}$$

$$W_{\mathrm{S}} = \frac{1}{2} \int_S \xi_n^2 \frac{\partial}{\partial n} \left(\frac{B_{0,\mathrm{ex}}^2}{2\mu_0} - \frac{B_{0,\mathrm{in}}^2}{2\mu_0} - p_0 \right) \mathrm{d}S, \tag{8.83}$$

$$W_{\mathrm{V}} = \frac{1}{2\mu_0} \int_{V_{\mathrm{ex}}} (\nabla \times \boldsymbol{A})^2 \mathrm{d}\boldsymbol{r} = \int_{V_{\mathrm{ex}}} \frac{\boldsymbol{B}_1^2}{2\mu_0} \mathrm{d}\boldsymbol{r}. \tag{8.84}$$

The stability condition is $W > 0$ for all possible $\boldsymbol{\xi}$. The frequency or growth rate of a perturbation can be obtained from the energy integral. When the perturbation varies as $\exp(-i\omega t)$, the equation of motion is

$$\widehat{\boldsymbol{K}}\boldsymbol{\xi} - \omega^2 \rho_{\mathrm{m0}} \boldsymbol{\xi} = 0. \tag{8.85}$$

The solution of the eigenvalue problem is the same as the solution based on the calculus of variations

$$\delta L = 0 \qquad L = \int \boldsymbol{\xi} \cdot \widehat{\boldsymbol{K}}\boldsymbol{\xi} \mathrm{d}\boldsymbol{r} - \omega^2 \int \rho_{\mathrm{m0}} \boldsymbol{\xi}^2 \mathrm{d}\boldsymbol{r}, \tag{8.86}$$

(refer to (4.48) and Figure 4.13).

As $\widehat{\boldsymbol{K}}$ is a Hermitian operator, ω^2 is real; that is

$$\omega^2 = \frac{\int \boldsymbol{\xi} \cdot \widehat{\boldsymbol{K}}\boldsymbol{\xi} \mathrm{d}\boldsymbol{r}}{\int \rho_{\mathrm{m0}} \boldsymbol{\xi}^2 \mathrm{d}\boldsymbol{r}}.$$

In the MHD analysis of an ideal plasma with zero resistivity, the perturbation either increases or decreases monotonically, or else the perturbed plasma oscillates with constant amplitude.

8.4.2 Energy Integral in Illuminating Form

The energy integral (8.82) derived in Section 8.4.1

$$W = \frac{1}{2} \int_V \left(\frac{\boldsymbol{B}_1^2}{\mu_0} + \gamma p (\nabla \cdot \boldsymbol{\xi})^2 + (\nabla \cdot \boldsymbol{\xi})(\boldsymbol{\xi} \cdot \nabla p) - \boldsymbol{\xi} \cdot (\boldsymbol{j} \times \boldsymbol{B}_1) \right) \mathrm{d}\boldsymbol{r}$$

can be further rearranged to the more illuminating form [8.9] of

$$W = \frac{1}{2} \int \left(\gamma p (\nabla \cdot \boldsymbol{\xi})^2 + \frac{1}{\mu_0} |\boldsymbol{B}_{1\perp}|^2 + \frac{1}{\mu_0} \left| \boldsymbol{B}_{1\parallel} - \boldsymbol{B} \frac{\mu_0 (\boldsymbol{\xi} \cdot \nabla p)}{B^2} \right|^2 \right.$$
$$\left. - \frac{(\boldsymbol{j} \cdot \boldsymbol{B})}{B^2} (\boldsymbol{\xi} \times \boldsymbol{B}) \cdot \boldsymbol{B}_1 - 2(\boldsymbol{\xi} \cdot \nabla p)(\boldsymbol{\xi} \cdot \boldsymbol{\kappa}) \right) \mathrm{d}\boldsymbol{r}. \tag{8.87}$$

The first term of the integrand of (8.98) is the term of sonic wave. The second and the third terms are of Alfv́en wave. The fourth term is of kink mode. The last one is the term of ballooning mode. $\boldsymbol{\kappa}$ is the vector of curvature of field line. The derivation of (8.87) is described in detail in Section 8.5.1 of [8.10].

8.4.3 Energy Integral of Axisymmetric Toroidal System

In any axisymmetric toroidal system, the energy integral may be reduced to a more convenient form. The axisymmetric magnetic field is expressed as

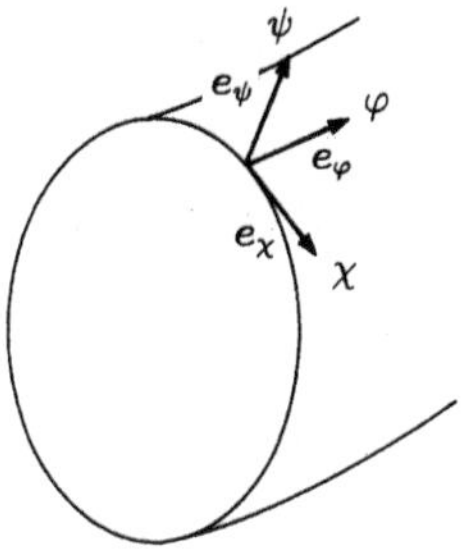

Figure 8.6 Orthogonal coordinate system (ψ, χ, φ) and $\boldsymbol{e}_\psi, \boldsymbol{e}_\chi, \boldsymbol{e}_\varphi$ are unit vectors of ψ, χ, φ directions, respectively.

$$\boldsymbol{B} = \frac{\hat{I}(\psi)}{R}\boldsymbol{e}_\varphi + B_\chi \boldsymbol{e}_\chi, \qquad \hat{I}(\psi) \equiv \frac{\mu_0 I(\psi)}{2\pi}, \tag{8.88}$$

where φ is the angle around the axis of torus and ψ is the flux function defined by

$$\psi = -RA_\varphi. \tag{8.89}$$

R is the distance from the axis of symmetry and A_φ is the φ component of the vector potential of the magnetic field. B_χ is the poloidal component of the magnetic field. $\boldsymbol{e}_\varphi$ and $\boldsymbol{e}_\chi$ are the unit vectors with the directions of toroidal angle and poloidal angle, respectively (see Figure 8.6). The R and Z components of the magnetic field are given by

$$RB_R = \frac{\partial\psi}{\partial Z}, \qquad RB_Z = -\frac{\partial\psi}{\partial R}.$$

Since $\nabla\varphi = (1/R)\boldsymbol{e}_\varphi$, the poloidal component of the magnetic field is expressed by $\boldsymbol{B}_\mathrm{p} = -\nabla\psi \times \nabla\varphi$. Therefore, we have

$$\boldsymbol{B} = -\nabla\psi \times \nabla\varphi + \hat{I}(\psi)\nabla\varphi. \tag{8.90}$$

We can introduce an orthogonal coordinate system (ψ, χ, φ), where $\psi =$ const. are the magnetic surfaces and χ, φ are poloidal and toroidal angles, respectively. The metric for these coordinates is

$$\mathrm{d}s^2 = \left(\frac{\mathrm{d}\psi}{RB_\chi}\right)^2 + (JB_\chi \mathrm{d}\chi)^2 + (R\mathrm{d}\varphi)^2,$$

where the volume element is $\mathrm{d}V = J(\psi)\mathrm{d}\psi\mathrm{d}\chi\mathrm{d}\varphi$. A field line is defined by $\psi=$const. and by

$$\frac{R\mathrm{d}\varphi}{JB_\chi \mathrm{d}\chi} = \frac{B_\varphi}{B_\chi} = \frac{\hat{I}(\psi)}{RB_\chi}.$$

Then the toroidal safety factor is given by

$$q(\psi) = \frac{1}{2\pi}\oint \frac{\mathrm{d}\varphi}{\mathrm{d}\chi}\mathrm{d}\chi = \frac{1}{2\pi}\oint \frac{J(\psi)\hat{I}(\psi)}{R^2}\mathrm{d}\chi.$$

Introducing

$$X \equiv RB_\chi \xi_\psi, \quad Y \equiv \xi_\chi/B_\chi, \quad U \equiv (B_\chi\xi_\varphi - B_\varphi\xi_\chi)/(RB_\chi),$$

the energy integral of axisymmetric toroidal system is given by [8.11].

$$W = \frac{1}{2}\int_V \left(\frac{|\boldsymbol{B}_1|^2}{\mu_0} + \gamma p|(\nabla\cdot\boldsymbol{\xi})|^2 + (\nabla\cdot\boldsymbol{\xi}^*)(\boldsymbol{\xi}\cdot\nabla p) - \boldsymbol{\xi}^*\cdot(\boldsymbol{j}\times\boldsymbol{B}_1)\right)\mathrm{d}\boldsymbol{r}$$

$$= \int_V \left(\frac{1}{2\mu_0}\frac{B^2k_\parallel^2}{B_\chi^2R^2}|X|^2 + \frac{1}{2\mu_0}\frac{R^2}{J^2}\left|\frac{\partial U}{\partial\chi} - I\left(\frac{JX}{R^2}\right)'\right|^2\right.$$

$$+\frac{B_\chi^2}{2\mu_0}\left|inU + X' - \frac{\mu_0 j_\varphi}{RB_\chi^2}X\right|^2 + \frac{1}{2}\gamma p\left|\frac{1}{J}(JX)' + iBk_\parallel Y + inU\right|^2$$

$$\left. -KXX^*\right)J\mathrm{d}\psi\mathrm{d}\chi\mathrm{d}\varphi. \tag{8.91}$$

This energy integral is important to analyze the ballooning instability with use of ballooning representation. Equation (8.91) is applied to analysis of ballooning mode by use of ballooning representation of $X(\psi,\chi)$ [8.11]. The derivation of (8.91) is described in detail in Section 8.5.2 of [8.10]. The notations in (8.91) are explained in the references.

8.5 Cylindrical Plasma with Diffuse Boundary Configurations

8.5.1 Energy Integral of Cylindrical Plasma with Diffuse Boundary Configuration

The sharp-boundary configuration treated in Section 8.3 is a special case; in most cases the plasma current decreases gradually at the boundary. Let us consider the case of a diffuse-boundary plasma whose parameters in the equilibrium state are

$$p_0(r), \qquad \boldsymbol{B}_0(r) = (0, B_\theta(r), B_z(r)).$$

The perturbation $\boldsymbol{\xi}$ is assumed to be

$$\boldsymbol{\xi} = \boldsymbol{\xi}(r)\exp(im\theta + ikz).$$

The perturbation of the magnetic field $\boldsymbol{B}_1 = \nabla \times (\boldsymbol{\xi} \times \boldsymbol{B}_0)$ is

$$B_{1r} = i(\boldsymbol{k}\cdot\boldsymbol{B}_0)\xi_r,$$

$$B_{1\theta} = ikA - \frac{\mathrm{d}}{\mathrm{d}r}(\xi_r B_\theta),$$

$$B_{1z} = -\left(\frac{imA}{r} + \frac{1}{r}\frac{\mathrm{d}}{\mathrm{d}r}(r\xi_r B_z),\right)$$

where

$$(\boldsymbol{k}\cdot\boldsymbol{B}_0) = kB_z + \frac{m}{r}B_\theta,$$

$$A = \xi_\theta B_z - \xi_z B_\theta = (\boldsymbol{\xi}\times\boldsymbol{B}_0)_r. \tag{8.92}$$

Since the pressure terms $\gamma p_0(\nabla\cdot\boldsymbol{\xi})^2 + (\nabla\cdot\boldsymbol{\xi})(\boldsymbol{\xi}\cdot\nabla p_0) = (\gamma-1)p_0(\nabla\cdot\boldsymbol{\xi})^2 + (\nabla\cdot\boldsymbol{\xi})(\nabla\cdot p_0\boldsymbol{\xi})$ in the energy integral are nonnegative, we examine the incompressible displacement $\nabla\cdot\boldsymbol{\xi} = 0$ again, i.e.,

$$\frac{1}{r}\frac{\mathrm{d}}{\mathrm{d}r}(r\xi_r) + \frac{im}{r}\xi_\theta + ik\xi_z = 0.$$

From this and (8.92) for A, ξ_θ and ξ_z are expressed in terms of ξ_r and A as

$$i(\boldsymbol{k}\cdot\boldsymbol{B})\xi_\theta = ikA - \frac{B_\theta}{r}\frac{\mathrm{d}}{\mathrm{d}r}(r\xi_r),$$

$$-i(\boldsymbol{k}\cdot\boldsymbol{B})\xi_z = \frac{imA}{r} + \frac{B_z}{r}\frac{\mathrm{d}}{\mathrm{d}r}(r\xi_r).$$

A is selected to minimize the integrand of (8.82). The detail process of derivation is found in Chapter 8 of [8.10]. Then total energy integral (8.81) $W = W_{\mathrm{p}} + W_{\mathrm{V}}$ is given by ($W_{\mathrm{S}} = 0$)

$$\left.\begin{aligned} W_{\mathrm{p}} &= \frac{\pi}{2\mu_0}\int_0^a \left(f\left|\frac{\mathrm{d}\xi_r}{\mathrm{d}r}\right|^2 + g|\xi_r|^2\right)\mathrm{d}r + W_a, \\ W_a &= \frac{\pi}{2\mu_0}\frac{k^2B_{\mathrm{s}}^2 - (m/a)^2B_{\mathrm{a}}^2}{k^2 + (m/a)^2}|\xi_r(a)|^2, \\ W_{\mathrm{V}} &= \frac{\pi}{2\mu_0}\int_a^b \left(\frac{1}{r(k^2+(m/r)^2)}\left|\frac{\mathrm{d}\zeta}{\mathrm{d}r}\right|^2 + \frac{1}{r}|\zeta|^2\right)\mathrm{d}r \end{aligned}\right\} \tag{8.93}$$

where

$$f = \frac{r(kB_z + (m/r)B_\theta)^2}{k^2 + (m/r)^2}, \tag{8.94}$$

$$g = \frac{1}{r}\frac{(kB_z - (m/r)B_\theta)^2}{k^2 + (m/r)^2} + r\left(kB_z + \frac{m}{r}B_\theta\right)^2$$

$$-\frac{2B_\theta}{r}\frac{\mathrm{d}(rB_\theta)}{\mathrm{d}r} - \frac{\mathrm{d}}{\mathrm{d}r}\left(\frac{k^2B_z^2 - (m/r)^2B_\theta^2}{k^2 + (m/r)^2}\right), \tag{8.95}$$

$$\zeta \equiv rB_{e1r}.$$

B_{e1r} is a radial component of perturbed magnetic field in vacuum between plasma boundary $r = a$ and the conductive wall $r = b$. (Note that B_{1r} in $r < a$ is $B_{1r} = i(\boldsymbol{k} \cdot \boldsymbol{B})\xi_r$). When the equation of equilibrium $(\mathrm{d}/\mathrm{d}r)(\mu_0 p + B^2/2) = -B_\theta^2/r$ is used, (8.95) is reduced to

$$g = \frac{2k^2}{k^2 + (m/r)^2}\mu_0\frac{\mathrm{d}p_0}{\mathrm{d}r} + r(kB_z + \frac{m}{r}B_\theta)^2\frac{k^2 + (m/r)^2 - (1/r)^2}{k^2 + (m/r)^2}$$

$$+\frac{(2k^2/r)(k^2B_z^2 - (m/r)^2B_\theta^2)}{(k^2 + (m/r)^2)^2}. \tag{8.96}$$

The functions ξ_r and ζ that will minimize W_p or W_V are the solutions of Euler's equation:

$$\frac{\mathrm{d}}{\mathrm{d}r}\left(\frac{r(\boldsymbol{k} \cdot \boldsymbol{B}_0)^2}{k^2 + (m/r)^2}\frac{\mathrm{d}\xi_r}{\mathrm{d}r}\right) - g\xi_r = 0, \quad r \le a, \tag{8.97}$$

$$\frac{\mathrm{d}}{\mathrm{d}r}\left(\frac{1}{r(k^2 + (m/r)^2)}\frac{\mathrm{d}\zeta}{\mathrm{d}r}\right) - \frac{1}{r}\zeta = 0, \quad r > a\ . \tag{8.98}$$

In the region $r \ll a$, (8.79) reduces to

$$\frac{\mathrm{d}}{\mathrm{d}r}\left(r^3\frac{\mathrm{d}\xi_r}{\mathrm{d}r}\right) - (m^2 - 1)r\xi_r = 0,$$

so that there are two solutions $\xi_r \propto r^{m-1}$, r^{-m-1} as $r \to 0$. Since ξ_r is finite at $r = 0$, ξ_r must satisfy the following boundary conditions:

$$\xi_r \propto r^{m-1} \qquad \text{as} \ \ r \to 0,$$

$$ia(kB_z(a) + \frac{m}{a}B_\theta(a))\xi_r(a) = \zeta(a), \qquad r = a$$

$$\zeta(b) = 0.$$

The solution of (8.98) for $b > r > a$ is

$$\zeta = i\frac{I'_m(kr)K'_m(kb) - K'_m(kr)I'_m(kb)}{I'_m(ka)K'_m(kb) - K'_m(ka)I'_m(kb)}r\left(kB_\mathrm{s} + \frac{m}{a}B_a\right)\xi_r(a). \tag{8.99}$$

8.5.2 Suydam's Criterion

The function f in the integrand of $W_{\rm p}$ in the previous section is always $f \geq 0$, so that the term in f is a stabilizing term. The first and second terms in (8.94) for g are stabilizing terms, but the third and fourth terms may contribute to the instabilities. When a singular point

$$f \propto (\boldsymbol{k} \cdot \boldsymbol{B}_0)^2 = 0$$

of Euler's equation (8.97) is located at some point $r = r_0$ within the plasma region, the contribution of the stabilizing term becomes small near $r = r_0$, so that a local mode near the singular point is dangerous. In terms of the notation

$$r - r_0 = x, \qquad f = \alpha x^2, \qquad g = \beta, \qquad \beta = \frac{2B_\theta^2}{B_0^2} \mu_0 \frac{\mathrm{d}p_0}{\mathrm{d}r}\bigg|_{r=r_0},$$

$$\alpha = \frac{r_0}{k^2 r_0^2 + m^2}\left(kr\frac{\mathrm{d}B_z}{\mathrm{d}r} + kB_z + m\frac{\mathrm{d}B_\theta}{\mathrm{d}r}\right)^2_{r=r_0} = \frac{rB_\theta^2 B_z^2}{B^2}\left(\frac{\tilde{\mu}'}{\tilde{\mu}}\right)^2_{r=r_0}, \quad \tilde{\mu} \equiv \frac{B_\theta}{rB_z}.$$

Euler's equation is reduced to

$$\alpha\frac{\mathrm{d}}{\mathrm{d}r}\left(x^2\frac{\mathrm{d}\xi_r}{\mathrm{d}x}\right) - \beta\xi_r = 0.$$

The solution is

$$\xi_r = c_1 x^{-n_1} + c_2 x^{-n_2},$$

where n_1 and n_2 are given by

$$n^2 - n - \frac{\beta}{\alpha} = 0, \qquad n_i = \frac{1 \pm (1 + 4\beta/\alpha)^{1/2}}{2} \qquad (i = 1, 2).$$

When $\alpha + 4\beta > 0$, n_1 and n_2 are real. The relation $n_1 + n_2 = 1$ holds always. For $n_1 < n_2$, we have the solution x^{-n_1}, called a *small solution.* When n is complex ($n = \gamma \pm i\delta$), ξ_r is in the form $\exp((-\gamma \mp i\delta)\ln x)$ and ξ_r is oscillatory.

Let us consider a local mode ξ_r, which is nonzero only in the neighborhood ε around $r = r_0$ and set

$$r - r_0 = \varepsilon t, \quad \xi_r(r) = \xi(t), \quad \xi(1) = \xi(-1) = 0.$$

Then $W_{\rm p}$ becomes

$$W_{\rm p} = \frac{\pi}{2\mu_0}\varepsilon \int_{-1}^{1}\left(\alpha t^2 \left|\frac{\mathrm{d}\xi}{\mathrm{d}t}\right|^2 + \beta|\xi|^2\right)\mathrm{d}t + O(\varepsilon^2).$$

Since Schwartz's inequality yields

$$\int_{-1}^{1} t^2|\xi'|^2 \mathrm{d}t \int_{-1}^{1} |\xi|^2 \mathrm{d}t \geq \left|\int_{-1}^{1} t\xi'\xi^* \mathrm{d}t\right|^2 = \left(\frac{1}{2}\int_{-1}^{1} |\xi|^2 \mathrm{d}t\right)^2,$$

W_{p} is

$$W_{\mathrm{p}} > \frac{\pi}{2\mu_0}\frac{1}{4}(\alpha + 4\beta)\int_{-1}^{1} |\xi|^2 \mathrm{d}t.$$

The stability condition is $\alpha + 4\beta > 0$, i.e.,

$$\frac{r}{4}\left(\frac{\tilde{\mu}'}{\tilde{\mu}}\right)^2 + \frac{2\mu_0}{B_z^2}\frac{\mathrm{d}p_0}{\mathrm{d}r} > 0. \tag{8.100}$$

$r(\tilde{\mu}'/\tilde{\mu})$ is called *shear parameter.* Usually the second term is negative, because, most often, $dp_0/dr < 0$. The first term $(\tilde{\mu}'/\tilde{\mu})^2$ represents the stabilizing effect of shear. This condition is called *Suydam's criterion* [8.12]. This is a necessary condition for stability; but it is not always a sufficient condition, as Suydam's criterion is derived from consideration of local-mode behavior only. Newcomb derived the necessary and sufficient conditions for the stability of a cylindrical plasma. His 12 theorems are described in [8.13].

8.5.3 Tokamak Configuration

In this case the longitudinal magnetic field B_{s} is much larger than the poloidal magnetic field B_θ. The plasma region is $r \leq a$ and the vacuum region is $a \leq r \leq b$ and an ideal conducting wall is at $r = b$. It is assumed that $ka \ll 1$, $kb \ll 1$. The function ζ of (8.99) for W_{V} becomes

$$\zeta = i\frac{(mB_a + kaB_{\mathrm{s}})}{1 - (a/b)^{2m}}\xi_r(a)\frac{a^m}{b^m}\left(\frac{b^m}{r^m} - \frac{r^m}{b^m}\right),$$

and W_{V} is

$$W_{\mathrm{V}} = \frac{\pi}{2\mu_0}\frac{(mB_a + kaB_{\mathrm{s}})^2}{m}\xi_r^2(a)\lambda, \qquad \lambda \equiv \frac{1 + (a/b)^{2m}}{1 - (a/b)^{2m}}.$$

From the periodic condition for a torus, it follows that

$$\frac{2\pi n}{k} = -2\pi R \quad (n \text{ is an integer}),$$

so that $(\boldsymbol{k}\cdot\boldsymbol{B})$ is given by

$$a(\boldsymbol{k}\cdot\boldsymbol{B}) = mB_a + kaB_{\mathrm{s}} = mB_a\left(1 - \frac{nq_a}{m}\right)$$

in terms of the safety factor. The W_a term in (8.93) is reduced to

$$k^2B_s^2 - \left(\frac{m}{a}\right)^2 B_a^2 = \left(kB_s + \frac{m}{a}B_a\right)^2 - 2\frac{m}{a}B_a\left(kB_s + \frac{m}{a}B_a\right)$$
$$= \left(\frac{nB_a}{a}\right)^2\left(\left(1-\frac{nq_a}{m}\right)^2 - 2\left(1-\frac{nq_a}{m}\right)\right).$$

Accordingly, the energy integral becomes

$$W_p + W_V = \frac{\pi}{2\mu_0}B_a^2\xi_r^2(a)\left(\left(1-\frac{nq_a}{m}\right)^2(1+m\lambda) - 2\left(1-\frac{nq_a}{m}\right)\right)$$
$$+\frac{\pi}{2\mu_0}\int\left(f\left(\frac{d\xi_r}{dr}\right)^2 + g\xi_r^2\right)dr. \quad (8.101)$$

The first term of (8.101) is negative when

$$1 - \frac{2}{1+m\lambda} < \frac{nq_a}{m} < 1. \quad (8.102)$$

The assumption $nq_a/m \sim 1$ corresponds to $ka \sim mB_a/B_s$. As $B_a/B_s \ll 1$, this is consistent with the assumption $ka \ll 1$. When $m = 1$, $(m^2-1)/m^2$ in the second term of (8.96) for g is zero. The magnitude of g is of the order of k^2r^2, which is very small since $kr \ll 1$. The term in $f(d\xi_r/dr)^2$ can be very small if ξ_r is nearly constant. Accordingly, the contribution of the integral term in (8.101) is negligible. When $m = 1$ and $a^2/b^2 < nq_a < 1$, the energy integral becomes negative ($W < 0$). The mode $m = 1$ is unstable in the region specified by (8.102) irrespective of the current distribution. The Kruskal-Shafranov condition for the mode $m = 1$ derived from the sharp-boundary configuration is also applicable to the diffuse-boundary plasma. The growth rate $\gamma^2 = -\omega^2$ is

$$\gamma^2 \simeq \frac{-W}{\int(\rho_{m0}|\boldsymbol{\xi}|^2/2)dr} = \frac{1}{\langle\rho_{m0}\rangle}\frac{B_a^2}{\mu_0 a^2}\left(2(1-nq_a) - \frac{2(1-nq_a)^2}{1-a^2/b^2}\right), \quad (8.103)$$

$$\langle\rho_{m0}\rangle = \frac{\int\rho_{m0}|\boldsymbol{\xi}|^2 2\pi r dr}{\pi a^2\xi_r^2(a)}.$$

The maximum growth rate is $\gamma_{max}^2 \sim (1-a^2/b^2)B_a^2/(\mu_0\langle\rho\rangle a^2)$. When $m \neq 1$, $(m^2-1)/m^2$ in the second term of (8.96) for g is large, and $g \sim 1$. Accordingly, the contribution of the integral term to W_p must be checked. The region $g < 0$ is given by $\chi_1 < \chi < \chi_2$, when $\chi \equiv -krB_z/B_\theta = nq(r)$ and

$$\chi_{1,2} = m - \frac{2}{m(m^2-1)}k^2r^2 \pm \frac{2k^2r^2}{m(m^2-1)}\left(1 - \frac{m^2(m^2-1)}{2k^2r^2}\frac{\mu_0 r p_0'}{B_\theta^2}\right)^{1/2}. \quad (8.104)$$

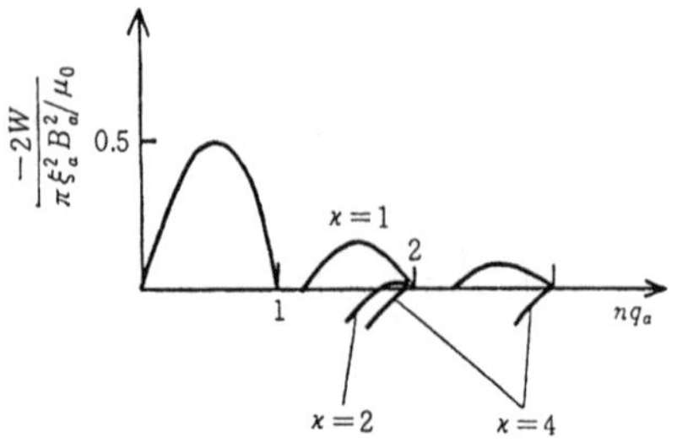

Figure 8.7 The relation of the growth rate γ and nq_a for kink instability $(-2W/(\pi\xi_a^2 B_a^2/\mu_0) = \gamma^2 a^2(\langle\rho_{m0}\rangle\mu_0/B_a^2))$.

Since $kr \ll 1$, the region $g < 0$ is narrow and close to the singular point $nq(r) = m$ and the contribution of the integral term to W_p can be neglected in the case of $m \neq 1$ as well. Therefore, if nq_a/m is in the range given by (8.102), the plasma is unstable due to the displacement $\xi_r(a)$ of the plasma boundary. When the current distribution is $j(r) = j_0 \exp(-\kappa^2 r^2/a^2)$ and the conducting wall is at infinity ($b = \infty$), γ^2 can be calculated from (8.103), using the solution of Euler's equation; and the dependence of γ^2 on q_a can be estimated. The result is shown in Figure 8.7 [8.6].
When the value of nq_a/m is outside the region given by (8.102), the effect of the displacement of the plasma boundary is not great and the contribution of the integral term in W_p is dominant. However, the growth rate γ^2 is k^2r^2 times as small as that given by (8.103), as is clear from consideration of (8.104).

The characteristics of the Reversed Field Pinch (RFP) (refer to Section 4.1) is that B_a and B_s are of the same order of magnitude, so that the approximation based upon $ka \ll 1$ or $B_a \ll B_s$ can no longer be used. Analysis of MHD instability in RFP is well analyzed in [8.14].

8.6 Hain-Lüst MHD Equation

When the displacement $\boldsymbol{\xi}$ is denoted by

$$\boldsymbol{\xi}(r,\theta,z,t) = \boldsymbol{\xi}(r)\exp i(m\theta + kz - \omega t),$$

and the equilibrium magnetic field $\boldsymbol{B}$ is expressed by

$$\boldsymbol{B}(r) = (0,\, B_\theta(r),\, B_z(r)),$$

the (r,θ,z) components of MHD equation of motion are given by

$$-\mu_0\rho_{\rm m}\omega^2\xi_r = \frac{\rm d}{{\rm d}r}\left(\mu_0\gamma p(\nabla\cdot\xi) + B^2\frac{1}{r}\frac{\rm d}{{\rm d}r}(r\xi_r) + iD(\xi_\theta B_z - \xi_z B_\theta)\right)$$

$$-\left(F^2 + r\frac{\rm d}{{\rm d}r}\left(\frac{B_\theta}{r}\right)^2\right)\xi_r - 2ik\frac{B_\theta}{r}(\xi_\theta B_z - \xi_z B_\theta), \qquad (8.105)$$

$$-\mu_0\rho_{\rm m}\omega^2\xi_\theta = i\frac{m}{r}\gamma\mu_0 p(\nabla\cdot\boldsymbol{\xi}) + iDB_z\frac{1}{r}\frac{\rm d}{{\rm d}r}(r\xi_r) + 2ik\frac{B_\theta B_z}{r}\xi_r - H^2 B_z(\xi_\theta B_z - \xi_z B_\theta), \qquad (8.106)$$

$$-\mu_0\rho_{\rm m}\omega^2\xi_z = ik\gamma\mu_0 p(\nabla\cdot\boldsymbol{\xi}) - iDB_\theta\frac{1}{r}\frac{\rm d}{{\rm d}r}(r\xi_r) - 2ik\frac{B_\theta^2}{r}\xi_r + H^2 B_\theta(\xi_\theta B_z - \xi_z B_\theta), \qquad (8.107)$$

where

$$F = \frac{m}{r}B_\theta + kB_z = (\boldsymbol{k}\cdot\boldsymbol{B}), \qquad D = \frac{m}{r}B_z - kB_\theta, \qquad H^2 = \left(\frac{m}{r}\right)^2 + k^2,$$

$$\nabla\cdot\boldsymbol{\xi} = \frac{1}{r}\frac{\rm d}{{\rm d}r}(r\xi_r) + \frac{im}{r}\xi_\theta + ik\xi_z.$$

When ξ_θ, ξ_z are eliminated by (8.106) and (8.107), we find

$$\frac{\rm d}{{\rm d}r}\left(\frac{(\mu_0\rho_{\rm m}\omega^2 - F^2)}{\Delta}(\mu_0\rho_{\rm m}\omega^2(\gamma\mu_0 p + B^2) - \gamma\mu_0 pF^2)\frac{1}{r}\frac{\rm d}{{\rm d}r}(r\xi_r)\right)$$

$$+\left[\mu_0\rho_{\rm m}\omega^2 - F^2 - 2B_\theta\frac{\rm d}{{\rm d}r}\left(\frac{B_\theta}{r}\right) - \frac{4k^2}{\Delta}\frac{B_\theta^2}{r^2}(\mu_0\rho_{\rm m}\omega^2 B^2 - \gamma\mu_0 pF^2)\right.$$

$$\left. + r\frac{\rm d}{{\rm d}r}\left(\frac{2kB_\theta}{\Delta\, r^2}\left(\frac{m}{r}B_z - kB_\theta\right)(\mu_0\rho_{\rm m}\omega^2(\gamma\mu_0 p + B^2) - \gamma\mu_0 pF^2)\right)\right]\xi_r$$

$$= 0, \qquad (8.108)$$

where Δ is

$$\Delta = \mu_0^2\rho_{\rm m}^2\omega^4 - \mu_0\rho_{\rm m}\omega^2 H^2(\gamma\mu_0 p + B^2) + \gamma\mu_0 pH^2F^2.$$

This equation was derived by Hain-Lüst [8.15]. The solution of (8.108) gives $\xi_r(r)$ in the region of $0 < r < a$. The equations for the vacuum region $a < r < a_w$ (a_w is the radius of wall) are

$$\nabla\times\boldsymbol{B}_1 = 0, \qquad \nabla\cdot\boldsymbol{B}_1 = 0,$$

so that we find

$$\boldsymbol{B}_1 = \nabla\psi, \qquad \Delta\psi = 0,$$

$$\psi = (bI_m(kr) + cK_m(kr))\exp(im\theta + ikz),$$

$$B_{1r} = \frac{\partial \psi}{\partial r} = (bI'_m(kr) + cK'_m(kr))\exp(im\theta + ikz).$$

B_{1r} in the plasma region is given by

$$B_{1r} = i(\boldsymbol{k} \cdot \boldsymbol{B})\xi_r = iF\xi_r,$$

and the boundary conditions at $r = a$ are

$$B_{1r}(a) = iF\xi_r(a),$$

$$B'_{1r}(a) = i(F'\xi_r(a) + F\xi'_r(a)),$$

and the coefficients b,c can be fixed.

To deal with this equation as an eigenvalue problem, boundary conditions must be imposed on ξ_r; one is that $\xi_r \propto r^{m-1}$ at $r = 0$, and the other is that the radial component of the perturbed magnetic field at the perfect conducting wall $B_{1r}(a_w) = 0$. After finding suitable ω^2 to satisfy these conditions, the growth rate $\gamma^2 \equiv -\omega^2$ is obtained [8.16].

8.7 Ballooning Instability

In interchange instability, the parallel component $k_\parallel = (\boldsymbol{k} \cdot \boldsymbol{B})/B$ of the propagation vector is zero and an average minimum-B condition may stabilize such an instability. Suydam's condition and the local-mode stability condition of toroidal system are involved in perturbations with $k_\parallel = 0$. In this section we will study perturbations where $k_\parallel \neq 0$ but $|k_\parallel / k_\perp| \ll 1$. Although the interchange instability is stabilized by an average minimum-B configuration, it is possible that the perturbation with $k_\parallel \neq 0$ can grow locally in the bad region of the average minimum-B field. This type of instability is called the *ballooning mode* (see Figure 8.8).

There is a beta limit for the stability of ballooning mode. It will be shown that average minimum-B and shear stabilize the ballooning mode. Therefore, measures of magnetic well and shear are important parameters of magnetic confinement configurations [8.17].

The energy integral δW is given by

$$\delta W = \frac{1}{2\mu_0} \int ((\nabla \times (\boldsymbol{\xi} \times \boldsymbol{B}_0))^2 - (\boldsymbol{\xi} \times (\nabla \times \boldsymbol{B}_0)) \cdot \nabla \times (\boldsymbol{\xi} \times \boldsymbol{B}_0)$$

$$+ \gamma \mu_0 p_0 (\nabla \cdot \boldsymbol{\xi})^2 + \mu_0 (\nabla \cdot \boldsymbol{\xi})(\boldsymbol{\xi} \cdot \nabla p_0))\mathrm{d}\boldsymbol{r}.$$

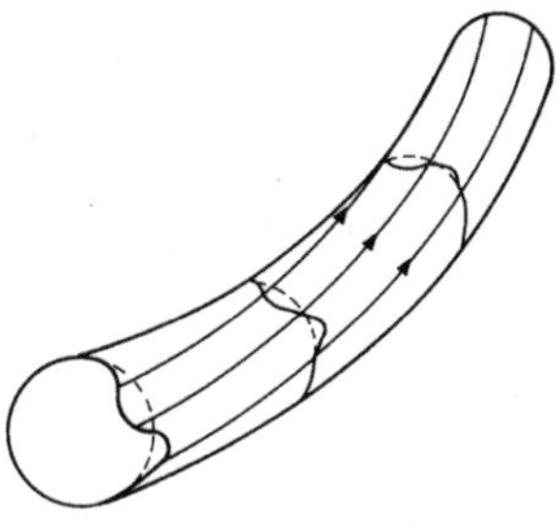

Figure 8.8 Ballooning mode.

Let us consider the case that $\boldsymbol{\xi}$ can be expressed by

$$\boldsymbol{\xi} = \frac{\boldsymbol{B}_0 \times \nabla\phi}{B_0^2},$$

where ϕ is considered to be the time integral of the scalar electrostatic potential of the perturbed electric field. Because of

$$\boldsymbol{\xi} \times \boldsymbol{B}_0 = \nabla_\perp \phi,$$

the energy integral is reduced to

$$\delta W = \frac{1}{2\mu_0}\int\left((\nabla \times \nabla_\perp\phi)^2 - \left(\frac{(\boldsymbol{B}_0 \times \nabla_\perp\phi)\times \mu_0 \boldsymbol{j}_0}{B_0^2}\right)\nabla\times\nabla_\perp\phi \right.$$
$$\left. + \gamma\mu_0 p_0 (\nabla\cdot\boldsymbol{\xi})^2 + \mu_0(\nabla\cdot\boldsymbol{\xi})(\boldsymbol{\xi}\cdot\nabla p_0)\right)\mathrm{d}\boldsymbol{r}.$$

$\nabla\cdot\boldsymbol{\xi}$ is given by

$$\nabla\cdot\boldsymbol{\xi} = \nabla\cdot\left(\frac{\boldsymbol{B}_0\times\nabla\phi}{B_0^2}\right) = \nabla\phi\cdot\nabla\times\left(\frac{\boldsymbol{B}_0}{B_0^2}\right) = \nabla\phi\cdot\left(\left(\nabla\frac{1}{B^2}\right)\times\boldsymbol{B} + \frac{1}{B^2}\nabla\times\boldsymbol{B}\right).$$

The second term in ( ) is negligible compared with the first term in the low beta case. By means of $\nabla p_0 = \boldsymbol{j}_0 \times \boldsymbol{B}_0$, δW is expressed by

$$\delta W = \frac{1}{2\mu_0}\int (\nabla\times\nabla_\perp\phi)^2 + \frac{\mu_0 \nabla p_0\cdot(\nabla_\perp\phi\times\boldsymbol{B}_0)}{B_0^2}\left(\frac{\boldsymbol{B}_0\cdot\nabla\times\nabla_\perp\phi}{B_0^2}\right)$$
$$-\frac{\mu_0(\boldsymbol{j}_0\cdot\boldsymbol{B}_0)}{B_0^2}\nabla_\perp\phi\cdot\nabla\times\nabla_\perp\phi + \gamma\mu_0 p_0\left(\nabla\left(\frac{1}{B_0^2}\right)\cdot(\boldsymbol{B}_0\times\nabla_\perp\phi)\right)^2$$
$$+\frac{\mu_0\nabla p_0\cdot(\boldsymbol{B}_0\times\nabla_\perp\phi)}{B_0^2}\left(\nabla\left(\frac{1}{B_0^2}\right)\cdot(\boldsymbol{B}_0\times\nabla_\perp\phi)\right)\mathrm{d}\boldsymbol{r}.$$

Let us use z coordinate as a length along a field line, r as radial coordinate of magnetic surfaces, and θ as poloidal angle in the perpendicular direction to field lines. The r, θ, z components of ∇p_0, $\boldsymbol{B}$, and $\nabla\phi$ are approximately given by

$$\nabla p_0 = (p_0', 0, 0), \qquad \boldsymbol{B} = (0,\ B_\theta(r),\ B_0(1 - rR_c^{-1}(z))),$$

$$\nabla\phi = (\partial\phi/\partial r, \partial\phi/r\partial\theta, \partial\phi/\partial z), \qquad \phi(r,\theta,z) = \phi(r,z)\mathrm{Re}(\exp im\theta).$$

$R_c(z)$ is the radius of curvature of the line of magnetic force:

$$\frac{1}{R_c(z)} = \frac{1}{R_0}\left(-w + \cos 2\pi\frac{z}{L}\right).$$

When $R_c(z) < 0$, the curvature is said to be good. If the configuration is average minimum-B, w must be $1 > w > 0$. Since $B_\theta/B_0, r/R_0, r/L$ are all small quantities, we find

$$\nabla_\perp\phi = \nabla\phi - \nabla_\parallel\phi \approx \mathrm{Re}\left(\frac{\partial\phi}{\partial r}, \frac{im}{r}\phi, 0\right),$$

$$\nabla\times(\nabla_\perp\phi) \approx \mathrm{Re}\left(\frac{-im}{r}\frac{\partial\phi}{\partial z}, \frac{\partial^2\phi}{\partial z\partial r}, 0\right),$$

$$\boldsymbol{B}_0 \times \nabla_\perp\phi \approx \mathrm{Re}\left(\frac{-im}{r}B_0\phi, B_0\frac{\partial\phi}{\partial r}, 0\right),$$

and δW is reduced to (in the case of complex $\phi(r,z)$)

$$\delta W = \frac{1}{2\mu_0}\int\frac{m^2}{r^2}\left(\left|\frac{\partial\phi(r,z)}{\partial z}\right|^2 - \frac{\beta}{r_p R_c(z)}|\phi(r,z)|^2\right)2\pi r\mathrm{d}r\mathrm{d}z,$$

where $-p_0/p_0' = r_p$ and $\beta = p_0/(B_0^2/2\mu_0)$. The first term comes from $(\nabla\times\nabla_\perp\phi)^2$ in the integrand of energy integral and the energy to bend the magnetic line of force. This is a stabilizing term, where

$$\frac{\partial\phi}{\partial z} \approx ik_\parallel\phi \sim i\frac{\phi}{q_s R}.$$

The second term comes from $\xi\cdot\nabla p_0$ in the integrand and is the energy to spend due to pressure driven motion. This is the destabilizing term. Therefore, the beta limit of the ballooning mode is

$$\beta < \frac{r_p}{Rq_s^2}. \tag{8.109}$$

In the configuration with magnetic shear, more rigorous treatment is necessary. For ballooning modes with large toroidal mode number $n \gg 1$ and $m - nq \sim 0$, the stable region in the shear parameter S and the measure

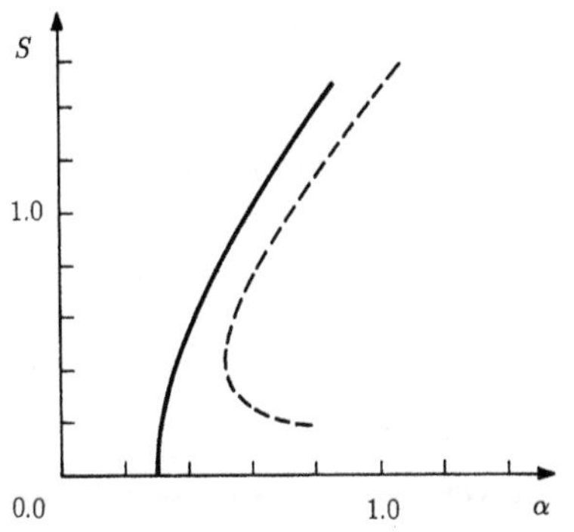

Figure 8.9 The maximum stable pressure gradient α as a function of the shear parameter S of ballooning mode. The dotted line is the stability boundary obtained by imposing a more restricted boundary condition on the perturbation [8.18].

of pressure gradient α of ballooning mode is shown in Figure 8.9. The shear parameter S is defined by

$$S = \frac{r}{q_s}\frac{dq_s}{dr},$$

where q is the safety factor ($q_s \equiv 2\pi/\iota$: ι rotational transform angle) and the measure of pressure gradient α is defined by

$$\alpha = -\frac{q_s^2 R}{B^2/2\mu_0}\frac{dp}{dr}.$$

The straight-line approximation of the maximum pressure gradient in the range of large positive shear ($S > 0.8$) is $\alpha \sim 0.6S$ as is shown in Figure 8.9. Since

$$\beta = \frac{1}{B_0^2/2\mu_0}\frac{1}{\pi a^2}\int_0^a p2\pi r dr = -\frac{1}{B_0^2/2\mu_0}\frac{1}{a^2}\int_0^a \frac{dp}{dr}r^2 dr,$$

the maximum ballooning stable beta is

$$\beta = 0.6\frac{a}{R}\left(\frac{1}{a^3}\int_0^a \frac{1}{q_s^3}\frac{dq_s}{dr}r^3 dr\right).$$

Under an optimum q profile, the maximum beta is given by [8.18]

$$\beta_{\max} \sim 0.28\frac{a}{Rq_a} \qquad (q_a > 2), \tag{8.110}$$

where q_a is the safety factor at the plasma boundary. In the derivation of (8.109) $q_a > 2$, $q_0 = 1$ are assumed. This beta limit is consistent with the Troyon result of computer simulation as is described in (4.51).

It must be noted that the ballooning mode is stable in the negative shear region of S, as is shown in Figure 8.9. When the shear parameter S is negative ($q_s(r)$ decreases outwardly), the outer lines of magnetic force rotate around the magnetic axis more quickly than the inner ones. When the pressure increases, the tokamak plasma tends to expand in a direction of major radius (Shafranov shift). This must be counter balanced by strengthening the poloidal field on the outside of tokamak plasma. In the region of strong pressure gradient, the necessary poloidal field increases outwardly, so on outer magnetic surfaces the magnetic field lines rotate around the magnetic axis faster than on inner ones and the shear parameter becomes more negative [8.19].

In reality, the shear parameter in a tokamak is positive in usual operations. However, the fact that the ballooning mode is stable in negative shear parameter region is very important to develop tokamak configuration stable against ballooning modes. Since

$$\frac{r}{Rq_s} = \frac{B_\theta}{B_0} = \frac{1}{B_0}\frac{\mu_0}{2\pi r}\int_0^r j(r)2\pi r\mathrm{d}r,$$

the profile of safety factor $q_s(r)$ is

$$\frac{1}{q_s(r)} = \frac{R}{2B_0}\left(\frac{\mu_0}{\pi r^2}\int_0^r j2\pi r\mathrm{d}r\right) \equiv \frac{\mu_0 R}{2B_0}\langle j(r)\rangle_r.$$

Therefore, a negative shear configuration can be realized by a hollow current profile. The MHD stability of tokamak with hollow current profiles is analyzed in detail in [8.20].

8.8 Eta-i Mode due to Density and Temperature Gradient

Let us consider a plasma with the density gradient $\mathrm{d}n_0/\mathrm{d}r$, and the temperature gradient $\mathrm{d}T_{e0}/\mathrm{d}r$, $\mathrm{d}T_{i0}/\mathrm{d}r$ in the magnetic field with the z direction. Assume that the ion's density becomes $n_i = n_{i0} + \tilde{n}_i$ by disturbance. The equation of continuity

$$\frac{\partial n_i}{\partial t} + \boldsymbol{v}_i \cdot \nabla n_i + n_i \nabla \cdot \boldsymbol{v}_i = 0$$

is reduced, by the linearization, to

$$-i\omega\tilde{n}_i + \tilde{v}_r\frac{\partial n_0}{\partial r} + n_0 i k_\parallel \tilde{v}_\parallel = 0. \tag{8.111}$$

It is assumed that the perturbation terms change as $\exp i(k_\theta r\theta + k_\parallel z - \omega t)$ and $k_\theta, k_\parallel$ are the θ and z components of the propagation vector. When the perturbed electrostatic potential is denoted by $\tilde{\phi}$, the $\boldsymbol{E} \times \boldsymbol{B}$ drift velocity is $\tilde{v}_{\mathrm{r}} = E_\theta / B = i k_\theta \tilde{\phi}/B$. Since the electron density follows Boltzmann distribution, we find

$$\frac{\tilde{n}_{\mathrm{e}}}{n_0} = \frac{e\tilde{\phi}}{T_{\mathrm{e}}}. \tag{8.112}$$

The parallel component of the equation of motion to the magnetic field

$$n_{\mathrm{i}} m_{\mathrm{i}} \frac{\mathrm{d}v_\parallel}{\mathrm{d}t} = -\nabla_\parallel p_{\mathrm{i}} - en\nabla_\parallel \phi$$

is reduced, by the linearization, to

$$-i\omega n_{\mathrm{i}} m_{\mathrm{i}} \tilde{v}_\parallel = -ik_\parallel (\tilde{p}_{\mathrm{i}} + en_0 \tilde{\phi}). \tag{8.113}$$

Similarly, the adiabatic equation

$$\frac{\partial}{\partial t}(p_{\mathrm{i}} n_{\mathrm{i}}^{-5/3}) + \boldsymbol{v} \cdot \nabla (p_{\mathrm{i}} n_{\mathrm{i}}^{-5/3}) = 0$$

is reduced to

$$-i\omega \left(\frac{\tilde{p}_{\mathrm{i}}}{p_{\mathrm{i}}} - \frac{5}{3}\frac{\tilde{n}_{\mathrm{i}}}{n_{\mathrm{i}}} \right) - \frac{ik_\theta \tilde{\phi}}{B} \left(\frac{\frac{\mathrm{d}T_{\mathrm{i}0}}{\mathrm{d}r}}{T_{\mathrm{i}0}} - \frac{2}{3}\frac{\frac{\mathrm{d}n_0}{\mathrm{d}r}}{n_0} \right) = 0. \tag{8.114}$$

Let us define the *electron drift frequencies* $\omega_{n\mathrm{e}}^*, \omega_{T\mathrm{ee}}^*$ and the *ion drift frequencies* $\omega_{n\mathrm{i}}^*, \omega_{T\mathrm{i}}^*$ by

$$\omega_{n\mathrm{e}}^* \equiv -\frac{k_\theta T_{\mathrm{e}}}{eBn_{\mathrm{e}}}\frac{\mathrm{d}n_{\mathrm{e}}}{\mathrm{d}r}, \qquad \omega_{n\mathrm{i}}^* \equiv \frac{k_\theta T_{\mathrm{i}}}{eBn_{\mathrm{i}}}\frac{\mathrm{d}n_{\mathrm{i}}}{\mathrm{d}r},$$

$$\omega_{T\mathrm{e}}^* \equiv -\frac{k_\theta}{eB}\frac{\mathrm{d}T_{\mathrm{e}}}{\mathrm{d}r}, \qquad \omega_{T\mathrm{i}}^* \equiv \frac{k_\theta}{eB}\frac{\mathrm{d}T_{\mathrm{i}}}{\mathrm{d}r}.$$

The ratio of the temperature gradient to the density gradient of electrons and ions is given by

$$\eta_{\mathrm{e}} \equiv \frac{\mathrm{d}T_{\mathrm{e}}/\mathrm{d}r}{T_{\mathrm{e}}}\frac{n_{\mathrm{e}}}{\mathrm{d}n_{\mathrm{e}}/\mathrm{d}r} = \frac{\mathrm{d}\ln T_{\mathrm{e}}}{\mathrm{d}\ln n_{\mathrm{e}}}, \qquad \eta_{\mathrm{i}} \equiv \frac{\mathrm{d}T_{\mathrm{i}}/\mathrm{d}r}{T_{\mathrm{i}}}\frac{n_{\mathrm{i}}}{\mathrm{d}n_{\mathrm{i}}/\mathrm{d}r} = \frac{\mathrm{d}\ln T_{\mathrm{i}}}{\mathrm{d}\ln n_{\mathrm{i}}},$$

respectively. There are, following relations among these values:

$$\omega_{n\mathrm{i}}^* = -\frac{T_{\mathrm{i}}}{T_{\mathrm{e}}}\omega_{n\mathrm{e}}^*, \qquad \omega_{T\mathrm{e}}^* = \eta_{\mathrm{e}}\omega_{n\mathrm{e}}^*, \qquad \omega_{T\mathrm{i}}^* = \eta_{\mathrm{i}}\omega_{n\mathrm{i}}^*.$$

Then Equations (8.111), (8.112), (8.113), and (8.114) are reduced to

$$\frac{\tilde{n}_{\mathrm{i}}}{n_0} = \frac{\tilde{v}_\parallel}{\omega/k_\parallel} + \frac{\omega_{n\mathrm{e}}^*}{\omega}\frac{e\tilde{\phi}}{T_{\mathrm{e}}},$$

$$\frac{\tilde{n}_e}{n_0} = \frac{e\tilde{\phi}}{T_e},$$

$$\frac{\tilde{v}_\parallel}{\omega/k_\parallel} = \frac{1}{m_i(\omega/k_\parallel)^2}\left(e\tilde{\phi} + \frac{\tilde{p}_i}{n_0}\right),$$

$$\left(\frac{\tilde{p}_i}{p_{i0}} - \frac{5}{3}\frac{\tilde{n}_i}{n_0}\right) = \frac{\omega_{ne}^*}{\omega}\left(\eta_i - \frac{2}{3}\right)\frac{e\tilde{\phi}}{T_e}.$$

Charge neutrality condition $\tilde{n}_i/n_0 = \tilde{n}_e/n_0$ yields the dispersion equation.

$$1 - \frac{\omega_{ne}^*}{\omega} - \left(\frac{v_{Ti}}{\omega/k_\parallel}\right)^2\left(\frac{T_e}{T_i} + \frac{5}{3} + \frac{\omega_{ne}^*}{\omega}\left(\eta_i - \frac{2}{3}\right)\right) = 0$$

($v_{Ti}^2 = T_i/m_i$). The solution in the case of $\omega \ll \omega_{n_e}^*$ is [8.21]

$$\omega^2 = -k_\parallel^2 v_{Ti}^2\left(\eta_i - \frac{2}{3}\right). \tag{8.115}$$

The dispersion equation shows that this type of perturbation is unstable when $\eta_i > 2/3$. This mode is called η_i *mode* or ion temperature gradient (ITG) mode.

When the propagation velocity $|\omega/k_\parallel|$ becomes the order of the ion thermal velocity v_{T_i}, the interaction (Landau damping) between ions and wave (perturbation) becomes important as will be described in Chapter 11 and MHD treatment must be modified. When the value of η_i is not large, the kinetic treatment is necessary and the threshold of η_i becomes $\eta_{i,cr} \sim 1.5$.

8.9 Alfvén Eigen Mode

Alfvén waves in homogeneous magnetic field in infinite plasma were analyzed in Section 8.2.2. Shear Alfvén wave, fast and slow magnetosonic waves appear. In the case of incompressible plasma ($\nabla \cdot \boldsymbol{\xi} = 0$ or ratio of specific heat $\gamma \to \infty$), only the shear Alfvén wave can exist.

In the case of cylindrical plasma in the axisymmetric magnetic field, the displacement of MHD perturbation $\boldsymbol{\xi}(r,\theta,z) = \boldsymbol{\xi}(r)\exp i(-m\theta + kz - \omega t)$ is given by Hain-Lüst equations (8.105) through (8.108) as was discussed in Section 8.6. In the case of incompressible plasma, Hain-Lüst equation (8.108) is reduced to [recalling that the perturbation is assumed to be $\boldsymbol{\xi}(r)\exp i(m\theta + kz - \omega t)$ in Section 8.6]

$$\frac{\mathrm{d}}{\mathrm{d}r}\left(\frac{F^2 - \mu_0\rho_m\omega^2}{\mathrm{m}^2/r^2 + k^2}\right)\frac{1}{r}\frac{\mathrm{d}}{\mathrm{d}r}(r\xi_r) + \left(-(F^2 - \mu_0\rho_m\omega^2) + 2B_\theta\frac{\mathrm{d}}{\mathrm{d}r}\left(\frac{B_\theta}{r}\right)\right.$$

$$+\frac{4k^2 B_\theta^2 F^2}{r^2(\mathrm{m}^2/r^2+k^2)(F^2-\mu_0\rho_\mathrm{m}\omega^2)}+2r\frac{\mathrm{d}}{\mathrm{d}r}\left(\frac{(\mathrm{m}/r)FB_\theta}{r^2(\mathrm{m}^2/r^2+k^2)}\right)\Bigg)\xi_r=0, \qquad (8.116)$$

where

$$F=(\boldsymbol{k}\cdot\boldsymbol{B})=\left(\frac{-\mathrm{m}}{r}B_\theta(r)+\frac{\mathrm{n}}{R}B_z(r)\right)=\frac{B_z}{R}\left(\mathrm{n}-\frac{\mathrm{m}}{q_\mathrm{s}(r)}\right), \qquad q_\mathrm{s}(r)=\frac{R}{r}\frac{B_z}{B_\theta}.$$

The position at which $F^2-\mu_0\rho_\mathrm{m}\omega^2=0\rightarrow\omega^2=\boldsymbol{k}_\parallel^2 v_\mathrm{A}^2$, $v_\mathrm{A}^2\equiv B^2/\mu_0\rho_\mathrm{m}$ holds is singular radius. It was shown by Hasegawa and Chen [8.22] that at this singular radius (resonant layer) shear Alfvén wave is mode converted to the kinetic Alfvén wave and absorbed by Landau damping. Therefore, Alfvén wave is stable in the cylindrical plasma.

Toroidal Alfv́en Eigenmode

Let us consider shear Alfvén waves in toroidal plasma and the perturbation of (-m, n) mode given by

$$\phi(r,\theta,z,t)=\phi(r)\exp i(-\mathrm{m}\theta+\mathrm{n}\frac{z}{R}-\omega t). \qquad (8.117)$$

R is the major radius of torus and $k_\parallel$ is

$$k_\parallel=\frac{\boldsymbol{k}\cdot\boldsymbol{B}}{B}=\frac{1}{R}\left(\mathrm{n}-\frac{\mathrm{m}}{q_\mathrm{s}(r)}\right).$$

The resonant conditions of m and m+1 modes in linear cylindrical plasma are

$$\frac{\omega^2}{v_\mathrm{A}^2}-k_{\parallel\mathrm{m}}^2=0, \qquad \frac{\omega^2}{v_\mathrm{A}^2}-k_{\parallel\mathrm{m+1}}^2=0.$$

However, wave of m mode can couple with m±1 in toroidal plasma since the magnitude of toroidal field changes as $B_z=B_{z0}(1-(r/R)\cos\theta)$. Then the resonant condition of m and m+1 modes in toroidal plasma becomes

$$\begin{vmatrix}\frac{\omega^2}{v_\mathrm{A}^2}-k_{\parallel\mathrm{m}}^2 & \alpha\epsilon\frac{\omega^2}{v_\mathrm{A}^2}\\ \alpha\epsilon\frac{\omega^2}{v_\mathrm{A}^2} & \frac{\omega^2}{v_\mathrm{A}^2}-k_{\parallel\mathrm{m+1}}^2\end{vmatrix}=0,$$

where $\epsilon=r/R$ and α is a constant with order of 1. Then the solutions are

$$\frac{\omega_\pm^2}{v_\mathrm{A}^2}=\frac{k_{\parallel\mathrm{m}}^2+k_{\parallel\mathrm{m+1}}^2\pm\left((k_{\parallel\mathrm{m}}^2-k_{\parallel\mathrm{m+1}}^2)^2+4\alpha\epsilon^2k_{\parallel\mathrm{m}}^2k_{\parallel\mathrm{m+1}}^2\right)^{1/2}}{2(1-\alpha^2\epsilon^2)}. \qquad (8.118)$$

The resonant condition (8.118) is plotted in Figure 8.10.

At the radius satisfying $k_{\parallel\mathrm{m}}^2=k_{\parallel\mathrm{m+1}}^2$, the difference of $\omega_\pm$ becomes minimum and the radius is given by

$$\frac{1}{R}\left(\mathrm{n}-\frac{\mathrm{m}}{q_\mathrm{s}(r)}\right)=-\frac{1}{R}\left(\mathrm{n}-\frac{\mathrm{m}+1}{q_\mathrm{s}(r)}\right), \qquad q_\mathrm{s}(r_0)=\frac{\mathrm{m}+1/2}{\mathrm{n}}, \qquad (8.119)$$

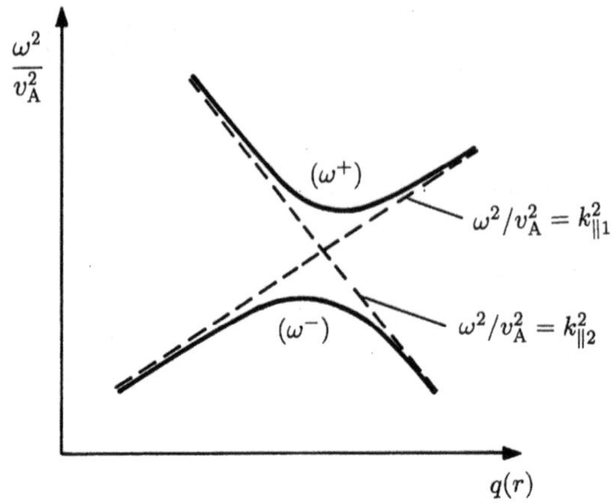

Figure 8.10 The Alfvén resonance frequency ω of toroidally coupled m and m+1 modes.

$$k_{\|\mathrm{m}} = -k_{\|\mathrm{m}+1} = \frac{1}{2q_\mathrm{s}(r_0)R}.$$

$q_\mathrm{s}(r_0) = 1.5$ for the case of $\mathrm{m}=1$ and $\mathrm{n}=1$. Therefore, Alfvén resonance does not exist in the frequency gap $\omega_- < \omega < \omega_+$.

The continuum Alfvén waves correspond to the excitation of shear Alfvén waves on a given flux surface where the mode frequency is resonant $\omega^2 = k_{\|\mathrm{m}}^2 v_\mathrm{A}^2(r)$ and such a resonance leads to wave damping. However, frequencies excited within the spectral gaps are not resonant with the continuum and hence will not damp in the gap region. This allows a discrete eigen-frequency of *toroidicty-induced Alfvén eigenmode* or *toroidal Alfvén eigenmode* (TAE) to be established. This TAE can easily be destabilized by the kinetic effect of energetic particles. The equations of TAE are described in detail in [8.23].

Growth Rate of Alfvén Wave by Energetic Particles

Dynamics of energetic particles must be treated by kinetic theory. The growth rate (ω_i) of Alfvén wave due to energetic particles is given by [8.24]

$$\frac{\omega_\mathrm{i}}{k_\| v_\mathrm{A}} = \sum_\mathrm{j} \frac{2\pi^2 \mu_0 m_\mathrm{j}^2 R q_0^3}{2B^2} \int \left(v_\|^2 + \frac{v_\perp^2}{2}\right)^2 \left(\omega_\mathrm{r} \frac{\partial F_\mathrm{j}}{\partial \epsilon} + \frac{\mathrm{n}}{q_\mathrm{j}} \frac{\partial F_\mathrm{j}}{\partial \psi}\right) v_\perp \mathrm{d}v_\perp \Bigg|_{v_\| = v_\mathrm{A}} + \cdots \Bigg|_{v_\| = v_\mathrm{A}/3}. \quad (8.120)$$

(The derivation process is described in detail in Section 13.2 of [8.10].) $F_j(\epsilon, p_\psi)$ is the zeroth order distribution function of each species. The term

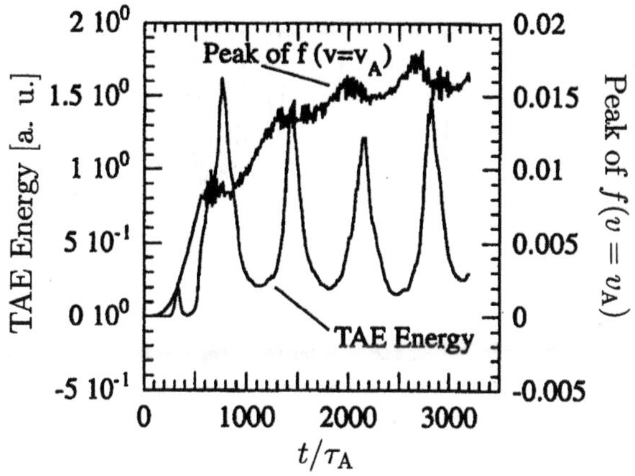

Figure 8.11 Time evolutions of total TAE energy and the energetic ion distribution peak value for $v = v_A$. After [8.25].

with subscript of $v_{\parallel} = v_A/3$ is the term of side band resonance due to toroidal coupling. The growth rate is the sum of contributions from bulk thermal ions and electrons and energetic ions such as α particles or fast ions produced by heating and is given by

$$\frac{\omega_i}{k_{\parallel} v_A} = -q_0^2 \left(\beta_i G_{mi}^T + \beta_e G_{me}^T + \beta_\alpha (G_{s\alpha}^T - nq_0 \delta_\alpha H_{s\alpha}^T)\right), \qquad (8.121)$$

where β_i, β_e, and β_α are beta ratio of ions, electrons, and α particles. Bulk thermal ions and electrons contribute to stabilization of Alfvén wave due to Landau dampings and energetic ions destabilize Alfvén wave. Therefore, there is a threshold of β_α to destabilize Alfvén eigen mode (AEM).

An Analysis by Computor Simulation

AEM is analyzed by computer simulation based on the MHD-kinetic hybrid model [8.25]. The bulk plasma is described by nonlinear full MHD equations (refer to (3.30)) and includes the effect of energetic ions through the current $\boldsymbol{j}'_\alpha$ due to the energetic ions as follows:

$$\rho \left(\frac{\partial \boldsymbol{v}}{\partial t} + (\boldsymbol{v} \cdot \nabla)\boldsymbol{v} \right) = \left(\frac{1}{\mu_0} \nabla \times \boldsymbol{B} - \boldsymbol{j}'_\alpha \right) \times \boldsymbol{B} - \nabla p,$$

$$\boldsymbol{j}'_\alpha = \int (v_{\parallel} \boldsymbol{b} + \boldsymbol{v}_d) f_\alpha \mathrm{d}^3 \boldsymbol{v} + \nabla \times \boldsymbol{M},$$

$$\boldsymbol{M} = -\int \mu \boldsymbol{b} f_\alpha \mathrm{d}^3 \boldsymbol{v},$$

where μ is the magnetic moment and $\boldsymbol{v}_d$ is curvature and ∇B drift velocity. The distribution function of energetic ions f_α is calculated by 4D (R, φ, z, v)

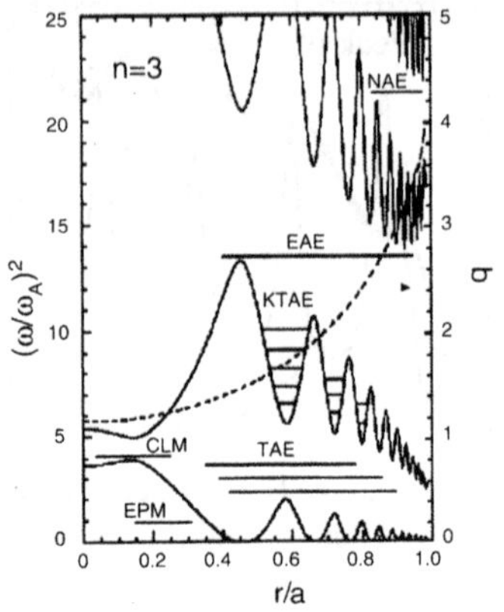

Figure 8.12 Representative shear Alfvén frequency continuum curves as a function of minor radius r [8.29]. Horizontal lines indicate the approximate radial location and mode width for toroidal Alfvén eigenmode (TAE), kinetic TAE mode (KTAE), core-localized TAE mode (CLM), ellipticity Alfvén eigenmode (EAE), noncircular triangularity Alfvén eigenmode (NAE), and energetic particle continuum mode (EPM). After [8.30].

Fokker-Planck equation (v is the parallel component of $\boldsymbol{v}$). This simulation reproduces the experimentally observed periodic burst of Alfvén eigenmodes in some aspect as is shown in Figure 8.11.

Various Alfvén Eigen Mode (AEM)

We discussed the excitation of AEM by super-Alfvénic energetic particles. There are various Alfvén modes. Typical ones are low n toroidal Alfvén wave [8.23] explained in the beginning of this section. Besides this, high n toroidal Alfvén waves [8.26] exist. In high-temperature plasmas, non-ideal effects such as finite Larmor radius of core plasma become important in gap region and cause the Alfvén continuum to split into a series of kinetic toroidal Alfvén eigenmodes (KTAE) at closely spaced frequencies above the ideal TAE frequency [8.27].

In the central region of the plasma, a low-shear version of TAE can arise, called the core-localized mode (CLM) [8.28].

Non-circular shaping of the plasma poloidal cross-section creates other gaps in the Alfvén continuum at high frequency. Ellipticity creats a gap, at about twice TAE frequency, within which exist ellipticity-induced Alfvén eigenmodes (EAE) [8.23] and similarly for triangularity-induced Alfvén eigenmodes (NAE) [8.23] at about three times the TAE frequency.

The ideal and kinetic TAE's are "cavity" mode, whose frequencies are determined by the bulk plasma. In addition, a "beam mode" can arise that is not a natural eigenmode of plasma but is supported by the presence of a

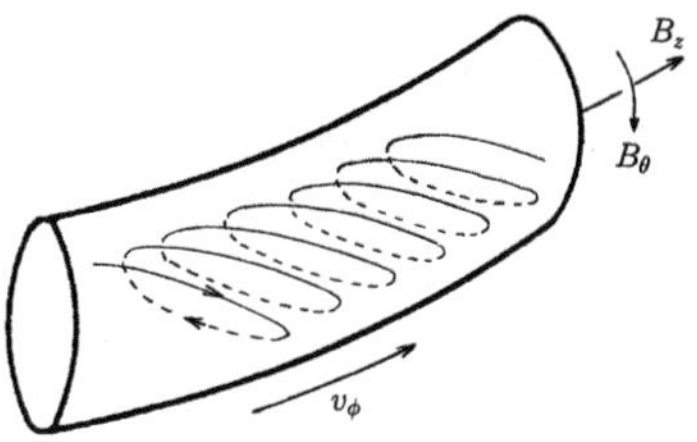

Figure 8.13 Toroidal precession of banana orbit of trapped ions.

population of energetic particles and is also destabilized by them. This so-called energetic particle mode (EPM) [8.29], which can also exit outside the TAE gaps, has frequencies related to the toroidal precession frequency and poloidal transit/bounce frequency of the fast ions. The schematic in Figure 8.11 illustrates these various modes. There are reviews on toroidal Alfvén eigenmode [8.30, 8.31].

Remark on Fishbone Instability
Trapped ions (bananas) in tokamak precess toroidally with the velocity * of $v_\phi = mv_\perp^2/(2eBr)$ and angular velocity $\omega_\phi = mv_\perp^2/(2eBRr)$ as is shown in Figure 8.13. Then the fishbone instability is excited by a resonant interaction between precessing energetic banana ions and $m = 1$, $n = 1$ MHD perturbation (inverse Landau damping) with frequency of $\omega \sim \omega_\phi$ and $\omega \sim (1/R)v_\phi$ [8.32]. More details on fishbone instability are described in Section 13.1 of [8.10].

Problems

1. Rayleigh-Taylor Instability Estimate the growth rate of Rayleigh-Taylor instability when the mass density $\rho_0(x)$ is $\rho_0(x) = \rho_2$ $(x > d)$, $\rho_0(x) = (\rho_2 - \rho_1)x/d$ $(d > x > 0)$ and $\rho_0(x) = \rho_1$ $(0 > x)$. Refer to (8.57).

2. Sausage Instability The stability condition of sausage instability ($m = 0$ mode) is given by $B_z^2 > B_\theta^2/2$ (refer to (8.69). Derive this stability condition from the following consideration. When a radius r of cylindrical plasma is constricted by the amount of $\delta r < 0$ near some z position adiabatically, the longitudinal magnetic flux inside the plasma is conserved $\delta(B_z\pi r^2) = 0$ and the plasma current does not change $\delta(B_\theta 2\pi r) = 0$.

*The toroidal vertical drift velocity of an ion is $v_\mathrm{d} = mv_\perp/(2eBR)$, so that the poloidal displacement of the ion between bounce is $r\mathrm{d}\theta \sim v_\mathrm{d}\tau_\mathrm{d}$, τ_d being the bounce period. Since $\mathrm{d}\phi/\mathrm{d}\theta = q_\mathrm{s}$ along the magnetic field line, the associated toroidal displacement between bounces is $R\mathrm{d}\phi = Rq_\mathrm{s}\mathrm{d}\theta = Rq_\mathrm{s}v_\mathrm{d}\tau_\mathrm{d}/r$, $q_\mathrm{s} = 1$. Thus, the toroidal precession velocity $v_\phi = R\mathrm{d}\phi/\tau_\mathrm{d} = Rv_\mathrm{d}/r = mv_\perp/(2eBr)$ is derived.

$\delta(B_z^2/2\mu_0 - B_\theta^2/2\mu_0) > 0$ for $\delta r < 0$ is necessary to be stable because the first term is stabilizing magnetic pressure and the second term is destabilizing one against sausage perturbation.

3. Kink Instability When a toroidal plasma with large aspect ratio is suffered from $m = -1, n = 1$ kink instability. Explain the geometrical configuration of the plasma displacement $\boldsymbol{\xi}(r)\exp(i(\theta - z/R))$ of the toroidal plasma. Consider incompressible displacement $1/r(\partial r\xi_r(r)/\partial r) + i/r\xi_\theta(r)$ $+(i/R)\xi_z(r) = 0$ and neglect ξ_z term, so that we have $\xi_r = \xi_r(r)\cos(\theta - z/R)$ and $\xi_\theta = (\partial/\partial r)(r\xi_r(r))\sin(\theta - z/R) \sim \xi_r(r)\sin(\theta - z/R)$.

4. Energy Integral Derive the formula (8.96) of g in the energy integral from (8.95).

5. Energy Integral The energy integral per unit length of cylindrical plasma with the boundary of conductor is given by (8.93) as $W_{\rm p} = (\pi/2)\int_0^a (f|{\rm d}\xi_r/{\rm d}r|^2 + g|\xi_r|^2){\rm d}r$. In the case of $m = -1$ perturbation with the singular radius $r = r_{\rm s}$, $q(r_{\rm s}) = 1/n$, the solution of ξ_r of Euler's equation (8.97) is $\xi_r(r) = \xi_0$ =const. for $0 < r < r_{\rm s}$ and $\xi_r(r) = 0$ for $r_{\rm s} < r < a$. This solution satisfies the boundary conditions at $r = 0,\, a$ (refer to Section 8.5.1). Confirm that the energy integral by use of $m = -1$, $n = 1$ solution becomes

$$W_{\rm p} = \frac{\pi B_{\theta\,{\rm s}}^2}{2\mu_0}|\xi_0|^2\left(\frac{r_{\rm s}}{R}\right)^2\left(-\frac{\beta_{\rm p}}{2} - \int_0^1 \rho^3\left(\frac{1}{q^2} + \frac{2}{q} - 3\right){\rm d}\rho\right),$$

where $\rho = r/r_{\rm s}$, $q(r_{\rm s}) = 1$, $B_{\theta\,{\rm s}} \equiv (r_s/R)B_z$, $\beta_{\rm p} \equiv 2\mu_0\langle p_0\rangle/B_{\theta\,{\rm s}}^2$, and $\langle p_0\rangle \equiv -\int_0^{r_{\rm s}}(r/r_{\rm s})^2({\rm d}p_0/{\rm d}r){\rm d}r = r_{\rm s}^{-2}\int_0^{r_{\rm s}}(p_0 - p_{\rm s})2r{\rm d}r$. Refer to [8.33].

6. Negative Shear Negative shear configuration of tokamak is more stable against ballooning mode than the positive shear configuration. State the reason qualitatively. Refer to Section 8.7.

7. Toroidal Precession of Banana in Tokamak The magnitude of the magnetic field of tokamak with large aspect ratio is given by $B = B_0(1 - \epsilon(r/R)\cos\theta)$, so that the longitudinal adiabatic invariant $J_\parallel$ of banana is identical to (7.9) of trapped particle in mirror. Generalized coordinates are $u^1 \approx r$, $u^2 \approx \theta - \varphi/q_{\rm s}$ and $u^3 \approx R\varphi$ in tokamak. Furthermore, there is constraint $\delta\theta = 0$ on banana center. Confirm the angular velocity of toroidal precession of banana ion is ${\rm d}\varphi/{\rm d}t = mv_\perp^2/(2eBRr)$, which is already derived intuitively in the end of Section 8.9. Refer to Section 3.5 of [4.25].

9

Resistive Instabilities

In the preceding chapter we discussed instabilities of plasmas with zero resistivity. In such a case, the conducting plasma is frozen to the line of magnetic force. However, the resistivity of a plasma is not generally zero and the plasma may hence deviate from the magnetic line of force. Modes that are stable in the ideal case may in some instances become unstable if a finite resistivity is introduced.

Ohm's law is

$$\eta \boldsymbol{j} = \boldsymbol{E} + \boldsymbol{V} \times \boldsymbol{B}. \tag{9.1}$$

For simplicity here we assume that $\boldsymbol{E}$ is zero. The current density is $\boldsymbol{j} = \boldsymbol{V} \times \boldsymbol{B}/\eta$ and the $\boldsymbol{j} \times \boldsymbol{B}$ force is

$$\boldsymbol{F}_{\mathrm{s}} = \boldsymbol{j} \times \boldsymbol{B} = \frac{\boldsymbol{B}(\boldsymbol{V} \cdot \boldsymbol{B}) - \boldsymbol{V}B^2}{\eta}. \tag{9.2}$$

When η tends to zero, this force becomes infinite and prevents the deviation of the plasma from the line of magnetic force. When the magnitude B of the magnetic field is small, this force does not become large, even if η is small, and the plasma can deviate from the line of magnetic force. When we consider a perturbation with the propagation vector $\boldsymbol{k}$, only the parallel (to $\boldsymbol{k}$) component of the zeroth-order magnetic field $\boldsymbol{B}$ affects the perturbation, as will be shown later. Even if shear exists, we can choose a propagation vector $\boldsymbol{k}$ perpendicular to the magnetic field $\boldsymbol{B}$:

$$(\boldsymbol{k} \cdot \boldsymbol{B}) = 0. \tag{9.3}$$

Accordingly, if there is any force $\boldsymbol{F}_{\mathrm{dr}}$ driving the perturbation, this driving force may easily exceed the force $\boldsymbol{F}_{\mathrm{s}}$, which is very small for a perturbation where $(\boldsymbol{k} \cdot \boldsymbol{B}) = 0$, and the plasma becomes unstable. This type of instability is called *resistive instability.*

9.1 Tearing Instability

Let us consider a slab model in which the zeroth-order magnetic field $\boldsymbol{B}_0$ depends on only x and $\boldsymbol{B}$ is given as follows:

$$\boldsymbol{B}_0 = B_{0y}(x)\boldsymbol{e}_y + B_{0z}(x)\boldsymbol{e}_z. \tag{9.4}$$

From Ohm's law (9.1) we find

$$\frac{\partial \boldsymbol{B}}{\partial t} = -\nabla \times \boldsymbol{E} = \nabla \times ((\boldsymbol{V} \times \boldsymbol{B}) - \eta \boldsymbol{j}) = \nabla \times (\boldsymbol{V} \times \boldsymbol{B}) + \frac{\eta}{\mu_0}\Delta \boldsymbol{B}, \tag{9.5}$$

where η is assumed to be constant. It is assumed that the plasma is incompressible. Since the growth rate of the resistive instability is small compared with the MHD characteristic rate (inverse of Alfvén wave transit time) and the movement is slower than the sound velocity, the assumption of incompressibility is justified and it follows that

$$\nabla \cdot \boldsymbol{V} = 0. \tag{9.6}$$

The magnetic field $\boldsymbol{B}$ always satisfies

$$\nabla \cdot \boldsymbol{B} = 0. \tag{9.7}$$

The equation of motion is

$$\begin{aligned} \rho_{\mathrm{m}} \frac{\mathrm{d}\boldsymbol{V}}{\mathrm{d}t} &= \frac{1}{\mu_0}(\nabla \times \boldsymbol{B}) \times \boldsymbol{B} - \nabla p \\ &= \frac{1}{\mu_0}\left((\boldsymbol{B}_0 \cdot \nabla)\boldsymbol{B}_1 + (\boldsymbol{B}_1 \cdot \nabla)\boldsymbol{B}_0 - \frac{\nabla B^2}{2}\right) - \nabla p. \end{aligned} \tag{9.8}$$

Let us consider the perturbation expressed by $f_1(\boldsymbol{r}, t) = f_1(x)\times$ $\exp\left(i(k_y y + k_z z) \ + \gamma t\right)$. Then (9.5) reduces to

$$\gamma B_{1x} = i(\boldsymbol{k} \cdot \boldsymbol{B})V_x + \frac{\eta}{\mu_0}\left(\frac{\partial^2}{\partial x^2} - k^2\right) B_{1x}, \tag{9.9}$$

where $k^2 = k_y^2 + k_z^2$. The first term in the right-hand side of (9.8) becomes $(\boldsymbol{B}_0 \cdot \nabla)\boldsymbol{B}_1 = i(\boldsymbol{k} \cdot \boldsymbol{B}_0)\boldsymbol{B}_1$. The rotation of (9.8) is

$$\mu_0 \rho_{\mathrm{m}} \gamma \nabla \times \boldsymbol{V} = \nabla \times \left(i(\boldsymbol{k} \cdot \boldsymbol{B}_0)\boldsymbol{B}_1 + \left(B_{1x}\frac{\partial}{\partial x}\right)\boldsymbol{B}_0\right). \tag{9.10}$$

Equations (9.6) and (9.7) reduce to

$$\frac{\partial B_{1x}}{\partial x} + ik_y B_{1y} + ik_z B_{1z} = 0, \tag{9.11}$$

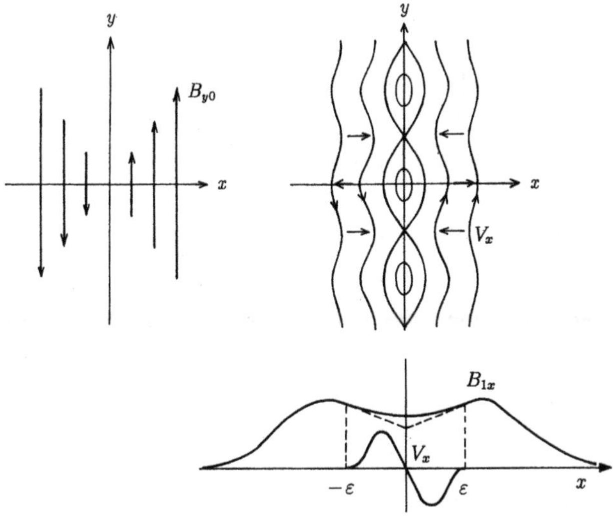

Figure 9.1 Zeroth-order magnetic configuration and magnetic islands due to tearing instability. Profiles of B_{1x} and V_x are also shown.

$$\frac{\partial V_x}{\partial x} + ik_y V_y + ik_z V_z = 0. \tag{9.12}$$

Multiply k_y and z component of (9.10) and multiply k_z and the y component and take the difference. Use the relations of (9.11) and (9.12); then we find [9.1]

$$\mu_0 \rho_{\rm m} \gamma \left(\frac{\partial^2}{\partial x^2} - k^2 \right) V_x = i(\boldsymbol{k} \cdot \boldsymbol{B}_0) \left(\frac{\partial^2}{\partial x^2} - k^2 \right) B_{1x} - i(\boldsymbol{k} \cdot \boldsymbol{B}_0)'' B_{1x}, \tag{9.13}$$

where $'$ is differentiation in x. Ohm's law and the equation of motion are reduced to (9.9) and (9.13). It must be noted that the zeroth-order magnetic field $\boldsymbol{B}_0$ appears only in the form of $(\boldsymbol{k} \cdot \boldsymbol{B}_0)$. When we introduce a function

$$F(x) \equiv (\boldsymbol{k} \cdot \boldsymbol{B}_0), \tag{9.14}$$

the location of $F(x) = 0$ is the position where resistive instabilities are likely to occur. We choose this position to be $x = 0$ (see Figure 9.1). $F(x)$ is equal to $(\boldsymbol{k} \cdot \boldsymbol{B}_0) \simeq (\boldsymbol{k} \cdot \boldsymbol{B}_0)' x$ near $x = 0$. As is clear from (9.9) and (9.13), B_{1x} is an even function and V_x is an odd function near $x = 0$. The term $|\Delta B_{1x}| \sim |\mu_0 k_y j_{1z}|$ can be large only in the region $|x| < \varepsilon$. Since the growth rate of resistive instability is much smaller than MHD growth rate, the left-hand side of the equation of motion (9.13) can be neglected in the region $|x| > \varepsilon$ and we have

$$\frac{{\rm d}^2 B_{1x}}{{\rm d}x^2} - k^2 B_{1x} = \frac{F''}{F} B_{1x}, \qquad |x| > \varepsilon. \tag{9.15}$$

The solution in the region $x > 0$ is

$$B_{1x} = e^{-kx}\left(\int_{-\infty}^{x} e^{2k\xi}\,\mathrm{d}\xi \int_{\infty}^{\xi} (F''/F)B_{1x}e^{-k\eta}\,\mathrm{d}\eta + A\right),$$

and the solution in the region $x < 0$ is

$$B_{1x} = e^{kx}\left(\int_{\infty}^{x} e^{-2k\xi}\,\mathrm{d}\xi \int_{-\infty}^{\xi} (F''/F)B_{1x}e^{k\eta}\,\mathrm{d}\eta + B\right).$$

Let us define Δ' as the difference between $B'_{1x}(+\varepsilon)$ at $x = +\varepsilon$ and $B'_{1x}(-\varepsilon)$ at $x = -\varepsilon$ as follows:

$$\Delta' = \frac{B'_{1x}(+\varepsilon) - B'_{1x}(-\varepsilon)}{B_{1x}(0)}. \tag{9.16}$$

Then the value of Δ' obtained from the solutions in the region $|x| > \varepsilon$ is given by

$$\Delta' = -2k - \frac{1}{B_{1x}(0)}\left(\int_{-\infty}^{-\varepsilon} + \int_{\varepsilon}^{\infty}\right)\exp(-k|x|)(F''/F)B_{1x}\,\mathrm{d}x. \tag{9.17}$$

For a trial function of

$$F(x) = F_{\mathrm{s}}x/L_{\mathrm{s}} \quad (|x| < L_{\mathrm{s}}), \qquad F(x) = F_{\mathrm{s}}x/|x| \quad (|x| > L_{\mathrm{s}}),$$

we can solve (9.15) and Δ' is reduced to

$$\Delta' = \left(\frac{2\alpha}{L_{\mathrm{s}}}\right)\frac{\mathrm{e}^{-2\alpha} + (1-2\alpha)}{\mathrm{e}^{-2\alpha} - (1-2\alpha)} \approx \frac{2}{L_{\mathrm{s}}}\left(\frac{1}{\alpha} - \alpha\right).$$

Here $\alpha \equiv kL_{\mathrm{s}}$ was used and L_{s} is *shear length* defined by $L_{\mathrm{s}} = (F_{\mathrm{s}}/F')_{x=0}$. For more general cases of $F(x)$, $B_{1x}(x)$ has logarithmic singularity at $x = 0$, since $F''/F \propto 1/x$ generally. Reference [9.2] describes the method to avoid difficulties arising from the corresponding logarithmic singularity.

Equations (9.9) and (9.13) in the region $|x| < \varepsilon$ reduce to

$$\frac{\partial^2 B_{1x}}{\partial x^2} - \left(k^2 + \frac{\gamma\mu_0}{\eta}\right)B_{1x} = -i\frac{\mu_0}{\eta}F'xV_x, \tag{9.18}$$

$$\frac{\partial^2 V_x}{\partial x^2} - \left(k^2 + \frac{(F')^2}{\rho_{\mathrm{m}}\eta\gamma}x^2\right)V_x = i\left(F'x\frac{1}{\rho_{\mathrm{m}}\eta} - \frac{F''}{\mu_0\rho_{\mathrm{m}}\gamma}\right)B_{1x}. \tag{9.19}$$

The value of Δ' obtained from the solution in the region $|x| < \varepsilon$ is given from (9.18) as follows:

$$\Delta' \times B_{1x}(0) = \frac{\partial B_{1x}(+\varepsilon)}{\partial x} - \frac{\partial B_{1x}(-\varepsilon)}{\partial x}$$

$$= \frac{\mu_0}{\eta} \int_{-\varepsilon}^{\varepsilon} \left(\left(\gamma + \frac{\eta}{\mu_0} k^2 \right) B_{1x} - iF' x V_x \right) \mathrm{d}x. \tag{9.20}$$

The value Δ' of (9.20) must be equal to the value of Δ' of (9.17). This requirement gives the eigenvalue γ and the growth rate of this resistive instability can be obtained [9.1]. However, we try to reduce the growth rate in a qualitative manner in this section. In the region $|x| < \varepsilon$, it is possible to write

$$\frac{\partial^2 B_{1x}}{\partial x^2} \sim \frac{\Delta' B_{1x}}{\varepsilon}.$$

It is assumed that the three terms of (9.9), namely the term of induced electric field (the left-hand side), the $\boldsymbol{V} \times \boldsymbol{B}$ term (the first term in the right-hand side), and Ohm's term (the second term) are the same order:

$$\gamma B_{1x} \sim \frac{\eta}{\mu_0} \frac{\Delta' B_{1x}}{\varepsilon}, \tag{9.21}$$

$$\gamma B_{1x} \sim iF' \varepsilon V_x. \tag{9.22}$$

Then (9.21) yields

$$\gamma \sim \frac{\eta}{\mu_0} \frac{\Delta'}{\varepsilon}. \tag{9.23}$$

Accordingly,

$$\Delta' > 0 \tag{9.24}$$

is the condition of instability. In order to get the value of γ, the evaluation of ε is necessary. Equation (9.13) reduces to

$$\mu_0 \rho_{\mathrm{m}} \gamma \left(\frac{-V_x}{\varepsilon^2} \right) \sim iF' \varepsilon \frac{\Delta' B_{1x}}{\varepsilon}. \tag{9.25}$$

If the terms V_x, B_{1x}, γ are eliminated by (9.21), (9.22), and (9.25), we find

$$\varepsilon^5 \sim \left(\frac{\eta}{\mu_0 a^2} \right)^2 (\Delta' a) \frac{\rho_{\mathrm{m}} \mu_0}{(F' a)^2} a^5,$$

$$\frac{\varepsilon}{a} \sim \left(\left(\frac{\tau_{\mathrm{A}}}{\tau_{\mathrm{R}}} \right)^2 (\Delta' a) \left(\frac{B_0}{F' a^2} \right)^2 \right)^{1/5} \sim S^{-2/5} (\Delta' a)^{1/5} \left(\frac{B_0}{(\boldsymbol{k} \cdot \boldsymbol{B}_0)' a^2} \right)^{2/5}, \tag{9.26}$$

where the physical quantities

$$\tau_{\mathrm{R}} = \frac{\mu_0 a^2}{\eta},$$

$$\tau_{\rm A} = \frac{a}{B_0/(\mu_0 \rho_{\rm m})^{1/2}}$$

are the resistive diffusion time and Alfvén transit time, respectively. A non-dimensional factor

$$S = \tau_{\rm R}/\tau_{\rm A}$$

is *magnetic Reynolds number* and a is a typical plasma size. Accordingly, the growth rate γ is given by

$$\gamma = \frac{\eta}{\mu_0 a^2}\frac{a}{\varepsilon}(\Delta' a) = \frac{(\Delta' a)^{4/5}\eta^{3/5}}{\mu_0^{4/5}\rho_{\rm m}^{1/5}a^{8/5}}\left(\frac{(\boldsymbol{k}\cdot\boldsymbol{B}_0)' a^2}{B_0}\right)^{2/5}$$

$$= \frac{(\Delta' a)^{4/5}}{\tau_{\rm R}^{3/5}\tau_{\rm A}^{2/5}}\left(\frac{(\boldsymbol{k}\cdot\boldsymbol{B}_0)' a^2}{B_0}\right)^{2/5} = \frac{(\Delta' a)^{4/5}}{S^{3/5}}\left(\frac{(\boldsymbol{k}\cdot\boldsymbol{B}_0)' a^2}{B_0}\right)^{2/5}\frac{1}{\tau_{\rm A}}. \quad (9.27)$$

Since this mode likely breaks up the plasma into a set of magnetic islands as is shown in Figure 9.1, this mode is called *tearing instability* [9.1].

The solution of (9.18) and (9.19) in the internal region $|x| \ll \varepsilon$ is solved in [9.3] and the explicit form of Δ' is derived. According to the results, γ is 0.55 times the value of (9.27) (refer to Problem 2 of this chapter).

The foregoing discussion has been based on the slab model. Let us consider this mode in a toroidal plasma. The poloidal and the toroidal components of the propagation vector $\boldsymbol{k}$ are m/r and $-n/R$, respectively. Accordingly, there are correspondences of $k_y \leftrightarrow m/r$, and $k_z \leftrightarrow -n/R$, and

$$(\boldsymbol{k}\cdot\boldsymbol{B}_0) = \frac{m}{r}B_\theta - \frac{n}{R}B_z = \frac{n}{r}B_\theta\left(\frac{m}{n} - q_{\rm s}\right), \qquad q_{\rm s} \equiv \frac{r}{R}\frac{B_z}{B_\theta}.$$

Therefore, weak positions for tearing instability are given by $(\boldsymbol{k}\cdot\boldsymbol{B}_0) = 0$ and these are rational surfaces satisfying $q(r_{\rm s}) = m/n$. The shear is given by

$$(\boldsymbol{k}\cdot\boldsymbol{B}_0)' = \frac{-n}{r}B_\theta\frac{{\rm d}q_{\rm s}}{{\rm d}r}, \qquad \frac{(\boldsymbol{k}\cdot\boldsymbol{B}_0)' r_{\rm s}^2}{B_0} = -n\left(\frac{r_{\rm s}}{R}\right)\frac{q_{\rm s}' r_{\rm s}}{q_{\rm s}}.$$

The tearing mode is closely related to the internal disruption in tokamak and plays an important role as is described in Section 4.3.

It has been assumed that the specific resistivity η and the mass density $\rho_{\rm m}$ are uniform and there is no gravitation (acceleration) $\boldsymbol{g} = 0$. If η depends on x, the resistive term in (9.5) becomes $\nabla\times(\eta\nabla\times\boldsymbol{B})/\mu_0$. When there is temperature gradient ($\eta' \neq 0$), rippling mode with short wavelength ($kL_{\rm s} \gg 1$) may appear in the smaller-resistivity-side (high-temperature-side) of $x = 0$ position. When there is gravitation, the term $\rho\boldsymbol{g}$ is added to the equation of motion (9.8). If the direction of $\boldsymbol{g}$ is opposite to $\nabla\rho_{\rm m}$ ($\boldsymbol{g}$ is toward low-density-side), gravitational interchange mode may appear [9.1].

9.2 Neoclassical Tearing Mode

Much attention has been focused on tokamak operational pressure limit imposed by non-ideal MHD instabilities, such as the effects of bootstrap current driven magnetic islands. At high β_p (poloidal beta) and low collisionality, the pressure gradient in the plasma gives rise to a bootstrap current (refer to Section 4.5.3). If an island develops, the pressure within the island tends to flatten out, thereby removing the drive for the bootstrap current. This gives rise to a helical 'hole' in the bootstrap current, which increases the size of the island.

Tearing instability was treated in slab model in the previous section. The zero order magnetic field $\boldsymbol{B}_0$ depends on only x and is given by $\boldsymbol{B}_0 = B_{0y}(x)\boldsymbol{e}_y + B_{0z}\boldsymbol{e}_z$, $|B_{0y}(x)| \ll |B_{0z}|$, $B_{0z} =$ const.. The basic equations are

$$\rho\left(\frac{\partial \boldsymbol{v}}{\partial t} + (\boldsymbol{v}\cdot\nabla)\boldsymbol{v}\right) = -\nabla p + \boldsymbol{j}\times\boldsymbol{B}, \tag{9.28}$$

$$-\boldsymbol{E} = \boldsymbol{v}\times\boldsymbol{B} - \eta\boldsymbol{j} = \frac{\partial \boldsymbol{A}}{\partial t}, \qquad \boldsymbol{A} = (0,0,-\psi), \tag{9.29}$$

$$B_x = -\frac{\partial\psi}{\partial y}, \qquad B_y = \frac{\partial\psi}{\partial x}$$

,

$$-\frac{\partial\psi}{\partial t} = (v_x B_y - v_y B_x) - \eta j_z = (\boldsymbol{v}\cdot\nabla)\psi - \eta j_z, \tag{9.30}$$

$$\nabla^2\psi = \mu_0 j_z. \tag{9.31}$$

Since

$$\boldsymbol{v} = \frac{\boldsymbol{E}\times\boldsymbol{B}}{B^2} = \left(\frac{E_y}{B_{0z}}, -\frac{E_x}{B_{0z}}, 0\right) = \left(-\frac{1}{B_{0z}}\frac{\partial\phi}{\partial y}, \frac{1}{B_{0z}}\frac{\partial\phi}{\partial x}, 0\right),$$

it is possible to introduce a stream function φ such as

$$v_x = -\frac{\partial\varphi}{\partial y}, \qquad v_y = \frac{\partial\varphi}{\partial x}.$$

Furthermore, z component of vorticity $w_z = (\nabla\times\boldsymbol{v})_z$ is introduced, then $w_z = \nabla^2\varphi$. Rotation of (9.28) yields

$$\rho\frac{\partial w_z}{\partial t} + (\boldsymbol{v}\cdot\nabla)w_z = (\nabla\times(\boldsymbol{j}\times\boldsymbol{B}))_z = (\boldsymbol{B}\cdot\nabla)j_z - (\boldsymbol{j}\cdot\nabla)B_z = (\boldsymbol{B}\cdot\nabla)j_z. \tag{9.32}$$

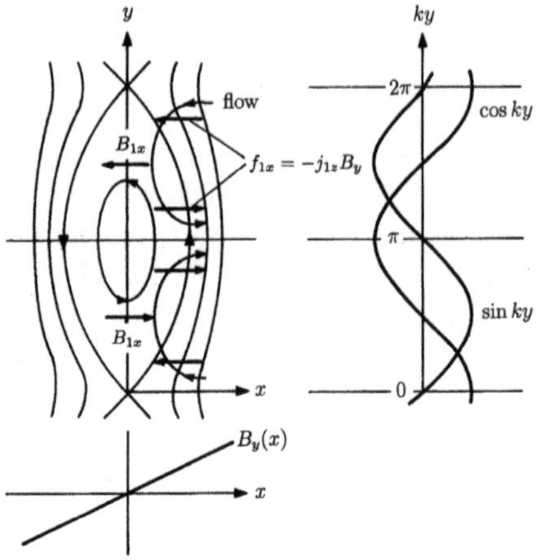

Figure 9.2 Tearing mode structure in the singular layer.

The relations $\nabla \cdot \boldsymbol{B} = 0$, $\nabla \cdot \boldsymbol{j} = 0$ were used here. The zeroth order flux function ψ_0 and the first order perturbation $\tilde{\psi}$ are

$$\psi_0(x) = B'_{0y}\frac{x^2}{2}, \qquad \boldsymbol{B}_0 = (0, B'_{0y}x, B_{0z}),$$

$$\tilde{\psi}(y,t) = \frac{B_{1x}(t)}{k}\cos ky, \qquad \boldsymbol{B}_1 = (B_{1x}(t))\sin ky, 0, 0), \qquad \tilde{\psi}_{\mathrm{A}}(t) \equiv \frac{B_{1x}(t)}{k},$$

$$\psi = \psi_0(x) + \tilde{\psi}(y,t) = B'_{0y}\frac{x^2}{2} + \frac{B_{1x}(t)}{k}\cos ky. \tag{9.33}$$

$x = 0$ is the location of the singular layer. The separatrix of islands is given by

$$B'_{0y}\frac{x^2}{2} + \frac{B_{1x}(t)}{k}\cos ky = \frac{B_{1x}(t)}{k}, \qquad x_{\mathrm{s}} = 2\left(\frac{B_{1x}}{kB'_{0y}}\right)^{1/2},$$

and the full width w of the island is

$$w = 4\left(\frac{B_{1x}}{kB'_{0y}}\right)^{1/2} = 4\left(\frac{\tilde{\psi}_{\mathrm{A}}(t)}{B'_{0y}}\right)^{1/2}. \tag{9.34}$$

The perturbation $B_{1x}(t)\sin ky$ growing with the growth rate γ induces a current $j_{1z} = E_{1z}/\eta = \gamma B_{1x}/\eta k$, which provides the x direction linear force $f_{1x} = -j_{1z}B'_{0y}x$ indicated in Figure 9.2. These drive the flow pattern of narrow vortices. Moving away from the resistive singular layer, the induced electric field produces a flow $v_x = -E_z/B_y = -\gamma B_{1x}\cos ky/(kB'_{0y}x)$. For incompressible flow (in strong equilibrium field B_{0z}), this requires a strongly sheared flow $v_y(x)$ over the layer $x \sim x_{\rm T}$, that is the narrow vortex pattern shown in Figure 9.2 and we have

$$v_y x_{\rm T} \sim v_x/k, \qquad v_y \sim v_x/kx_{\rm T} \sim \gamma B_{1x}/(k^2 B'_{0y} x_{\rm T}).$$

That this shear flow be driven against inertia by the torque produced by the linear forces requires

$$x_{\rm T} j_{1z} B_{0y} = \gamma\rho v_y/k, \quad B_{0y} = B'_{0y} x_{\rm T}, \quad \rightarrow \quad x_{\rm T}^4 = \frac{\gamma\rho}{j_{1z} k B'_{0y}} \frac{\gamma B_{1x}}{k^2 B'_{0y}} = \frac{\gamma\rho\eta}{(kB'_{0y})^2}$$

since $j_{1z} = E_z/\eta = \gamma B_{1x}/\eta k$. Thus, determined width of perturbation is

$$x_{\rm T} = \frac{(\gamma\rho\eta)^{1/4}}{(kB'_{0y})^{1/2}}. \tag{9.35}$$

This is consistent with the results (9.26) obtained by linear theory of tearing instability, which were described in Section 9.1 (Notation ε was used in Section 9.1 instead of $x_{\rm T}$).

Rutherford showed that the growth of the mode is drastically slowed and perturbation grows only linearly in time when non-linear effects are taken into account [9.4]. The vortex flow will induce the second order y-independent eddy current $\delta j_{1z} = -v_y B_{1x}/\eta \sim \gamma B_{1x}^2/(\eta k^2 B'_{0y} x_{\rm p}^2)$. The y-direction third order nonlinear forces $\delta f_y \sim \delta j_z B_{1x}$ indicated on Figure 9.3 provide a torque opposing vortex flow and decelerate v_y flow.

We restrict ourselves to the case where the inertia may be neglected in (9.32).

$$(\boldsymbol{B}\cdot\nabla) j_z = -\frac{\partial\psi}{\partial y}\frac{\partial j_z}{\partial x} + \frac{\partial\psi}{\partial x}\frac{\partial j_z}{\partial y} = 0 \qquad \rightarrow \qquad j_z = j_z(\psi).$$

Equation (9.30) yields

$$\frac{\partial\psi_0}{\partial t} + \frac{\partial\tilde{\psi}}{\partial t} = -v_x B'_{0y} x + \eta j_{1z}, \qquad \frac{\partial\psi_0}{\partial t} = \eta j_{0z},$$

$$\rightarrow \quad \frac{\partial\tilde{\psi}}{\partial t} = -\frac{\partial\varphi}{\partial y} B'_{0y} x + \eta j_{1z} - \eta j_{0z}. \tag{9.36}$$

We may eliminate φ from (9.36) by dividing by x and averaging over y at constant ψ. From (9.33), x is given by

$$x = \left(\frac{2}{B'_{0y}}(\psi - \tilde{\psi})\right)^{1/2} = \left(\frac{2}{B'_{0y}}\right)^{1/2} \tilde{\psi}_{\rm A}^{1/2}(W - \cos ky)^{1/2}, \quad W \equiv \frac{\psi}{\tilde{\psi}_{\rm A}}, \tag{9.37}$$

and

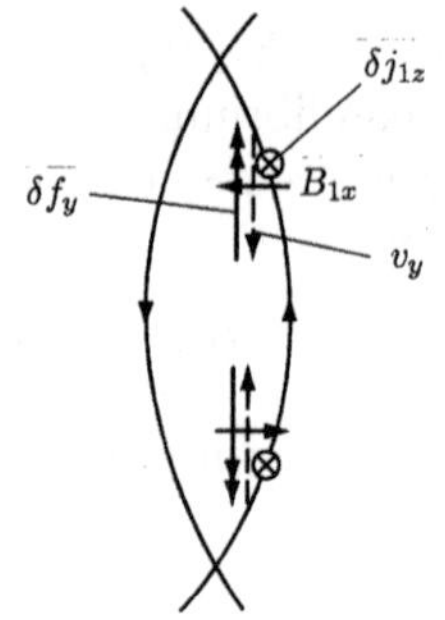

Figure 9.3 Nonlinear forces decelerating v_y flow in tearing mode.

$$\left\langle \frac{1}{(\psi - \tilde{\psi})^{1/2}} \right\rangle (\eta j_{1z}(\psi) - \eta j_{0z}(\psi)) = \left\langle \frac{\partial \tilde{\psi}(y,t)/\partial t}{(\psi - \tilde{\psi}(y,t))^{1/2}} \right\rangle,$$

$$j_{1z}(\psi) = j_{0z}(\psi) + \frac{1}{\eta} \left\langle \frac{\partial \tilde{\psi}/\partial t}{(\psi - \tilde{\psi})^{1/2}} \right\rangle \left\langle (\psi - \tilde{\psi})^{-1/2} \right\rangle^{-1}, \tag{9.38}$$

where

$$\langle f \rangle \equiv \frac{k}{2\pi} \int_0^{2\pi/k} f \mathrm{d}y.$$

For the outer solution we require the discontinuity in the logarithmic derivatives across singularity. We must match the logarithmic discontinuity from the solution within the singular layer to that of the outer solution:

$$\Delta' \equiv \left(\left. \frac{\partial \tilde{\psi}_{\mathrm{A}}}{\partial x} \right|_{+0} - \left. \frac{\partial \tilde{\psi}_{\mathrm{A}}}{\partial x} \right|_{-0} \right) \frac{1}{\tilde{\psi}_{\mathrm{A}}} = \left. \frac{\partial}{\partial x} \ln \tilde{\psi}_{\mathrm{A}} \right|_{-0}^{+0}.$$

We utilize $\nabla^2 \tilde{\psi} = \mu_0 j_{1z}$ and $\partial^2 \tilde{\psi}/\partial x^2 \approx \mu_0 j_{1z}$ and

$$\Delta' \tilde{\psi}_{\mathrm{A}} = \frac{\partial \tilde{\psi}_{\mathrm{A}}}{\partial x} = 2\mu_0 \left\langle \cos ky \int_{-\infty}^{\infty} j_{1z} \mathrm{d}x \right\rangle, \tag{9.39}$$

$$\mathrm{d}x = \left(\frac{1}{2B'_{0y}} \right)^{1/2} \frac{\mathrm{d}\psi}{(\psi - \tilde{\psi})^{1/2}}.$$

Insertion of (9.38) into (9.39) yields

$$\Delta' \tilde{\psi}_{\mathrm{A}} = 2 \frac{\mu_0}{\eta (2B'_{0y})^{1/2}} \int_{x=-\infty}^{x=\infty} \left\langle \frac{\partial \tilde{\psi}/\partial t}{(\psi - \tilde{\psi})^{1/2}} \right\rangle \left\langle (\psi - \tilde{\psi})^{-1/2} \right\rangle^{-1} \left\langle \frac{\cos ky}{(\psi - \tilde{\psi})^{1/2}} \right\rangle \mathrm{d}\psi$$

$$= \frac{4\mu_0}{\eta(2B'_{0y})^{1/2}} \int_{\psi_{\min}}^{\infty} \mathrm{d}\psi \left\langle \frac{\partial \tilde{\psi}_{\mathrm{A}} \cos ky/\partial t}{(\psi - \tilde{\psi})^{1/2}} \right\rangle \left\langle (\psi - \tilde{\psi})^{-1/2} \right\rangle^{-1} \left\langle \frac{\cos ky}{(\psi - \tilde{\psi})^{1/2}} \right\rangle .$$

Since

$$\int \mathrm{d}\psi \left\langle \frac{\cos ky}{(\psi - \tilde{\psi})^{1/2}} \right\rangle^2 \frac{1}{\left\langle (\psi - \tilde{\psi})^{-1/2} \right\rangle}$$

$$= \int \left\langle \frac{\cos ky}{(W - \cos ky)^{1/2}} \right\rangle^2 \frac{\mathrm{d}W \tilde{\psi}_{\mathrm{A}}^{1/2}}{\langle (W - \cos ky)^{-1/2} \rangle} \equiv A \tilde{\psi}_{\mathrm{A}}^{1/2},$$

we obtain

$$\Delta' \tilde{\psi}_{\mathrm{A}} = \frac{4\mu_0 A}{\eta(2B'_{0y})^{1/2}} \frac{\partial \tilde{\psi}_{\mathrm{A}}}{\partial t} \tilde{\psi}_{\mathrm{A}}^{1/2},$$

and

$$\frac{\partial}{\partial t} \tilde{\psi}_{\mathrm{A}}^{1/2} = \frac{\eta(2B'_{0y})^{1/2}}{8\mu_0 A} \Delta' .$$

Taking note (9.34), the time variation of the island width is reduced to

$$\frac{\mathrm{d}w}{\mathrm{d}t} = \frac{1}{2^{1/2} A} \frac{\eta}{\mu_0} \Delta' \approx \frac{\eta}{\mu_0} \Delta', \qquad \tau_{\mathrm{R}} \frac{\mathrm{d}}{\mathrm{d}t} \frac{w}{r_{\mathrm{s}}} = \Delta' r_{\mathrm{s}} \qquad \tau_{\mathrm{R}} \equiv \frac{\mu_0 r_{\mathrm{s}}^2}{\eta}. \tag{9.40}$$

Let us consider a toroidal plasma as is shown in Figure 9.4. The magnetic field

$$B_{\mathrm{p}} - \frac{\mathrm{n}r}{\mathrm{m}R} B_{\mathrm{t}} = \left(\frac{1}{q_{\mathrm{s}}(r)} - \frac{1}{q_{\mathrm{s}}} \right) \frac{r}{R} B_{\mathrm{t}}, \qquad \left(q_{\mathrm{s}} = \frac{\mathrm{m}}{\mathrm{n}} \right)$$

corresponds to B_{0y} in slab model near singular radius. The coordinates (x, y, z) in slab model correspond to radial direction $(r - r_{\mathrm{s}})$, poloidal direction $(\sim r\theta)$, and the direction of magnetic field at the rational surface in toroidal plasma, respectively. The flux function is

$$\psi(x, y) = \int_0^{r - r_{\mathrm{s}}} \left(\frac{1}{q_{\mathrm{s}}(r)} - \frac{1}{q_{\mathrm{s}}} \right) \frac{r}{R} B_{\mathrm{t}} \mathrm{d}x + \frac{B_{1x}}{k} \cos ky, \tag{9.41}$$

and the magnetic field is given by

$$B_{1x} = -\frac{\partial \psi}{\partial y} = B_{1x} \sin ky,$$

$$B_{0y} = \frac{\partial \psi}{\partial x} = \left(\frac{1}{q_{\mathrm{s}}(r)} - \frac{1}{q_{\mathrm{s}}} \right) \frac{r}{R} B_{\mathrm{t}} = -\frac{q'_{\mathrm{s}}}{q_{\mathrm{s}}} B_{\mathrm{p}} x = B'_{0y} x. \tag{9.42}$$

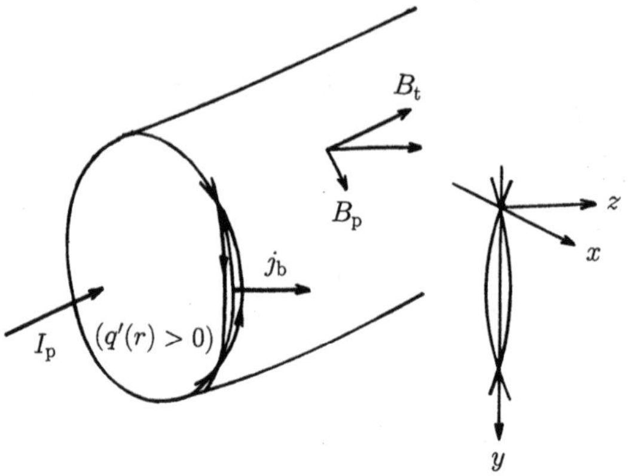

Figure 9.4 The coordinates in slab model and the coordinates in toroidal plasma. The coordinates (x, y, z) correspond radial direction $(r - r_s)$, poloidal direction $(\sim r\theta)$, and the direction of the magnetic field at the rational surface in toroidal plasma, respectively. The arrows in the island indicate the direction of magnetic field $B_p - (nr/mR)B_t$ (refer to (9.42)).

Equation (9.41) is reduced to

$$\psi(x, y) = B'_{0y}\frac{x^2}{2} + \frac{B_{1x}}{k}\cos ky. \tag{9.43}$$

The change of bootstrap current δj^b_{1z} induces the change in flux function $\delta\psi_b$ and electric field E_z

$$E_z = \frac{\partial\psi_b}{\partial t} = \eta\delta j^b_{1z}.$$

Discontinuity of logarithmic derivative due to δj^b_{1z} is

$$\Delta'_b = \frac{1}{\tilde{\psi}_A}\left(\left.\frac{\partial\tilde{\psi}^b_A}{\partial r}\right|_{r_s+} - \left.\frac{\partial\tilde{\psi}^b_A}{\partial r}\right|_{r_s-}\right) = \frac{1}{\tilde{\psi}_A}\int_{r_s-}^{r_s+}\mu_0\delta j^b_{1z}\mathrm{d}r,$$

where

$$\tilde{\psi}_A = \frac{B_{1x}}{k} = \frac{w^2 B'_{0y}}{16},$$

so that

$$\Delta'_b = \frac{16}{w^2 B'_{0y}}\int_{r_s-}^{r_s+}\mu_0\delta j^b_{1z}\mathrm{d}r.$$

Because of flattening of pressure profile due to the formation of island, δj^b_{1z} is given by (refer to (4.90))

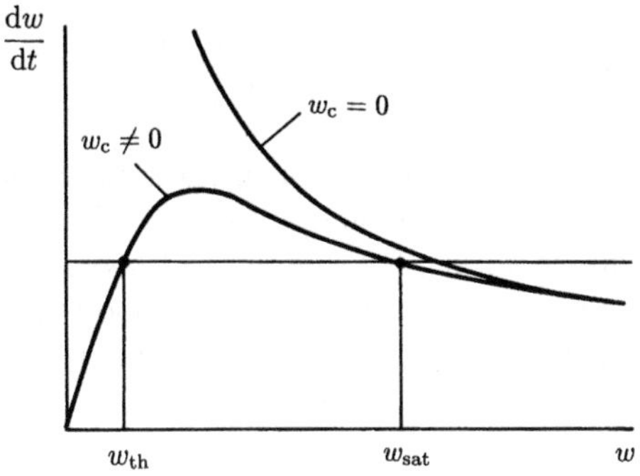

Figure 9.5 The curve of (9.46). w_{th} is the threshold width of island for the onset of neoclassical tearing mode and w_{sat} is the saturated width.

$$\delta j_{1z}^{b} = 0 - \left(-\frac{\epsilon_s^{1/2}}{B_p} \frac{dp}{dr} \right) = \frac{\epsilon_s^{1/2}}{B_p} \frac{dp}{dr}. \tag{9.44}$$

This is called the helical hole of bootstrap current. Thus, discontinuity of logarithmic derivative due to δj_{1z}^{b} is reduced to

$$\Delta_b' r_s = \frac{16\mu_0}{w^2 B_{0y}'} \left(\frac{\epsilon_s^{1/2}}{B_p} \frac{dp}{dr} \right)_{r_s} w r_s = \frac{8 r_s}{w} \frac{p}{B_p^2/2\mu_0} \epsilon_s^{1/2} \frac{L_q}{L_p},$$

$$B_{0y}' = -\frac{q'}{q} B_p \equiv -\frac{B_p}{L_q}, \qquad \frac{dp}{dr} \equiv -\frac{p}{L_p}.$$

Then the time variation of the island's width is given by

$$\tau_R \frac{d}{dt} \frac{w}{r_s} = \Delta' r_s + a \epsilon_s^{1/2} \beta_p \frac{L_q}{L_p} \frac{r_s}{w} \qquad a \sim 8. \tag{9.45}$$

The first term of the right-hand side of (9.45) is the Rutherford term and the second is the destabilizing term of bootstrap current. This is the equation of neoclassical tearing mode. When the transport and the effect of ion polarization current across the island are taken into account, a reduction in the bootstrap current takes place. Then the term due to bootstrap current in (9.45) is modified to

$$\tau_R \frac{d}{dt} \frac{w}{r_s} = \Delta' r_s + a_1 \beta_p \epsilon_s^{1/2} \frac{L_q}{L_p} \frac{r_s w}{w^2 + w_c^2} + a_2 \beta_p \left(\frac{L_q}{L_p} \right)^2 \frac{r_s \rho^2}{w^3}, \tag{9.46}$$

where w_c in the second term of the right-hand side is the effect of transport across the island parameterizing the magnitude of the contribution of the $\chi_\perp/\chi_\parallel$ model [9.5] and being given by the relation

$$w_c = 1.8r_s \left(\frac{8RL_q}{r_s^2 n}\right)^{1/2} \left(\frac{\chi_\perp}{\chi_\parallel}\right)^{1/4}.$$

The third term of ion polarization current in the right-hand side of (9.46) is given in [9.6]. Figure 9.5 shows the curve of (9.46). When the effect of w_c is included, there is a threshold w_{th} for the onset of neoclassical tearing mode. When w becomes large, the destabilizing term of bootstap current becomes weak and the island width is saturated. It is possible to control neoclassical tearing mode by local current drive in rational (singular) surface [9.7].

9.3 Resistive Drift Instability

A finite density and temperature gradient always exists at a plasma boundary. Configurations including a gradient may be unstable under certain conditions. Let us consider a slab model. The directon of the uniform magnetic field is taken in the z direction and $\boldsymbol{B}_0 = (0, 0, B_0)$. The x axis is taken in the direction of the density gradient with the positive direction outward from the plasma. The pressure is $p_0(x)$, (see Figure 9.6). The zeroth-order plasma current is $\boldsymbol{j}_0 = (0, p_0'/B_0, 0)$ and we assume that the flow velocity and the electric field are zero $\boldsymbol{V}_0 = 0$, $\boldsymbol{E}_0 = 0$ in the zeroth order. The flow velocity due to classical diffusion is neglected here. Electron inertia is also neglected. The usual relations in this configuration are

$$Mn\frac{\partial \boldsymbol{V}}{\partial t} = \boldsymbol{j} \times \boldsymbol{B} - \nabla p, \tag{9.47}$$

$$\boldsymbol{E} + \boldsymbol{V} \times \boldsymbol{B} = \eta \boldsymbol{j} + \frac{1}{en}\left(\boldsymbol{j} \times \boldsymbol{B} - \nabla p_e\right), \tag{9.48}$$

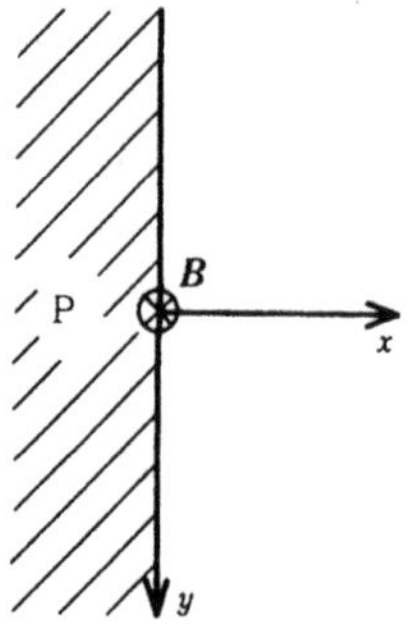

Figure 9.6 Slab model of resistive drift wave.

$$\frac{\partial n}{\partial t} + \nabla \cdot (n\boldsymbol{V}) = 0, \tag{9.49}$$

$$\nabla \cdot \boldsymbol{j} = 0, \tag{9.50}$$

where M is the ion mass. In this configuration, electrostatic perturbations are considered here. The first order electric field $\boldsymbol{E}_1$ is expressed by the electrostatic potential $\boldsymbol{E}_1 = -\nabla\phi_1$ and the first order magnetic field perturbation is zero $\boldsymbol{B}_1 = 0$ ($\partial B/\partial t = \nabla \times \boldsymbol{E}$). The characteristics of electrostatic perturbation will be explained in Chapter 10 in detail. For simplicity the ion temperature is assumed to be zero $T_{\mathrm{i}} = 0$. Let us consider the mode

$$n_1 = n_1(x) \exp i(ky + k_\parallel z - \omega t),$$
$$\phi_1 = \phi_1(x) \exp i(ky + k_\parallel z - \omega t).$$

Equations (9.47) and (9.48) reduce to

$$-i\omega M n_0 \boldsymbol{V}_1 = \boldsymbol{j}_1 \times \boldsymbol{B}_0 - \kappa T_{\mathrm{e}} \nabla n_1, \tag{9.51}$$

$$\boldsymbol{j}_1 \times \boldsymbol{B}_0 - \kappa T_{\mathrm{e}} \nabla n_1 = e n_0(-\nabla\phi_1 + \boldsymbol{V}_1 \times \boldsymbol{B}_0 - \eta \boldsymbol{j}_1). \tag{9.52}$$

Equations (9.51) and (9.52) yield

$$i\omega \left(\frac{M}{e}\right) \boldsymbol{V}_1 = \nabla\phi_1 - \boldsymbol{V}_1 \times \boldsymbol{B}_0 + \eta \boldsymbol{j}_1. \tag{9.53}$$

When η is small ($\nu_{\mathrm{ei}} \ll \Omega_{\mathrm{e}}$), the contribution of $\eta \boldsymbol{j}$ can be neglected in (9.53), i.e., we may write

$$V_x = -ik\frac{\phi_1}{B_0}, \qquad V_y = \left(\frac{\omega}{\Omega_{\mathrm{i}}}\right)\frac{k\phi_1}{B_0}, \qquad V_z = \left(-\frac{\Omega_{\mathrm{i}}}{\omega}\right)\frac{k_\parallel \phi_1}{B_0}.$$

Ω_{i} is the ion cyclotron frequency ($\Omega_{\mathrm{i}} = -eB/M$). The wave frequency ω was assumed to be low $(\omega/\Omega_{\mathrm{i}})^2 \ll 1$. The x, y component of (9.51) and the z component of (9.52) are

$$j_x = -ik\frac{T_{\mathrm{e}} n_1}{B_0}, \qquad j_y = k n_0 \left(\frac{\omega}{\Omega_{\mathrm{i}}}\right)\frac{e\phi_1}{B_0}, \qquad j_z = \frac{ik_\parallel}{e\eta}\left(\kappa T_{\mathrm{e}}\frac{n_1}{n_0} - e\phi_1\right).$$

Since (9.50) is $j_x' + ikj_y + ik_\parallel j_z = 0$, and (9.49) is $-i\omega n_1 + n_0' V_x + n_0 ik V_y + n_0 ik_\parallel V_z = 0$, we find

$$\frac{n_1}{n_0} - \left(1 + i\left(\frac{k}{k_\parallel}\right)^2 \frac{\omega}{\Omega_{\mathrm{i}}}\frac{e n_0 \eta}{B_0}\right)\frac{e\phi_1}{T_{\mathrm{e}}} = 0, \tag{9.54}$$

$$\frac{n_1}{n_0} - \left(\left(\frac{k_\parallel}{\omega} \right)^2 \frac{T_\mathrm{e}}{M} + k^2 \frac{T_\mathrm{e}}{eB_0 \Omega_\mathrm{i}} + \frac{k(-n_0'/n_0)T_\mathrm{e}}{eB_0} \frac{1}{\omega} \right) \frac{e\phi_1}{\kappa T_\mathrm{e}} = 0. \qquad (9.55)$$

The dispersion equation is given by the determinant of the coefficients of (9.54) and (9.55):

$$1 + i \left(\frac{k}{k_\parallel} \right)^2 \frac{\omega}{\Omega_\mathrm{i}} \frac{\nu_\mathrm{ei}}{\Omega_\mathrm{e}} - \left(\frac{k_\parallel}{\omega} \right)^2 c_\mathrm{s}^2 + (k\rho_\Omega)^2 - \frac{\omega_\mathrm{e}^*}{\omega} = 0, \qquad (9.56)$$

where $\eta = m_\mathrm{e}\nu_\mathrm{ei}/ne^2$, $(n_0 e\eta)/B_0 = \nu_\mathrm{ei}/\Omega_\mathrm{e}$. $c_\mathrm{s}^2 \equiv T_\mathrm{e}/M$, $\rho_\Omega \equiv c_\mathrm{s}/|\Omega_\mathrm{i}|$ and $\omega_\mathrm{e}^* \equiv k(-n_0'/n_0)(T_\mathrm{e}/eB_0)$. The *drift velocities* $\boldsymbol{v}_\mathrm{di}, \boldsymbol{v}_\mathrm{de}$ of ions and electrons due to the density gradient ∇n_0 are given by

$$\boldsymbol{v}_\mathrm{di} = \frac{-(T_\mathrm{i}\nabla n_0/n_0) \times \boldsymbol{b}}{eB_0} = \frac{-T_\mathrm{i}}{eB_0} \left(\frac{-n_0'}{n_0} \right) \boldsymbol{e}_y,$$

$$\boldsymbol{v}_\mathrm{de} = \frac{(T_\mathrm{e}\nabla n_0/n_0) \times \boldsymbol{b}}{eB_0} = \frac{T_\mathrm{e}}{eB_0} \left(\frac{-n_0'}{n_0} \right) \boldsymbol{e}_y.$$

These drift velocities did not appear in the drift velocities of a single particle. These drift velocities are called ion and electron diamagnetic velocities, respectively. The origin of diamagnetic drift is due to the macroscopic effect of density gradient (refer to Chapter 2 of [2.4], Section 3.4 of [9.8]). The diamagnetic current (4.41) that appeared in MHD equilibrium is expressed by $\boldsymbol{j}_\perp = ne(\boldsymbol{v}_\mathrm{di} - \boldsymbol{v}_\mathrm{de})$. The *drift frequencies* of ions and electrons are defined by $\omega_\mathrm{i}^* \equiv kv_\mathrm{di} = -k(-n_0'/n_0)(T_\mathrm{i}/eB_0)$ and $\omega_\mathrm{e}^* \equiv kv_\mathrm{de} = k(-n_0'/n_0)(T_\mathrm{e}/eB_0)$, respectively. As $n_0'/n_0 < 0$, $\omega_\mathrm{e}^* > 0$ and $\omega_\mathrm{i}^* = -(T_\mathrm{i}/T_\mathrm{e})\omega_\mathrm{e}^* < 0$. The dispersion equation is reduced to

$$\left(\frac{\omega}{\omega_\mathrm{e}^*} \right)^2 - i \left(1 + (k\rho_\Omega)^2 - \frac{k_\parallel^2 c_\mathrm{s}^2}{\omega^2} \right) \frac{\Omega_\mathrm{e}\Omega_\mathrm{i}}{\nu_\mathrm{ei}\omega_\mathrm{e}^*} \left(\frac{k_\parallel}{k} \right)^2 \left(\frac{\omega}{\omega_\mathrm{e}^*} \right) + i \frac{\Omega_\mathrm{e}\Omega_\mathrm{i}}{\nu_\mathrm{ei}\omega_\mathrm{e}^*} \left(\frac{k_\parallel}{k} \right)^2 = 0. \qquad (9.57)$$

Denote $\omega/\omega_\mathrm{e}^* = x + iz$, and $-(\Omega_\mathrm{e}\Omega_\mathrm{i}\ /\nu_\mathrm{ei}\omega_\mathrm{e}^*)(k_\parallel/k)^2 = y^2$ and assume $(k\rho_\Omega)^2 - (k_\parallel c_\mathrm{s}/\omega)^2 \ll 1$. The dispersion equation is then

$$(x + iz)^2 + iy^2(x + iz) - iy^2 = 0. \qquad (9.58)$$

The dependence of the two solutions $x_1(y), z_1(y)$ and $x_2(y), z_2(y)$ on $y \propto (k_\parallel/k)$, is shown in Figure 9.7. As $z_2(y) < 0$, the mode corresponding to $x_2(y), z_2(y)$ is stable. This wave propagates in the direction of the ion drift. The solution $x_1, z_1 > 0$ propagates in the direction of the electron drift and it is unstable. If the value of $(k_\parallel/k)$ is adjusted to be $y \simeq 1.3$, the z_1 value becomes maximum to be $z_1 \approx 0.25$ and the growth rate is $\mathrm{Im}\,\omega \approx 0.25\,\omega_\mathrm{e}^*$. If η is small, the wavelength of the most unstable wave becomes long and the necessary number of collisions to interrupt the electron motion along the

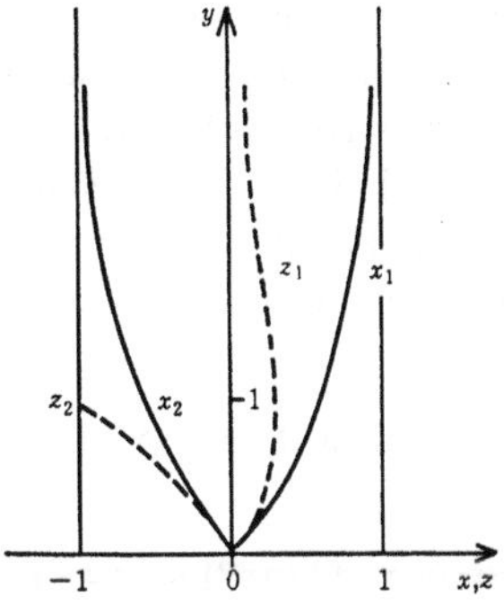

Figure 9.7 Dependence of $\omega/\omega_e^* = x + iz$ on $y \propto k_\parallel/k$ for resistive drift instability.

magnetic line of force is maintained. If the lower limit of $k_\parallel$ is fixed by an appropriate method, the growth rate is

$$\mathrm{Im}(\omega/\omega_e^*) \approx y^{-2} = \frac{\nu_{ei}\omega_e^*}{\Omega_e|\Omega_i|}\left(\frac{k}{k_\parallel}\right)^2,$$

and the growth rate is proportional to $\eta \propto \nu_{ei}$. This instability is called *resistive drift instability* or *dissipative drift instability* [9.9, 9.10].

If the ion's inertia term can be neglected ($M \to 0$, $|\Omega_i| \to \infty$ in (9.56)), the dispersion equation becomes $\omega^2 - \omega\omega_e^* - (k_\parallel c_s)^2 = 0$. The instability does not appear. This instability originates in the charge separation between electrons and ions due to ion inertia. The charge separation thus induced is neutralized by the electrons motion along the lines of magnetic force. However, if the parallel motion of electrons is interrupted by collision, i.e., resistivity, the charge separation grows and the wave becomes unstable. This instability is therefore also called *collisional drift instability.*

The instability does not appear in the collisionless case in the framework of MHD theory. However, the instability may occur even in the collisionless case when it is analyzed by the kinetic theory. This instability is called *collisionless drift instability* (refer to Section 11.6.1).

9.4 Resistive Wall Mode

MHD kink instabilities in tokamak are of major importance because they have a beta limit. In the absence of a conducting wall, the results obtained in [9.11] with a variety of current and pressure profiles show that this beta limit is of the form $\beta(\%)/(I_p(\mathrm{MA})/a(\mathrm{m})B(\mathrm{T})) \equiv \beta_N < 2.8$ (refer to (4.49)).

However, the external kink can be stabilized at a higher value of β_N by including a closely fitting conducting wall. The situation is complicated by the existence of a resistive wall in the case when an ideal MHD instability is stabilized by a perfect conducting wall, but unstable if the wall is removed. In this situation, there is a resistive wall mode that grows on the resistive time of the wall. Furthermore, it is an interesting issue whether this resistive wall mode is stabilized by plasma rotation or not. In this section, the analysis of resistive wall mode is described.

9.4.1 Growth Rate of Resistive Wall Mode

Basic equations of motion in slab model are already given by (9.9) and (9.13). In the case of $\eta = 0$ and $|\mathrm{d}^2/\mathrm{d}x^2| \gg k^2$, we have

$$B_{1x} = i(\boldsymbol{k}\cdot\boldsymbol{B})\xi_x,$$

$$\mu_0\gamma^2\nabla(\rho_{\mathrm{m}}\nabla\xi_x) = i(\boldsymbol{k}\cdot\boldsymbol{B})\Delta B_{1x} - i(\boldsymbol{k}\cdot\boldsymbol{B})''B_{1x}.$$

For cylindrical coordinates, the same mathematical process as for the slab model leads to

$$(rB_{1r}) = iF(r\xi_r), \tag{9.59}$$

$$\frac{\mu_0\gamma^2}{F}\nabla(\rho_{\mathrm{m}}\nabla(ir\xi_r)) = -\nabla^2(rB_{1r}) + \frac{F''}{F}(rB_{1r}), \tag{9.60}$$

where

$$F = (\boldsymbol{k}\cdot\boldsymbol{B}) = \left(\frac{-\mathrm{m}}{r}B_\theta + \frac{\mathrm{n}}{R}B_z\right) = \frac{B_z}{R}\left(\mathrm{n} - \frac{\mathrm{m}}{q(r)}\right) = \frac{B_\theta}{r}(\mathrm{n}q(r) - \mathrm{m}).$$

The flux function $\psi = A_z(r,\theta) = A_z(r)\exp(-\mathrm{m}\theta)$, z component of vector potential, is introduced and then

$$B_{1r} = \frac{1}{r}\frac{\partial A_z}{\partial\theta} = -i\frac{\mathrm{m}}{r}\psi, \qquad B_z = -\frac{\partial A_z}{\partial r} = -\frac{\partial\psi}{\partial r}.$$

Equation (9.60) is reduced to

$$\frac{\mu_0\gamma^2}{F}\nabla\left(\rho_{\mathrm{m}}\nabla_\perp\frac{\psi}{F}\right) = -\nabla^2\psi - \frac{\mathrm{m}\mu_0}{F}\frac{j'}{r}\psi, \qquad F'' \approx -\frac{\mathrm{m}\mu_0 j'}{r}. \tag{9.61}$$

First, a step function model of cylindrical plasma presented by Finn [9.12] is used for analysis. The mass density and plasma current profiles are flat within the plasma $r < a$ as is shown in Figure 9.8, that is,

$$j(r) = j_0, \quad \rho(r) = \rho_0, \quad q(r) = q, \qquad \text{for } r < a,$$

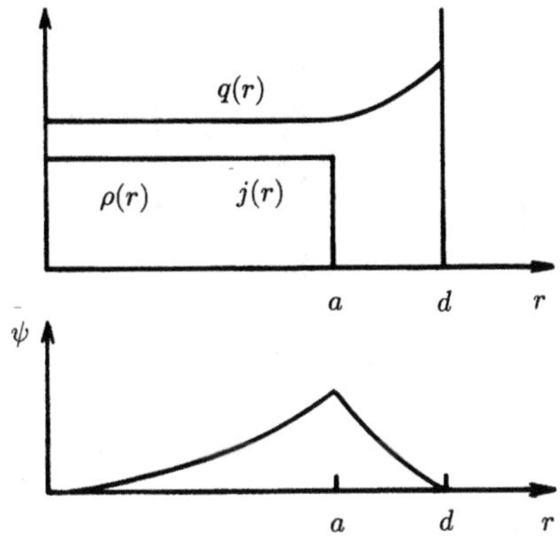

Figure 9.8 Upper figure: Profiles of mass density $\rho(r)$, current density $j(r)$, and q profile $q(r)$. The plasma radius is $r = a$ and the wall is located at $r = d$. Lower figure: Profile of flux function $\psi(r)$ in the case of a conducting wall.

and $j(r) = 0,\ \ \rho(r) = 0,\ \ q(r) = q(r)$ for $r < a$. Then (9.61) for $r < a$ yields

$$\left(1 + \frac{\mu_0 \rho_0 \gamma^2}{F^2}\right) \nabla^2 \psi = 0, \qquad \psi(r) = \psi_0 \left(\frac{r}{a}\right)^{\rm m}, \qquad \frac{\psi(a_-)'}{\psi(a)} = \frac{\rm m}{a}, \tag{9.62}$$

and (9.61) for $r > a$ yields

$$\nabla^2 \psi = 0, \qquad \psi(r) = \frac{\psi(a)}{1 - \alpha} \left(\left(\frac{r}{a}\right)^{-\rm m} - \alpha \left(\frac{r}{a}\right)^{\rm m} \right).$$

When a conducting wall is located at $r = d$, $\psi(d) = 0$ must be satisfied and $\alpha = (a/d)^{2\rm m}$. Then

$$\frac{\psi'(a_+)}{\psi(a)} = -\frac{\rm m}{a} \frac{1 + (a/d)^{2\rm m}}{1 - (a/d)^{2\rm m}}. \tag{9.63}$$

At the plasma boundary, (9.61) yields

$$\frac{\psi'(a_+)}{\psi(a)} - \left(1 + \frac{\mu_0 \rho_0 \gamma^2}{B_\theta^2 ({\rm n}q - {\rm m})^2 / a^2}\right) \frac{\psi'(a_-)}{\psi(a)} = \frac{{\rm m}\mu_0}{Fa} j_0 = \frac{\rm m}{a} \frac{2}{({\rm n}q - {\rm m})}, \tag{9.64}$$

since $\mu_0 j_0 = 2B_\theta / a$ for flat current profile. The growth rate $\gamma_{\rm c}(d)$ of the MHD perturbation in the case of a conducting wall at $r = d$ is reduced from (9.62), (9.63) and (9.64) as follows:

$$\gamma_{\rm c}(d)^2 \tau_{{\rm A}\theta}^2 = -2({\rm n}q - {\rm m}) \left(1 + \frac{({\rm n}q - {\rm m})}{1 - (a/d)^{2\rm m}}\right). \tag{9.65}$$

The stable region in d/a-q diagram is shown in Figure (9.9).

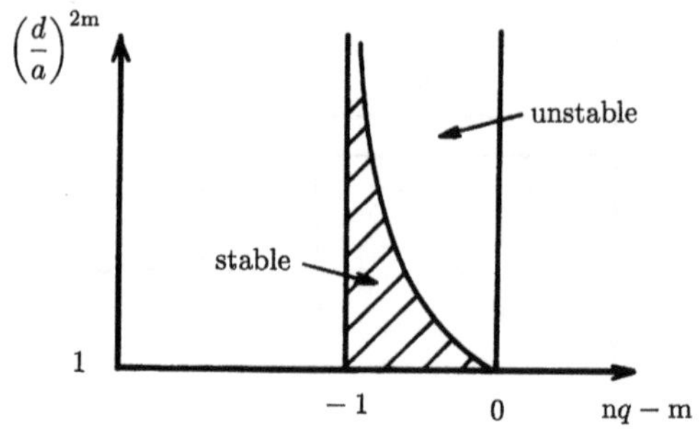

Figure 9.9 Stable region in $nq-$m-d/a-diagram.

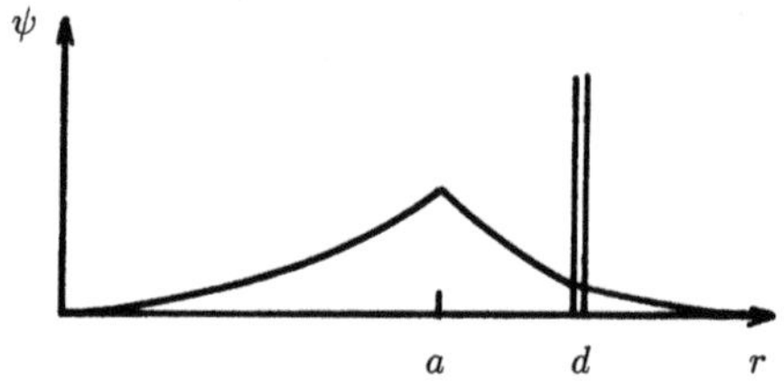

Figure 9.10 Profile of $\psi(r)$ in the case of a resistive wall at $r = d$.

When a thin resistive wall is located at $r = d$ instead of a conducting wall, the external solution of ψ is modified and is given by (see Figure (9.10))

$$\psi(r) = \psi(d)(r/d)^{-\mathrm{m}}, \qquad (r > d),$$

$$\psi(r) = \frac{\psi(d)}{1-\alpha_{\mathrm{res}}}\left((r/d)^{-\mathrm{m}} - \alpha_{\mathrm{res}}(r/d)^{2\mathrm{m}}\right), \qquad (d > r > a). \qquad (9.66)$$

When the wall current and wall specific resistivity are denoted by j_{w} and η_{w}, there are the following relations:

$$\nabla^2\psi = -\mu_0 j_{\mathrm{w}}, \qquad j_{\mathrm{w}} = E_z/\eta_{\mathrm{w}} = -\frac{\gamma}{\eta_{\mathrm{w}}}\psi.$$

Then the discontinuity of logarithmic derivative at $r = d$ is

$$\frac{\psi(d_+)' - \psi(d_-)'}{\psi(d)} = -\frac{\mu_0 \int j_{\mathrm{w}} \mathrm{d}r}{\psi} = \frac{\mu_0\gamma/\eta_{\mathrm{w}} \int \psi \mathrm{d}r}{\psi} = \frac{\mu_0\gamma\delta_{\mathrm{w}}}{\eta_{\mathrm{w}}} = \frac{\gamma\tau_{\mathrm{w}}}{d},$$

$$\tau_{\mathrm{w}} \equiv \mu_0 d\delta_{\mathrm{w}}/\eta_{\mathrm{w}},$$

where $\delta_{\rm w}$ is the wall thickness and we obtain

$$\frac{\psi(d_-)'}{\psi(d)} = -\frac{\rm m}{d} - \frac{\gamma_{\rm res}\tau_{\rm w}}{d}.$$

Thus, $\alpha_{\rm res}$ in (9.66) is given by

$$\alpha_{\rm res} = \frac{\gamma_{\rm res}\tau_{\rm w}/(2{\rm m})}{1+\gamma_{\rm res}\tau_{\rm w}/(2{\rm m})}.$$

We have already $\psi'(a_+)/\psi(a)$ from (9.64) as follows:

$$\frac{\psi'(a_+)}{\psi(a)} = \frac{\rm m}{a}\left(1+\frac{\gamma_{\rm res}^2\tau_{\rm A\theta}^2}{({\rm n}q-{\rm m})^2}+\frac{2}{({\rm n}q-{\rm m})}\right).$$

On the other hand, $\psi'(a_+)/\psi(a)$ is also given by (9.66) as follows:

$$\frac{\psi'(a_+)}{\psi(a)} = -\frac{\rm m}{a}\frac{1+\alpha_{\rm res}(a/d)^{2{\rm m}}}{1-\alpha_{\rm res}(a/d)^{2{\rm m}}}. \tag{9.67}$$

Therefore, the growth rate of the mode in the resistive wall is given by

$$\gamma_{\rm res}(d)^2\tau_{\rm A\theta}^2 = -2({\rm n}q-{\rm m})\left(1+\frac{({\rm n}q-{\rm m})}{1-\alpha_{\rm res}(a/d)^{2{\rm m}}}\right). \tag{9.68}$$

Since

$$\frac{1}{1-\alpha_{\rm res}(a/d)^{2{\rm m}}} = \frac{1}{1+R}+\frac{R}{1+R}\frac{1}{1-(a/d)^{2{\rm m}}}, \quad R = (1-(a/d)^{2{\rm m}})\frac{\gamma_{\rm res}(d)\tau_{\rm w}}{2{\rm m}},$$

(9.68) is reduced to

$$\gamma_{\rm res}(d)^2 = \frac{\gamma_{\rm c}^2(\infty)+R\gamma_{\rm c}^2(d)}{1+R}. \tag{9.69}$$

Let us consider the case where the mode is stable with the conducting wall at $r=d$ and is unstable without the wall; that is, $\gamma_{\rm c}(d)<0$ and $\gamma_{\rm c}(\infty)>0$. Then the growth rate of the mode with thin resistive wall at $r=d$ is (under the assumption of $\gamma_{\rm res}(d)^2 \ll \gamma_{\rm c}^2(d), \gamma_{\rm c}^2(\infty)$)

$$R = -\frac{\gamma_{\rm c}^2(\infty)}{\gamma_{\rm c}^2(d)}, \qquad \gamma_{\rm res}(d)\tau_{\rm w} = \frac{2{\rm m}}{1-(a/d)^{2{\rm m}}}\left(-\frac{\gamma_{\rm c}^2(\infty)}{\gamma_{\rm c}^2(d)}\right). \tag{9.70}$$

Therefore, the growth rate is the order of inverse resistive wall time constant. For $d \to a$, $\gamma_{\rm res}(d)\tau_{\rm w} \to -2{\rm m}(1+{\rm n}q-{\rm m})/({\rm n}q-{\rm m})$ remains finite. This mode is called resistive wall mode (RWM). When wall position d approaches to critical one $d_{\rm cr}$, where ideal MHD mode becomes unstable even with conducting wall $\gamma_{\rm c}(d_{\rm cr})=0$, the growth rate of RWM becomes infinity as is seen in (9.70) and connects to ideal MHD mode.

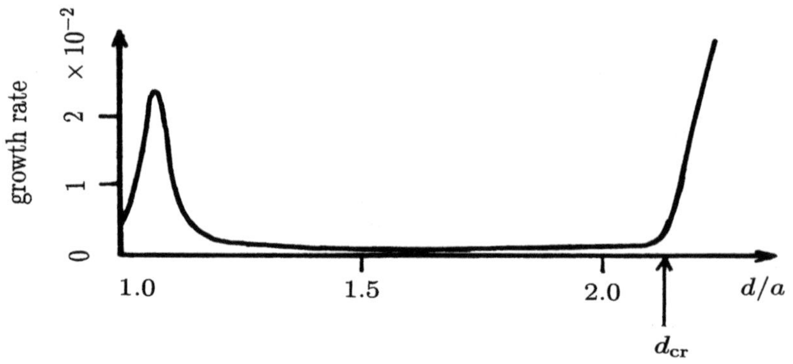

Figure 9.11 The growth rate $\gamma_{\rm res}(d)$ versus resistive wall position d/a in the case of $\omega_{\rm rot} = 0.5$. $\gamma_{\rm res}(d)$ and $\omega_{\rm rot}$ are in units of $\tau_{\rm A}^{-1} = B/(a(\mu_0\rho)^{1/2}) = (B/B_\theta)\tau_{{\rm A}\theta}^{-1}$, $d_{\rm cr} = 2.115$. ($R/a = 5, q_0 = 1.05$, m=2, n=1, $\tau_{\rm A}/\tau_{\rm w} = 5\times10^{-4}$) [9.12].

When plasmas rotate rigidly and if perturbations also rotate without slip in the plasmas, the effect of rigid rotation is included by adding a Doppler shift

$$\gamma \rightarrow \gamma + i\boldsymbol{k}\cdot\boldsymbol{v} = \gamma + i\left(\frac{\rm n}{R}v_z - {\rm m}\omega_\theta\right) = \gamma + \omega_{\rm rot}$$

on the left-hand side of (9.69),but not on the right-hand side of (9.69). Figure 9.11 shows the dependence of the growth rate $\gamma_{\rm res}(d)$ on the resistive wall position d/a while Ω is fixed for the case of $\omega_{\rm rot}\tau_{\rm A} = 0.5$, $\tau_{\rm A}^{-1} \equiv B/(a(\mu_0\rho)^{1/2}) = (B/B_\theta)\tau_{{\rm A}\theta}^{-1}$, $R/a = 5, q_0 = 1.05$, m=2, n=1, $\tau_{\rm A}/\tau_{\rm w} = 5\times10^{-4}$. When d/a increases beyond $d_{\rm cr}/a$, plasma becomes unstable in ideal MHD time scale. For d/a approaching near unity, there is an enhancement in the growth rate due to the inductance factor $(1-(a/d)^{2{\rm m}})$. Due to the fact the effective flux decay time becomes shorter, the resistive wall behaves as if the wall is more resistive. There is an initial increase in the growth rate with $\omega_{\rm rot}$, after which the growth rate decreases, but does not go to zero as $\omega_{\rm rot}\rightarrow\infty$.
Ward and Bondeson [9.13] analyzed the full toroidal plasma with resistive wall by numerical code. Numerical analysis indicates that there are two modes. One is the mode that has zero frequency in the frame of plasma and perturbation hardly penetrates the resistive wall, the "plasma mode". In other words, resistive wall behaves as if the wall is ideal when $\omega_{\rm rot} \gg \tau_{\rm w}^{-1}$. The other one is the mode that the perturbation rotates slowly with resistive wall, the "resistive wall mode". In other words, perturbation rotates with respect to the plasma. The two modes are influenced in opposite ways by the wall distance. The plasma mode is destabilized as the wall is moved further from the plasma, while the resistive wall mode is stabilized.

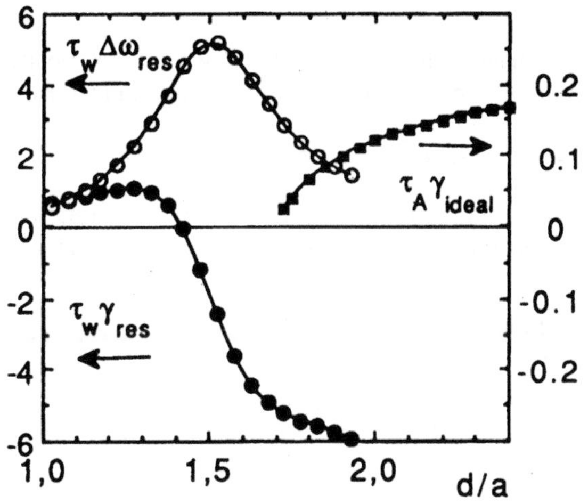

Figure 9.12 The growth rate $\gamma_{\rm res}$ and slip frequency $\Delta\omega_{\rm slip} = \omega_{\rm rot} - \omega_{\rm res}$ of resistive wall mode and the growth rate $\gamma_{\rm ideal}$ of plasma mode versus resistive wall radius d/a for n=1. $\omega_{\rm rot} = 0.06$ is in units of $\tau_{\rm A}^{-1} = B/(a(\mu_0\rho)^{1/2}) = (B/B_\theta)\tau_{{\rm A}\theta}^{-1}$. After [9.13].

Therefore, there can be a finite window for the wall position such that both modes are stable (see Figure 9.12). They suggest that important aspects of stabilization mechanism are that inertial effects become important near the resonant layers in the plasma where the rotation speed exceeds the local Alfvén frequency $k_\parallel v_{\rm A}$ and that coupling to sound waves contributes to the stabilization.

These numerical results can be interpreted by analytical consideration. In the case of resistive wall at $r = d$, $\psi'(a_+)/\psi(a)$ is already given by (9.67). Formally $\psi'(a_-)/\psi(a)$ can be expressed by

$$\frac{\psi'(a_-)}{\psi(a)} = -\frac{\rm m}{a}(1+Z).$$

Z should be calculated by solving ψ for a given plasma model. Then the dispersion relation is given by

$$(1+Z) = \frac{1+\alpha_{\rm res}(a/d)^{2\rm m}}{1-\alpha_{\rm res}(a/d)^{2\rm m}} = \frac{(1+\gamma'_{\rm res}) + \gamma'_{\rm res}(a/d)^{2\rm m}}{(1+\gamma'_{\rm res}) - \gamma'_{\rm res}(a/d)^{2\rm m}}, \qquad \gamma'_{\rm res} = \frac{\gamma_{\rm res}\tau_{\rm w}}{2\rm m}$$

and the growth rate is given by

$$\frac{\gamma_{\rm res}\tau_{\rm w}}{2\rm m}\left(1-(a/d)^{2\rm m}\right) = \frac{Z}{w-Z}, \qquad w \equiv \frac{2(a/d)^{2\rm m}}{1-(a/d)^{2\rm m}} = \frac{2}{(d/a)^{2\rm m}-1}.$$

(9.71)

In the case when plasma does not rotate, Z is real and positive and the resistive wall mode is unstable for $w > Z$ or equivalently for

$$(1+2/Z) > (d/a)^{2\mathrm{m}}, \qquad a < d < d_{\mathrm{ideal}} = a(1+2/Z)^{1/2\mathrm{m}}.$$

As the wall radius approaches to d_{ideal} ($w \to Z$), γ_{res} tends to increase to infinity and the resistive wall mode connects to ideal MHD instability, which is ideally unstable for $d > d_{\mathrm{ideal}}$.

When plasma rotates, the logarithmic derivative has non-zero imaginary part $Z = Z_{\mathrm{r}} + Z_{\mathrm{i}} i$ and the growth rate is

$$\gamma_{\mathrm{res}}\tau_{\mathrm{w}} 2\mathrm{m}\left(1-(a/d)^{2\mathrm{m}}\right) = \frac{wZ_{\mathrm{r}} - (Z_{\mathrm{r}}^2 + Z_{\mathrm{i}}^2) + iwZ_{\mathrm{i}}}{(w-Z_{\mathrm{r}})^2 + Z_{\mathrm{i}}^2}. \tag{9.72}$$

This eliminates the zero in the demominator of (9.69) and γ_{res} remains finite and complex for all wall distances. The resistive wall mode does not connect to the ideal instability. When $wZ_{\mathrm{r}} < (Z_{\mathrm{r}}^2 + Z_{\mathrm{i}}^2)$, the resistive wall mode becomes stable. This condition becomes

$$d > d_{\mathrm{res}}, \qquad d_{\mathrm{res}} \equiv \left(1 + \frac{2Z_{\mathrm{r}}}{Z_{\mathrm{r}}^2 + Z_{\mathrm{i}}^2}\right)^{1/2\mathrm{m}}.$$

These results are consistent with the numerical results.

9.4.2 Feedback Stabilization of Resistive Wall Mode

Feedback stabilization of resistive wall mode is discussed according to [9.14]. We begin with the eigenmode equation used to determine the stability of a large aspect ratio tokamak with low beta.

$$\frac{\mathrm{d}}{\mathrm{d}r}\left((\gamma\tau_{\mathrm{A}}^2 + F^2) r \frac{\mathrm{d}}{\mathrm{d}r}(r\xi_r)\right) - \left(\mathrm{m}^2(\gamma\tau_{\mathrm{A}}^2 + F^2) + r\frac{\mathrm{d}F^2}{\mathrm{d}r}\right)\xi_r = 0, \tag{9.73}$$

$$F = \left(-\frac{\mathrm{m}}{r}B_\theta + \frac{\mathrm{n}}{R}B_z\right) = \frac{B_\theta}{r}(\mathrm{n}q - \mathrm{m}).$$

This formula can be derived from (8.116) under the assumption $\epsilon = r/R \ll 1$. In the vacuum, the perturbation of magnetic field $\boldsymbol{B}_1 = \nabla\phi$ is the solution of

$$\nabla^2\phi = 0, \qquad \phi = A\left((r/b_{\mathrm{w}})^{-\mathrm{m}} + \alpha_{\mathrm{w}}(r/b_{\mathrm{w}})^{\mathrm{m}}\right)\exp(-i\mathrm{m}\theta + \mathrm{n}z/R). \tag{9.74}$$

A plasma-vacuum boundary condition at plasma edge $r = a$ is of the form of

$$(\gamma^2\tau^2 + f^2)\frac{1}{\xi_r}\frac{\mathrm{d}(r\xi_r)}{\mathrm{d}r} = f^2\left(\frac{a\psi'(a_+)}{\psi(a)} - \frac{2\mathrm{m}}{f}\right), \tag{9.75}$$

where $f = nq - \mathrm{m}$ and ψ is the flux function of external perturbation $\boldsymbol{B}_{1\mathrm{ex}}$. The flux function $\psi = \psi(r)\exp(-im\theta)$ (z component of vector potential A_z) in the vacuum is given by $\psi = -rB_{1r}/\mathrm{m}$. This formula can be derived from the plasma-vacuum boundary condition (8.33) and (8.38). The boundary condition (8.38) $\boldsymbol{n}\cdot\boldsymbol{B}_{1\mathrm{ex}} = \boldsymbol{n}\cdot\nabla\times(\boldsymbol{\xi}\times\boldsymbol{B})$ becomes
$\partial\phi/\partial r = r^{-1}\partial(\xi_r B_\theta/\partial\theta + \partial(\xi_r B_z)/\partial z$ and determines A in (9.74) as follows:

$$A = -i\frac{\xi_r F}{(\mathrm{m}/a)\left((a/b_\mathrm{w})^{-\mathrm{m}} - \alpha_\mathrm{w}(a/b_\mathrm{w})^{\mathrm{m}}\right)}.$$

The constant α_w is to be determined by the boundary condition on the wall at $r = b_\mathrm{w}$. The boundary condition (8.33) becomes

$$B_\theta B_{1\theta\mathrm{in}} + B_z B_{1z\mathrm{in}} = B_\theta B_{1\theta\mathrm{ex}} + B_z B_{1z\mathrm{ex}},$$

where $\boldsymbol{B}_{1\mathrm{in}} = \nabla\times(\boldsymbol{\xi}\times\boldsymbol{B})$ is given in Section 8.5.1 and $\boldsymbol{B}_{1\mathrm{ex}}$ is given by (9.74). $\boldsymbol{\xi}$ is given by Hain-Lüst equation (8.105-107), in which low β and incompressivility are assumed. Two boundary conditions yield (9.75).

The boundary condition (9.75) is used to provide a circuit equation for the plasma by defining $\beta_0 \equiv (1/\xi_r)(\mathrm{d}(r\xi_r)/\mathrm{d}r)|_a$. In principle, β_0 would be determined by (9.73) and (9.75) consistently. The flux function $2\pi R\psi(a_+)$ is the perturbed poloidal flux of (m, n) mode in the vacuum region ($B_{1\theta} = -(\partial\psi/\partial r)$. $2\pi R\psi(a_+)$ consists of contributions from the perturbed current I_1, resistive wall current I_2 and circuit current I_3 corresponding to active feedback coil for (m, n) mode; that is,

$$2\pi R\psi(a_+) = L_1 I_1 + M_{12}I_2 + M_{13}I_3,$$

$$2\pi R\psi'(a_+) = L_1' I_1 + M_{12}' I_2 + M_{13}' I_3.$$

Therefore, (9.75) is reduced to

$$\left((\gamma^2\tau^2 + f^2)\beta_0 - f^2\frac{aL_1'}{L_1} + 2\mathrm{m}\right)L_1 I_1 + \left((\gamma^2\tau^2 + f^2)\beta_0 - f^2\frac{aM_{12}'}{M_{12}}\right.$$

$$\left.+2\mathrm{m}\right)M_{12}I_2 + \left((\gamma^2\tau^2 + f^2)\beta_0 - f^2\frac{aM_{13}'}{M_{13}} + 2\mathrm{m}\right)M_{13}I_3 = 0.$$

For a circuit corresponding to the resistive wall, the flux function at the resistive wall ($r = r_\mathrm{w}$) is given by

$$2\pi R\frac{\partial\psi(r_\mathrm{w})}{\partial t} = L_2\frac{\partial I_2}{\partial t} + M_{21}\frac{\partial I_1}{\partial t} + M_{23}\frac{\partial I_3}{\partial t},$$

and

$$2\pi R\frac{\partial\psi(r_\mathrm{w})}{\partial t} = -2\pi R\eta_\mathrm{w} j_\mathrm{w} = -R_2 I_2,$$

where

$$I_2 \equiv 2\pi r_{\rm w}\delta_{\rm w} j_{\rm w}/(2{\rm m}), \qquad R_2 \equiv 2{\rm m}\frac{\eta_{\rm w} 2\pi R}{2\pi r_{\rm w}\delta_{\rm w}}.$$

For the circuit corresponding to the active feedback control, a voltage term to drive the feedback current must be included and

$$M_{31}\frac{\partial I_1}{\partial t} + M_{32}\frac{\partial I_2}{\partial t} + M_{31}\frac{\partial I_3}{\partial t} + R_3 I_3 = V_3.$$

The form of the feedback voltage V_3 should be applied to minimize $\psi(a_+)$ by use of appropriate sensors of the perturbations.

Problems

1. Resistive Tearing Instability Confirm that Δ' defined by (9.16) calculated by external solution is given by
$\Delta' = (2\alpha/L_{\rm s})(({\rm e})^{-2\alpha} + (1-2\alpha))/(({\rm e})^{-2\alpha} - (1-2\alpha)), \ \ \alpha \equiv {\rm kL_s}$, in the case of $F(x) = F_{\rm s}x/L_{\rm s}$ $(|x| < L_{\rm s})$ and $F(x) = F_{\rm s}x/|x|$ $(|x| > L_{\rm s})$. Note that F'' is delta function at $x = \pm L_{\rm s}$. Refer to Section 9.1.

2. Δ' by Internal Solution To solve the internal solution of (9.18) and (9.19), we assume that $F'' \to 0$, $\partial^2/\partial y^2 = k^2 \ll \partial^2/\partial x^2$ and B_{1x}=const. in $x < \varepsilon$. Then (9.19) reduces to

$$\frac{\partial^2 V_x}{\partial x^2} = i\alpha x B_{1x} + \beta x^2 V_x, \quad \alpha = \frac{F'}{\rho_{\rm m}\eta}, \quad \beta = \frac{F'^2}{\rho_{\rm m}\eta\gamma}.$$

When x, V_x are transformed to $x = \beta^{1/4}X$, $V_x = i\alpha B_{1x}\beta^{-3/4}U_x$, the equation becomes $\partial^2 U_x/\partial x^2 = X(1 + XU_x)$ and the solution becomes [9.3]

$$U_x(X) = -\frac{X}{2}\int_0^{\pi/2} \exp\left(-\frac{X^2}{2}\cos\theta\right)\sin^{1/2}\theta {\rm d}\theta.$$

Since $(\partial B_{1x}/\partial x) = (\gamma\mu_0/\eta)\int^x (B_{1x} - (i/\gamma)F' x V_x){\rm d}x$ from (9.18), we have

$$\Delta' = \frac{B'_{1x}(+\varepsilon) - B'_{1x}(-\varepsilon)}{B_{1x}(0)} = \frac{\mu_0\gamma}{\eta B_{1x}}\int_{-\infty}^{\infty}(1 - \frac{iF'}{\gamma B_{1x}}xV_x){\rm d}x$$

$$= \frac{\mu_0\gamma}{\eta B_{1x}}\int_{-\infty}^{\infty}(1 - XU_x){\rm d}X\left(\frac{F'^2}{\rho_{\rm m}\eta\gamma}\right)^{-1/4} = \frac{\gamma^{5/4}\rho_m^{1/4}\mu_0}{\eta^{3/4}F'^{1/2}}\int_{-\infty}^{\infty}\frac{1}{X}\frac{\partial^2 U_x}{\partial X^2}{\rm d}X.$$

The definite integral is 2.12. Then the growth rate is given by

$$\gamma_{\rm Rutherford} = 0.55(\Delta' a)^{4/5}\left(\frac{\eta}{\mu_0 a^2}\right)^{3/5}\left(\frac{B_0^2}{\rho_{\rm m}\mu_0 a^2}\right)^{1/5}\left(\frac{(\boldsymbol{k}\cdot\boldsymbol{B}_0)' a^2}{B_0^2}\right)^{2/5}.$$

The growth rate of (9.27) is different from this value by a factor of 0.55. Confirm that $U_x(X)$ is the solution of $\partial^2 U_x/\partial x^2 = X(1 + XU_x)$.

3. Island Width Estimate the ratio of the island width of $m = 2,\ n = -1$ mode to the minor radius a of tokamak with the safety factor $q_s(r) = 1 + 2(r/a)^2$ and aspect ratio $A = 3$, when the radial component of the perturbed magnetic field δB_r is $\delta B_r/B_t = 10^{-4}$, where B_t is the magnitude of the toroidal field. Calculate the island width in the case of $a = 2\,\mathrm{m}$.

4. Drift Frequency Electron drift frequncy is given by $\omega_e^* = k_\theta \kappa_n T_e/(eB)$, $\kappa_n \approx 1/a$, $k_\theta = 1/a \sim 1/\rho_i$. Calculate the range of drift frequncy in the case of $a = 2\,\mathrm{m}$, $B = 5\mathrm{T}$, $T_e = T_i = 10\,keV$ and D-T plasma. Calculate the phase velocity of electron drift wave. Is the direction of the phase velocity the same as the direction of electron diamagnetic drift or ion diamagnetic drift?

10

Boltzmann Equation and Landau Damping

A plasma consists of many ions and electrons, but the individual behavior of each particle can hardly be observed. What can be observed instead are statistical averages. In order to describe the properties of a plasma, it is necessary to define a distribution function that indicates particle number density in the phase space whose ordinates are the particle positions and velocities. The distribution function is not necessarily stationary with respect to time. In Section 10.1, the equation governing the distribution function $f(q_i, p_i, t)$ is derived by means of Liouville's theorem. Boltzmann equation for the distribution function $f(\boldsymbol{x}, \boldsymbol{v}, t)$ is formulated in Section 10.2. When the collision term is neglected, Boltzmann's equation is called Vlasov equation. In Section 10.3, the Fokker-Planck collision term is described.

10.1 Boltzmann Equation

10.1.1 Phase Space and Distribution Function

A particle can be specified by its coordinates (x, y, z), velocity (v_x, v_y, v_z), and time t. More generally, the particle can be specified by *canonical variables* $q_1, q_2, q_3, p_1, p_2, p_3$, and t in phase space. When canonical variables are used, an infinitesimal volume in phase space $\Delta = \delta q_1 \delta q_2 \delta q_3 \delta p_1 \delta p_2 \delta p_3$ is conserved (Liouville's theorem). The motion of a particle in phase space is described by Hamilton's equations:

$$\frac{\mathrm{d}q_i}{\mathrm{d}t} = \frac{\partial H(q_j, p_j, t)}{\partial p_i}, \qquad \frac{\mathrm{d}p_i}{\mathrm{d}t} = -\frac{\partial H(q_j, p_j, t)}{\partial q_i}, \qquad (i, j = 1, 2, 3). \tag{10.1}$$

An infinitesimal volume Δ in phase space is invariant

$$\Delta = \text{invariant} \tag{10.2}$$

according to Liouville's theorem (refer to Figure 10.1).

Let the number of particles in a small volume of phase space be δN

$$\delta N = F(q_i, p_i, t)\delta \boldsymbol{q} \delta \boldsymbol{p}, \tag{10.3}$$

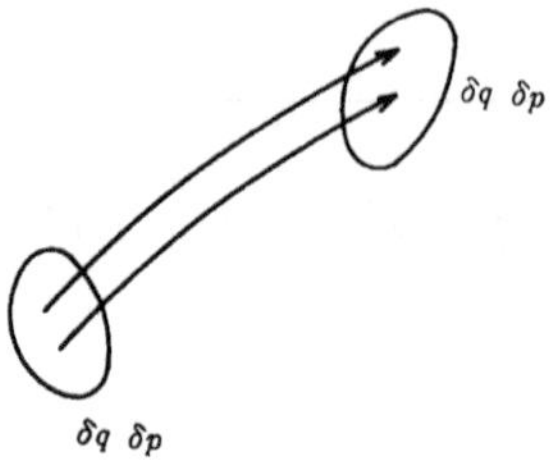

Figure 10.1 Movement of particles in phase space.

where $\delta \boldsymbol{q} = \delta q_1 \delta q_2 \delta q_3$, $\delta \boldsymbol{p} = \delta p_1 \delta p_2 \delta p_3$, and $F(q_i, p_i, t)$ is the *distribution function in phase space.* If the particles move according to the equation of motion and are not scattered by collisions, the small volume in phase space is conserved. As the particle number δN within the small phase space is conserved, the distribution function ($F = \delta N/\Delta$) is also constant, i.e.,

$$\frac{\mathrm{d}F}{\mathrm{d}t} = \frac{\partial F}{\partial t} + \sum_i \left(\frac{\partial F}{\partial q_i} \frac{\mathrm{d}q_i}{\mathrm{d}t} + \frac{\partial F}{\partial p_i} \frac{\mathrm{d}p_i}{\mathrm{d}t} \right) = \frac{\partial F}{\partial t} + \sum_i \left(\frac{\partial H}{\partial p_i} \frac{\partial F}{\partial q_i} - \frac{\partial H}{\partial q_i} \frac{\partial F}{\partial p_i} \right) = 0. \tag{10.4}$$

In the foregoing discussion we did not take collisions into account. If we denote the variation of F due to the collisions by $(\delta F/\delta t)_{\mathrm{coll}}$, (10.4) becomes

$$\frac{\partial F}{\partial t} + \sum_i \left(\frac{\partial H}{\partial p_i} \frac{\partial F}{\partial q_i} - \frac{\partial H}{\partial q_i} \frac{\partial F}{\partial p_i} \right) = \left(\frac{\delta F}{\delta t} \right)_{\mathrm{coll.}} . \tag{10.5}$$

10.1.2 Boltzmann Equation and Vlasov Equation

Let us use the space and velocity-space coordinates x_1, x_2, x_3, v_1, v_2, v_3 instead of canonical coordinates. Hamiltonian is

$$H = \frac{1}{2m} (\boldsymbol{p} - q\boldsymbol{A})^2 + q\phi, \tag{10.6}$$

$$p_i = mv_i + qA_i, \tag{10.7}$$

$$q_i = x_i, \tag{10.8}$$

and

$$\frac{\mathrm{d}x_i}{\mathrm{d}t} = \frac{\partial H}{\partial p_i} = v_i, \tag{10.9}$$

$$\frac{\mathrm{d}p_i}{\mathrm{d}t} = -\frac{\partial H}{\partial x_i} = \sum_k \frac{(p_k - qA_k)}{m} q \frac{\partial A_k}{\partial x_i} - q\frac{\partial \phi}{\partial x_i}. \tag{10.10}$$

Consequently, (10.5) becomes

$$\frac{\partial F}{\partial t} + \sum_i v_k \frac{\partial F}{\partial x_k} + q \sum_i \left(\sum_k v_k \frac{\partial A_k}{\partial x_i} - \frac{\partial \phi}{\partial x_i} \right) \frac{\partial F}{\partial p_i} = \left(\frac{\delta F}{\delta t} \right)_{\mathrm{coll}}. \tag{10.11}$$

By use of (10.7) and (10.8), independent variables are transformed from (q_i, p_i, t) to (x_j, v_j, t) and

$$\frac{\partial v_j(x_k, p_k, t)}{\partial p_i} = \frac{1}{m}\delta_{ij},$$

$$\frac{\partial v_j(x_k, p_k, t)}{\partial x_i} = -\frac{q}{m}\frac{\partial A_j}{\partial x_i},$$

$$\frac{\partial v_j(x_k, p_k, t)}{\partial t} = -\frac{q}{m}\frac{\partial A_j}{\partial t}.$$

We denote $F(x_i, p_i, t) = F(x_i, p_i(x_j, v_j, t), t) \equiv f(x_j, v_j, t)/m^3$. Then we have $m^3 F(x_i, p_i, t) = f(x_j, v_j(x_i, p_i, t), t)$ and

$$m^3 \frac{\partial}{\partial p_i} F(x_h, p_h, t) = \frac{\partial}{\partial p_i} f(x_j, v_j(x_h, p_h, t), t) = \sum_j \frac{\partial f}{\partial v_j}\frac{\partial v_j}{\partial p_i} = \frac{\partial f}{\partial v_i}\frac{1}{m},$$

$$\begin{aligned} m^3 \frac{\partial}{\partial x_k} F(x_h, p_h, t) &= \frac{\partial}{\partial x_k} f(x_i, v_i(x_h, p_h, t), t) = \frac{\partial f}{\partial x_k} + \sum_i \frac{\partial f}{\partial v_i}\frac{\partial v_i}{\partial x_k} \\ &= \frac{\partial f}{\partial x_k} + \sum_i \frac{\partial f}{\partial v_i}\left(\frac{-q}{m}\right)\frac{\partial A_i}{\partial x_k}, \end{aligned}$$

$$m^3 \frac{\partial}{\partial t} F(x_h, p_h, t) = \frac{\partial}{\partial t} f(x_i, v_i(x_h, p_h, t), t) = \frac{\partial f}{\partial t} + \sum_i \frac{\partial f}{\partial v_i}\left(\frac{-q}{m}\right)\frac{\partial A_i}{\partial t}.$$

Accordingly, (10.11) is reduced to

$$\begin{aligned} &\frac{\partial f}{\partial t} + \sum_i \frac{\partial f}{\partial v_i}\left(\frac{-q}{m}\right)\frac{\partial A_i}{\partial t} + \sum_k v_k \left(\frac{\partial f}{\partial x_k} + \sum_i \frac{\partial f}{\partial v_i}\left(\frac{-q}{m}\right)\frac{\partial A_i}{\partial x_k} \right) \\ &\qquad + \sum_i \left(\sum_k v_k \frac{\partial A_k}{\partial x_i} - \frac{\partial \phi}{\partial x_i} \right) \frac{q}{m}\frac{\partial f}{\partial v_i} = \left(\frac{\delta f}{\delta t} \right)_{\mathrm{coll}}, \end{aligned}$$

$$\frac{\partial f}{\partial t} + \sum_k v_k \frac{\partial f}{\partial x_k} + \sum_i \left(-\frac{\partial A_i}{\partial t} - \sum_k v_k \frac{\partial A_i}{\partial x_k} + \sum_k v_k \frac{\partial A_k}{\partial x_i} - \frac{\partial \phi}{\partial x_i} \right) \frac{q}{m}\frac{\partial f}{\partial v_i} = \left(\frac{\delta f}{\delta t} \right)_{\mathrm{coll}}.$$

Since the following relation holds,

$$\sum_k v_k \frac{\partial A_k}{\partial x_i} = \sum_k v_k \frac{\partial A_i}{\partial x_k} + (\boldsymbol{v} \times (\nabla \times \boldsymbol{A}))_i = \sum_k v_k \frac{\partial A_i}{\partial x_k} + (\boldsymbol{v} \times \boldsymbol{B})_i,$$

we have

$$\frac{\partial f}{\partial t} + \sum_i v_i \frac{\partial f}{\partial x_i} + \sum_i \frac{q}{m} (\boldsymbol{E} + \boldsymbol{v} \times \boldsymbol{B})_i \frac{\partial f}{\partial v_i} = \left(\frac{\delta f}{\delta t} \right)_{\text{coll}}. \tag{10.12}$$

This equation is called *Boltzmann equation*. The electric charge density ρ and the electric current $\boldsymbol{j}$ are expressed by

$$\rho = \sum_{\text{i,e}} q \int f \mathrm{d}v_1 \mathrm{d}v_2 \mathrm{d}v_3, \tag{10.13}$$

$$\boldsymbol{j} = \sum_{\text{i,e}} q \int \boldsymbol{v} f \mathrm{d}v_1 \mathrm{d}v_2 \mathrm{d}v_3. \tag{10.14}$$

Accordingly, Maxwell equations are given by

$$\nabla \cdot \boldsymbol{E} = \frac{1}{\epsilon_0} \sum_{\text{i,e}} q \int f \mathrm{d}\boldsymbol{v}, \tag{10.15}$$

$$\frac{1}{\mu_0} \nabla \times \boldsymbol{B} = \epsilon_0 \frac{\partial \boldsymbol{E}}{\partial t} + \sum_{\text{i,e}} q \int \boldsymbol{v} f \mathrm{d}\boldsymbol{v}, \tag{10.16}$$

$$\nabla \times \boldsymbol{E} = -\frac{\partial \boldsymbol{B}}{\partial t}, \tag{10.17}$$

$$\nabla \cdot \boldsymbol{B} = 0. \tag{10.18}$$

When the plasma is rarefied, the collision term $(\delta f / \delta t)_{\text{coll}}$ may be neglected. However, the interactions of the charged particles are still included through the internal electric and magnetic field which are calculated from the charge and current densities by means of Maxwell equations. The charge and current densities are expressed by the distribution functions for the electron and the ion. This equation is called *collisionless Boltzmann equation* or *Vlasov equation*.

When *Fokker-Planck collision term* is adopted as the collision term of Boltzmann equation, this equation is called *Fokker-Planck equation* (see Section 10.3).

10.1.3 Fokker-Planck Collision Term

In the case of Coulomb collision, scattering into small angles has a large cross-section and a test particle interacts with many field particles at the same time, since the Coulomb force is a long-range interaction. Consequently it is appropriate to treat Coulomb collision statistically. Assume that the velocity $\boldsymbol{v}$ of a particle is changed to $\boldsymbol{v}+\Delta\boldsymbol{v}$ after the time Δt by Coulomb collisions; denote the probability of this process by $W(\boldsymbol{v},\Delta\boldsymbol{v})$. Then the distribution function $f(\boldsymbol{r},\boldsymbol{v},t)$ satisfies

$$f(\boldsymbol{r},\boldsymbol{v},t+\Delta t)=\int f(\boldsymbol{r},\boldsymbol{v}-\Delta\boldsymbol{v},t)W(\boldsymbol{v}-\Delta\boldsymbol{v},\Delta\boldsymbol{v})\mathrm{d}(\Delta\boldsymbol{v}). \qquad (10.19)$$

In this process the state at $t+\Delta t$ depends only on the state at t. Such a process (i.e., one independent of the history of the process) is called the *Markoff process.* The change of the distribution function by virtue of Coulomb collision is

$$\left(\frac{\delta f}{\delta t}\right)_{\mathrm{coll}}\Delta t=f(\boldsymbol{r},\boldsymbol{v},t+\Delta t)-f(\boldsymbol{r},\boldsymbol{v},t).$$

Taylor expansion of the integrand of (10.19) gives

$$f(\boldsymbol{r},\boldsymbol{v}-\Delta\boldsymbol{v},t)W(\boldsymbol{v}-\Delta\boldsymbol{v},\Delta\boldsymbol{v})$$

$$=f(\boldsymbol{r},\boldsymbol{v},t)W(\boldsymbol{v},\Delta\boldsymbol{v})-\sum_r\frac{\partial(fW)}{\partial v_r}\Delta v_r+\sum_{r\,s}\frac{1}{2}\frac{\partial^2(fW)}{\partial v_r\partial v_s}\Delta v_r\Delta v_s+\cdots. \qquad (10.20)$$

From the definition of $W(\boldsymbol{v},\Delta\boldsymbol{v})$, the intgral of W is

$$\int W\mathrm{d}(\Delta\boldsymbol{v})=1.$$

Introducing the quantities

$$\int W\Delta\boldsymbol{v}\mathrm{d}(\Delta\boldsymbol{v})=\langle\Delta\boldsymbol{v}\rangle_t\Delta t, \qquad \int W\Delta v_r\Delta v_s\mathrm{d}(\Delta\boldsymbol{v})=\langle\Delta v_r\Delta v_s\rangle_t\Delta t,$$

we find

$$\left(\frac{\delta f}{\delta t}\right)_{\mathrm{coll}}=-\nabla_v(\langle\Delta\boldsymbol{v}\rangle_t f)+\sum\frac{1}{2}\frac{\partial^2}{\partial\Delta_r\partial\Delta_s}(\langle\Delta v_r\Delta v_s\rangle_t f). \qquad (10.21)$$

This term is called the *Fokker-Planck collision term* and $\langle\Delta\boldsymbol{v}\rangle_t\rangle$, $\Delta v_r\Delta v_s$ are called *Fokker-Planck coefficients.* $\int W\Delta v_r\Delta v_s\mathrm{d}(\Delta\boldsymbol{v})$ is proportional to Δt. Δv_r is the sum of Δv_r^i, which is, the change of v_r due to the i-th collisons during Δt, i.e., $\Delta v_r=\sum_i\Delta v_r^i$, so that $\Delta v_r\Delta v_s=\sum_i\sum_j\Delta v_r^i\Delta v_s^j$. When the collisions are statistically independent, statistical average of $\langle\Delta v_r^1\Delta v_s^2\rangle_t$ $(i\neq j)$ is zero and

$$\int W\Delta v_r\Delta v_s\mathrm{d}(\Delta\boldsymbol{v})=\sum_i\int\Delta v_r^i\Delta v_s^i\mathrm{d}(\Delta\boldsymbol{v}).$$

This expression is proportinal to Δt.

The Fokker-Planck equation can be expressed in the form [10.1]

$$\frac{\partial f}{\partial t} + \boldsymbol{v} \cdot \nabla_r f + \frac{\boldsymbol{F}}{m} \nabla_v f + \nabla_v \cdot \boldsymbol{J} = 0, \tag{10.22}$$

where

$$J_i = A_i f - \sum_j D_{ij} \frac{\partial f}{\partial v_j},$$

$$A_i = \langle \Delta v_i \rangle_t - \frac{1}{2} \sum_j \frac{\partial}{\partial v_j} \langle \Delta v_i \Delta v_j \rangle_t,$$

$$D_{ij} = \frac{1}{2} \langle \Delta v_i \Delta v_j \rangle_t.$$

The tensor D is called the *diffusion tensor* in the velosity space and $\boldsymbol{A}$ is called the *coefficient of dynamic friction*. For convenience we consider the components of $\boldsymbol{J}$ parallel and perpendicular to the velocity $\boldsymbol{v}$ of the test particle. When the distribution function of field particles is isotropic, we find

$$\left.\begin{aligned} \boldsymbol{J}_\parallel &= -D_\parallel \nabla_\parallel f + \boldsymbol{A} f, \\ \boldsymbol{J}_\perp &= -D_\perp \nabla_\perp f. \end{aligned}\right\} \tag{10.23}$$

$\boldsymbol{A}$ is parallel to $\boldsymbol{v}$ and the diffusion tensor becomes diagonal. When the distribution function of field particles is Maxwellian, $\boldsymbol{A}$ and D are given by [10.1]

$$m v D_\parallel = -T^* A, \tag{10.24}$$

$$D_\parallel = \frac{(qq^*)^2 n^* \ln \Lambda}{8\pi\varepsilon_0^2 v m^2} \frac{\Phi_1(b^*)}{b^{*2} v^2}, \tag{10.25}$$

$$D_\perp = \frac{(qq^*)^2 n^* \ln \Lambda}{8\pi\varepsilon_0^2 v m^2} \left(\Phi(b^* v) - \frac{\Phi_1(b^* v)}{2b^{*2} v^2} \right). \tag{10.26}$$

q^*, n^*, b^*, and T^* are those of field particles and q, m, and v are those of test particles. $\Phi(x)$ and $\Phi_1(x)$ are

$$\Phi(x) = \frac{2}{\pi^{1/2}} \int_0^x \exp(-\xi^2 \mathrm{d}\xi),$$

$$\Phi_1(x) = \Phi(x) - \frac{2x}{\pi^{1/2} \exp(-x^2)}.$$

When $x > 2$, then $\Phi(x) \approx \Phi_1(x) \approx 1$.

10.2 Landau Damping

The existence of a damping mechanism by which plasma particles absorb wave energy even in a collisionless plasma was found by L. D. Landau, under the condition that the plasma is not cold and the velocity distribution is of finite extent. Energy-exchange processes between particles and wave are possible even in a collisionless plasma and play important roles in plasma heating by waves (wave absorption) and in the mechanism of instabilities (wave amplification). These important processes will be explained in terms of simplified physical models in this chapter. In Chapter 12, these processes will be described systematically. In hot plasma models, pressure term and particle-wave interaction term appear in the dielectric tensor that are absent in the dielectric tensor for a cold plasma. Let us assume that many particles drift with different velocities in the direction of the lines of magnetic force. When an electrostatic wave (a longitudinal wave with $\boldsymbol{k} \parallel \boldsymbol{E}$) propagates along the lines of magnetic force, there appears an interaction between the wave and a group of particles (see Figure 10.2). Take the z axis in the direction of the magnetic field and denote the unit vector in this direction by $\hat{\boldsymbol{z}}$. Then the electric field and the velocity $\boldsymbol{v} = v\hat{\boldsymbol{z}}$ satisfy

$$\boldsymbol{E} = \hat{\boldsymbol{z}} E \cos(kz - \omega t), \tag{10.27}$$

$$m\frac{\mathrm{d}v}{\mathrm{d}t} = qE\cos(kz - \omega t). \tag{10.28}$$

The electric field $\boldsymbol{E}$ is a quantity of the first order. The zeroth order solution of (10.28) is

$$z = v_0 t + z_0,$$

and the first order equation is

$$m\frac{\mathrm{d}v_1}{\mathrm{d}t} = qE\cos(kz_0 + kv_0 t - \omega t). \tag{10.29}$$

The solution of (10.29) for the initial condition $v_1 = 0$ at $t = 0$ is

$$v_1 = \frac{qE}{m}\frac{\sin(kz_0 + kv_0 t - \omega t) - \sin kz_0}{kv_0 - \omega}. \tag{10.30}$$

The kinetic energy of the particle becomes

$$\frac{\mathrm{d}}{\mathrm{d}t}\frac{mv^2}{2} = v\frac{\mathrm{d}}{\mathrm{d}t}mv = v_1\frac{\mathrm{d}}{\mathrm{d}t}mv_1 + v_0\frac{\mathrm{d}}{\mathrm{d}t}mv_2 + \cdots. \tag{10.31}$$

From (10.28) and (10.30), we have the relation

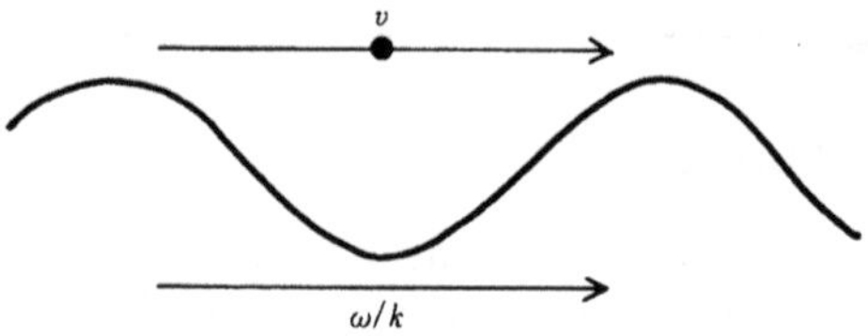

Figure 10.2 Propagation of wave and motion of particles in the process of Landau damping.

$$m\frac{\mathrm{d}(v_1+v_2)}{\mathrm{d}t} = qE\cos(k(z_0+v_0t+z_1)-\omega t)$$
$$= qE\cos(kz_0+\alpha t) - qE\sin(kz_0+\alpha t)kz_1,$$
$$z_1 = \int_0^t v_1\,\mathrm{d}t = \frac{qE}{m}\left(\frac{-\cos(kz_0+\alpha t)+\cos kz_0}{\alpha^2} - \frac{t\sin kz_0}{\alpha}\right),$$

where

$$\alpha \equiv kv_0 - \omega.$$

Using these, we may put (10.31) into the form

$$\frac{\mathrm{d}}{\mathrm{d}t}\frac{mv^2}{2} = \frac{q^2E^2}{m}\left(\frac{\sin(kz_0+\alpha t)-\sin kz_0}{\alpha}\right)\cos(kz_0+\alpha t)$$
$$-\frac{kv_0q^2E^2}{m}\left(\frac{-\cos(kz_0+\alpha t)+\cos kz_0}{\alpha^2} - \frac{t\sin kz_0}{\alpha}\right)\sin(kz_0+\alpha t).$$

The average of the foregoing quantitiy with respect to the initial position z_0 is

$$\left\langle\frac{\mathrm{d}}{\mathrm{d}t}\frac{mv^2}{2}\right\rangle_{z_0} = \frac{q^2E^2}{2m}\left(\frac{-\omega\sin\alpha t}{\alpha^2} + t\cos\alpha t + \frac{\omega t\cos\alpha t}{\alpha}\right). \quad (10.32)$$

When we take the velocity average of (10.32) over v_0 with the weighting factor, i.e., distribution function $\hat{f}(v_0)$ (defining $\alpha \equiv kv_0 - \omega$)

$$\hat{f}(v_0) = \hat{f}\left(\frac{\alpha+\omega}{k}\right) = g(\alpha),$$

the rate of increase of the kinetic energy of the particles is obtained. The distribution function is normalized:

$$\int_{-\infty}^{\infty}\hat{f}(v_0)\,\mathrm{d}v_0 = \frac{1}{k}\int g(\alpha)\,\mathrm{d}\alpha = 1.$$

The integral of the second term of (10.32)

$$\frac{1}{k}\int g(\alpha)t\cos\alpha t\,\mathrm{d}\alpha = \frac{1}{k}\int g\left(\frac{x}{t}\right)\cos x\,\mathrm{d}x \tag{10.33}$$

approaches zero as $t \to \infty$. The integral of the third term of (10.32) becomes

$$\frac{\omega}{k}\int \frac{g(\alpha)t\cos\alpha t}{\alpha}\mathrm{d}\alpha = \frac{\omega}{k}\int \frac{t}{x}g\left(\frac{x}{t}\right)\cos x\,\mathrm{d}x. \tag{10.34}$$

The function $g(\alpha)$ can be considered to be the sum of an even and an odd function. The even function does not contribute to the integral. The contribution of the odd function approaches zero when $t \to \infty$ if $g(\alpha)$ is continuous at $\alpha = 0$. Therefore, only the contribution of the first term in (10.32) remains and we find

$$\left\langle \frac{\mathrm{d}}{\mathrm{d}t}\frac{mv^2}{2}\right\rangle_{z_0,v_0} = -\frac{\omega q^2E^2}{2mk}\mathrm{P}\int \frac{g(\alpha)\sin\alpha t}{\alpha^2}\mathrm{d}\alpha. \tag{10.35}$$

P denotes Cauchy's principal value of integral. The main contribution to the integral comes from near $\alpha = 0$, so that $g(\alpha)$ may be expanded around $\alpha = 0$:

$$g(\alpha) = g(0) + \alpha g'(0) + \frac{\alpha^2}{2}g''(0) + \cdots.$$

As $\sin\alpha t/\alpha^2$ is an odd function, only the second term of the foregoing equation contributes to the integral [10.2] and we find for large t that

$$\begin{aligned}\left\langle \frac{\mathrm{d}}{\mathrm{d}t}\frac{mv^2}{2}\right\rangle_{z_0,v_0} &= -\frac{\omega q^2E^2}{2m|k|}\int_{-\infty}^{\infty}\frac{g'(0)\sin\alpha t}{\alpha}\mathrm{d}\alpha \\ &= \frac{-\pi q^2E^2}{2m|k|}\left(\frac{\omega}{k}\right)\left(\frac{\partial\hat{f}(v_0)}{\partial v_0}\right)_{v_0=\omega/k}.\end{aligned} \tag{10.36}$$

If the number of particles slightly slower than the phase velocity of the wave is larger than the number slightly faster, i.e., if $v_0\partial\hat{f}_0/\partial v_0 < 0$, the group of particles as a whole gains energy from the wave and the wave is damped. On the contrary, when $v_0\partial\hat{f}_0/\partial v_0 > 0$ at $v_0 = \omega/k$, the particles give their energy to the wave and the amplitude of the wave increases (Figure 10.3). This mechanism is called *Landau damping* or *amplification* [10.3]. Experimental verification of Landau damping of waves in a collisionless plasma was demonstrated by Malemberg et al. [10.4] in 1965, 20 years after Landau's prediction.

The growth rate (10.36) of the kinetic energy of particles must be equal to the damping rate of wave energy. Therefore, the growth rate γ of the amplitude of the wave field is obtained by ($\gamma < 0$ in the case of damping)

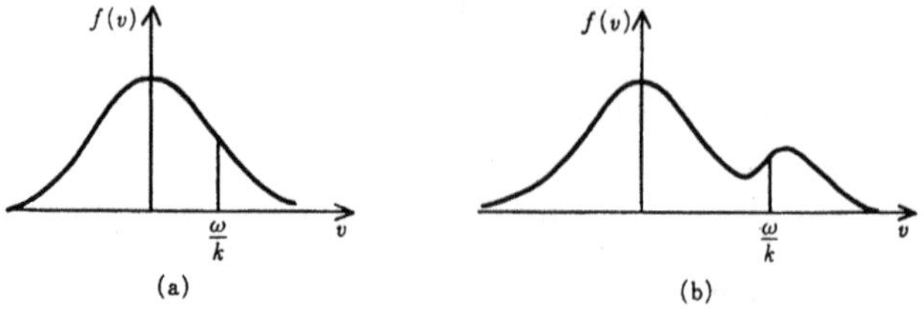

Figure 10.3 (a) Landau damping and (b) Landau amplification.

$$n\left\langle \frac{\mathrm{d}}{\mathrm{d}t}\frac{mv^2}{2} \right\rangle_{z_0 v_0} = -2\gamma W,$$

and the growth rate γ is given by

$$\frac{\gamma}{\omega} = \frac{\pi}{2}\left(\frac{\Pi}{\omega}\right)^2 \left(\frac{\omega}{|k|}\right)\left(v_0 \frac{\partial \hat{f}(v_0)}{\partial v_0}\right)_{v_0=\omega/k}, \tag{10.37}$$

where $\Pi^2 = nq^2/\epsilon_0 m,\ W \approx 2\epsilon_0 E^2/4,\ \int f(v)\mathrm{d}v = 1.$

There is a restriction on the applicability of linear Landau damping. When this phenomenon runs its course before the particle orbit deviates from the linear-approximation solution, the reductions leading to linear Landau damping are justified. The period of oscillation in the potential well of the electric field of the wave gives the time for the particle orbit to deviate from the linear approximation ($\omega^2 \sim eEk/m$ from $m\omega^2 x = eE$). The period of oscillation is

$$\tau_{\mathrm{osc}} = \frac{1}{\omega_{\mathrm{osc}}} \approx \left(\frac{m}{ekE}\right)^{1/2}.$$

Consequently the condition for the applicability of linear Landau damping is that the Landau damping time $1/\gamma$ is shorter than τ_{osc} or the collision time $1/\nu_{\mathrm{coll}}$ is shorter than τ_{osc}.

$$|\gamma\tau_{\mathrm{osc}}| > 1, \tag{10.38}$$

$$|\nu_{\mathrm{coll}}\tau_{\mathrm{osc}}| > 1. \tag{10.39}$$

On the other hand, it was assumed that particles are collisionless. The condition that the collision time $1/\nu_{\mathrm{coll}}$ is longer than λ/v_{rms} is necessary for the asymptotic approximation of the integral (10.35) as $t \to \infty$, where λ is the wavelength of the wave and v_{rms} is the spread in the velocity distribution;

$$\frac{1}{\nu_{\mathrm{coll}}} > \frac{2\pi}{k v_{\mathrm{rms}}}. \tag{10.40}$$

Transit Time Damping
In the low-frequency region, the magnetic moment $\mu_{\rm m}$ is conserved and the equation of motion along the field lines is

$$m\frac{{\rm d}v_z}{{\rm d}t} = -\mu_{\rm m}\frac{\partial B_{1z}}{\partial z}. \tag{10.41}$$

This equation is the same as that for Landau damping if $-\mu_{\rm m}$ and $\partial B_{1z}/\partial z$ are replaced by the electric charge and the electric field, respectively. The rate of change of the kinetic energy is derived similarly, and is

$$\left\langle \frac{\rm d}{{\rm d}t}\frac{mv^2}{2}\right\rangle_{z_0,v_0} = -\frac{\pi\mu_{\rm m}^2|k|}{2m}|B_{1z}|^2\left(\frac{\omega}{k}\right)\left(\frac{\partial\hat{f}(v_0)}{\partial v_0}\right)_{v_0=\omega/k}. \tag{10.42}$$

This phenomena is called *transit time damping.*
Cyclotron Damping
The mechanism of cyclotron damping is different from that of Landau damping. Here the electric field of the wave is perpendicular to the direction of the magnetic field and accelerates or decelerates the particle perpendicularly when the Doppler shifted frequency is nearly harmonic of cyclotron frequency

$$\omega - kV + n\Omega = 0 \qquad n = \pm 1, \pm 2, \cdots.$$

This phenomena is called *cyclotron damping* or *cyclotron amplification.*

10.3 Quasi-Linear Theory of Evolution in the Distribution Function

It has been assumed that the perturbation is small and the zeroth-order terms do not change. Under these assumption, the linearized equations on the perturbations are analyzed. However, if the perturbations grow, then the zeroth-order quantities may change and the growth rate of the perturbations may change due to the evolution of the zeroth order quantities. Finally the perturbations saturate (growth rate becomes zero) and shift to steady state. Let us consider a simple case of $\boldsymbol{B} = 0$ and one-dimensional electrostatic perturbation ($\boldsymbol{B}_1 = 0$). Ions are uniformly distributed. Then the distribution function $f(x, v, t)$ of electrons obeys the following Vlasov equation:

$$\frac{\partial f}{\partial t} + v\frac{\partial f}{\partial x} - \frac{e}{m}E\frac{\partial f}{\partial v} = 0. \tag{10.43}$$

Let the distribution function f be divided into two parts

$$f(x,v,t) = f_0(v,t) + f_1(x,v,t). \tag{10.44}$$

where f_0 is slowly changing zeroth order term and f_1 is the oscillatory first order term. It is assumed that the time derivative of f_0 is the second order term. When (10.44) is substituted into (10.43), the first and second terms satisfy the following equations:

$$\frac{\partial f_1}{\partial t} + v\frac{\partial f_1}{\partial x} = \frac{e}{m}E\frac{\partial f_0}{\partial v}, \tag{10.45}$$

$$\frac{\partial f_0}{\partial t} = \frac{e}{m}E\frac{\partial f_1}{\partial v}. \tag{10.46}$$

f_1 and E may be expressed by Fourier integrals:

$$f_1(x,v,t) = \frac{1}{(2\pi)^{1/2}}\int f_k(v)\exp(i(kx-\omega(k)t))\mathrm{d}k, \tag{10.47}$$

$$E(x,t) = \frac{1}{(2\pi)^{1/2}}\int E_k \exp(i(kx-\omega(k)t))\mathrm{d}k. \tag{10.48}$$

Since f_1 and E are real, $f_{-k} = f_k^*$, $E_{-k} = E_k^*$, $\omega(-k) = -\omega^*(k)$ $(\omega(k) = \omega_{\mathrm{r}}(k) + i\gamma(k))$. The substitution of (10.47) and (10.48) into (10.45) yields

$$f_k(v) = \frac{e}{m}\left(\frac{i}{\omega(k)-kv}\right)E_k\frac{\partial f_0}{\partial v}. \tag{10.49}$$

If (10.48) and (10.49) are substituted into (10.46), we find

$$\frac{\partial f_0(v,t)}{\partial t} = \left(\frac{e}{m}\right)^2\frac{\partial}{\partial v}\left\langle\frac{1}{2\pi}\int E_{k'}\exp(i(k'x-\omega(k')t))\mathrm{d}k'\right.$$
$$\left.\times\frac{i}{\omega(k)-kv}E_k\frac{\partial f_0(v,t)}{\partial v}\exp(i(kx-\omega(k)t))\mathrm{d}k\right\rangle. \tag{10.50}$$

The statistical average of (10.50) (integration by x) is reduced to

$$\frac{\partial f_0(v,t)}{\partial t} = \frac{\partial}{\partial v}\left(D_{\mathrm{v}}(v)\frac{\partial f_0(v,t)}{\partial v}\right), \tag{10.51}$$

$$D_{\mathrm{v}}(v) = \left(\frac{e}{m}\right)^2\frac{1}{L}\int_{-\infty}^{\infty}\frac{i|E_k|^2\exp(2\gamma(k)t)}{\omega_{\mathrm{r}}(k)-kv+i\gamma(k)}\mathrm{d}k$$

$$= \left(\frac{e}{m}\right)^2\frac{1}{L}\int_{-\infty}^{\infty}\frac{\gamma(k)|E_k|^2\exp(2\gamma(k)t)}{(\omega_{\mathrm{r}}(k)-kv)^2+\gamma(k)^2}\mathrm{d}k,$$

Figure 10.4 Integral path of (10.53) in the case of $k > 0$.

where L is the range of x in the statistical average integral. When $|\gamma(k)| \ll |\omega_{\rm r}(k)|$, the diffusion coefficient in velocity space is

$$D_{\rm v}(v) = \left(\frac{e}{m}\right)^2 \frac{\pi}{L} \int |E_k|^2 \exp(2\gamma(k)t)\, \delta(\omega_{\rm r}(k) - kv){\rm d}k$$

$$= \left(\frac{e}{m}\right)^2 \frac{\pi}{|v|} \frac{1}{L} |E_k|^2 \exp(2\gamma(k)t)\Big|_{\omega/k=v}. \tag{10.52}$$

From Poisson's equation and (10.49), the dispersion equation can be derived:

$$\nabla \cdot \boldsymbol{E} = -\frac{e}{\epsilon_0} \int f_1 {\rm d}v, \qquad ikE_k = -\frac{e}{\epsilon_0} \int f_k {\rm d}v,$$

$$1 + \frac{\Pi_{\rm e}^2}{k} \frac{1}{n_0} \int \left(\frac{1}{\omega(k) - kv}\right) \frac{\partial f_0}{\partial v} {\rm d}v = 0. \tag{10.53}$$

The point $\omega - kv = 0$ is a singular point and the integral path must be chosen in the case of $k > 0$ as is shown in Figure 10.4 (refer to Section 11.5.5). Under the assumption of $|\gamma| \ll |\omega_{\rm r}|$ ($\omega = \omega_{\rm r} + i\gamma$) and $k > 0$, (10.53) becomes

$$D(k, \omega_r + \gamma i) \equiv 1 - \frac{\Pi^2}{k^2} {\rm P} \int_{-\infty}^{\infty} \frac{\frac{1}{n_0}\frac{\partial f_0}{\partial v}}{v - \omega_r/k} {\rm d}v - \frac{\Pi^2}{k^2} \int_{\text{half circle}} \frac{\frac{1}{n_0}\frac{\partial f_0}{\partial v}}{v - \omega_r/k} {\rm d}v = 0, \tag{10.54}$$

where P means Cauchy's principal value. Then (10.54) reduces to

$$D(k, \omega_r + \gamma i) = D(k, \omega_r) + \frac{\Pi^2}{k^2} i\pi \frac{1}{n_0} \frac{\partial f_0}{\partial v}\Big|_{\omega_r/k}, \tag{10.55}$$

where $D(k, \omega_r)$ is the real part of $D(k, \omega_r + \gamma i)$ and the sum of the first two terms in the right-hand side of (10.54). When $|\gamma| \ll |\omega_r|$, we have $D(k, \omega_r + \gamma i) = D(k, \omega_r) + (\partial D/\partial \omega)\gamma i$ and the growth rate γ is

$$\gamma = \pi \frac{\Pi^2}{k^2} \frac{1}{n_0} \frac{\partial f_0}{\partial v}\Big|_{\omega_r/k} \left(\frac{\partial D(k, \omega_r)}{\partial \omega_r}\right)^{-1} \approx \omega \frac{\pi}{2} \left(\frac{\omega}{k}\right)^2 \frac{1}{n_0} \frac{\partial f_0}{\partial v}\Big|_{\omega_r/k}. \tag{10.56}$$

When the phase velocity ω_r/k is much larger than v in the region where $\partial f_0/\partial v$ is not small, we have $D(k, \omega_r) \approx 1 - \Pi_{\rm e}^2/\omega^2$, and (10.56) is almost the same equation as (10.37).

Equation (10.51) is the diffusion equation in the velocity space. When the distribution function of electrons is given by the profile shown in Figure 10.3(b), in which the positive gradient of $v\,\partial f/\partial v > 0$ exists near v_1, waves with the phase velocity of $\omega/k \approx v_1$ grow due to Landau amplification, and the amplitude of $|E_k|$ increases. The diffusion coefficient $D_{\rm v}$ in velocity space becomes large and anomalous diffusion takes place at $v_1 = \omega/k$ in velocity space due to (10.52) and (10.56). The positive gradient of $\partial f/\partial v$ near $\sim v_1$ decreases and finally the profile of the distribution function becomes flat near $v \sim v_1$.

Let us consider the other case. When a wave (lower hybrid wave) is externally exited (by antenna) in a plasma with Maxwellian distribution function (Figure 10.3(a)), electron diffusion coefficient $D_{\rm v}$ at $v = \omega/k$ is increased. The gradient of the distribution function near $v = \omega/k$ becomes flat as will be seen in Figure 12.7 and drives electron current in the direction of wave propagation (refer to Section 12.3).

Problems

1. Electron Plasma Wave Dispersion relation of electron plasma wave is given by the real part of (10.53). Expand $(\omega - kv)^{-1} = \omega^{-1}(1 + (kv/\omega) + (kv/\omega)^2 + (kv/\omega)^3 + \cdots)$ and the electron distribution function is Maxwellian $f_0 = n_0(1/2\pi)^{1/2}v_{\rm T}\exp(v^2/(2v_{\rm T}^2))$. Confirm the dispersion relation is given by $1 - (\Pi_{\rm e}/\omega)^2(1 + 3(k/\omega)^2(T_{\rm e}/m_{\rm e}) \approx 1 - (\Pi_{\rm e}/\omega)^2 - 3(k/\omega)^2(T_{\rm e}/m_{\rm e}) = 0$, $v_{\rm T}^2 \equiv T_{\rm e}/m_{\rm e}$.

2. Landau Damping Rate Formulate the ratio of Landau damping rate to the frequency ω of electron plasma wave. What condition makes Landau damping effective?

11

Plasma as Medium of Waves

A plasma is an ensemble of an enormous number of moving ions and electrons interacting with each other. In order to describe the behavior of such an ensemble, the distribution function was introduced in Chapter 10; and Boltzmann and Vlasov equations were derived with respect to the distribution function. A plasma viewed as an ensemble of a large number of particles has a large number of degrees of freedom; thus, the mathematical description of plasma behavior is feasible only for simplified analytical models.

In Chapter 3, statistical averages in velocity space, such as mass density, flow velocity, pressure, etc. were introduced and the MHD equations for these averages were derived. We have thus obtained a mathematical description of the MHD fluid model; and we have studied the equilibrium conditions, stability problems, etc. for this model. Since the fluid model considers only average quantities in velocity space, it is not capable of describing instabilities or damping phenomena, in which the profile of the distribution function plays a significant role. The phenomena which can be handled by means of the fluid model are of low frequency (less than the ion or electron cyclotron frequency); high-frequency phenomena are not describable in terms of it.

In the first part of this chapter, we will focus on a model which allows us to study wave phenomena while retaining the essential features of plasma dynamics, while at the same time maintaining relative simplicity in its mathematical form. A model is given by a homogeneous plasma of ions and electrons at 0 K in a uniform magnetic field. In the unperturbed state, both the ions and electrons of this plasma are motionless. Any small deviation from the unperturbed state induces an electric field and a time-dependent component of the magnetic field, and consequently movements of ions and electrons are excited. The movements of the charged particles induce electric and magnetic fields which are themselves consistent with the previously induced small perturbations. This is called the kinetic model of a *cold plasma*. We will use it in this chapter to derive the dispersion relation which characterizes wave phenomena in the cold plasma.

Although this model assumes uniformity of the magnetic field, and the density and as well as the zero temperature, this cold plasma model is applicable for a nonuniform, warm plasma if the typical length of variation of the magnetic field and the density is much larger than the wavelength and the phase velocity of wave is much larger than the thermal velocity of the particles.

It is possible to consider the plasma as a medium of electromagnetic wave propagation with a dielectric tensor $\boldsymbol{K}$. This dielectric tensor $\boldsymbol{K}$ is a function of the magnetic field and the density which may change with the position. Accordingly, plasma is in general a nonuniform, anisotropic, and dispersive medium.

In the later part of this chapter, the general mathematical analysis of the hot-plasma wave will be discussed. When the temperature of plasma is finite and the thermal velocity of the particles is comparable to the phase velocity of the propagating wave, the interaction of the particles and the wave becomes important. A typical interaction is Landau damping, which is included in the dielectric tensor $\boldsymbol{K}$ of hot plasma. Reference [11.1] describes the plasma wave in more detail.

11.1 Dielectric Tensor of Cold Plasma

In an unperturbed cold plasma, the particle density n and the magnetic field $\boldsymbol{B}_0$ are both homogeneous in space and constant in time. The ions and electrons are motionless.

Now assume that the first-order perturbation term $\exp i(\boldsymbol{k}\cdot\boldsymbol{r}-\omega t)$ is applied. In this case the phase velocity $v_{\rm ph}$ is equal to $v_{\rm ph}=\omega/k$ $(k=|\boldsymbol{k}|)$ propagating to the direction of $\boldsymbol{k}$. The ions and electrons are forced to move by the perturbed electric field $\boldsymbol{E}$ and the induced magnetic field $\boldsymbol{B}_1$. Let us denote velocity by $\boldsymbol{v}_k$, where the suffix k indicates the species of particle (electrons or ions of various kinds). The current $\boldsymbol{j}$ due to the particle motion is given by

$$\boldsymbol{j}=\sum_k n_k q_k \boldsymbol{v}_k. \tag{11.1}$$

n_k and q_k are the density and charge of the kth species, respectively. The electric displacement $\boldsymbol{D}$ is

$$\boldsymbol{D}=\epsilon_0\boldsymbol{E}+\boldsymbol{P}, \tag{11.2}$$

$$\boldsymbol{j}=\frac{\partial \boldsymbol{P}}{\partial t}=-i\omega\boldsymbol{P}, \tag{11.3}$$

where $\boldsymbol{E}$ is the electric intensity, $\boldsymbol{P}$ is the electric polarization, and ϵ_0 is the dielectric constant of vacuum. Consequently $\boldsymbol{D}$ is expressed by

$$\boldsymbol{D}=\epsilon_0\boldsymbol{E}+\frac{i}{\omega}\boldsymbol{j}\equiv\epsilon_0\boldsymbol{K}\cdot\boldsymbol{E}. \tag{11.4}$$

$\boldsymbol{K}$ is called the *dielectric tensor*. In the case of isotropic medium with dielectric constant ϵ ($\boldsymbol{D} = \epsilon \boldsymbol{E}$), $\boldsymbol{K}$ is equal to $\boldsymbol{K} = \hat{\epsilon}\boldsymbol{I}$, where $\hat{\epsilon} = \epsilon/\epsilon_0$ and $\boldsymbol{I}$ is the unit matrix. The equation of motion of a single particle of the kth kind is

$$m_k \frac{\mathrm{d}\boldsymbol{v}_k}{\mathrm{d}t} = q_k(\boldsymbol{E} + \boldsymbol{v}_k \times \boldsymbol{B}). \tag{11.5}$$

Here $\boldsymbol{B}$ consists of $\boldsymbol{B} = \boldsymbol{B}_0 + \boldsymbol{B}_1$, where $\boldsymbol{v}_k$, $\boldsymbol{E}$, $\boldsymbol{B}_1$ are the first-order quantities. The linearized equation in these quantities is

$$-i\omega m_k \boldsymbol{v}_k = q_k(\boldsymbol{E} + \boldsymbol{v}_k \times \boldsymbol{B}_0). \tag{11.6}$$

When the z axis is taken along the direction of $\boldsymbol{B}_0$, the solution is given by

$$\left.\begin{aligned} v_{k,x} &= \frac{-iE_x}{B_0}\frac{\Omega_k \omega}{\omega^2 - \Omega_k^2} - \frac{E_y}{B_0}\frac{\Omega_k^2}{\omega^2 - \Omega_k^2}, \\ v_{k,y} &= \frac{E_x}{B_0}\frac{\Omega_k^2}{\omega^2 - \Omega_k^2} - \frac{iE_y}{B_0}\frac{\Omega_k \omega}{\omega^2 - \Omega_k^2}, \\ v_{k,z} &= \frac{-iE_z}{B_0}\frac{\Omega_k}{\omega}, \end{aligned}\right\} \tag{11.7}$$

where Ω_k is the cyclotron frequency of the charged particle of the kth kind:

$$\Omega_k = \frac{-q_k B_0}{m_k} \tag{11.8}$$

($\Omega_{\mathrm{e}} > 0$ for electrons and $\Omega_{\mathrm{i}} < 0$ for ions). The components of $\boldsymbol{v}_k$ are the linear functions of $\boldsymbol{E}$ given by (11.7); and $\boldsymbol{j}$ of (11.1) and the electric displacement $\boldsymbol{D}$ of (11.4) are also the linear function of $\boldsymbol{E}$, so that the *dielectric tensor* is given by

$$\boldsymbol{K}\cdot\boldsymbol{E} = \begin{bmatrix} K_\perp & -iK_\times & 0 \\ iK_\times & K_\perp & 0 \\ 0 & 0 & K_\parallel \end{bmatrix} \begin{bmatrix} E_x \\ E_y \\ E_z \end{bmatrix} \tag{11.9}$$

where

$$K_\perp \equiv 1 - \sum_k \frac{\Pi_k^2}{\omega^2 - \Omega_k^2}, \tag{11.10}$$

$$K_\times \equiv -\sum_k \frac{\Pi_k^2}{\omega^2 - \Omega_k^2}\frac{\Omega_k}{\omega}, \tag{11.11}$$

$$K_\parallel \equiv 1 - \sum_k \frac{\Pi_k^2}{\omega^2}, \tag{11.12}$$

$$\Pi_k^2 \equiv \frac{n_k q_k^2}{\epsilon_0 m_k}. \tag{11.13}$$

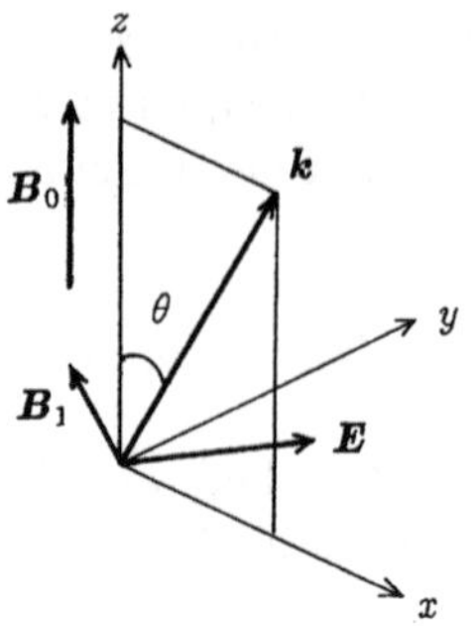

Figure 11.1 Propagation vector $\boldsymbol{k}$ and x, y, z coordinates.

According to the Stix notation, the following quantities are introduced:

$$\left.\begin{aligned} R &\equiv 1 - \sum_k \frac{\Pi_k^2}{\omega^2} \frac{\omega}{\omega - \Omega_k} = K_\perp + K_\times, \\ L &\equiv 1 - \sum_k \frac{\Pi_k^2}{\omega^2} \frac{\omega}{\omega + \Omega_k} = K_\perp - K_\times. \end{aligned}\right\} \tag{11.14}$$

From Maxwell's equation

$$\nabla \times \boldsymbol{E} = -\frac{\partial \boldsymbol{B}}{\partial t}, \tag{11.15}$$

$$\nabla \times \boldsymbol{H} = \boldsymbol{j} + \epsilon_0 \frac{\partial \boldsymbol{E}}{\partial t} = \frac{\partial \boldsymbol{D}}{\partial t}, \tag{11.16}$$

it follows that

$$\boldsymbol{k} \times \boldsymbol{E} = \omega \boldsymbol{B}_1,$$

$$\boldsymbol{k} \times \boldsymbol{H}_1 = -\omega\epsilon_0 \boldsymbol{K} \cdot \boldsymbol{E}$$

and

$$\boldsymbol{k} \times (\boldsymbol{k} \times \boldsymbol{E}) + \frac{\omega^2}{c^2} \boldsymbol{K} \cdot \boldsymbol{E} = 0. \tag{11.17}$$

Let us define a dimensionless vector

$$\boldsymbol{N} \equiv \frac{\boldsymbol{k}c}{\omega}$$

(c is light velocity in a vacuum). The absolute value $N = |\boldsymbol{N}|$ is the ratio of the light velocity to the phase velocity of the wave, i.e., N is the refractive index. Using $\boldsymbol{N}$, we may write (11.17) as

$$\boldsymbol{N}\times(\boldsymbol{N}\times\boldsymbol{E})+\boldsymbol{K}\cdot\boldsymbol{E}=0. \tag{11.18}$$

In the case of isotropic medium $\boldsymbol{K}=\hat{\epsilon}\boldsymbol{I}$, (11.18) reduces $N^2=\hat{\epsilon}$ and $\boldsymbol{k}\cdot\boldsymbol{E}=0$. Plasmas are not isotropic and if the angle between $\boldsymbol{N}$ and $\boldsymbol{B}_0$ is denoted by θ (Figure 11.1) and the x axis is taken so that $\boldsymbol{N}$ lies in the z,x plane, then (11.18) may be expressed by

$$\begin{bmatrix} K_\perp - N^2\cos^2\theta & -iK_\times & N^2\sin\theta\cos\theta \\ iK_\times & K_\perp - N^2 & 0 \\ N^2\sin\theta\cos\theta & 0 & K_\parallel - N^2\sin^2\theta \end{bmatrix} \begin{bmatrix} E_x \\ E_y \\ E_z \end{bmatrix} = 0. \tag{11.19}$$

For a nontrivial solution to exist, the determinant of the matrix must be zero; that is,

$$AN^4 - BN^2 + C = 0, \tag{11.20}$$

$$A = K_\perp\sin^2\theta + K_\parallel\cos^2\theta, \tag{11.21}$$

$$B = (K_\perp^2 - K_\times^2)\sin^2\theta + K_\parallel K_\perp(1+\cos^2\theta), \tag{11.22}$$

$$C = K_\parallel(K_\perp^2 - K_\times^2) = K_\parallel RL. \tag{11.23}$$

Equation (11.20) determines the relationship between the propagation vector $\boldsymbol{k}$ and the frequency ω, and it is called the *dispersion equation.* The solution of (11.20) is

$$\begin{aligned} N^2 &= \frac{B \pm (B^2-4AC)^{1/2}}{2A} \\ &= \big((K_\perp^2 - K_\times^2)\sin^2\theta + K_\parallel K_\perp(1+\cos^2\theta) \\ &\quad \pm[(K_\perp^2 - K_\times^2 - K_\parallel K_\perp)^2\sin^4\theta + 4K_\parallel^2K_\times^2\cos^2\theta]^{1/2}\big) \\ &\quad \times\big(2(K_\perp\sin^2\theta + K_\parallel\cos^2\theta)\big)^{-1}. \end{aligned} \tag{11.24}$$

When the wave propagates along the line of magnetic force ($\theta=0$), the dispersion equation (11.20) is

$$K_\parallel(N^4 - 2K_\perp N^2 + (K_\perp^2 - K_\times^2)) = 0, \tag{11.25}$$

and the solutions are

$$K_\parallel = 0, \quad N^2 = K_\perp + K_\times = R, \quad N^2 = K_\perp - K_\times = L. \tag{11.26}$$

For the wave propagating in the direction perpendicular to the magnetic field ($\theta=\pi/2$), the dispersion equation and the solutions are given by

$$K_\perp N^4 - (K_\perp^2 - K_\times^2 + K_\parallel K_\perp)N^2 + K_\parallel(K_\perp^2 - K_\times^2) = 0, \tag{11.27}$$

$$N^2 = \frac{K_\perp^2 - K_\times^2}{K_\perp} = \frac{RL}{K_\perp}, \qquad N^2 = K_\parallel. \tag{11.28}$$

11.2 Properties of Waves

11.2.1 Polarization and Particle Motion

The dispersion relation for waves in a cold plasma was derived in the previous section. We consider here the electric field of the waves and the resultant particle motion. The y component of (11.19) is

$$iK_\times E_x + (K_\perp - N^2)E_y = 0,$$

$$\frac{iE_x}{E_y} = \frac{N^2 - K_\perp}{K_\times}. \tag{11.29}$$

The relation between the components of the particle velocity is

$$\frac{iv_{k,x}}{v_{k,y}} = \frac{i\left(\dfrac{-iE_x}{E_y}\dfrac{\omega}{\omega^2 - \Omega_k^2} - \dfrac{\Omega_k}{\omega^2 - \Omega_k^2}\right)}{\dfrac{E_x}{E_y}\dfrac{\Omega_k}{\omega^2 - \Omega_k^2} - i\dfrac{\omega}{\omega^2 - \Omega_k^2}}$$

$$= \frac{(\omega + \Omega_k)(N^2 - L) + (\omega - \Omega_k)(N^2 - R)}{(\omega + \Omega_k)(N^2 - L) - (\omega - \Omega_k)(N^2 - R)}. \tag{11.30}$$

The wave satisfying $N^2 = R$ at $\theta = 0$ has $iE_x/E_y = 1$ and the electric field is right-circularly polarized. In other words, the electric field rotates in the direction of the electron Larmor motion. The motion of ions and electrons is also right-circular motion. In the wave satisfying $N^2 = L$ at $\theta \to 0$, the relation $iE_x/E_y = -1$ holds and the electric field is left-circularly polarized. The motion of ions and electrons is also left-circular motion. The waves with $N^2 = R$ and $N^2 = L$ as $\theta \to 0$ are called the R *wave* and the L *wave*, respectively. The solution of the dispersion equation (11.25) at $\theta = 0$ is

$$N^2 = \frac{1}{2}\left(R + L \pm \frac{|K_\parallel|}{K_\parallel}|R - L|\right), \tag{11.31}$$

so that R and L waves are exchanged when $K_\parallel$ changes sign. When $K_\times = R - L$ changes sign, R and L waves are also exchanged.

When $\theta = \pi/2$, the electric field of the wave satisfying $N^2 = K_\parallel$ is $E_x = E_y = 0$, $E_z \neq 0$. For the wave satisfying $N^2 = RL/K_\perp$, the electric field satisfies the relations $iE_x/E_y = -(R - L)/(R + L) = -K_\times/K_\perp$, $E_z = 0$. The waves with $N^2 = K_\parallel$ and $N^2 = RL/K_\perp$ as $\theta \to \pi/2$ are called the ordinary wave (O) and the extraordinary wave (X), respectively. It should be pointed out that the electric field of the extraordinary wave at $\theta = \pi/2$ is perpendicular to the magnetic field ($E_z = 0$) and the electric field of the ordinary wave at $\theta = \pi/2$ is parallel to the magnetic field ($E_x = E_y = 0$). The dispersion relation (11.24) at $\theta = \pi/2$ is

$$N^2 = \frac{1}{2K_\perp}(K_\perp^2 - K_\times^2 + K_\parallel K_\perp + |K_\perp^2 - K_\times^2 - K_\parallel K_\perp|)$$
$$= \frac{1}{2K_\perp}(RL + K_\parallel K_\perp \pm |RL - K_\parallel K_\perp|), \tag{11.32}$$

so that the ordinary wave and the extraordinary wave are exchanged at $RL - K_\parallel K_\perp = 0$.

Besides the classification into R and L waves, and O and X waves, there is another classification, namely, that of *fast* wave and *slow* wave, following the difference in the phase velocity. Since the term inside the square root of the equation $N^2 = (B \pm (B^2 - 4AC)^{1/2})/2A$ is always positive, as is clear from (11.24), the fast wave and slow wave do not exchange between $\theta = 0$ and $\theta = \pi/2$.

11.2.2 Cutoff and Resonance

The refractive index (11.24) may become infinity or zero. When $N^2 = 0$, the wave is said to be *cutoff*; at cutoff the phase velocity

$$v_{\text{ph}} = \frac{\omega}{k} = \frac{c}{N} \tag{11.33}$$

becomes infinity. As is clear from (11.20) and (11.23), cutoff occurs when

$$K_\parallel = 0, \qquad R = 0, \qquad L = 0. \tag{11.34}$$

When $N^2 = \infty$, the wave is said to be at *resonance*; here the phase velocity becomes zero. The wave will be absorbed by the plasma at resonance. The resonance condition is

$$\tan^2\theta = -\frac{K_\parallel}{K_\perp}. \tag{11.35}$$

When $\theta = 0$, the resonance condition is $K_\perp = (R + L)/2 \to \pm\infty$. The condition $R \to \pm\infty$ is satisfied at $\omega = \Omega_e$, Ω_e being the electron cyclotron frequency. This is called *electron cyclotron resonance*. The condition $L \to \pm\infty$ holds when $\omega = |\Omega_i|$, and this is called *ion cyclotron resonance*. When $\theta = \pi/2$, $K_\perp = 0$ is the resonance condition. This is called *hybrid resonance*. When waves approach a cutoff region, the wave path is curved according to Snell's refraction law and the waves are reflected (Figure 11.2a). When waves approach a resonance region, the waves propagate perpendicularly toward the resonance region. The phase velocities tend to zero and the wave energy will be absorbed (Figure 11.2b).

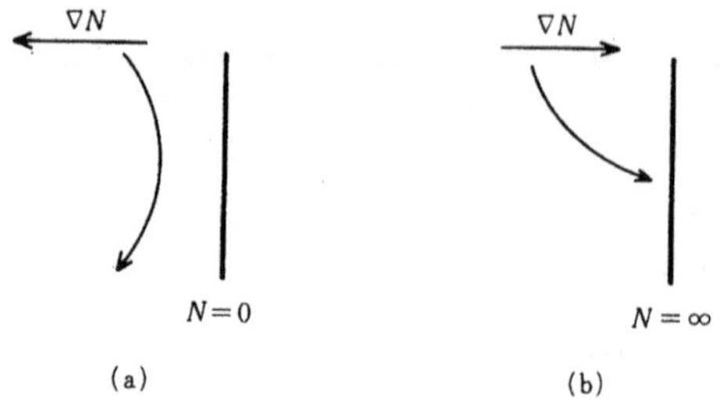

Figure 11.2 Wave propagation (a) near cutoff region and (b) near a resonance region.

11.3 Waves in a Two-Component Plasma

Let us consider a plasma that consists of electrons and of one kind of ion. Charge neutrality is

$$n_{\rm i}Z_{\rm i} = n_{\rm e}. \tag{11.36}$$

A dimensionless parameter is introduced for convenience:

$$\delta = \frac{\mu_0(n_{\rm i}m_{\rm i} + n_{\rm e}m_{\rm e})c^2}{B_0^2}. \tag{11.37}$$

The quantity defined by (11.13), which was also introduced in Section 2.2,

$$\Pi_{\rm e}^2 = n_{\rm e}e^2/(\epsilon_0 m_{\rm e}) \tag{11.38}$$

is called electron plasma frequency. Then we have the relations

$$\begin{aligned} &\Pi_{\rm e}^2/\Pi_{\rm i}^2 = m_{\rm i}/m_{\rm e} \gg 1, \\ &\frac{\Pi_{\rm i}^2 + \Pi_{\rm e}^2}{|\Omega_{\rm i}|\Omega_{\rm e}} = \delta \approx \frac{\Pi_{\rm i}^2}{\Omega_{\rm i}^2}. \end{aligned} \tag{11.39}$$

$K_\perp$, $K_\times$, $K_\parallel$, and R, L are given by

$$\left.\begin{aligned} K_\perp &= 1 - \frac{\Pi_{\rm i}^2}{\omega^2 - \Omega_{\rm i}^2} - \frac{\Pi_{\rm e}^2}{\omega^2 - \Omega_{\rm e}^2}, \\ K_\times &= -\frac{\Pi_{\rm i}^2}{\omega^2 - \Omega_{\rm i}^2}\frac{\Omega_{\rm i}}{\omega} - \frac{\Pi_{\rm e}^2}{\omega^2 - \Omega_{\rm e}^2}\frac{\Omega_{\rm e}}{\omega}, \\ K_\parallel &= 1 - \frac{\Pi_{\rm e}^2 + \Pi_{\rm i}^2}{\omega^2} \simeq 1 - \frac{\Pi_{\rm e}^2}{\omega^2}, \end{aligned}\right\} \tag{11.40}$$

$$R = 1 - \frac{\Pi_{\rm e}^2 + \Pi_{\rm i}^2}{(\omega - \Omega_{\rm i})(\omega - \Omega_{\rm e})} \simeq \frac{\omega^2 - (\Omega_{\rm i} + \Omega_{\rm e})\omega + \Omega_{\rm i}\Omega_{\rm e} - \Pi_{\rm e}^2}{(\omega - \Omega_{\rm i})(\omega - \Omega_{\rm e})}, \tag{11.41}$$

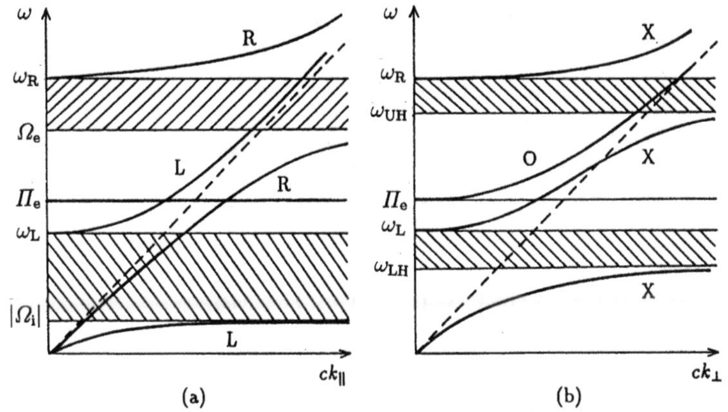

Figure 11.3 (a) Dispersion relations (ω - $ck_\parallel$) for R and L waves propagating parallel to the magnetic field ($\theta = 0$). (b) Dispersion relations (ω - $ck_\parallel$) for O and X waves propagating perpendicular to the magnetic field ($\theta = \pi/2$).

$$L = 1 - \frac{\Pi_e^2 + \Pi_i^2}{(\omega + \Omega_i)(\omega + \Omega_e)} \simeq \frac{\omega^2 + (\Omega_i + \Omega_e)\omega + \Omega_i\Omega_e - \Pi_e^2}{(\omega + \Omega_i)(\omega + \Omega_e)}. \tag{11.42}$$

The dispersion relations for the waves propagating parallel to $\boldsymbol{B}_0$ ($\theta = 0$) are found by setting $K_\parallel = 0$, $N^2 = R$, and $N^2 = L$. Then

$$\omega^2 = \Pi_e^2, \tag{11.43}$$

$$\frac{\omega^2}{c^2k_\parallel^2} = \frac{1}{R} = \frac{(\omega - \Omega_i)(\omega - \Omega_e)}{\omega^2 - \omega\Omega_e + \Omega_e\Omega_i - \Pi_e^2} = \frac{(\omega + |\Omega_i|)(\omega - \Omega_e)}{(\omega - \omega_R)(\omega + \omega_L)}, \tag{11.44}$$

where ω_R, ω_L are given by

$$\omega_R = \frac{\Omega_e}{2} + \left(\left(\frac{\Omega_e}{2}\right)^2 + \Pi_e^2 + |\Omega_e\Omega_i|\right)^{1/2} > 0, \tag{11.45}$$

$$\omega_L = -\frac{\Omega_e}{2} + \left(\left(\frac{\Omega_e}{2}\right)^2 + \Pi_e^2 + |\Omega_e\Omega_i|\right)^{1/2} > 0, \tag{11.46}$$

$$\frac{\omega^2}{c^2k_\parallel^2} = \frac{1}{L} = \frac{(\omega + \Omega_i)(\omega + \Omega_e)}{\omega^2 + \omega\Omega_e + \Omega_e\Omega_i - \Pi_e^2} = \frac{(\omega - |\Omega_i|)(\omega + \Omega_e)}{(\omega - \omega_L)(\omega + \omega_R)}. \tag{11.47}$$

Note that $\Omega_e > 0$, $\Omega_i < 0$, and $\omega_R > \Omega_e$. Plots of the dispersion relations $\omega - ck_\parallel$ in the case of $\Omega_e > \Pi_e$ are shown in Figure 11.3a. The dispersion

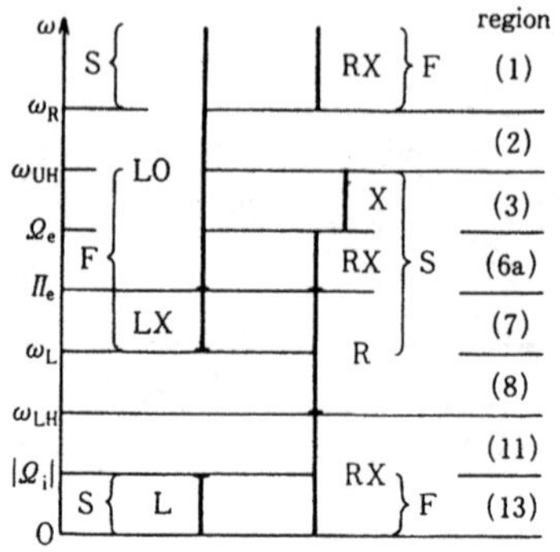

Figure 11.4 The ω regions of R and L waves at $\theta = 0$; O and X waves at $\theta = \pi/2$; F and S waves; in the case of $(\omega_L < \Pi_e < \Omega_e)$. The numbers on the right identify regions shown in the CMA diagram (Figure 11.5).

relations for the waves propagating perpendicular to $\boldsymbol{B}_0$ are found by setting $N^2 = K_\parallel$ (ordinary wave) and $N^2 = (K_\perp^2 - K_\times^2)/K_\perp$ (extraordinary wave). Then

$$\frac{\omega^2}{c^2k_\perp^2} = \frac{1}{K_\parallel} = \left(1 - \frac{\Pi_e^2}{\omega^2}\right)^{-1} = 1 + \frac{\Pi_e^2}{c^2k_\perp^2}, \tag{11.48}$$

$$\frac{\omega^2}{c^2k_\perp^2} = \frac{K_\perp}{K_\perp^2 - K_\times^2} = \frac{K_\perp}{RL}$$

$$= \frac{2(\omega^2-\Omega_i^2)(\omega^2-\Omega_e^2)-\Pi_e^2((\omega+\Omega_i)(\omega+\Omega_e)+(\omega-\Omega_i)(\omega-\Omega_e))}{2(\omega^2-\omega_L^2)(\omega^2-\omega_R^2)}$$

$$= \frac{\omega^4 - (\Omega_i^2 + \Omega_e^2 + \Pi_e^2)\omega^2 + \Omega_i^2\Omega_e^2 - \Pi_e^2\Omega_i\Omega_e}{(\omega^2-\omega_L^2)(\omega^2-\omega_R^2)}. \tag{11.49}$$

Equation (11.48) is the dispersion equation of electron plasma wave (Langmuir wave). Let us define ω_{UH} and ω_{LH} by

$$\omega_{UH}^2 \equiv \Omega_e^2 + \Pi_e^2, \tag{11.50}$$

$$\frac{1}{\omega_{LH}^2} \equiv \frac{1}{\Omega_i^2 + \Pi_i^2} + \frac{1}{|\Omega_i|\Omega_e}. \tag{11.51}$$

ω_{UH} is called *upper hybrid resonant frequency* and ω_{LH} is called *lower hybrid resonant frequency*. Using these, we may write (11.49) as

$$\frac{\omega^2}{c^2k_\perp^2} = \frac{(\omega^2-\omega_{LH}^2)(\omega^2-\omega_{UH}^2)}{(\omega^2-\omega_L^2)(\omega^2-\omega_R^2)}. \tag{11.52}$$

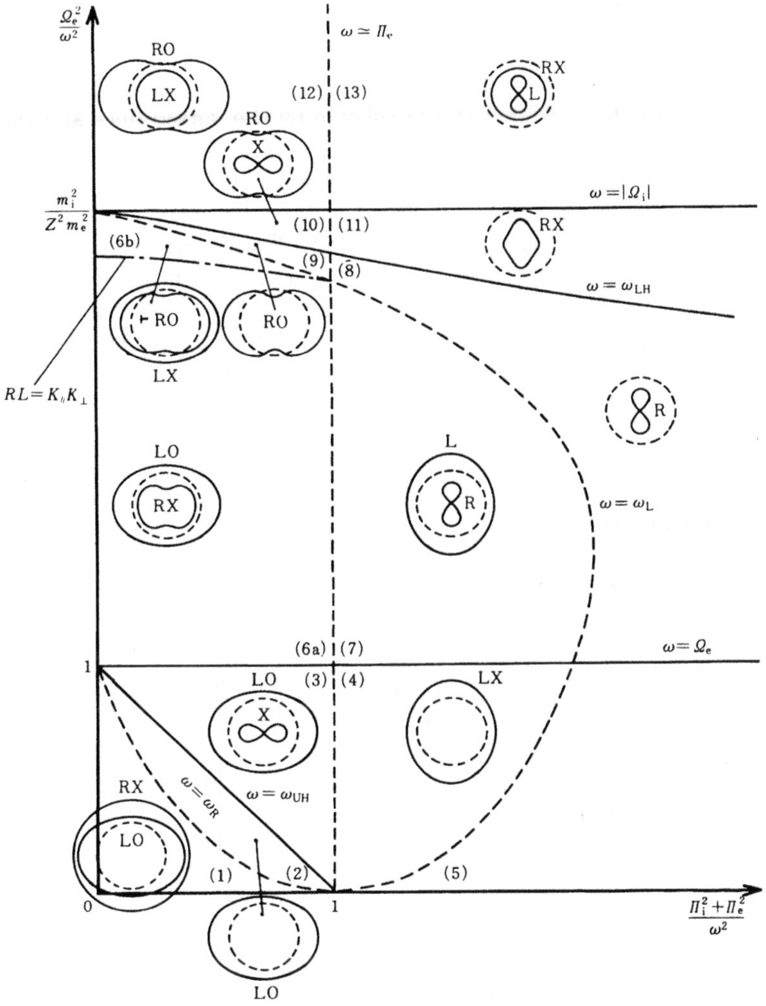

Figure 11.5 CMA diagram of a two-component plasma. The surfaces of constant phase are drawn in each region. The dotted circles give the wave front in vacuum. The magnetic field is directed toward the top of the diagram.

We have $\omega_R > \omega_{UH} > \Pi_e, \Omega_e$ and $\omega_{LH}^2 < \Omega_e|\Omega_i|, \Omega_i^2 + \Pi^2$. Plots of the dispersion relation ω - $ck_\perp$ in the case of $\Omega_e > \Pi_e$ are shown in Figure 11.3b. The gradient ω/ck in ω - $ck_\perp$ diagram is the ratio of the phase velocity v_{ph} to c. The steeper the gradient, the greater the phase velocity. The regions (in terms of ω) of R and L waves at $\theta = 0$, and O and X waves at $\theta = \pi/2$, and F and S waves are shown in Figure 11.4, for the case of $\omega_L < \Pi_e < \Omega_e$.

We explain here the CMA diagram (Figure 11.5), which was introduced by P. C. Clemmow and R. F. Mullaly and later modified by W. P. Allis [11.2]. The quantities Ω_e^2/ω^2 and $(\Pi_i^2 + \Pi_e^2)/\omega^2$ are plotted along the vertical and horizontal ordinates, respectively. The cutoff conditions $R = 0\,(\omega = \omega_R)$, $L = 0\,(\omega = \omega_L)$, $K_\parallel = 0\,(\omega = \Pi_e)$ are shown by the dotted lines and the resonance conditions $R = \infty\,(\omega = \Omega_e)$, $L = \infty\,(\omega = \Omega_i)$, $K_\perp = 0\,(\omega = \Omega_{LH},\ \omega = \Omega_{UH})$ are shown by solid lines. The cutoff and the resonance contours form the boundaries of the different regions. The boundary $RL = K_\parallel K_\perp$, at which O wave and X wave are exchanged, is also shown by broken and dotted lines in Figure 11.5. The surfaces of constant phase for R, L and O, X waves are shown for the different regions. As the vertical and horizontal ordinates correspond to the magnitude of B and the density n_e, one can easily assign waves to the corresponding regions simply by giving their frequencies ω.

11.4 Various Waves

11.4.1 Alfvén Wave

When the frequency ω is smaller than the ion cyclotron frequency $(\omega \ll |\Omega_i|)$, the dielectric tensor $\boldsymbol{K}$ is expressed by

$$\left.\begin{aligned} K_\perp &= 1 + \delta, \\ K_\times &= 0, \\ K_\parallel &= 1 - \frac{\Pi_e^2}{\omega^2}, \end{aligned}\right\} \tag{11.53}$$

where $\delta = \mu_0 n_i m_i c^2/B_0^2$. As $\Pi_e^2/\omega^2 = (m_i/m_e)(\Omega_i^2/\omega^2)\delta$, we find $\Pi_e^2/\omega^2 \gg \delta$. Assuming that $\Pi_e^2/\omega^2 \gg 1$, we have $|K_\parallel| \gg |K_\perp|$; then A, B, C of (11.20) are given by

$$\left.\begin{aligned} A &\approx -\frac{\Pi_e^2}{\omega^2}\cos^2\theta, \\ B &\approx -\frac{\Pi_e^2}{\omega^2}(1+\delta)(1+\cos^2\theta), \\ C &\approx -\frac{\Pi_e^2}{\omega^2}(1+\delta)^2, \end{aligned}\right\} \tag{11.54}$$

and the dispersion relations are

$$\frac{c^2}{N^2} = \frac{\omega^2}{k^2} = \frac{c^2}{1+\delta} = \frac{c^2}{1+\dfrac{\mu_0\rho_\mathrm{m}c^2}{B_0^2}} \simeq \frac{B_0^2}{\mu_0\rho_\mathrm{m}}, \tag{11.55}$$

$$\frac{c^2}{N^2} = \frac{\omega^2}{k^2} = \frac{c^2}{1+\delta}\cos^2\theta \tag{11.56}$$

(ρ_m is the mass density). The wave satisfying this dispersion relation is called the *Alfvén wave.* We define the *Alfvén velocity* by

$$v_\mathrm{A}^2 = \frac{c^2}{1+\delta} = \frac{c^2}{1+\dfrac{\mu_0\rho_\mathrm{m}c^2}{B_0^2}} \simeq \frac{B_0^2}{\mu_0\rho_\mathrm{m}}. \tag{11.57}$$

Equations (11.55) and (11.56) correspond to modes appearing in region (13) of the CMA diagram. Substitution of (11.55) and (11.56) into (11.19) shows that E_z for either mode is $E_z = 0$; $E_x = 0$ for the mode (11.55) (R wave, F wave, X wave) and $E_y = 0$ for mode (11.56) (L wave, S wave). From (11.6), we find for $\omega \ll |\Omega_\mathrm{i}|$ that

$$\boldsymbol{E} + \boldsymbol{v}_\mathrm{i} \times \boldsymbol{B}_0 = 0, \tag{11.58}$$

and $\boldsymbol{v}_\mathrm{i} = (\boldsymbol{E} \times \boldsymbol{B}_0)/\boldsymbol{B}_0^2$, so that $\boldsymbol{v}_\mathrm{i}$ of the mode (11.55) is

$$\boldsymbol{v}_\mathrm{i} \propto \hat{\boldsymbol{x}} \cos(k_x x + k_z z - \omega t), \tag{11.59}$$

and $\boldsymbol{v}_\mathrm{i}$ of the mode (11.56) is

$$\boldsymbol{v}_\mathrm{i} \propto \hat{\boldsymbol{y}} \cos(k_x x + k_z z - \omega t), \tag{11.60}$$

where $\hat{\boldsymbol{x}}$, $\hat{\boldsymbol{y}}$ are unit vectors along x and y axes, respectively. From these last equations, the fast mode (11.55) and (11.59) is called the *compressional mode* and the slow mode (11.56) and (11.60) is called the *torsional* or *shear mode.* The R wave (11.55) still exists, though it is deformed in the transition from region (13) to regions (11) and (8), but the L wave (11.56) disappears in these transitions.

As is clear from (11.58), the plasma is frozen to the magnetic field. There is tension $B^2/2\mu_0$ along the magnetic-field lines and the pressure $B^2/2\mu_0$ exerted perpendicularly to the magnetic field. As the plasma of mass density ρ_m sticks to the field lines, the wave propagation speed in the direction of the field is $B_0^2/(\mu_0\rho_\mathrm{m})$.

11.4.2 Ion Cyclotron Wave and Fast Wave

Let us consider the case where the frequency ω is shifted from low frequency toward the ion cyclotron frequency and $\Pi_\mathrm{e}^2/\omega^2 \gg 1$. The corresponding waves are located in regions (13) and (11) of the CMA diagram. When $|\omega| \ll \Omega_e$, $\delta \gg 1$, and $\Pi_e^2/\omega^2 \gg 1$, the values of $K_\perp$, $K_\times$, and $K_\parallel$ are

$$K_\perp = \frac{-\delta\Omega_i^2}{\omega^2 - \Omega_i^2}, \qquad K_\times = \frac{-\delta\omega\Omega_i^2}{\omega^2 - \Omega_i^2}, \qquad K_\parallel = -\frac{\Pi_e^2}{\omega^2}. \tag{11.61}$$

Since $\Pi_e^2/\omega^2 = (m_i/m_e)(\Omega_i^2/\omega^2)\delta \gg \delta$, the coefficients A, B, C are

$$\left.\begin{aligned} A &= -\frac{\Pi_e^2}{\omega^2}\cos^2\theta, \\ B &= \frac{\Pi_e^2}{\omega^2}\frac{\delta\Omega_i^2}{\omega^2 - \Omega_i^2}(1+\cos^2\theta), \\ C &= \frac{\Pi_e^2}{\omega^2}\frac{\delta^2\Omega_i^2}{\omega^2 - \Omega_i^2}. \end{aligned}\right\} \tag{11.62}$$

The dispersion equation becomes ($\Pi_i^2 = \Omega_i^2\delta$, $v_A^2 = c^2/\delta$)

$$N^4\cos^2\theta - N^2\frac{\delta\Omega_i^2}{\Omega_i^2 - \omega^2}(1+\cos^2\theta) + \frac{\delta^2\Omega_i^2}{\Omega_i^2 - \omega^2} = 0. \tag{11.63}$$

Setting $N^2\cos^2\theta = c^2k_\parallel^2/\omega^2$, and $N^2\sin^2\theta = c^2k_\perp^2/\omega^2$, we may write (11.63) as

$$k_\perp^2 c^2 = \frac{\omega^4\delta^2\Omega_i^2 - \omega^2(2\delta\Omega_i^2k_\parallel^2c^2 + k_\parallel^4c^4) + \Omega_i^2k_\parallel^4c^4}{\omega^2(\delta\Omega_i^2 + k_\parallel^2c^2) - \Omega_i^2k_\parallel^2c^2}, \tag{11.64}$$

$$\frac{k_\perp^2}{k_\parallel^2} = \frac{(\omega/v_Ak_\parallel)^4 - 2(\omega/v_Ak_\parallel)^2 - (\omega/\Omega_i)^2 + 1}{(\omega/v_Ak_\parallel)^2 - (1 - \omega^2/\Omega_i^2)}$$

$$= \frac{((\omega/v_Ak_\parallel)^2 - (1 - \omega/\Omega_i))((\omega/v_Ak_\parallel)^2 - (1 + \omega/\Omega_i))}{(\omega/v_Ak_\parallel)^2 - (1 - \omega^2/\Omega_i^2)}. \tag{11.64'}$$

Therefore, resonance occurs at

$$\omega^2 = \Omega_i^2\frac{k_\parallel^2c^2}{k_\parallel^2c^2 + \delta\Omega_i^2} = \Omega_i^2\frac{k_\parallel^2c^2}{k_\parallel^2c^2 + \Pi_i^2}, \tag{11.65}$$

$$\left(\frac{\omega}{v_Ak_\parallel}\right)^2 = 1 - \left(\frac{\omega}{\Omega_i}\right)^2. \tag{11.65'}$$

When $|\omega|$ approaches $|\Omega_i|$, the dispersion equation (11.63) approaches

$$N^2 \approx \frac{\delta}{1+\cos^2\theta}, \tag{11.66}$$

$$N^2 \cos^2 \theta \approx \delta(1 + \cos^2 \theta) \frac{\Omega_i^2}{\Omega_i^2 - \omega^2}. \tag{11.67}$$

The mode (11.66) corresponds to the compressional Alfvén mode (fast wave) and is not affected by the ion cyclotron resonance. The dispersion relation (11.67) is that of the *ion cyclotron wave*, and can be expressed by

$$\omega^2 = \Omega_i^2 \left(1 + \frac{\Pi_i^2}{k_\parallel^2 c^2} + \frac{\Pi_i^2}{k_\parallel^2 c^2 + k_\perp^2 c^2}\right)^{-1}. \tag{11.68}$$

Note that here ω^2 is always less than Ω_i^2.

The ions move in a left circular motion (i.e., in the direction of the ion Larmor motion) at $\omega \simeq |\Omega_i|$ for both waves (see (11.30)). The mode (11.66) satisfies $iE_x/E_y = 1$, i.e., it is circularly polarized, with the electric field rotating opposite to the ion Larmor motion.

The ion cyclotron wave satisfies

$$\frac{iE_x}{E_y} \approx -\frac{\omega}{|\Omega_i|} \frac{1}{\left(1 + \frac{k_\perp^2}{k_\parallel^2}\right)}, \tag{11.69}$$

i.e., the electric field is elliptically polarized, rotating in the same direction as the ion Larmor motion.

11.4.3 Lower Hybrid Resonance

The frequency at lower hybrid resonance at $\theta = \pi/2$ is given by

$$\omega^2 = \omega_{LH}^2,$$

$$\frac{1}{\omega_{LH}^2} = \frac{1}{\Omega_i^2 + \Pi_i^2} + \frac{1}{|\Omega_i|\Omega_e}, \qquad \frac{\omega_{LH}^2}{|\Omega_i|\Omega_e} = \frac{\Pi_i^2 + \Omega_i^2}{\Pi_i^2 + |\Omega_i|\Omega_e + \Omega_i^2}. \tag{11.70}$$

When the density is high and $\Pi_i^2 \gg |\Omega_i|\Omega_e$, it follows that $\omega_{LH} = (|\Omega_i|\Omega_e)^{1/2}$. When $\Pi_i^2 \ll |\Omega_i|\Omega_e$, then $\omega_{LH}^2 = \Pi_i^2 + \Omega_i^2$. At lower hybrid resonance, we have $E_y = E_z = 0$ and $E_x \neq 0$.

When the density is high (that is, $\Pi_i^2 > |\Omega_i|\Omega_e$), then $|\Omega_i| \ll \omega_{LH} \ll \Omega_e$ and the analysis of the motions of ions and electrons becomes simple. From (11.7), the velocity is given by

$$v_{k,x} = \frac{i\epsilon_k E_x}{B_0} \frac{\omega|\Omega_k|}{\omega^2 - \Omega_k^2}, \tag{11.71}$$

and $v_{k,x} = \mathrm{d}x_k/\mathrm{d}t = -i\omega x_k$ yields

$$x_k = \frac{-\epsilon_k E_x}{B_0} \frac{|\Omega_k|}{\omega^2 - \Omega_k^2}. \tag{11.72}$$

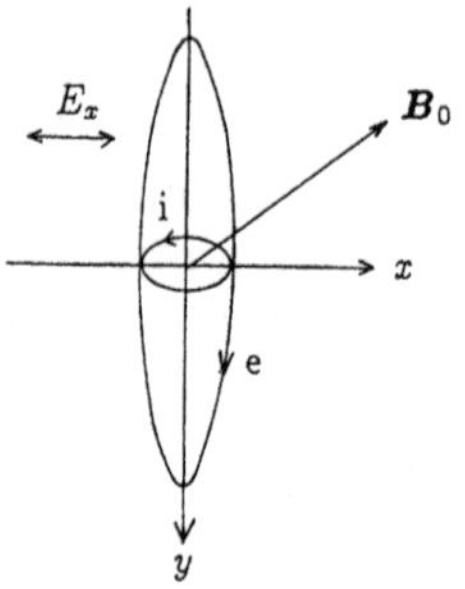

Figure 11.6 Orbits of ions and electrons at lower hybrid resonance.

At $\omega^2 = |\Omega_i|\Omega_e$, we find that $x_i \approx -E_x/B_0\Omega_e$ and $x_e \simeq -E_x/B_0\Omega_e$, or $x_i \approx x_e$ (see Figure 11.6). Consequently, charge separation does not occur and the lower hybrid wave can exist.

We have been discussing lower hybrid resonance at $\theta = \pi/2$. Let us consider the case in which θ is slightly different from $\theta = \pi/2$. The resonance condition is obtained from (11.24) as follows:

$$K_\perp \sin^2\theta + K_\parallel \cos^2\theta = 0. \tag{11.73}$$

$K_\perp$ is given by use of (11.46),(11.50), and (11.51). Equation (11.73) is reduced to

$$\frac{(\omega^2-\omega_{LH}^2)(\omega^2-\omega_{UH}^2)}{(\omega^2-\Omega_i^2)(\omega^2-\Omega_e^2)}\sin^2\theta + \left(1-\frac{\Pi_e^2}{\omega^2}\right)\cos^2\theta = 0. \tag{11.74}$$

When θ is near $\pi/2$ and ω is not much different than ω_{LH}, we find that

$$\omega^2-\omega_{LH}^2 = \frac{(\omega_{LH}^2-\Omega_e^2)(\omega_{LH}^2-\Omega_i^2)}{\omega_{LH}^2-\omega_{UH}^2}\frac{\Pi_e^2-\omega_{LH}^2}{\omega_{LH}^2}\cos^2\theta$$
$$\approx \frac{\Omega_e^2\Pi_e^2}{\omega_{UH}^2}\left(1-\left(\frac{\Omega_i}{\omega_{LH}}\right)^2\right)\left(1-\left(\frac{\omega_{LH}}{\Pi_e}\right)^2\right)\cos^2\theta.$$

As $\omega_{UH}^2\omega_{LH}^2 = \Omega_i^2\Omega_e^2 + \Pi_e^2|\Omega_i|\Omega_e$, ω^2 is expressed by

$$\omega^2 = \omega_{LH}^2\left[1+\frac{m_i}{Zm_e}\cos^2\theta\frac{\left(1-\left(\frac{\Omega_i}{\omega_{LH}}\right)^2\right)\left(1-\left(\frac{\omega_{LH}}{\Pi_e}\right)^2\right)}{\left(1+\frac{|\Omega_i|\Omega_e}{\Pi_e^2}\right)}\right]. \tag{11.75}$$

When $\Pi_e^2/|\Omega_i|\Omega_e \approx \delta = c^2/v_A^2 \gg 1$, (11.75) becomes

$$\omega^2 = \omega_{LH}^2\left(1+\frac{m_i}{Zm_e}\cos^2\theta\right). \tag{11.76}$$

Even if θ is different from $\pi/2$ by a slight amount of $(Zm_{\rm e}/m_{\rm i})^{1/2}$, ω^2 becomes $\omega^2 \approx 2\omega_{\rm LH}^2$, so that (11.76) holds only in the region very near $\theta = \pi/2$.

11.4.4 Upper Hybrid Resonance

The upper hybrid resonant frequency $\omega_{\rm UH}$ is given by

$$\omega_{\rm UH}^2 = \Pi_{\rm e}^2 + \Omega_{\rm e}^2. \tag{11.77}$$

Since this frequency is much larger than $|\Omega_{\rm i}|$, ion motion can be neglected.

11.4.5 Electron Cyclotron Wave

Let us consider high-frequency waves, so that ion motion can be neglected. When $\omega \gg |\Omega_{\rm i}|$, we find

$$\left.\begin{aligned} K_\perp &\approx 1 - \frac{\Pi_{\rm e}^2}{\omega^2 - \Omega_{\rm e}^2}, \\ K_\times &\approx -\frac{\Pi_{\rm e}^2}{\omega^2 - \Omega_{\rm e}^2}\frac{\Omega_{\rm e}}{\omega}, \\ K_\parallel &= 1 - \frac{\Pi_{\rm e}^2}{\omega^2}. \end{aligned}\right\} \tag{11.78}$$

The solution of dispersion equation $AN^4 - BN^2 + C = 0$

$$N^2 = \frac{B \pm (B^2 - 4AC)^{1/2}}{2A}$$

may be modified to

$$\begin{aligned} N^2 - 1 &= \frac{-2(A - B + C)}{2A - B \pm (B^2 - 4AC)^{1/2}} \\ &= \frac{-2\Pi_{\rm e}^2(1 - \Pi_{\rm e}^2/\omega^2)}{2\omega^2(1 - \Pi_{\rm e}^2/\omega^2) - \Omega_{\rm e}^2 \sin^2\theta \pm \Omega_{\rm e}\Delta}, \end{aligned} \tag{11.79}$$

$$\Delta = \left(\Omega_{\rm e}^2 \sin^4\theta + 4\omega^2\left(1 - \frac{\Pi_{\rm e}^2}{\omega^2}\right)^2 \cos^2\theta\right)^{1/2}. \tag{11.80}$$

The ordinary wave and extraordinary wave will be obtained by taking the plus and minus sign, respectively, in (11.79).

In the case of quasi-transverse propagaton

$$\Omega_{\rm e}^2 \sin^4\theta \gg 4\omega^2\left(1 - \frac{\Pi_{\rm e}^2}{\omega^2}\right)^2 \cos^2\theta, \tag{11.81}$$

we find

$$N^2 = \frac{1 - \Pi_e^2/\omega^2}{1 - (\Pi_e^2/\omega^2)\cos^2\theta}, \tag{11.82}$$

$$N^2 = \frac{(1 - \Pi_e^2/\omega^2)^2\omega^2 - \Omega_e^2 \sin^2\theta}{(1 - \Pi_e^2/\omega^2)\omega^2 - \Omega_e^2 \sin^2\theta}. \tag{11.83}$$

The ordinary wave (11.82) becomes $N^2 = K_\parallel = 1 - \Pi_e^2/\omega^2$ at $\theta \sim \pi/2$ and does not depend on the magnitude of the magnetic field. This wave is used for density measurements by microwave interferometry.

In the case of logitudinal propagation

$$\Omega_e^2 \sin^4\theta \ll 4\omega^2 \left(1 - \frac{\Pi_e^2}{\omega^2}\right)^2 \cos^2\theta \tag{11.84}$$

with the additional condition

$$\Omega_e^2 \sin^2\theta \ll \left| 2\omega^2 \left(1 - \frac{\Pi_e^2}{\omega^2}\right) \right| \tag{11.85}$$

the dispersion relations become

$$N^2 = 1 - \frac{\Pi_e^2}{(\omega + \Omega_e \cos\theta)\omega}, \tag{11.86}$$

$$N^2 = 1 - \frac{\Pi_e^2}{(\omega - \Omega_e \cos\theta)\omega}. \tag{11.87}$$

Equation (11.86) corresponds to the L wave, and (11.87) to the R wave. R-wave resonance occurs near the electron cyclotron frequency. This wave can propagate in regions (7) and (8) of the CMA diagram, where the frequency is less than the plasma frequency. This wave is called the *electron cyclotron wave.* It must be noticed that the assumptions (11.84) and (11.85) are not satisfied near $K_\parallel = 1 - \Pi_e^2/\omega^2 \simeq 0$. The electron cyclotron wave is also called the *whistler wave.*

Electromagnetic disturbances initiated by lightning flashes propagate through the ionosphere along magnetic-field lines. The frequency of a lightning-induced whistler wave falls in the audio region, and its group velocity increases with frequency so that this wave is perceived as a whistle of descending tone. This is why it is called the whistler wave.

11.4.6 Dispersion Relation of Electrostatic Wave

When the electric field $\boldsymbol{E}$ can be expressed by an electrostatic potential ϕ

$$\boldsymbol{E} = -\nabla\phi = -i\boldsymbol{k}\phi \tag{11.88}$$

the resultant wave is called an *electrostatic wave.* The electric field $\boldsymbol{E}$ is always parallel to the propagation vector $\boldsymbol{k}$, so that the electrostatic wave is longitudinal. The magnetic field $\boldsymbol{B}_1$ of the electrostatic wave is always zero:

$$\boldsymbol{B}_1 = \boldsymbol{k} \times \boldsymbol{E}/\omega = 0. \tag{11.89}$$

Alfvén waves are not electrostatic waves. Here we will discuss the conditions for electrostatic waves. Since the dispersion relation is

$$\boldsymbol{N} \times (\boldsymbol{N} \times \boldsymbol{E}) + \boldsymbol{K} \cdot \boldsymbol{E} = 0$$

the scalar product with $\boldsymbol{N}$ becomes

$$\boldsymbol{N} \cdot \boldsymbol{K} \cdot (\boldsymbol{E}_{\parallel} + \boldsymbol{E}_{\perp}) = 0,$$

where $\boldsymbol{E}_{\parallel}$ and $\boldsymbol{E}_{\perp}$ are the components of the electric field parallel and perpendicular to $\boldsymbol{k}$. If $|\boldsymbol{E}_{\parallel}| \gg |\boldsymbol{E}_{\perp}|$, the wave is electrostatic and the dispersion relation becomes

$$\boldsymbol{N} \cdot \boldsymbol{K} \cdot \boldsymbol{N} = 0. \tag{11.90}$$

Rewriting the general dispersion relation as

$$(N^2 - \boldsymbol{K}) \cdot \boldsymbol{E}_{\perp} = \boldsymbol{K} \cdot \boldsymbol{E}_{\parallel}$$

shows that $|\boldsymbol{E}_{\parallel}| \gg |\boldsymbol{E}_{\perp}|$ holds when

$$|N^2| \gg |K_{ij}| \tag{11.91}$$

is satisfied for all K_{ij}. The dispersion relation (11.90) for the electrostatic wave is then

$$k_x^2 K_{xx} + 2k_x k_z K_{xz} + k_z^2 K_{zz} = 0. \tag{11.92}$$

The condition (11.91) for the electrostatic wave indicates that the phase velocity $\omega/k = c/N$ of this wave is low. The K_{ij} have already been given by Equations (11.9) through (11.12) for cold plasmas, and the general formula for hot plasma will be discussed in Section 11.5. We have stated that magnetic field $\boldsymbol{B}_1$ of the electrostatic wave is zero. Disturbances of the magnetic field propagate with the Alfvén velocity $v_{\mathrm{A}} \simeq B_0^2/(\mu_0 n_{\mathrm{i}} m_{\mathrm{i}})$. If the phase velocity of the wave is much lower than v_{A}, the disturbance of the magnetic field will be damped within a few cycles of the wave and the propagated magnetic-field disturbance becomes zero. When the electron thermal velocity v_{Te} is taken as a typical phase velocity for electrostatic waves, then the condition of $v_{\mathrm{A}} > v_{\mathrm{Te}}$ reduces to

$$\frac{B_0^2}{\mu_0 n_{\mathrm{i}} m_{\mathrm{i}} v_{\mathrm{Te}}^2} = \frac{2m_{\mathrm{e}}}{\beta_{\mathrm{e}} m_{\mathrm{i}}} > 1,$$

$$\beta_{\rm e} < \frac{2m_{\rm e}}{m_{\rm i}}.$$

This is a measure that a wave is electrostatic.

At resonance the refractive index N becomes infinite. As the K_{ij} are finite for lower hybrid and upper hybrid resonance, the condition (11.91) is satisfied so that these hybrid waves are electrostatic. Since some of the K_{ij} become infinite for the ion or electron cyclotron waves, these cyclotron waves are not always electrostatic.

11.5 Dielectric Tensor of Hot Plasma

When the temperature of plasma becomes hot and the thermal velocities $v_{\rm T} = (T/m)^{1/2}$ of electrons or ions become comparable to the phase velocity of waves, Landau damping or amplification may occur as is described in Chapter 10. In order to study the propagation and absorption (damping) or excitation (amplification) of waves and perturbation systematically, the dielectric tensor of hot plasma must be used.

In the process of wave absorption by hot plasma, Landau damping or cyclotron damping are most important as was described in Chapter 10. These damping processes are due to the interaction between the wave and so-called *resonant particles* satisfying

$$\omega - k_z v_z - {\rm n}\Omega = 0. \qquad {\rm n} = 0, \pm 1, \pm 2, \cdots.$$

In the coordinates running with the same velocity, the electric field is static ($\omega = 0$) or of cyclotron harmonic frequency ($\omega = n\Omega$). The case of $\rm n = 0$ corresponds to Landau damping, the case of $\rm n = 1$ corresponds to electron cyclotron damping, and the case of $\rm n = -1$ corresponds to ion cyclotron damping ($\omega > 0$ is assumed).

Although nonlinear or stochastic processes accompany wave heating in many cases, the experimental results of wave heating or absorption can usually well-described by linear or quasi-linear theories, because the wave heating is the damping phenomena of waves. The basis of the linear theory is the dispersion relation with the dielectric tensor $\boldsymbol{K}$ of finite-temperature plasma.

11.5.1 Formulation of Dispersion Relation in Hot Plasma

In previous sections, dispersion relation of cold plasma was derived. In the unperturbed state, both the electrons and ions are motionless in cold plasma. However, in hot plasma, the electron and ions move along spiral trajectories even in the unperturbed state. The motion of charged particles in a uniform magnetic field $\boldsymbol{B}_0 = B_0\hat{\boldsymbol{z}}$ may be described by

$$\frac{\mathrm{d}\boldsymbol{r}'}{\mathrm{d}t'} = \boldsymbol{v}', \quad \frac{\mathrm{d}\boldsymbol{v}'}{\mathrm{d}t'} = \frac{q}{m}\boldsymbol{v}' \times \boldsymbol{B}_0. \tag{11.93}$$

Assuming that $\boldsymbol{r}' = \boldsymbol{r}$, $\boldsymbol{v}' = \boldsymbol{v} = (v_\perp \cos\theta, v_\perp \sin\theta, v_z)$ at $t' = t$, the solution of (11.93) is obtained as follows:

$$v'_x(t') = v_\perp \cos(\theta + \Omega(t' - t)),$$

$$v'_y(t') = v_\perp \sin(\theta + \Omega(t' - t)), \tag{11.94}$$

$$v'_z(t') = v_z,$$

$$x'(t') = x + \frac{v_\perp}{\Omega}(\sin(\theta + \Omega(t' - t)) - \sin\theta),$$

$$y'(t') = y - \frac{v_\perp}{\Omega}(\cos(\theta + \Omega(t' - t)) - \cos\theta), \tag{11.95}$$

$$z'(t') = z + v_z(t' - t),$$

where $\Omega = -qB_0/m$ and $v_x = v_\perp \cos\theta, v_y = v_\perp \sin\theta$. The analysis of the behavior due to a perturbation of this system must be based on Boltzmann's equation. The distribution function $f_k(\boldsymbol{r}, \boldsymbol{v}, t)$ of kth kind of particles is given by

$$\frac{\partial f_k}{\partial t} + \boldsymbol{v} \cdot \nabla_{\mathrm{r}} f_k + \frac{q_k}{m_k}(\boldsymbol{E} + \boldsymbol{v} \times \boldsymbol{B}) \cdot \nabla_{\mathrm{v}} f_k = 0. \tag{11.96}$$

Maxwell equation is

$$\nabla \cdot \boldsymbol{E} = \frac{1}{\epsilon_0}\sum_k q_k \int \boldsymbol{v} f_k \mathrm{d}\boldsymbol{v}, \tag{11.97}$$

$$\frac{1}{\mu_0}\nabla \times \boldsymbol{B} = \epsilon_0 \frac{\partial \boldsymbol{E}}{\partial t} + \sum_k q_k \int \boldsymbol{v} f_k \mathrm{d}\boldsymbol{v}, \tag{11.98}$$

$$\nabla \times \boldsymbol{E} = -\frac{\partial \boldsymbol{B}}{\partial t}, \tag{11.99}$$

$$\nabla \cdot \boldsymbol{B} = 0. \tag{11.100}$$

As usual, we indicate zeroth order quantities (the unperturbed state) by a subscript 0 and the first order perturbation terms by a subscript 1. The first order terms are expressed in the form of $\exp i(\boldsymbol{k} \cdot \boldsymbol{r} - \omega t)$. Using

$$f_k = f_{k0}(\boldsymbol{r}, \boldsymbol{v}) + f_{k1}, \tag{11.101}$$

$$\boldsymbol{B} = \boldsymbol{B}_0 + \boldsymbol{B}_1, \tag{11.102}$$

$$\boldsymbol{E} = 0 + \boldsymbol{E}_1, \tag{11.103}$$

we can linearize (11.96)∼(11.100) as follows:

$$\boldsymbol{v} \cdot \nabla_{\mathrm{r}} f_{k0} + \frac{q_k}{m_k}(\boldsymbol{v} \times \boldsymbol{B}_0) \cdot \nabla_{\mathrm{v}} f_{k0} = 0, \tag{11.104}$$

$$\sum_k q_k \int f_{k0} \mathrm{d}\boldsymbol{v} = 0, \tag{11.105}$$

$$\frac{1}{\mu_0} \nabla \times \boldsymbol{B}_0 = \sum_k q_k \int \boldsymbol{v} f_{k0} \mathrm{d}\boldsymbol{v} = \boldsymbol{j}_0, \tag{11.106}$$

$$\frac{\partial f_{k1}}{\partial t} + \boldsymbol{v} \cdot \nabla_{\mathrm{r}} f_{k1} + \frac{q_k}{m_k}(\boldsymbol{v} \times \boldsymbol{B}_0) \cdot \nabla_{\mathrm{v}} f_{k1} = -\frac{q_k}{m_k}(\boldsymbol{E}_1 + \boldsymbol{v} \times \boldsymbol{B}_1) \cdot \nabla_{\mathrm{v}} f_{k0}, \tag{11.107}$$

$$i\boldsymbol{k} \cdot \boldsymbol{E}_1 = \frac{1}{\epsilon_0} \sum_k q_k \int f_{k1} \mathrm{d}\boldsymbol{v}, \tag{11.108}$$

$$\frac{1}{\mu_0} \boldsymbol{k} \times \boldsymbol{B}_1 = -\omega \left(\epsilon_0 \boldsymbol{E}_1 + \frac{i}{\omega} \sum_k q_k \int \boldsymbol{v} f_{k1} \mathrm{d}\boldsymbol{v} \right), \tag{11.109}$$

$$\boldsymbol{B}_1 = \frac{1}{\omega}(\boldsymbol{k} \times \boldsymbol{E}_1). \tag{11.110}$$

The right-hand side of (11.108) is a linear equation in $\boldsymbol{E}_1$ as is clear from (11.110), so that f_{k1} is given as a linear function in $\boldsymbol{E}_1$. The electric tensor of the hot plasma as defined by $\boldsymbol{K}(\boldsymbol{D} = \epsilon_0 \boldsymbol{K} \cdot \boldsymbol{E})$ is given by

$$\boldsymbol{E}_1 + \frac{i}{\epsilon \omega} \boldsymbol{j} = \boldsymbol{E}_1 + \frac{i}{\epsilon_0 \omega} \sum_k q_k \int \boldsymbol{v} f_{k1} \mathrm{d}\boldsymbol{v} \equiv \boldsymbol{K} \cdot \boldsymbol{E}_1. \tag{11.111}$$

The linear relation of $\boldsymbol{E}_1$ is derived from (11.109) and (11.110):

$$\boldsymbol{k} \times (\boldsymbol{k} \times \boldsymbol{E}_1) + \frac{\omega^2}{c^2} \boldsymbol{K} \cdot \boldsymbol{E}_1 = 0, \tag{11.112}$$

and the dispersion relation is obtained by equating the determinant of the coefficient matrix of the linear equation to zero. Consequently, if f_{k1} can be solved from (11.107), then $\boldsymbol{K}$ can be obtained. As for cold plasmas, the properties of waves in hot plasmas can be studied by the dispersion relation of hot plasma.

11.5.2 Solution of Linearized Vlasov Equation

When the right-hand side of (11.107) is time-integrated along the particle orbit (11.94) and (11.95) in the unperturbed state, we find

$$f_{k1}(\boldsymbol{r},\boldsymbol{v},t) = -\frac{q_k}{m_k}\int_{-\infty}^{t}\left(\boldsymbol{E}_1(\boldsymbol{r}'(t'),t') + \frac{1}{\omega}\boldsymbol{v}'(t')\times(\boldsymbol{k}\times\boldsymbol{E}_1(\boldsymbol{r}'(t'),t'))\right)$$
$$\cdot\nabla'_{\mathrm{v}} f_{k0}(\boldsymbol{r}'(t'),\boldsymbol{v}'(t'))\mathrm{d}t'. \qquad (11.113)$$

Substitution of (11.113) into (11.107) yields

$$-\frac{q_k}{m_k}\left(\boldsymbol{E}_1 + \frac{1}{\omega}\boldsymbol{v}\times(\boldsymbol{k}\times\boldsymbol{E}_1)\right)\cdot\nabla_{\mathrm{v}} f_{k0}$$
$$-\frac{q_k}{m_k}\int_{-\infty}^{t}\left(\frac{\partial}{\partial t} + \boldsymbol{v}\cdot\nabla_{\mathrm{r}} + \frac{q_k}{m_k}(\boldsymbol{v}\times\boldsymbol{B}_0)\cdot\nabla_{\mathrm{v}}\right)[\text{Integrand of (11.113)}]\,\mathrm{d}t'$$
$$= -\frac{q_k}{m_k}(\boldsymbol{E}_1 + \boldsymbol{v}\times\boldsymbol{B}_1)\cdot\nabla_{\mathrm{v}} f_{k0}. \qquad (11.114)$$

Therefore, if it is proven that the second term of the left-hand side of (11.114) is zero, (11.113) is confirmed to be the solution of (11.107). When the variables $(\boldsymbol{r},\boldsymbol{v},t)$ are changed to $(\boldsymbol{r}',\boldsymbol{v}',t')$ by use of (11.94) and (11.95), the differential operators in the second term of the left-hand side of (11.114) are reduced to

$$\frac{\partial}{\partial t} = \frac{\partial t'}{\partial t}\frac{\partial}{\partial t'} + \frac{\partial \boldsymbol{r}'}{\partial t}\cdot\nabla'_{\mathrm{r}} + \frac{\partial \boldsymbol{v}'}{\partial t}\cdot\nabla'_{\mathrm{v}}$$
$$= \frac{\partial(t'-t)}{\partial t}\left(\frac{\partial \boldsymbol{r}'}{\partial(t'-t)}\cdot\nabla'_{\mathrm{r}} + \frac{\partial \boldsymbol{v}'}{\partial(t'-t)}\cdot\nabla'_{\mathrm{v}}\right)$$
$$= -\boldsymbol{v}'\cdot\nabla'_{\mathrm{r}} - \frac{q_k}{m_k}(\boldsymbol{v}'\times\boldsymbol{B}_0)\cdot\nabla'_{\mathrm{v}},$$

$$\boldsymbol{v}\cdot\nabla_{\mathrm{r}} = \boldsymbol{v}\cdot\nabla'_{\mathrm{r}},$$

$$\frac{\partial}{\partial v_x} = \frac{\partial \boldsymbol{r}'}{\partial v_x}\cdot\nabla'_{\mathrm{r}} + \frac{\partial \boldsymbol{v}'}{\partial v_x}\cdot\nabla'_{\mathrm{v}}$$
$$= \frac{1}{\Omega}\left(\sin\Omega(t'-t)\frac{\partial}{\partial x'} + [-\cos\Omega(t'-t)+1]\frac{\partial}{\partial y'}\right)$$
$$+\left(\cos\Omega(t'-t)\frac{\partial}{\partial v'_x} + \sin\Omega(t'-t)\frac{\partial}{\partial v'_y}\right),$$

$$\frac{\partial}{\partial v_y} = \frac{1}{\Omega}\left((\cos\Omega(t'-t)-1)\frac{\partial}{\partial x'} + \sin\Omega(t'-t)\frac{\partial}{\partial y'}\right)$$

$$+\left(-\sin\Omega(t'-t)\frac{\partial}{\partial v'_x}+\cos\Omega(t'-t)\frac{\partial}{\partial v'_y}\right),$$

$$\frac{q}{m}(\boldsymbol{v}\times\boldsymbol{B}_0)\cdot\nabla_{\rm v}=-\Omega\left(v_y\frac{\partial}{\partial v_x}-v_x\frac{\partial}{\partial v_y}\right)$$

$$=v'_x\frac{\partial}{\partial x'}+v'_y\frac{\partial}{\partial y'}-\left(v_x\frac{\partial}{\partial x'}+v_y\frac{\partial}{\partial y'}\right)-\Omega\left(v'_y\frac{\partial}{\partial v'_x}-v'_x\frac{\partial}{\partial v'_y}\right)$$

$$=(\boldsymbol{v}'-\boldsymbol{v})\cdot\nabla'_{\rm r}+\frac{q}{m}(\boldsymbol{v}'\times\boldsymbol{B}_0)\cdot\nabla'_{\rm v}.$$

Therefore, the second term of the left-hand side of (11.114) is zero.

Since the first order terms vary as $\exp(-i\omega t)$, the integral (11.113) converges when the imaginary part of ω is positive. When the imaginary part of ω is negative, the solution can be given by analytic continuation from the region of the positive imaginary part.

11.5.3 Dielectric Tensor of Hot Plasma

The zeroth-order distribution function f_0 must satisfiy (11.104), or

$$f_0(\boldsymbol{r},\boldsymbol{v})=f(v_\perp,v_z),\quad v_\perp^2=v_x^2+v_y^2.$$

Let us consider

$$\boldsymbol{E}_1(\boldsymbol{r}',t')=\boldsymbol{E}\exp i(\boldsymbol{k}\cdot\boldsymbol{r}'-\omega t').$$

The z axis is taken along $\boldsymbol{B}_0$ direction and the x axis is taken in the plane spanned by $\boldsymbol{B}_0$ and the propagation vector $\boldsymbol{k}$, so that the y component of the propagation vector is zero ($k_y=0$); that is:

$$\boldsymbol{k}=k_x\hat{\boldsymbol{x}}+k_z\hat{\boldsymbol{z}}.$$

Then (11.113) is reduced to

$$f_1(\boldsymbol{r},\boldsymbol{v},t)=-\frac{q}{m}\exp i(k_xx+k_zz-\omega t)\int_\infty^t\left(\left(1-\frac{\boldsymbol{k}\cdot\boldsymbol{v}'}{\omega}\right)\boldsymbol{E}+(\boldsymbol{v}'\cdot\boldsymbol{E})\frac{\boldsymbol{k}}{\omega}\right)\cdot\nabla'_{\rm v}f_0$$

$$\times\exp\left(i\frac{k_xv_\perp}{\Omega}\sin(\theta+\Omega(t'-t))-i\frac{k_xv_\perp}{\Omega}\sin\theta+i(k_zv_z-\omega)(t'-t)\right){\rm d}t'.$$

We introduce $\tau=t'-t$ and use the following formulas of Bessel function:

$$\exp(ia\sin\theta)=\sum_{\rm m=-\infty}^{\infty}J_{\rm m}(a)\exp i{\rm m}\theta,$$

$$J_{-\rm m}(a)=(-1)^{\rm m}J_{\rm m}(a),$$

$$\exp\Big(\ \Big) = \sum_{\mathrm{m}=-\infty}^{\infty}\sum_{\mathrm{n}=-\infty}^{\infty} J_{\mathrm{m}}\exp(-i\mathrm{m}\theta)J_{\mathrm{n}}\exp\left(i\mathrm{n}(\theta+\Omega\tau)\right)\exp i(k_z v_z-\omega)\tau.$$

Since

$$\left(\left(1-\frac{\boldsymbol{k}\cdot\boldsymbol{v}'}{\omega}\right)\boldsymbol{E}+(\boldsymbol{v}'\cdot\boldsymbol{E})\frac{\boldsymbol{k}}{\omega}\right)\cdot\nabla_{\mathrm{v}}' f_0 = \frac{\partial f_0}{\partial v_z}\left(\left(1-\frac{k_x v_x'}{\omega}\right)E_z\right.$$

$$\left.+(v_x' E_x + v_y' E_y)\frac{k_z}{\omega}\right)+\frac{\partial f_0}{\partial v_\perp}\left(\left(1-\frac{k_z v_z'}{\omega}\right)\left(E_x\frac{v_x'}{v_\perp}+E_y\frac{v_y'}{v_\perp}\right)+v_z E_z\frac{k_x}{\omega}\frac{v_x'}{v_\perp}\right)$$

$$=\left(\frac{\partial f_0}{\partial v_\perp}\left(1-\frac{k_z v_z}{\omega}\right)+\frac{\partial f_0}{\partial v_z}\frac{k_z v_\perp}{\omega}\right)\left(\frac{E_x}{2}(e^{i(\theta+\Omega\tau)}+e^{-i(\theta+\Omega\tau)})+\frac{E_y}{2i}(e^{i(\theta+\Omega\tau)}\right.$$

$$\left.-e^{-i(\theta+\Omega\tau)})\right)+\left(\frac{\partial f_0}{\partial v_\perp}\frac{k_x v_z}{\omega}-\frac{\partial f_0}{\partial v_z}\frac{k_x v_\perp}{\omega}\right)\frac{E_z}{2}(e^{i(\theta+\Omega\tau)}+e^{-i(\theta+\Omega\tau)})+\frac{\partial f_0}{\partial v_z}E_z,$$

we find

$$f_1(\boldsymbol{r},\boldsymbol{v},t)$$

$$=-\frac{q}{m}\exp i(k_x x+k_z z-\omega t)\sum_{\mathrm{m,n}}\left(U\left(\frac{J_{\mathrm{n}-1}+J_{\mathrm{n}+1}}{2}\right)E_x-iU\left(\frac{J_{\mathrm{n}-1}-J_{\mathrm{n}+1}}{2}\right)E_y\right.$$

$$\left.+\left(W\frac{J_{\mathrm{n}-1}+J_{\mathrm{n}+1}}{2}+\frac{\partial f_0}{\partial v_z}J_{\mathrm{n}}\right)E_z\right)\cdot\frac{J_{\mathrm{m}}(a)\exp(-i(\mathrm{m}-\mathrm{n})\theta)}{i(k_z v_z-\omega+\mathrm{n}\Omega)},$$

where

$$U=\left(1-\frac{k_z v_z}{\omega}\right)\frac{\partial f_0}{\partial v_\perp}+\frac{k_z v_\perp}{\omega}\frac{\partial f_0}{\partial v_z}, \tag{11.115}$$

$$W=\frac{k_x v_z}{\omega}\frac{\partial f_0}{\partial v_\perp}-\frac{k_x v_\perp}{\omega}\frac{\partial f_0}{\partial v_z}, \tag{11.116}$$

$$a=\frac{k_x v_\perp}{\Omega},\qquad \Omega=\frac{-qB}{m}, \tag{11.117}$$

and

$$\frac{J_{\mathrm{n}-1}(a)+J_{\mathrm{n}+1}(a)}{2}=\frac{\mathrm{n}J_{\mathrm{n}}(a)}{a},\qquad \frac{J_{\mathrm{n}-1}(a)-J_{\mathrm{n}+1}(a)}{2}=\frac{\mathrm{d}}{\mathrm{d}a}J_{\mathrm{n}}(a).$$

Since f_1 is obtained, the dielectric tensor $\boldsymbol{K}$ of hot plasma is reduced from (11.111) to

$$(\boldsymbol{K}-\boldsymbol{I})\cdot\boldsymbol{E}=\frac{i}{\epsilon_0\omega}\sum_j q_j\int \boldsymbol{v}f_{j1}\mathrm{d}\boldsymbol{v}. \tag{11.118}$$

Since $v_x = v_\perp \cos\theta$, $v_y = v_\perp \sin\theta$, $v_z = v_z$, only the terms of $e^{i(\mathrm{m}-\mathrm{n})\theta} = e^{\pm i\theta}$ in f_{j1} can contribute to x, y components of the integral (11.118) and only the term of $e^{i(\mathrm{m}-\mathrm{n})\theta} = 1$ in f_{j1} can contribute to z component of the integral (11.118) and we find:

$$\boldsymbol{K} = \boldsymbol{I} - \sum_j \frac{\Pi_j^2}{\omega n_{j0}} \sum_{\mathrm{n}=-\infty}^{\infty} \int \mathrm{d}\boldsymbol{v} \frac{S_{j\mathrm{n}}}{k_z v_z - \omega + \mathrm{n}\Omega_j}, \tag{11.119}$$

$$S_{j\mathrm{n}} = \begin{bmatrix} v_\perp(\mathrm{n}\frac{J_\mathrm{n}}{a})^2 U & -iv_\perp(\mathrm{n}\frac{J_\mathrm{n}}{a})J_\mathrm{n}' U & v_\perp(\mathrm{n}\frac{J_\mathrm{n}}{a})J_\mathrm{n}(\frac{\partial f_0}{\partial v_z} + \frac{\mathrm{n}}{a}W) \\ iv_\perp J_\mathrm{n}'(\mathrm{n}\frac{J_\mathrm{n}}{a})U & v_\perp(J_\mathrm{n}')^2 U & iv_\perp J_\mathrm{n}' J_\mathrm{n}(\frac{\partial f_0}{\partial v_z} + \frac{\mathrm{n}}{a}W) \\ v_z J_\mathrm{n}(\mathrm{n}\frac{J_\mathrm{n}}{a})U & -iv_z J_\mathrm{n} J_\mathrm{n}' U & v_z J_\mathrm{n}^2(\frac{\partial f_0}{\partial v_z} + \frac{\mathrm{n}}{a}W) \end{bmatrix}$$

where

$$\Pi_j^2 = \frac{n_j q_j^2}{\epsilon_0 m_j}.$$

When we use the relations

$$\frac{v_z U - v_\perp(\frac{\partial f_0}{\partial v_z} + \frac{\mathrm{n}\Omega}{k_x v_\perp}W)}{k_z v_z - \omega + \mathrm{n}\Omega} = -\frac{v_z}{\omega}\frac{\partial f_0}{\partial v_\perp} + \frac{v_\perp}{\omega}\frac{\partial f_0}{\partial v_z},$$

$$\sum_{\mathrm{n}=-\infty}^{\infty} J_\mathrm{n}^2 = 1, \qquad \sum_{\mathrm{n}=-\infty}^{\infty} J_\mathrm{n} J_\mathrm{n}' = 0, \qquad \sum_{\mathrm{n}=-\infty}^{\infty} \mathrm{n} J_\mathrm{n}^2 = 0 \quad (J_{-\mathrm{n}} = (-1)^\mathrm{n} J_\mathrm{n}),$$

and replace n by $-$n, then (11.119) is reduced to

$$\boldsymbol{K} = \boldsymbol{I} - \sum_j \frac{\Pi_j^2}{\omega} \sum_{\mathrm{n}=-\infty}^{\infty} \int T_{j\mathrm{n}} \frac{v_\perp^{-1} U_j n_{j0}^{-1}}{k_z v_z - \omega - \mathrm{n}\Omega_j} \mathrm{d}\boldsymbol{v}$$
$$- \boldsymbol{L} \sum_j \frac{\Pi_j^2}{\omega^2}\left(1 + \frac{1}{n_{j0}} \int \frac{v_z^2}{v_\perp}\frac{\partial f_{j0}}{\partial v_\perp} \mathrm{d}\boldsymbol{v}\right),$$

$$T_{j\mathrm{n}} = \begin{bmatrix} v_\perp^2(\mathrm{n}\frac{J_\mathrm{n}}{a})(\mathrm{n}\frac{J_\mathrm{n}}{a}) & iv_\perp^2(\mathrm{n}\frac{J_\mathrm{n}}{a})J_\mathrm{n}' & -v_\perp v_z(\mathrm{n}\frac{J_\mathrm{n}}{a})J_\mathrm{n} \\ -iv_\perp^2 J_\mathrm{n}'(\mathrm{n}\frac{J_\mathrm{n}}{a}) & v_\perp^2 J_\mathrm{n}' J_\mathrm{n}' & iv_\perp v_z J_\mathrm{n}' J_\mathrm{n} \\ -v_\perp v_z J_\mathrm{n}(\mathrm{n}\frac{J_\mathrm{n}}{a}) & -iv_\perp v_z J_\mathrm{n} J_\mathrm{n}' & v_z^2 J_\mathrm{n} J_\mathrm{n} \end{bmatrix}$$

where all the components of matrix $\boldsymbol{L}$ are zero except $L_{zz} = 1$. From the relations

$$\frac{U_j}{k_z v_z - \omega - \mathrm{n}\Omega_j} = -\frac{1}{\omega}\frac{\partial f_{j0}}{\partial v_\perp} + \frac{1}{\omega(k_z v_z - \omega - \mathrm{n}\Omega_j)}\left(-\mathrm{n}\Omega_j \frac{\partial f_{j0}}{\partial v_\perp} + k_z v_\perp \frac{\partial f_{j0}}{\partial v_z}\right),$$

$$\sum_{\mathrm{n}=-\infty}^{\infty} (J_\mathrm{n}')^2 = \frac{1}{2}, \qquad \sum_{\mathrm{n}=-\infty}^{\infty} \frac{\mathrm{n}^2 J_\mathrm{n}^2(a)}{a^2} = \frac{1}{2},$$

another expression of the dielectric tensor is obtained:

$$\boldsymbol{K}=\left(1-\frac{\Pi_j^2}{\omega^2}\right)\boldsymbol{I}-\sum_{j,\mathrm{n}}\frac{\Pi_j^2}{\omega^2}\int\frac{T_{j\mathrm{n}}}{k_z v_z-\omega-\mathrm{n}\Omega_j}\left(\frac{-\mathrm{n}\Omega_j}{v_\perp}\frac{\partial f_{j0}}{\partial v_\perp}+k_z\frac{\partial f_{j0}}{\partial v_z}\right)\frac{1}{n_{j0}}\mathrm{d}\boldsymbol{v}. \tag{11.120}$$

Using

$$\boldsymbol{N}\equiv\frac{\boldsymbol{k}}{\omega}c,$$

(11.112) is

$$(K_{xx}-N_\parallel^2)E_x+K_{xy}E_y+(K_{xz}+N_\perp N_\parallel)E_z=0,$$

$$K_{yx}E_x+(K_{yy}-N^2)E_y+K_{yz}E_z=0,$$

$$(K_{zx}+N_\perp N_\parallel)E_x+K_{zy}E_y+(K_{zz}-N_\perp^2)E_z=0,$$

where $N_\parallel$ is the z component of $\boldsymbol{N}$ (parallel to $\boldsymbol{B}$) and $N_\perp$ is the x component of $\boldsymbol{N}$ (perpendicular to $\boldsymbol{B}$). The dispersion relation is given by equating the determinent of the coefficient matrix to zero.

11.5.4 Dielectric Tensor of bi-Maxwellian Plasma

When the zeroth-order distribution function is bi-Maxwellian,

$$f_0(v_\perp,v_z)=n_0F_\perp(v_\perp)F_z(v_z), \tag{11.121}$$

$$F_\perp(v_\perp)=\frac{m}{2\pi T_\perp}\exp\left(-\frac{mv_\perp^2}{2T_\perp}\right), \tag{11.122}$$

$$F_z(v_z)=\left(\frac{m}{2\pi T_z}\right)^{1/2}\exp\left(-\frac{m(v_z-V)^2}{2T_z}\right), \tag{11.123}$$

we find

$$\left(-\frac{\mathrm{n}\Omega_j}{v_\perp}\frac{\partial f_0}{\partial v_\perp}+k_z\frac{\partial f_0}{\partial v_z}\right)\frac{1}{n_0}=m\left(\frac{\mathrm{n}\Omega_j}{T_\perp}-\frac{k_z(v_z-V)}{T_z}\right)F_\perp(v_\perp)F_z(v_z).$$

Integration over v_z can be done by use of the *plasma dispersion function* $Z(\zeta)$. Plasma dispersion function $Z(\zeta)$ is defined by:

$$Z(\zeta)\equiv\frac{1}{\pi^{1/2}}\int_{-\infty}^{\infty}\frac{\exp(-\beta^2)}{\beta-\zeta}\mathrm{d}\beta. \tag{11.124}$$

Using the following relations

$$\int_{-\infty}^{\infty} \frac{F_z}{k_z(v_z - V) - \omega_n} dv_z = \frac{1}{\omega_n} \zeta_n Z(\zeta_n),$$
$$\int_{-\infty}^{\infty} \frac{k_z(v_z - V)F_z}{k_z(v_z - V) - \omega_n} dv_z = 1 + \zeta_n Z(\zeta_n),$$
$$\int_{-\infty}^{\infty} \frac{(k_z(v_z - V))^2 F_z}{k_z(v_z - V) - \omega_n} dv_z = \omega_n(1 + \zeta_n Z(\zeta_n)),$$
$$\int_{-\infty}^{\infty} \frac{(k_z(v_z - V))^3 F_z}{k_z(v_z - V) - \omega_n} dv_z = \frac{k_z^2(\kappa T_z)}{m} + \omega_n^2(1 + \zeta_n Z(\zeta_n)),$$

$$\omega_n \equiv \omega - k_z V + n\Omega,$$
$$\zeta_n \equiv \frac{\omega - k_z V + n\Omega}{k_z(2T_z/m)^{1/2}},$$

$$\int_0^{\infty} J_n^2(b^{1/2}x) \exp\left(-\frac{x^2}{2\alpha}\right) x dx = \alpha I_n(\alpha b) e^{-b\alpha},$$

$$\sum_{n=-\infty}^{\infty} I_n(b) = e^b, \quad \sum_{n=-\infty}^{\infty} n I_n(b) = 0, \quad \sum_{n=-\infty}^{\infty} n^2 I_n(b) = b e^b,$$

(where $I_n(x)$ is nth modified Bessel function (refer to Section 12.1)), the formula for the dielectric tensor of a bi-Maxwellian plasma is obtained as follows [11.3]:

$$\boldsymbol{K} = \boldsymbol{I} + \sum_{i,e} \frac{\Pi^2}{\omega^2} \left(\sum_n \left(\zeta_0 Z(\zeta_n) - \left(1 - \frac{1}{\lambda_T}\right)(1 + \zeta_n Z(\zeta_n)) \right) e^{-b} \boldsymbol{X}_n + 2\eta_0^2 \lambda_T \boldsymbol{L} \right), \tag{11.125}$$

$$\boldsymbol{X}_n = \begin{bmatrix} n^2 I_n/b & in(I_n' - I_n) & -(2\lambda_T)^{1/2}\eta_n \frac{n}{\alpha} I_n \\ -in(I_n' - I_n) & (n^2/b + 2b)I_n - 2bI_n' & i(2\lambda_T)^{1/2}\eta_n \alpha (I_n' - I_n) \\ -(2\lambda_T)^{1/2}\eta_n \frac{n}{\alpha} I_n & -i(2\lambda_T)^{1/2}\eta_n \alpha (I_n' - I_n) & 2\lambda_T \eta_n^2 I_n \end{bmatrix} \tag{11.126}$$

$$\eta_n \equiv \frac{\omega + n\Omega}{2^{1/2} k_z v_{Tz}}, \quad \lambda_T \equiv \frac{T_z}{T_\perp}, \quad b \equiv \left(\frac{k_x v_{T\perp}}{\Omega}\right)^2,$$
$$\alpha \equiv \frac{k_x v_{T\perp}}{\Omega}, \quad v_{Tz}^2 \equiv \frac{T_z}{m}, \quad v_{T\perp}^2 \equiv \frac{T_\perp}{m},$$

$\boldsymbol{L}$ matrix components are $L_{zz} = 1$ and all others are 0.

When the plasma is isotropic Maxwellian ($T_z = T_\perp$) and $V = 0$, then $\eta_n = \zeta_n$, and $\lambda_T = 1$, and (11.125) reduces to

$$\boldsymbol{K} = \boldsymbol{I} + \sum_{i,e} \frac{\Pi^2}{\omega^2} \left[\sum_{n=-\infty}^{\infty} \zeta_0 Z(\zeta_n) e^{-b} \boldsymbol{X}_n + 2\zeta_0^2 \boldsymbol{L} \right]. \tag{11.127}$$

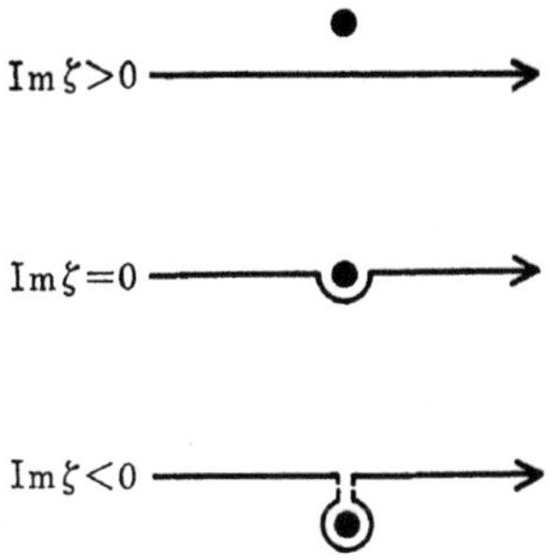

Figure 11.7 Integral paths of (11.128) for $\mathrm{Im}\zeta > 0$, $\mathrm{Im}\zeta = 0$, and $\mathrm{Im}\zeta < 0$ in the case of $k_z > 0$.

11.5.5 Plasma Dispersion Function

Let us examine the property of *plasma dispersion function* $Z_{\rm p}(\zeta)$ defined by

$$Z_{\rm p}(\zeta) \equiv \frac{1}{\pi^{1/2}} \int_{-\infty}^{\infty} \frac{\exp(-\beta^2)}{\beta - \zeta} {\rm d}\beta, \qquad (11.128)$$

in the case of $\mathrm{Im}\zeta > 0$. The solution (11.113) of the first order Vlasov equation is obtained by time integration of $\exp(-\omega t)$ along the particle orbit from $-\infty$ to t. Therefore, the imaginary part $\omega_{\rm i}$ must be positive. First, let us consider the case $k_z > 0$, so that $\mathrm{Im}\zeta > 0$. For $Z_{\rm p}(\zeta)$ in the case of $\mathrm{Im}\zeta < 0$, analytic continuation must be used. The integral paths for $\mathrm{Im}\zeta > 0$, $\mathrm{Im}\zeta = 0$, and $\mathrm{Im}\zeta < 0$ are shown in Figure 11.7. When ζ is real and equal to x, $Z_{\rm p}(x)$ is

$$Z_{\rm p}(\zeta) \equiv \frac{1}{\pi^{1/2}} {\rm P} \int_{-\infty}^{\infty} \frac{\exp(-\beta^2)}{\beta - x} {\rm d}x + i\pi^{1/2} \exp(-x^2),$$

where P means Cauchy's principal value $\lim_{\varepsilon \to 0} \left(\int_{-\infty}^{x-\varepsilon} + \int_{x+\varepsilon}^{\infty} \right)$. Setting $\beta - \zeta = \gamma$ in (11.128) and using the relation

$$\int_{-\infty}^{\infty} \frac{\exp(-\gamma^2 - 2\zeta\gamma)}{\gamma} {\rm d}\gamma = -2\pi^{1/2} \int_{+i\infty}^{\zeta} \exp(t^2) {\rm d}t$$

(if both sides are differentiated by ζ, it is clear that they are equal), we find

$$Z_{\rm p}(\zeta) = 2i \exp(-\zeta^2) \int_{-\infty}^{i\zeta} \exp(-t^2) {\rm d}t = i\pi^{1/2} \exp(-\zeta^2)(1 + {\rm erf}(i\zeta)),$$

$${\rm erf}(\eta) \equiv \frac{2}{\pi^{1/2}} \int_{-\infty}^{\eta} \exp(-t^2) {\rm d}t - 1 = 2\Phi(2^{1/2}\eta) - 1,$$

$$\Phi(x) = \frac{1}{(2\pi)^{1/2}} \int_{-\infty}^{x} \exp\left(-\frac{t^2}{2}\right) \mathrm{d}t.$$

This expression is valid regardless of the sign of $\mathrm{Im}\zeta$. The series expansion of $Z_{\mathrm{p}}(\zeta)$ is

$$Z_{\mathrm{p}}(\zeta) = i\pi^{1/2} \exp(-\zeta^2) - \zeta \sum_{n=0}^{\infty} \frac{(-\zeta^2)^n \pi^{1/2}}{\Gamma(n+3/2)}$$

$$= i\pi^{1/2} \exp(-\zeta^2) - 2\zeta \left(1 - \frac{2\zeta^2}{2} + \frac{4\zeta^4}{15} - \cdots\right),$$

and asymptotic expansion of $Z_{\mathrm{p}}(\zeta)$ is given by [11.1, 11.3]

$$Z_{\mathrm{p}}(\zeta) = i\sigma\pi^{1/2} \exp(-\zeta^2) - \sum_{n=0}^{\infty} \zeta^{-(2n+1)} \frac{\Gamma(n+1/2)}{\pi^{1/2}}$$

$$= i\sigma\pi^{1/2} \exp(-\zeta^2) - \frac{1}{\zeta}\left(1 + \frac{1}{2\zeta^2} + \frac{3}{4\zeta^4 \cdots}\right),$$

$$\sigma = 0 \text{ for } |\mathrm{Im}\zeta| > 0,\ |\mathrm{Im}\zeta| > |\mathrm{Re}\zeta|$$

$$\sigma = 0 \text{ for } |\mathrm{Im}\zeta| < |\mathrm{Re}\zeta|,$$

$$\sigma = 2 \text{ for } |\mathrm{Im}\zeta| < 0,\ |\mathrm{Im}\zeta| > |\mathrm{Re}\zeta|.$$

The curves for real and imaginary parts of $Z_{\mathrm{p}}(x)$ (x is real) are shown in Figure 11.8.

The function $Z(\zeta)$ of dielectric tensor (11.126) is defined for $\mathrm{Im}\omega > 0$ because we solved the linear Vlasov equation by time integral along the particle path from $-\infty$ to t ($|\exp(-i\omega t)| = \exp \mathrm{Im}\omega t$). Accordingly, in the case of $k > 0$, the function $Z(\zeta)$ of dielectric tensor (11.126) is the same as the plasma dispersion function $Z_{\mathrm{p}}(\zeta)$ defined by (11.128); that is

$$Z(\zeta) = Z_{\mathrm{p}}(\zeta) \quad \text{when } k > 0.$$

Let us consider the case for $k < 0$. Since the function $Z(\zeta)$ of dielectric tensor (11.126) was defined for $\mathrm{Im}\omega > 0$, this function $Z(\zeta)$ was defined when $\mathrm{Im}\zeta = \mathrm{Im}\omega/(2^{1/2}k_z v_{\mathrm{Tz}}) < 0$. Analytic continuation must be used for $\mathrm{Im}\zeta = \mathrm{Im}\omega/(2^{1/2}k_z v_{\mathrm{Tz}}) > 0$. The relation of $Z(\zeta)$ and $Z_{\mathrm{p}}(\zeta)$ is

$$Z(\zeta) = Z_{\mathrm{p}}(\zeta) - 2\pi i \exp(-\zeta^2) \quad \text{when } k < 0.$$

Therefore, the series expansion of $Z(\zeta)$ used for the dielectric tensor is [11.1, 11.3], Chapter 11 of [2.3]

$$Z(\zeta) = i\pi^{1/2} \frac{k_z}{|k_z|} \exp(-\zeta^2) - 2\zeta \left(1 - \frac{2\zeta^2}{2} + \frac{4\zeta^4}{15} - \cdots\right), \tag{11.129}$$

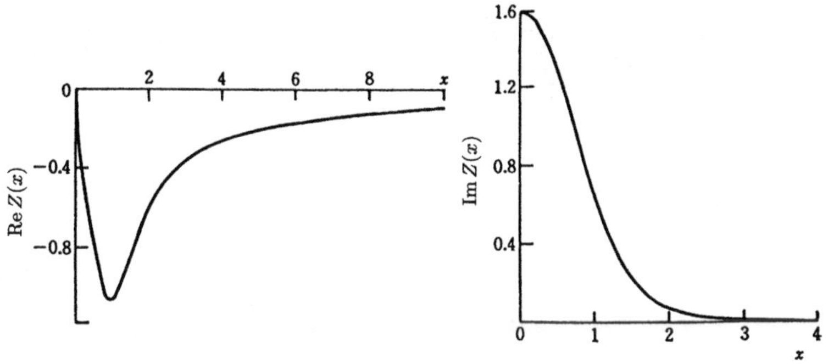

Figure 11.8 Real part $\mathrm{Re}\, Z(x)$ and imaginary part $\mathrm{Im}\, Z(x)$ of $Z(x)$.

and asymptotic expansion of $Z_{\mathrm{p}}(\zeta)$ is given by

$$Z(\zeta) = i\sigma\pi^{1/2}\frac{k_z}{|k_z|}\exp(-\zeta^2) - \frac{1}{\zeta}\left(1 + \frac{1}{2\zeta^2} + \frac{3}{4\zeta^4 \cdots}\right). \qquad (11.130)$$

The imaginary part of $Z(\zeta)$ represents the terms of Landau damping and cyclotron damping.

When $T \to 0$, that is, $\zeta_{\mathrm{n}} \to \pm\infty$, $b \to 0$, the dielectric tensor of hot plasma is reduced to the dielectric tensor (11.9)$\sim$(11.13) of cold plasma.

11.5.6 Dispersion Relation of Electrostatic Wave

When the electric field $\boldsymbol{E}$ of wave is expressed by electrostatic potential $\boldsymbol{E} = -\nabla\phi$, the wave is called an electrostatic wave. The dispersion relation of the electrostatic wave in hot plasma is reduced from (11.92) to

$$k_x^2 K_{xx} + 2k_x k_z K_{xz} + k_z^2 K_{zz} = 0. \qquad (11.131)$$

When K_{xx}, K_{xz}, K_{zz} given by (11.125) and (11.126) are substitued in (11.131), we find

$$k_x^2 + k_z^2 + \sum_{\mathrm{i,e}} \frac{\Pi^2}{\omega^2}\left(k_z^2 2\eta_0^2 \lambda_{\mathrm{T}} + \sum_{\mathrm{n}=-\infty}^{\infty} \left[\frac{\mathrm{n}^2 I_{\mathrm{n}}}{b} k_x^2 - (2\lambda_{\mathrm{T}})^{1/2}\eta_{\mathrm{n}}\frac{\mathrm{n}}{\alpha} I_{\mathrm{n}} 2k_x k_z + 2\lambda_{\mathrm{T}}\eta_{\mathrm{n}}^2 I_{\mathrm{n}} k_z^2 \right] \right.$$
$$\left. \times \left(\zeta_0 Z(\zeta_{\mathrm{n}}) - \left(1 - \frac{1}{\lambda_{\mathrm{T}}}\right)(1 + \zeta_{\mathrm{n}} Z(\zeta_{\mathrm{n}}))\right) e^{-b} \right) = 0.$$

Since $[\ \ ] = m\omega^2/T_\perp$, we have

$$k_x^2 + k_z^2 + \sum_{\mathrm{i,e}} \frac{\Pi^2}{\omega^2}\left(\frac{m\omega^2}{T_\perp} + \sum_{\mathrm{n}=-\infty}^{\infty} \frac{m\omega^2}{T_\perp} I_{\mathrm{n}} \left(\zeta_0 Z(\zeta_{\mathrm{n}}) - \left(1 - \frac{1}{\lambda_{\mathrm{T}}}\right)(1 + \zeta_{\mathrm{n}} Z(\zeta_{\mathrm{n}}))\right) e^{-b} \right) = 0,$$

that is,

$$k_x^2 + k_z^2 + \sum_{\mathrm{i,e}} \Pi^2 \frac{m}{T_z}\left(1 + \sum_{\mathrm{n}=-\infty}^{\infty}\left(1 + \frac{T_z}{T_\perp}\left(\frac{-\mathrm{n}\Omega}{\omega_\mathrm{n}}\right)\right)\zeta_\mathrm{n} Z(\zeta_\mathrm{n}) I_\mathrm{n} e^{-b}\right) = 0. \tag{11.132}$$

The relation of $\sum_{-\infty}^{\infty} I_\mathrm{n}(b) = e^b$ was used.

When the frequency of the wave is much higher than the cyclotron frequency ($|\omega| \gg |\Omega|$) or the magnetic field is very weak ($B \to 0$), then we find $\zeta_n \to \zeta_0$, $n\Omega/\omega_n \to 0$, $\sum I_n(b)e^{-b} = 1$, so that the dispersion equation is reduced to

$$k_x^2 + k_z^2 + \sum_{\mathrm{i,e}} \Pi^2 \frac{m}{T_z}(1 + \zeta_0 Z(\zeta_0)) = 0 \qquad (|\omega| \gg |\Omega|). \tag{11.133}$$

The dispersion equation in the case of $B = 0$ is given by

$$k^2 + \sum_{\mathrm{i,e}} \Pi^2 \frac{m}{T}(1 + \zeta Z(\zeta)) = 0 \qquad \left(\zeta = \frac{\omega - kV}{2^{1/2} k v_\mathrm{T}},\quad B = 0\right). \tag{11.134}$$

When $\zeta_n \to \infty$ ($n \neq 0$), then $\zeta_n Z_n \to -1$ ($\sum I_n(b)e^{-b} = 1$) and (11.132) yields

$$k_x^2 + k_z^2 + \sum_{\mathrm{i,e}} \Pi^2 \frac{m}{T_z}\left(I_0 e^{-b}(1 + \zeta_0 Z(\zeta_0)) + \frac{T_z}{T_\perp}\sum_{n=1}^{\infty}\left(I_n(b)e^{-b}\frac{2n^2\Omega^2}{n^2\Omega^2 - \omega^2}\right)\right) = 0. \tag{11.135}$$

This wave is called *Bernstein wave.*

Drift Waves in Inhomogenous Plasma

When the density and temperature of the zeroth-order state change in the direction of y, we must resort to (11.108) and (11.113). Since the electrostatic wave $\boldsymbol{E} = -\nabla\phi_1$, $\boldsymbol{B}_1 = 0$ is considered in this subsection, (11.108) and (11.113) reduce to

$$-\nabla^2\phi_1 = \frac{1}{\epsilon_0}\sum_k q_k \int f_{k1}\mathrm{d}\boldsymbol{v}, \tag{11.136}$$

$$f_{k1} = \frac{q_k}{m_k}\int_{-\infty}^{t} \nabla'_\mathrm{r}\phi_1(\boldsymbol{r}', t') \cdot \nabla'_\mathrm{v} f_{k0}(\boldsymbol{r}', \boldsymbol{v}')\mathrm{d}t'. \tag{11.137}$$

The zeroth order distribution function f_{k0} must satisfy (11.104); that is,

$$v_y \frac{\partial f_0}{\partial y} - \Omega\left(v_y \frac{\partial}{\partial v_x} - v_x \frac{\partial}{\partial v_y}\right) f_0 = 0. \tag{11.138}$$

Since $v_\perp^2 = \alpha$, $(v_z - V)^2 = \beta$, and $y + v_x/\Omega = \gamma$ are the solutions of the equation for particle motion, $f_0(\alpha, \beta, \gamma)$ satisfies (11.138) and we adopt the following zeroth order distribution function:

$$f_0\left(v_\perp^2, (v_z - V)^2, y + \frac{v_x}{\Omega}\right)$$

$$= \frac{n_0(1 - \epsilon(y + v_x/\Omega))\exp\left(-\frac{v_\perp^2}{2v_{\mathrm{T}\perp}^2(1-\delta_\perp(y+v_x/\Omega))} - -\frac{v_z^2}{2v_{\mathrm{T}z}^2(1-\delta_z(y+v_x/\Omega))}\right)}{2\pi v_{\mathrm{T}\perp}^2(1 - \delta_\perp(y + v_x/\Omega))(2\pi)^{1/2}v_{\mathrm{T}z}(1 - \delta_z(y + v_x/\Omega))^{1/2}}$$

$$= n_0\left(1 - \left((\epsilon - \delta_\perp - \frac{\delta_z}{2}) + \delta_\perp\frac{v_\perp^2}{2v_{\mathrm{T}\perp}^2} + \delta_z\frac{(v_z - V)^2}{2v_{\perp z}^2}\right)\left(y + \frac{v_x}{\Omega}\right)\right)$$

$$\times\left(\frac{1}{2\pi v_{\mathrm{T}\perp}^2}\right)\left(\frac{1}{2\pi v_{\mathrm{T}z}^2}\right)^{1/2}\exp\left(-\frac{v_\perp^2}{2v_{\mathrm{T}\perp}^2} - \frac{(v_z - V)^2}{2v_{\mathrm{T}z}^2}\right). \quad (11.139)$$

The density gradient and temperature gradient of this distribution function are

$$-\frac{1}{n_0}\frac{\mathrm{d}n_0}{\mathrm{d}y} = \epsilon, \qquad -\frac{1}{T_\perp}\frac{\mathrm{d}T_\perp}{\mathrm{d}y} = \delta_\perp, \qquad -\frac{1}{T_z}\frac{\mathrm{d}T_z}{\mathrm{d}y} = \delta_z.$$

Let us consider the following perturbation $\phi_1(\boldsymbol{r}, t) = \phi_1(y)\exp(ik_x x + ik_z z - i\omega t)$ and assume $|(k_x^2 + k_z^2)\phi_1| \gg |\partial^2\phi_1/\partial y^2|$. By the same way as in Section 11.5.3, (11.136) is reduced to the following dispersion relation of electrostatic wave in inhomogenous plasma (refer to [11.4], Chapter 11 of [2.3]):

$$(k_x^2 + k_z^2) - \sum_j \Pi_j^2\frac{1}{n_{0j}}\int\int\int[\quad]_j \mathrm{d}\theta\mathrm{d}v_\perp\mathrm{d}v_\perp\mathrm{d}v_z = 0,$$

$$k_x^2 + k_z^2 + \sum_j \Pi_j^2\Bigg\{\frac{1}{v_{\mathrm{T}z}^2} + \sum_{\mathrm{n}=-\infty}^{\infty} I_{\mathrm{n}}(b)e^{-b}\Bigg[\left(\frac{1}{v_{\mathrm{T}z}^2} - \frac{1}{v_{\mathrm{T}\perp}^2}\frac{\mathrm{n}\Omega}{\omega_{\mathrm{n}}}\right)\zeta_{\mathrm{n}}Z(\zeta_{\mathrm{n}})$$

$$-\frac{1}{v_{\mathrm{T}\perp}^2}\frac{\mathrm{n}}{k_x}\left((\epsilon' - \delta_\perp + f_{\mathrm{n}}(b)\delta_\perp)\left(1 + \frac{\mathrm{n}\Omega}{\omega_{\mathrm{n}}}\zeta_{\mathrm{n}}Z(\zeta_{\mathrm{n}})\right) + \frac{\delta_z}{2}\left(1 + \frac{\mathrm{n}\Omega\omega_{\mathrm{n}}}{k_z^2 v_{\mathrm{T}z}^2}(1 + \zeta_{\mathrm{n}}Z(\zeta_{\mathrm{n}}))\right)\right)$$

$$+\frac{1}{v_{\mathrm{T}z}^2}\frac{\mathrm{n}}{k_x}\left((\epsilon' - \delta_z + f_{\mathrm{n}}(b)\delta_\perp)(1 + \zeta_{\mathrm{n}}Z(\zeta_{\mathrm{n}})) + \frac{\delta_z}{2}\left(1 + \frac{\omega_{\mathrm{n}}^2}{k_z^2 v_{\mathrm{T}z}^2}(1 + \zeta_{\mathrm{n}}Z(\zeta_{\mathrm{n}}))\right)\right)$$

$$+\frac{k_x}{\Omega}\left((\epsilon' + f_{\mathrm{n}}(b)\delta_\perp)\frac{\zeta_n}{\omega_{\mathrm{n}}}Z(\zeta_{\mathrm{n}}) + \frac{\delta_z}{2}\frac{\omega_{\mathrm{n}}}{k_z^2 v_{\mathrm{T}z}^2}(1 + \zeta_{\mathrm{n}}Z(\zeta_{\mathrm{n}}))\right)\Bigg]_{\mathrm{n}}\Bigg\}_j = 0, \quad (11.140)$$

where $f_{\mathrm{n}}(b) \equiv (1 - b) + bI'_{\mathrm{n}}(b)/I_{\mathrm{n}}(b)$, $\epsilon' \equiv \epsilon - \delta_\perp - \delta_z/2$ and we used the following relation:

$$\int_{-\infty}^{\infty} J_{\mathrm{n}}^2(b^{1/2}x)\exp\left(-\frac{x^2}{2}\right)\cdot\frac{x^2}{2}x\mathrm{d}x = f_{\mathrm{n}}(b)I_{\mathrm{n}}(b)e^{-b}.$$

It is possible to include the effect of acceleration parallel to the y axis $\boldsymbol{g} = g\hat{\boldsymbol{y}}$ (the effect of curvature drift $g = v_{\parallel}^2/R$) in the dispersion equation (11.140) (refer to [11.4], Chapter 11 of [2.3]).

In the case of low frequency ($\omega \ll |\Omega|$), we have the following relations: $\zeta_n \gg 1$ ($n \neq 0$), $\zeta_n Z(\zeta_n) \to -1$ ($n \neq 0$) and $1 + \zeta_n Z(\zeta_n) \to -(1/2)\zeta_n^{-2}$ ($n \neq 0$) and (11.140) is reduced to

$$k_x^2 + k_z^2 + \sum_j \Pi_j^2 \left(\frac{1}{v_{\mathrm{T}z}^2} + I_0(b)e^{-b} \left(\frac{1}{v_{\mathrm{T}z}^2}(1 + \zeta_0 Z(\zeta_0)) - \frac{1}{v_{\mathrm{T}\perp}^2} \right.\right.$$
$$\left.\left. + \frac{k_x}{\Omega\omega_0}(\epsilon' + f_0(b)\delta_\perp)\zeta_0 Z(\zeta_0) + \frac{k_x}{\Omega\omega_0}\delta_z\zeta_0^2(1 + \zeta_0 Z(\zeta_0)) \right) \right)_j = 0, \qquad (11.141)$$

where $\sum_{-\infty}^{\infty} I_n(b)e^{-b} = 1$ is used.

When the gradient of temperature does not exist $\delta_\perp = \delta_z = 0$ in an isotropic plasma with $v_{\mathrm{T}\perp} = v_{\mathrm{T}z} = v_{\mathrm{T}}$ and $V = 0$, we have familiar dispersion relation of drift wave due to the density gradient as follows:

$$k_x^2 + k_z^2 + \sum_j \Pi_j^2 \left(\frac{1}{v_{\mathrm{T}}^2} + I_0(b)e^{-b} \left(\frac{1}{v_{\mathrm{T}}^2}\zeta_0 Z(\zeta_0) + \frac{k_x}{\Omega\omega_0}\epsilon\zeta_0 Z(\zeta_0) \right) \right)_j = 0.$$

We can usually assume $b_{\mathrm{e}} = 0$ for electrons. Then it reduces to

$$0 = (k_x^2 + k_z^2)\frac{v_{\mathrm{Te}}^2}{\Pi_{\mathrm{e}}^2} + 1 + \zeta_{\mathrm{e}} Z(\zeta_{\mathrm{e}}) \left(1 - \frac{\omega_{\mathrm{e}}^*}{\omega} \right)$$
$$+ \frac{ZT_{\mathrm{e}}}{T_{\mathrm{i}}} \left(1 + I_0(b)e^{-b}\zeta_{\mathrm{i}} Z(\zeta_{\mathrm{i}}) \left(1 - \frac{\omega_{\mathrm{i}}^*}{\omega} \right) \right), \qquad (11.142)$$

where

$$\omega_{\mathrm{e}}^* = -\frac{k_x \epsilon v_{\mathrm{Te}}^2}{\Omega_{\mathrm{e}}} = -\frac{k_x \epsilon T_{\mathrm{e}}}{eB} \quad \text{and} \quad \omega_{\mathrm{i}}^* = -\frac{k_x \epsilon v_{\mathrm{Ti}}^2}{\Omega_{\mathrm{i}}} = \frac{k_x \epsilon T_{\mathrm{i}}}{ZeB}.$$

ω_{e}^* and ω_{i}^* are called drift freqencies.

Note that the x direction is opposite the electron drift velocity $\boldsymbol{v}_{\mathrm{de}}$, the y direction is the direction of negative density gradient (outward), and the z direction is of the magnetic field in this subsection.

11.6 Velocity Space Instabilities

When the velocity distribution function is deviated from the stable Maxwell distribution, the plasma may be unstable due to Landau and cyclotron amplification. This type of instability is called *velocity space instability* or *microinstability*.

11.6.1 Drift Instability (Collisionless)

The dispersion relation of drift instability in the case of $\delta_\perp = \delta_\parallel = 0$ is given by (11.142). When the growth rate is much smaller than the real frequency, we have $\zeta Z(\zeta) = \zeta Z_r(\zeta) + ik_z/|k_z|\pi^{1/2}\exp(-\zeta^2)$. The solution of (11.142) is given by (Z=1)

$$\frac{\zeta_e Z(\zeta_e)\omega_e^* + I_0 e^{-b}\zeta_i Z(\zeta_i)(T_e/T_i)\omega_i^*}{\omega_r + i\gamma} = 1 + \zeta_e Z(\zeta_e) + \frac{T_e}{T_i}(1 + I_0 e^{-b}\zeta_i Z(\zeta_i)),$$

$$\frac{\omega_e^*}{\omega_r} = \frac{\frac{T_e}{T_i}(1 + e^{-b}I_0\zeta_i Z_r(\zeta_i)) + 1 + \zeta_e Z_r(\zeta_e)}{\zeta_e Z_r(\zeta_e) - e^{-b}I_0\zeta_i Z_r(\zeta_i)},$$

$$\frac{\gamma}{\omega_r} = \pi^{1/2}\frac{k_z}{|k_z|}\frac{\zeta_e\left(1 - \frac{\omega_r}{\omega_e^*}\right)\exp(-\zeta_e^2) - e^{-b}I_0\zeta_i\left((1 + \frac{\omega_r}{\omega_e^*}\frac{T_e}{T_i}\right)\exp(-\zeta_i)^2}{\zeta_e Z_r(\zeta_e) - e^{-b}I_0\zeta_i Z_r(\zeta_i)}$$

The second term in the numerator of γ/ω_r is the contribution of ion Landau damping. The first term contributes to instability when $\omega_r/\omega_e^* < 1$. When k_z/ω is in the region $v_{Ti} < \omega/k_z < v_{Te}$, it follows that $|\zeta_e| < 1$, $|\zeta_i| > 1$, the ion Landau damping decreases, and the first term in the numerator becomes dominant. When $b < 1$,

$$\frac{\omega_e^*}{\omega_r} = \frac{1 + (T_e/T_i)b}{1 - b} \approx 1 + \left(1 + \frac{T_e}{T_i}\right)b,$$

$$\frac{\gamma}{\omega_r} = \pi^{1/2}\frac{k_z}{|k_z|}\left(1 + \frac{T_e}{T_i}\right)b\zeta_e\exp(-\zeta_e^2) \approx \frac{b\omega_e^*}{|k_z|v_{Te}}.$$

The growth rate of collisionless instability becomes large when ion Larmor radius is large. The kinetic effect of electron Landau amplification destabilizes drift wave. In the analysis of drift instability in MHD (Section 9.3), collisionless drift instability was stable within resistive MHD time scale.

11.6.2 Various Velocity Space Instabilities

In the previous subsection, a simple case of drift instability was described. There are various velocity space instabilities.

The distribution function of a plasma confined in a mirror field is zero for loss cone region $(v_\perp/v)^2 < 1/R_M$ (R_M is mirror ratio; refer to Section 2.4). The instability associated with this is called *loss-cone instability* [11.5].

Plasmas heated by ICRF have higher ion temperature in perpendicular direction than that of parallel direction. In this case, instabilities with the higher harmonic ion cyclotron frequencies may occur. This type of instability is called *Harris instability* [11.6, 11.7]. Harris instability is electrostatic and is analyzed by the dispersion relation (11.132).

In general, plasmas are hot and dense in the center and are cold and low density in the edge. The instabilities driven by temperature gradient and density gradient are called *drift instabilities*. The electrostatic drift instability of inhomogeneous plasma can be analyzed by the dispersion equation (11.140) and (11.141) of inhomogenous plasma [11.4]. The simple case of drift instability was just analyzed.

In toroidal field, trapped particles always exist in the outside where the magnetic field is weak. The instabilities induced by the trapped particles are called *trapped particle instabilities* [11.8].

Problems

1. Density Measurement The refractive index of quasi-transverse ($\theta = \pi/2$) ordinary wave is $N = (1 - \Pi_e^2/\omega^2)^{1/2}$ given by (11.82). When the microwave or laser light with the high frequency $\omega^2 \gg \Pi_e^2$ is used to measure the electron density by interferometer, the change of the optical path Δ due to the plasma is $\Delta = \int \Delta N \mathrm{d}l \approx -(1/2)(\Pi_e^2/\omega^2)l$ where l is plasma size along the optical path and is proportional to the average electron density. When the light of CO_2 laser (vacuum wavelegth λ_0=10.6 μm) is used and the fringe shift due to the change of optical path is 0.2, that is, $\Delta = 0.2\lambda_0$, calculate the average electron density in the case of plasma size $l = 2\,\mathrm{m}$.

2. Dielectric Tensor of Cold Plasma and Hot Plasma Dielectric tensor of hot plasma is given by (11.125) and (11.126). Prove that the dielectric tensor of hot plasma becomes the dielectric tensor of cold plasma (11.9) when the temperatures $T_\perp$ and T_z tend to be zero. Use (12.13) in Chapter 12.

3. Bernstein Wave Drive the dispersion relation of Berstein wave when the Bernstein wave (11.135) propagates in the perpendicular direction to the magnetic field ($k_z = 0$).

4. Drift Instability Drift instability requires $v_{\mathrm{Te}} \gg \omega/k_z \gg v_{\mathrm{Ti}}$. Explain this reason. What condition is required for $k_\parallel/k_\perp$? Note $\omega \sim \omega_e^* = \kappa_n k_\perp T_e$ $/eB$.

5. Ion Temperature Gradient Instability Dispersion equation (11.141) is applicable to ion temperature gradient instability (eta-i mode). Assume that ions are cold and electrons are hot, that is, $\zeta_e = \omega/k_\parallel v_{\mathrm{Te}} \gg 1$ and $\zeta_e = \omega/k_\parallel v_{\mathrm{Te}} \ll 1$. Prove that the dispersion equation of ITG instability is reduced to

$$1 - \frac{\omega_e^*}{\omega} + (k_\perp \rho_s)^2 \left(1 - \frac{\omega_i^*}{\omega}\right) - \frac{k_\parallel^2 c_s^2}{\omega^2}\left(1 - \frac{\omega_{ip}^*}{\omega}\right) = 0$$

where $\omega_e^* \equiv \epsilon k_\perp T_e/(eB)$, $\omega_i^* \equiv -\epsilon k_\perp T_i/(eB)$, $\omega_{ip}^* \equiv -(\epsilon + \delta_i)k_\perp T_i/(eB)$, $c_s^2 = T_s/m_i$ and $\rho_s^2 = c_s^2/\Omega_i^2$.

When $k_\perp\rho_s \ll 1$, $\omega_e^*/\omega \ll 1$, $|\omega_{ip}|/\omega \gg 1$ ($\epsilon \ll \delta_i$), the dispersion relation is reduced to $\omega^3 = k_\parallel^2 c_s^2 |\omega_{ip}|$, that is $\omega = (k_\parallel^2 c_s^2 |\omega_{ip}|)^{1/3}\exp(2\pi/3)i$. Refer to [11.9].

12

Wave Heating and Non-Inductive Current Drive

First we discuss the wave heatings; i.e., wave heating in ion cyclotron range of frequency (ICRF) in Section 12.2, lower hybrid heating (LHH) in Section 12.3, and electron cyclotron heating (ECH) in Section 12.4. The physical processes of wave heating are not simple and the interactions of waves and plasmas have a lot of variety, so that various applications are possible depending on the development of wave heating methods.

Waves are excited in the plasma by antennas or waveguides located outside the plasma (*excitation of wave, antenna-plasma coupling*). When the electric field of the excited wave is parallel to the confining magnetic field of the plasma, the electron, which can move along the magnetic field, may cancel the electric field. However, if the frequency of the wave is larger than the plasma frequency the electron cannot follow the change in the electric field, and the wave then propagates through the plasma. When the electric field of the excited wave is perpendicular to the magnetic field, the electrons move in the direction of $\boldsymbol{E} \times \boldsymbol{B}$ (under the condition $\omega < \Omega_{\rm e}$) and thus they cannot cancel the electric field. In this case, the wave can propagate through the plasma even if the wave frequency is smaller than the plasma frequency. Excitation consists of pumping high-frequency electromagnetic wave into plasma through the coupling system. If the structure of the coupling system has the same periodicity as the eigenmode wave, the wave can be excited resonantly (*resonant excitation*). Electron cyclotron heating (and neutral beam injection) can be launched in vacuum and propagate directly into the plasma without attenuation or interaction with the edge. Consequently the launching structures do not have to be in close proximity to the plasma and have advantage against thermal load and erosion by plasma.

Excited waves may propagate and pass through the plasma center without damping (heating) in some cases and may refract and turn back to the external region without passing the plasma center or may be reflected by the cutoff layer (see Figure 12.1). The wave may be converted to the other mode by the *mode conversion* (*wave propagation*).

The waves propagating in the plasma are absorbed and damped at the locations where Landau damping and cyclotron damping occur and heat the plasma. Therefore, it is necessary for heating the plasma center that the waves be able to propagate into the plasma center without absorption and

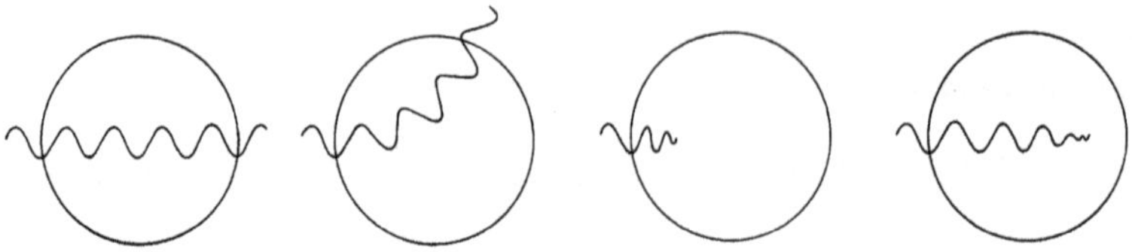

Figure 12.1 Passing through, refraction and reflection, absorption near boundary, and absorption at the center of plasma.

that they be absorbed when they reach the plasma center (*wave heating*).

12.1 Energy Flow

Energy transport and the propagation of waves in the plasma medium are very important in the wave heating of plasmas. The equation of energy flow is derived from Maxwell equations.

$$\nabla \times \boldsymbol{E} = -\frac{\partial \boldsymbol{B}}{\partial t}, \tag{12.1}$$

$$\nabla \times \boldsymbol{H} = \boldsymbol{j} + \epsilon_0 \frac{\partial \boldsymbol{E}}{\partial t} = \frac{\partial \boldsymbol{D}}{\partial t}. \tag{12.2}$$

By taking the difference between the scalar product of $\boldsymbol{H}$ and (12.1) and the scalar product of $\boldsymbol{E}$ and (12.2):

$$\nabla \cdot (\boldsymbol{E} \times \boldsymbol{H}) + \boldsymbol{E} \cdot \frac{\partial \boldsymbol{D}}{\partial t} + \boldsymbol{H} \cdot \frac{\partial \boldsymbol{B}}{\partial t} = 0. \tag{12.3}$$

$\boldsymbol{P} \equiv \boldsymbol{E} \times \boldsymbol{H}$ is called *Poynting vector* and represents the energy flow of the electromagnetic field. This Poynting equation does not include the effect of electric resistivity by electron-ion collision.

Plasmas are dispersive medium and the dielectric tensors are dependent on the propagation vector $\boldsymbol{k}$ and the frequency ω. Denote the Fourier components of $\boldsymbol{E}(\boldsymbol{r},t)$ and $\boldsymbol{D}(\boldsymbol{r},t)$ by $\boldsymbol{E}_{k\,\omega}(\boldsymbol{k},\omega)$ and $\boldsymbol{D}_{k\,\omega}(\boldsymbol{k},\omega)$, respectively. Then we find

$$\boldsymbol{D}_{k\,\omega} = \frac{1}{(2\pi)^2} \int \boldsymbol{D}(\boldsymbol{r},t) \exp(-i(\boldsymbol{k} \cdot \boldsymbol{r} - \omega t)) \, \mathrm{d}\boldsymbol{r} \, \mathrm{d}t,$$

$$\boldsymbol{E}_{k\,\omega} = \frac{1}{(2\pi)^2} \int \boldsymbol{E}(\boldsymbol{r},t) \exp(-i(\boldsymbol{k} \cdot \boldsymbol{r} - \omega t)) \, \mathrm{d}\boldsymbol{r} \, \mathrm{d}t.$$

There is the following relation between them:

$$\boldsymbol{D}_{k\,\omega}(\boldsymbol{k},\omega) = \epsilon_0 \boldsymbol{K}(\boldsymbol{k},\omega) \cdot \boldsymbol{E}_{k\,\omega}(\boldsymbol{k},\omega),$$

and we have

$$\boldsymbol{D}(\boldsymbol{r},t) = \frac{1}{(2\pi)^2}\epsilon_0 \int \boldsymbol{K}(\boldsymbol{k},\omega) \cdot \boldsymbol{E}_{k\,\omega}(\boldsymbol{k},\omega) \exp(i(\boldsymbol{k}\cdot\boldsymbol{r} - \omega t))\, \mathrm{d}\boldsymbol{k}\, \mathrm{d}\omega,$$

$$\boldsymbol{E}(\boldsymbol{r},t) = \frac{1}{(2\pi)^2} \int \boldsymbol{E}_{k\,\omega}(\boldsymbol{k},\omega) \exp(i(\boldsymbol{k}\cdot\boldsymbol{r} - \omega t))\, \mathrm{d}\boldsymbol{k}\, \mathrm{d}\omega.$$

From the formula of Fourier integral, the following equations are derived:

$$\boldsymbol{D}(\boldsymbol{r},t) = \epsilon_0 \int \widehat{\boldsymbol{K}}(\boldsymbol{r}-\boldsymbol{r}', t-t') \cdot \boldsymbol{E}(\boldsymbol{r}',t')\, \mathrm{d}\boldsymbol{r}'\, \mathrm{d}t',$$

where $\widehat{\boldsymbol{K}}(\boldsymbol{r},t)$ is

$$\widehat{\boldsymbol{K}}(\boldsymbol{r},t) = \frac{1}{(2\pi)^4} \int \boldsymbol{K}(\boldsymbol{k},\omega) \exp(i(\boldsymbol{k}\cdot\boldsymbol{r} - \omega t))\, \mathrm{d}\boldsymbol{k}\, \mathrm{d}\omega.$$

Therefore, analysis of general electromagnetic fields in dispersive medium is not simple. However, if the electric field consists of a Fourier component in narrow region near $\boldsymbol{k}_0$, ω_0 and $\boldsymbol{K}$ changes slowly as $\boldsymbol{k}$, ω change, then we can use the following relation:

$$\boldsymbol{D}(\boldsymbol{r},t) = \epsilon_0 \boldsymbol{K}(\boldsymbol{k}_0,\omega_0) \cdot \boldsymbol{E}(\boldsymbol{r},t).$$

Now we will discuss this simple case. The relation between the magnetic induction $\boldsymbol{B}$ and the magnetic intensity $\boldsymbol{H}$ is $\boldsymbol{B} = \mu_0 \boldsymbol{H}$.
The quasi-periodic functions A, B may be expressed by

$$A = A_0 \exp\left(-i \int_{-\infty}^{t} (\omega_\mathrm{r} + i\omega_\mathrm{i}) \mathrm{d}t'\right) = A_0 \exp(-i\phi_\mathrm{r} + \phi_\mathrm{i}),$$

$$B = B_0 \exp\left(-i \int_{-\infty}^{t} (\omega_\mathrm{r} + i\omega_\mathrm{i}) \mathrm{d}t'\right) = B_0 \exp(-i\phi_\mathrm{r} + \phi_\mathrm{i}),$$

where ϕ_r and ϕ_i are real. When the average of the multiplication of the real parts of A with the real part of B is denoted by $\overline{AB}$, then $\overline{AB}$ is given by

$$\overline{AB} = \frac{1}{2} \cdot \frac{1}{2} \langle (A_0 \exp(-i\phi_\mathrm{r} + \phi_\mathrm{i}) + A_0^* \exp(i\phi_\mathrm{r} + \phi_\mathrm{i}))$$

$$\times (B_0 \exp(-i\phi_\mathrm{r} + \phi_\mathrm{i}) + B_0^* \exp(i\phi_\mathrm{r} + \phi_\mathrm{i})) \rangle$$

$$= \frac{1}{4}(A_0 B_0^* + A_0^* B_0) \exp(2\phi_\mathrm{i}) = \frac{1}{2}\mathrm{Re}(AB^*)\,. \tag{12.4}$$

The averaging of the Poynting equation becomes

$$\nabla \cdot \boldsymbol{P} + \frac{\partial W}{\partial t} = 0, \tag{12.5}$$

$$\boldsymbol{P} = \frac{1}{2\mu_0}\mathrm{Re}(\boldsymbol{E}_0 \times \boldsymbol{B}_0^*)\exp 2\int_{-\infty}^{t} \omega_\mathrm{i}\mathrm{d}t', \tag{12.6}$$

$$\begin{aligned}\frac{\partial W}{\partial t} &= \frac{1}{2}\mathrm{Re}\left(\left(\frac{\boldsymbol{B}^*}{\mu_0}\cdot\frac{\partial \boldsymbol{B}}{\partial t}\right) + \epsilon_0 \boldsymbol{E}^* \cdot \frac{\partial}{\partial t}(\boldsymbol{K}\cdot\boldsymbol{E})\right) \\ &= \frac{1}{2}\mathrm{Re}\left(-i\omega\frac{\boldsymbol{B}^*\cdot\boldsymbol{B}}{\mu_0} + \epsilon_0(-i\omega)\boldsymbol{E}^*\cdot\boldsymbol{K}\cdot\boldsymbol{E}\right) \\ &= \frac{1}{2}\omega_\mathrm{i}\frac{\boldsymbol{B}\cdot\boldsymbol{B}^*}{\mu_0} + \frac{\epsilon_0}{2}\left(\omega_\mathrm{i}\mathrm{Re}(\boldsymbol{E}^*\cdot\boldsymbol{K}\cdot\boldsymbol{E}) + \omega_\mathrm{r}\mathrm{Im}(\boldsymbol{E}^*\cdot\boldsymbol{K}\cdot\boldsymbol{E})\right).\end{aligned} \tag{12.7}$$

From the relations

$$\boldsymbol{E}^*\cdot\boldsymbol{K}\cdot\boldsymbol{E} = \sum_i E_i^* \sum_j K_{ij}E_j,$$

$$\begin{aligned}\boldsymbol{E}\cdot\boldsymbol{K}^*\cdot\boldsymbol{E}^* &= \sum_i E_i \sum_j K_{ij}^* E_j^* = \sum_j E_j^* \sum_i (K_{ji}^\mathrm{T})^* E_i \\ &= \sum_i E_i^* \sum_j (K_{ij}^\mathrm{T})^* E_j,\end{aligned}$$

we find

$$\mathrm{Re}(\boldsymbol{E}^*\cdot\boldsymbol{K}\cdot\boldsymbol{E}) = \boldsymbol{E}^*\cdot\frac{\boldsymbol{K}+(\boldsymbol{K}^\mathrm{T})^*}{2}\cdot\boldsymbol{E},$$

$$\mathrm{Im}(\boldsymbol{E}^*\cdot\boldsymbol{K}\cdot\boldsymbol{E}) = \boldsymbol{E}^*\cdot\frac{(-i)[\boldsymbol{K}-(\boldsymbol{K}^\mathrm{T})^*]}{2}\cdot\boldsymbol{E}.$$

$(\boldsymbol{K}^\mathrm{T})^*$ is the complex conjugate of transpose matrix $\boldsymbol{K}^\mathrm{T}$ (lines and rows of components are exchanged) of $\boldsymbol{K}$, i.e., $K_{ij}^\mathrm{T} \equiv K_{ji}$. When a matrix $\boldsymbol{M}$ and $(\boldsymbol{M}^\mathrm{T})^*$ are equal, this kind of matrix is called a *Hermite matrix*. For the Hermite matrix, $(\boldsymbol{E}^*\cdot\boldsymbol{M}\cdot\boldsymbol{E})$ is always real. The dielectric tensor may be decomposed to

$$\boldsymbol{K}(\boldsymbol{k},\omega) = \boldsymbol{K}_\mathrm{H}(\boldsymbol{k},\omega) + i\boldsymbol{K}_\mathrm{I}(\boldsymbol{k},\omega).$$

As will be described in Section 12.3, $\boldsymbol{K}_\mathrm{H}$ and $\boldsymbol{K}_\mathrm{I}$ are Hermite, when $\boldsymbol{k}$, ω are real. It will be proved that the term $i\boldsymbol{K}_\mathrm{I}$ corresponds to Landau damping and cyclotron damping. When the imaginary part of ω is much smaller than the real part ($\omega = \omega_\mathrm{r} + i\omega_\mathrm{i}$, $|\omega_\mathrm{i}| \ll |\omega_\mathrm{r}|$), we may write

$$\boldsymbol{K}(\boldsymbol{k},\omega_{\rm r}+i\omega_{\rm i}) \approx \boldsymbol{K}_{\rm H}(\boldsymbol{k},\omega_{\rm r}) + i\omega_{\rm i}\frac{\partial}{\partial\omega_{\rm r}}\boldsymbol{K}_{\rm H}(\boldsymbol{k},\omega_{\rm r}) + i\boldsymbol{K}_{\rm I}(\boldsymbol{k},\omega_{\rm r}),$$

$$\frac{\boldsymbol{K}+(\boldsymbol{K}^{\rm T})^*}{2} = \boldsymbol{K}_{\rm H}, \qquad \frac{-i[\boldsymbol{K}-(\boldsymbol{K}^{\rm T})^*]}{2} = \omega_{\rm i}\frac{\partial}{\partial\omega_{\rm r}}\boldsymbol{K}_{\rm H} + \boldsymbol{K}_{\rm I}.$$

When the Hermite component of W (the term associated to $\boldsymbol{K}_{\rm H}$ in W) is denoted by W_0, W_0 is given by

$$W_0 = \frac{1}{2}{\rm Re}\left(\frac{\boldsymbol{B}_0^*\cdot\boldsymbol{B}_0}{2\mu_0} + \frac{\epsilon_0}{2}\boldsymbol{E}_0^*\cdot\boldsymbol{K}_{\rm H}\cdot\boldsymbol{E}_0 + \frac{\epsilon_0}{2}\boldsymbol{E}_0^*\cdot\left(\omega_{\rm r}\frac{\partial}{\partial\omega_{\rm r}}\boldsymbol{K}_{\rm H}\right)\cdot\boldsymbol{E}_0\right)$$

$$= \frac{1}{2}{\rm Re}\left(\frac{\boldsymbol{B}_0^*\cdot\boldsymbol{B}_0}{2\mu_0} + \frac{\epsilon_0}{2}\boldsymbol{E}_0^*\cdot\left(\frac{\partial}{\partial\omega}(\omega\boldsymbol{K}_{\rm H})\right)\cdot\boldsymbol{E}_0\right). \qquad (12.8)$$

Then Equations (12.5) and (12.7) yield

$$\frac{\partial W_0}{\partial t} = -\omega_{\rm r}\frac{1}{2}\epsilon_0\boldsymbol{E}_0^*\cdot\boldsymbol{K}_{\rm I}\cdot\boldsymbol{E}_0 - \nabla\cdot\boldsymbol{P}. \qquad (12.9)$$

The first term in (12.8) is the energy density of the magnetic field and the second term is the energy density of the electric field which includes the kinetic energy of coherent motion associated with the wave. Equation (12.8) gives the energy density of the wave in a dispersive media. The first term in the right-hand side of (12.9) represents the Landau and cyclotron dampings and the second term is the divergence of the flow of wave energy.

Therefore, the absorbed power per unit volume of plasma $P^{\rm ab}$ is given by the first term in (12.9):

$$P^{\rm ab} = \omega_{\rm r}\left(\frac{\epsilon_0}{2}\right)\boldsymbol{E}^*\cdot\boldsymbol{K}_{\rm I}\cdot\boldsymbol{E}.$$

Since $\boldsymbol{K}_{\rm H}$, $\boldsymbol{K}_{\rm I}$ is a Hermit matrix for real $\boldsymbol{k}$, ω as was shown in the previous chapter, the absorbed power $P^{\rm ab}$ is given by

$$P^{\rm ab} = \omega_{\rm r}\left(\frac{\epsilon_0}{2}\right){\rm Re}\left(\boldsymbol{E}^*\cdot(-i)\boldsymbol{K}\cdot\boldsymbol{E}\right)_{\omega=\omega_{\rm r}}. \qquad (12.10)$$

As is clear from the expression of $\boldsymbol{K}$, the absorbed power $P^{\rm ab}$ reduces to

$$P^{\rm ab} = \omega\frac{\epsilon_0}{2}\left(|E_x|^2{\rm Im}K_{xx} + |E_y|^2{\rm Im}K_{yy} + |E_z|^2{\rm Im}K_{zz}\right.$$

$$\left. +2{\rm Im}(E_x^*E_y){\rm Re}K_{xy} + 2{\rm Im}(E_y^*E_z){\rm Re}K_{yz} + 2{\rm Im}(E_x^*E_z){\rm Re}K_{xz}\right). \quad (12.11)$$

Since (12.3) gives $\boldsymbol{j} = -i\omega\boldsymbol{P} = -i\epsilon_0\omega(\boldsymbol{K}-\boldsymbol{I})\cdot\boldsymbol{E}$, (12.10) may be described by

$$P^{\rm ab} = \frac{1}{2}{\rm Re}(\boldsymbol{E}^*\cdot\boldsymbol{j})_{\omega=\omega_{\rm r}}. \qquad (12.12)$$

In the case of $b = (k_x\rho_\Omega)^2 \ll 1$ ($\rho_\Omega = v_{T\perp}/\Omega$ is Larmor radius), it is possible to expand $e^{-b}\boldsymbol{X}_\mathrm{n}$ by b using

$$I_\mathrm{n}(b) = \left(\frac{b}{2}\right)^\mathrm{n} \sum_{l=0}^{\infty} \frac{1}{l!(\mathrm{n}+l)!}\left(\frac{b}{2}\right)^{2l}$$
$$= \left(\frac{b}{2}\right)^\mathrm{n}\left(\frac{1}{\mathrm{n}!} + \frac{1}{1!(\mathrm{n}+1)!}\left(\frac{b}{2}\right)^2 + \frac{1}{2!(\mathrm{n}+2)!}\left(\frac{b}{2}\right)^4 + \cdots\right).$$

The expansion in b and the inclusion of terms up to the second harmonic in $\boldsymbol{K}$ gives

$$K_{xx} = 1 + \sum_j \left(\frac{\Pi_j}{\omega}\right)^2 \zeta_0\left((Z_1 + Z_{-1})\left(\frac{1}{2} - \frac{b}{2} + \cdots\right) + (Z_2 + Z_{-2})\left(\frac{b}{2} - \frac{b^2}{2} + \cdots\right) + \cdots\right)_j$$

$$K_{yy} = 1 + \sum_j \left(\frac{\Pi_j}{\omega}\right)^2 \zeta_0\left(Z_0(2b + \cdots) + (Z_1 + Z_{-1})\left(\frac{1}{2} - \frac{3b}{2} + \cdots\right)\right.$$
$$\left. + (Z_2 + Z_{-2})\left(\frac{b}{2} - b^2 + \cdots\right) + \cdots\right)_j$$

$$K_{zz} = 1 - \sum_j \left(\frac{\Pi_j}{\omega}\right)^2 \zeta_0\left(2\zeta_0 W_0(1 - b + \cdots) + (\zeta_1 W_1 + \zeta_{-1}W_{-1})(b + \cdots)\right.$$
$$\left. + (\zeta_2 W_2 + \zeta_{-2}W_{-2})\left(\frac{b^2}{4} + \cdots\right) + \cdots\right)_j$$

$$K_{xy} = i\sum_j \left(\frac{\Pi_j}{\omega}\right)^2 \zeta_0\left((Z_1 - Z_{-1})\left(\frac{1}{2} - b + \cdots\right) + (Z_2 - Z_{-2})\left(\frac{b}{2} + \cdots\right) + \cdots\right)_j$$

$$K_{xz} = 2^{1/2}\sum_j \left(\frac{\Pi_j}{\omega}\right)^2 b^{1/2}\zeta_0\left((W_1 - W_{-1})\left(\frac{1}{2} + \cdots\right) + (W_2 - W_{-2})\left(\frac{b}{4} + \cdots\right) + \cdots\right)_j$$

$$K_{yz} = -2^{1/2} i\sum_j \left(\frac{\Pi_j}{\omega}\right)^2 b^{1/2}\zeta_0\left(W_0\left(-1 + \frac{3}{2}b + \cdots\right)\right.$$
$$\left. + (W_1 + W_{-1})\left(\frac{1}{2} + \cdots\right) + (W_2 - W_{-2})\left(\frac{b}{4} + \cdots\right) + \cdots\right)_j$$

$$K_{yx} = -K_{xy}, \quad K_{zx} = K_{xz}, \quad K_{zy} = -K_{zy}, \tag{12.13}$$

where

$$Z_{\pm\mathrm{n}} \equiv Z(\zeta_{\pm\mathrm{n}}), \qquad W_\mathrm{n} \equiv -\left(1 + \zeta_\mathrm{n} Z(\zeta_\mathrm{n})\right),$$
$$\zeta_\mathrm{n} = (\omega + \mathrm{n}\Omega)/(2^{1/2}k_z(T_z/m)^{1/2}).$$

When $x \gg 1$, Re$W(x)$ is

$$\mathrm{Re}\, W(x) = (1/2)x^{-2}(1 + (3/2)x^{-2} + \cdots).$$

The absorbed power by Landau damping (including transit time damping) may be estimated by the terms associated with the imaginary part G_0 of $\zeta_0 Z(\zeta_0)$ in (12.13) of K_{ij}:

$$G_0 \equiv \mathrm{Im}\zeta_0 Z(\zeta_0) = (k_z/|k_z|)\pi^{1/2}\zeta_0 \exp(-\zeta_0^2).$$

Since

$$(\mathrm{Im}K_{yy})_0 = (\Pi_j/\omega)^2 2bG_0,$$
$$(\mathrm{Im}K_{zz})_0 = (\Pi_j/\omega)^2 2\zeta_0^2 G_0,$$
$$(\mathrm{Re}K_{yz})_0 = (\Pi_j/\omega)^2 2^{1/2} b^{1/2}\zeta_0 G_0,$$

the contribution of these terms to the absorption power (12.11) is

$$P_0^{\mathrm{ab}} = \omega(\Pi_j/\omega)^2 G_0 \epsilon_0(|E_y|^2 b + |E_z|^2\zeta_0^2 + \mathrm{Im}(E_y^* E_z)(2b)^{1/2}\zeta_0). \qquad (12.14)$$

The first term is of transit time damping and is equal to (11.42). The second term is of Landau damping and is equal to (11.36). The third one is the term of the interference of both.

The absorption powers due to cyclotron damping and the harmonic cyclotron damping are obtained by the contribution from the terms

$$G_{\pm \mathrm{n}} \equiv \mathrm{Im}\zeta_0 Z_{\pm \mathrm{n}} = (k_z/|k_z|)\pi^{1/2}\zeta_0 \exp(-\zeta_{\pm \mathrm{n}}^2),$$

and for the case $b \ll 1$,

$$(\mathrm{Im}K_{xx})_{\pm \mathrm{n}} = (\mathrm{Im}K_{yy})_{\pm \mathrm{n}} = (\Pi_j/\omega)^2 G_{\pm \mathrm{n}}\alpha_{\mathrm{n}},$$
$$(\mathrm{Im}K_{zz})_{\pm \mathrm{n}} = (\Pi_j/\omega)^2 2\zeta_{\pm \mathrm{n}}^2 G_{\pm \mathrm{n}} b\alpha_{\mathrm{n}}\mathrm{n}^{-2},$$
$$(\mathrm{Re}K_{xy})_{\pm \mathrm{n}} = -(\Pi_j/\omega)^2 G_{\pm \mathrm{n}}(\pm\alpha_{\mathrm{n}}),$$
$$(\mathrm{Re}K_{yz})_{\pm \mathrm{n}} = -(\Pi_j/\omega)^2 (2b)^{1/2}\zeta_{\pm \mathrm{n}} G_{\pm \mathrm{n}}\alpha_{\mathrm{n}}\mathrm{n}^{-1},$$
$$(\mathrm{Im}K_{xz})_{\pm \mathrm{n}} = -(\Pi_j/\omega)^2 (2b)^{1/2}\zeta_{\pm \mathrm{n}} G_{\pm \mathrm{n}}(\pm\alpha_{\mathrm{n}})\mathrm{n}^{-1},$$

$$\alpha_{\mathrm{n}} = \mathrm{n}^2(2\cdot \mathrm{n}!)^{-1}(b/2)^{\mathrm{n}-1}.$$

The contribution of these terms to the absorbed power (12.11) is

$$P_{\pm \mathrm{n}}^{\mathrm{ab}} = \omega\left(\frac{\Pi_j}{\omega}\right)^2 G_{\mathrm{n}}\left(\frac{\epsilon_0}{2}\right)\alpha_{\mathrm{n}}|E_x \pm iE_y|^2. \qquad (12.15)$$

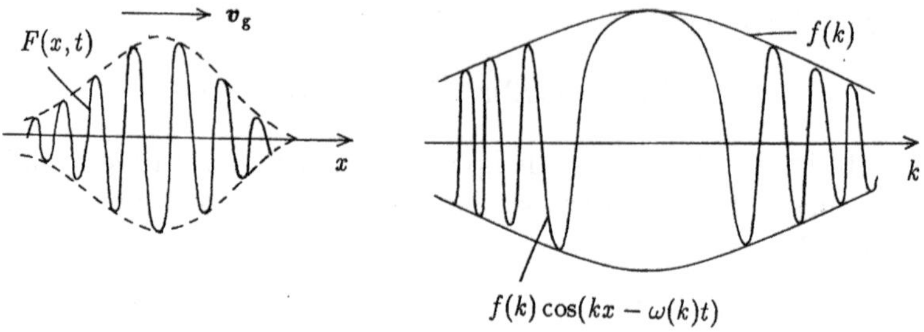

Figure 12.2 $F(x,t)$ and $f(k)\cos(kx-w(k)t)$.

Since

$$\zeta_{\rm n} = (\omega + {\rm n}\Omega_{\rm i})/(2^{1/2}k_z v_{\rm Ti}) = (\omega - {\rm n}|\Omega_{\rm i}|)/(2^{1/2}k_z v_{\rm Ti}),$$

the term of +n is dominant for the *ion cyclotron damping* ($\omega > 0$). The term of −n is dominant for *electron cyclotron damping* ($\omega > 0$), since

$$\zeta_{-{\rm n}} = (\omega - {\rm n}\Omega_{\rm e})/(2^{1/2}k_z v_{\rm Te}).$$

The relative ratio of $\boldsymbol{E}$ components can be estimated from the following equations:

$$(K_{xx} - N_\parallel^2)E_x + K_{xy}E_y + (K_{xz} + N_\perp N_\parallel)E_z = 0,$$

$$-K_{xy}E_x + (K_{yy} - N_\parallel^2 - N_\perp^2)E_y + K_{yz}E_z = 0, \tag{12.16}$$

$$(K_{xz} + N_\perp N_\parallel)E_x - K_{yz}E_y + (K_{zz} - N_\perp^2)E_z = 0.$$

For cold plasmas, $K_{xx} \to K_\perp$, $K_{yy} \to K_\perp$, $K_{zz} \to K_\parallel$, $K_{xy} \to -iK_\times$, $K_{xz} \to 0$, $K_{yz} \to 0$ can be substituted into (12.16), and the relative ratio is $E_x : E_y : E_z = (K_\perp - N^2) \times (K_\parallel - N_\perp^2) : -iK_x(K_\parallel - N_\perp^2) : -N_\parallel N_\perp (K_\perp - N^2)$.

In order to obtain the magnitude of the electric field, it is necessary to solve Maxwell equation with the dielectric tensor of (12.125) and (12.126). In this case the density, temperature, and magnetic field are functions of the coordinates. Therefore, the simplified model must be used for analytical solutions; otherwise, numerical calculations are necessary to derive the wave field.

Group Velocity

Let us consider the velocity of movement of the wave packet given by

$$F(\boldsymbol{r},t) = \int_{-\infty}^{\infty} f(\boldsymbol{k}) \exp i(\boldsymbol{k}\cdot\boldsymbol{r} - \omega(\boldsymbol{k})t) {\rm d}\boldsymbol{k}, \tag{12.17}$$

when

$$\omega = \omega(\boldsymbol{k})$$

is given. If $f(\boldsymbol{k})$ varies slowly, the position of the maximum of $F(\boldsymbol{r}, t)$ at t is the position of the stationary phase of

$$\frac{\partial}{\partial k_\alpha}(\boldsymbol{k}\cdot\boldsymbol{r} - \omega(\boldsymbol{k})t) = 0, \qquad (\alpha = x, y, z),$$

because the main contribution to the integral (12.8) comes from the region near the stationary phase as is seen in Figure 12.2. Consequently, the velocity of the wave packet is

$$\left(\frac{x}{t} = \frac{\partial\omega(\boldsymbol{k})}{\partial k_x}, \qquad \frac{y}{t} = \frac{\partial\omega(\boldsymbol{k})}{\partial k_y}, \qquad \frac{z}{t} = \frac{\partial\omega(\boldsymbol{k})}{\partial k_z}\right),$$

that is

$$\boldsymbol{v}_{\mathrm{g}} = \left(\frac{\partial\omega}{\partial k_x}, \frac{\partial\omega}{\partial k_y}, \frac{\partial\omega}{\partial k_z}\right). \tag{12.18}$$

This velocity is called *group velocity* and represents the velocity of energy flow.

Ray Tracing

When the wavelength of waves in the plasma is much less than the characteristic length (typically the minor radius a), the WKB approximation (geometrical optical approximation) can be applied. Let the dispersion relation be $D(\boldsymbol{k}, \omega, \boldsymbol{r}, t) = 0$. The direction of wave energy flow is given by the group velocity $\boldsymbol{v}_{\mathrm{g}} = \partial\omega/\partial\boldsymbol{k} \equiv (\partial\omega/\partial k_x, \partial\omega/\partial k_y, \partial\omega/\partial k_z)$, so that the optical ray can be given by $\mathrm{d}\boldsymbol{r}/\mathrm{d}t = \boldsymbol{v}_{\mathrm{g}}$. Although the quantities $(\boldsymbol{k}, \omega)$ change according to the change of $\boldsymbol{r}$, they always satisfy $D = 0$. Then the optical ray can be obtained by

$$\frac{\mathrm{d}\boldsymbol{r}}{\mathrm{d}s} = \frac{\partial D}{\partial \boldsymbol{k}}, \qquad \frac{\mathrm{d}\boldsymbol{k}}{\mathrm{d}s} = -\frac{\partial D}{\partial \boldsymbol{r}}, \tag{12.19}$$

$$\frac{\mathrm{d}t}{\mathrm{d}s} = -\frac{\partial D}{\partial \omega}, \qquad \frac{\mathrm{d}\omega}{\mathrm{d}s} = \frac{\partial D}{\partial t}. \tag{12.20}$$

Here s is a measure of the length along the optical ray. Along the optical ray the variation δD becomes zero,

$$\delta D = \frac{\partial D}{\partial \boldsymbol{k}}\cdot\delta\boldsymbol{k} + \frac{\partial D}{\partial \omega}\cdot\delta\omega + \frac{\partial D}{\partial \boldsymbol{r}}\cdot\delta\boldsymbol{r} + \frac{\partial D}{\partial t}\cdot\delta t = 0, \tag{12.21}$$

and $D(\boldsymbol{k}, \omega, \boldsymbol{r}, t) = 0$ is satisfied. Equations (12.19) and (12.20) reduce to

$$\frac{\mathrm{d}\boldsymbol{r}}{\mathrm{d}t} = \frac{\mathrm{d}\boldsymbol{r}}{\mathrm{d}s}\left(\frac{\mathrm{d}t}{\mathrm{d}s}\right)^{-1} = -\frac{\partial D}{\partial \boldsymbol{k}}\left(\frac{\partial D}{\partial \omega}\right)^{-1} = \left(\frac{\partial\omega}{\partial\boldsymbol{k}}\right)_{\boldsymbol{r}, t=\mathrm{const.}} = \boldsymbol{v}_{\mathrm{g}}.$$

Equation (12.19) has the same formula as the equation of motion with Hamiltonian D. When D does not depend on t explicitly, $D = \text{const.} = 0$ corresponds to the energy conservation law. If the plasma medium does not depend on z, k_z =const. corresponds to the momentum conservation law and is the same as the Snell law, $N_\parallel = \text{const.}$

When $\boldsymbol{k} = \boldsymbol{k}_\mathrm{r} + i\boldsymbol{k}_\mathrm{i}$ is a solution of $D = 0$ for a given real ω and $|\boldsymbol{k}_\mathrm{i}| \ll |\boldsymbol{k}_\mathrm{r}|$ is satisfied, we have

$$D(\boldsymbol{k}_\mathrm{r} + i\boldsymbol{k}_\mathrm{i}, \omega) = \mathrm{Re}D(\boldsymbol{k}_\mathrm{r}, \omega) + \frac{\partial \mathrm{Re}D(\boldsymbol{k}_\mathrm{r}, \omega)}{\partial \boldsymbol{k}_\mathrm{r}} \cdot i\boldsymbol{k}_\mathrm{i} + i\mathrm{Im}D(\boldsymbol{k}_\mathrm{r}, \omega) = 0,$$

then

$$\mathrm{Re}D(\boldsymbol{k}_\mathrm{r}, \omega) = 0,$$

$$\boldsymbol{k}_\mathrm{i} \cdot \frac{\partial \mathrm{Re}D(\boldsymbol{k}_\mathrm{r}, \omega)}{\partial \boldsymbol{k}_\mathrm{r}} = -\mathrm{Im}D(\boldsymbol{k}_\mathrm{r}, \omega). \tag{12.22}$$

Then the wave intensity $I(\boldsymbol{r})$ becomes

$$I(\boldsymbol{r}) = I(\boldsymbol{r}_0) \exp\left(-2 \int_{\boldsymbol{r}_0}^{\boldsymbol{r}} \boldsymbol{k}_\mathrm{i} \cdot \mathrm{d}\boldsymbol{r} \right), \tag{12.23}$$

$$\int \boldsymbol{k}_\mathrm{i} \cdot \mathrm{d}\boldsymbol{r} = \int \boldsymbol{k}_\mathrm{i} \cdot \frac{\partial D}{\partial \boldsymbol{k}} \mathrm{d}s = -\int \mathrm{Im}D(\boldsymbol{k}_\mathrm{r}, \omega)\mathrm{d}s = -\int \frac{\mathrm{Im}D(\boldsymbol{k}_\mathrm{r}, \omega)}{|\partial D/\partial \boldsymbol{k}|}\mathrm{d}l, \tag{12.24}$$

where dl is the length along the optical ray. Therefore, the wave absorption can be estimated from (12.23) and (12.25) by tracing many optical rays. The geometrical optical approximation can provide the average wave intensity with a space resolution of, say, two or three times the wavelength.

12.2 Wave Heating in Ion Cyclotron Range of Frequency (ICRF)

The dispersion relation of waves in ICRF is given by (12.64) and is reduced to

$$N_\parallel^2 =$$

$$\frac{N_\perp^2}{2[1-(\omega/\Omega_\mathrm{i})^2]} \left(-\left(1 - \left(\frac{\omega}{\Omega_\mathrm{i}}\right)^2\right) + \frac{2\omega^2}{k_\perp^2 v_\mathrm{A}^2} \pm \left[\left(1 - \left(\frac{\omega}{\Omega_\mathrm{i}}\right)^2\right)^2 + 4\left(\frac{\omega}{\Omega_\mathrm{i}}\right)^2 \left(\frac{\omega}{k_\perp v_\mathrm{A}}\right)^4 \right]^{1/2} \right).$$

The plus sign corresponds to the slow wave (L wave, ion cyclotron wave), and the minus sign corresponds to the fast wave (R wave, extraordinary wave). When $1 - \omega^2/\Omega_\mathrm{i}^2 \ll 2(\omega/k_\perp v_\mathrm{A})^2$, the dispersion relation becomes

$$k_z^2 = 2\left(\frac{\omega^2}{v_A^2}\right)\left(1 - \frac{\omega^2}{\Omega_i^2}\right)^{-1} \qquad \text{(for slow wave)}$$

$$k_z^2 = -\frac{k_\perp^2}{2} + \frac{\omega^2}{2v_A^2}. \qquad \text{(for fast wave)}$$

Since the externally excited waves have propagation vectors with $0 < k_z^2 < (\pi/a)^2$, $k_\perp^2 > (\pi/a)^2$ usually, there are constraints for slow wave

$$\frac{\omega^2}{v_A^2}\frac{2}{(1-\omega^2/\Omega_i^2)} < \left(\frac{\pi}{a}\right)^2,$$

$$n_{20}a^2 < 1.3 \times 10^{-3}\frac{A}{Z^2}\frac{\Omega_i^2}{\omega^2}\left(1 - \frac{\omega^2}{\Omega_i^2}\right),$$

and for fast wave [12.1]

$$\frac{\omega^2}{2v_A^2} > \left(\frac{\pi}{a}\right)^2,$$

$$n_{20}a^2 > 0.5 \times 10^{-2}\frac{A}{Z^2}\frac{\Omega_i^2}{\omega^2},$$

where n_{20} is the ion density in $10^{20}\,\mathrm{m}^{-3}$, a is the plasma radius in meters, and A is the atomic number.

An ion cyclotron wave (slow wave) can be excited by a Stix coil [12.1] and can propagate and heat ions in a low-density plasma. But it cannot be excited in a high-density plasma like that of a tokamak.

The fast wave is an extraordinary wave in this frequency range and can be excited by a loop antenna, which generates a high-frequency electric field perpendicular to the magnetic field (see Section 12.4.2). The fast wave can propagate in a high-density plasma. The fast wave in a plasma with a single ion species has $E_x + iE_y = 0$ at $\omega = |\Omega_i|$ in cold plasma approximation, so that it is not absorbed by the ion cyclotron damping. However, the electric field of the fast wave in a plasma with two ion species is $E_x + iE_y \neq 0$, so that the fast wave can be absorbed; that is, the fast wave can heat the ions in this case.

Let us consider the heating of a plasma with two ion species, M and m, by a fast wave. The masses, charge numbers, and densities of the M ion and m ion are denoted by m_M, Z_M, n_M and m_m, Z_m, n_m, respectively. When we use

$$\eta_M \equiv \frac{Z_M^2 n_M}{n_e}, \qquad \eta_m \equiv \frac{Z_m^2 n_m}{n_e},$$

we have $\eta_M/Z_M + \eta_m/Z_m = 1$ since $n_e = Z_M n_M + Z_m n_m$. Since $(\Pi_e/\omega)^2 \gg 1$ in ICRF wave, the dispersion relation in the cold plasma approximation is given by (10.2) as follows:

$$N_\perp^2 = \frac{(R - N_\parallel^2)(L - N_\parallel^2)}{K_\perp - N_\parallel^2},$$

$$R = -\frac{\Pi_\mathrm{i}^2}{\omega^2}\left(\frac{(m_\mathrm{M}/m_\mathrm{m})\eta_\mathrm{m}\omega}{\omega+|\Omega_\mathrm{m}|} + \frac{\eta_\mathrm{M}\omega}{\omega+|\Omega_\mathrm{M}|} - \frac{\omega}{|\Omega_\mathrm{M}|/Z_\mathrm{M}}\right),$$

$$L = -\frac{\Pi_\mathrm{i}^2}{\omega^2}\left(\frac{(m_\mathrm{M}/m_\mathrm{m})\eta_\mathrm{m}\omega}{\omega-|\Omega_\mathrm{m}|} + \frac{\eta_\mathrm{M}\omega}{\omega-|\Omega_\mathrm{M}|} + \frac{\omega}{|\Omega_\mathrm{M}|/Z_\mathrm{M}}\right),$$

$$K_\perp = -\frac{\Pi_\mathrm{i}^2}{\omega^2}\left(\frac{(m_\mathrm{M}/m_\mathrm{m})\eta_\mathrm{m}\omega^2}{\omega^2-\Omega_\mathrm{m}^2} + \frac{\eta_\mathrm{M}\omega^2}{\omega^2-\Omega_\mathrm{M}^2}\right),$$

$$\Pi_\mathrm{i}^2 \equiv \frac{n_\mathrm{e}e^2}{\epsilon_0 m_\mathrm{M}}.$$

Therefore, ion-ion hybrid resonance occurs at $K_\perp - N_\parallel^2 = 0$; that is,

$$\frac{\eta_\mathrm{m}(m_\mathrm{M}/m_\mathrm{m})\omega^2}{\omega^2-\Omega_\mathrm{m}^2} + \frac{\eta_\mathrm{M}\omega^2}{\omega^2-\Omega_\mathrm{M}^2} \approx -\frac{\omega^2}{\Pi_\mathrm{i}^2}N_\parallel^2 \approx 0,$$

$$\omega^2 \approx \omega_\mathrm{IH} \equiv \frac{\eta_\mathrm{M}+\eta_\mathrm{m}(\mu^2/\mu')}{\eta_\mathrm{M}+\eta_\mathrm{m}/\mu'}\Omega_\mathrm{m}^2,$$

$$\mu' \equiv \frac{m_\mathrm{m}}{m_\mathrm{M}}, \qquad \mu \equiv \frac{\Omega_\mathrm{M}}{\Omega_\mathrm{m}} = \frac{m_\mathrm{m}Z_\mathrm{M}}{m_\mathrm{M}Z_\mathrm{m}}.$$

Figure 12.3 shows the ion-ion hybrid resonance layer; $K_\perp - N_\parallel^2 = 0$, the L cutoff layer; $L - N_\parallel^2 = 0$, and R cutoff layer; $R - N_\parallel^2 = 0$, of a tokamak plasma with the two ion species D^+ (M ion) and H^+ (m ion). Since the K_{zz} component of the dielectric tensor is much larger than the other component, even in a hot plasma, the dispersion relation of a hot plasma is [12.2]

$$\begin{vmatrix} K_{xx}-N_\parallel^2 & K_{xy} \\ -K_{xy} & K_{yy}-N_\parallel^2-N_\perp^2 \end{vmatrix} = 0. \tag{12.25}$$

When we use the relation $K_{yy} \equiv K_{xx} + \Delta K_{yy}$, $|\Delta K_{yy}| \ll |K_{xx}|$,

$$N_\perp^2 = \frac{(K_{xx}-N_\parallel^2)(K_{xx}+\Delta K_{yy}-N_\parallel^2)+K_{xy}^2}{K_{xx}-N_\parallel^2}$$

$$\approx \frac{(K_{xx}+iK_{xy}-N_\parallel^2)(K_{xx}-iK_{xy}-N_\parallel^2)}{K_{xx}-N_\parallel^2}.$$

When ω^2 is near ω_IH^2, K_{xx} is given by (refer to (12.13))

$$K_{xx} = -\frac{\Pi_\mathrm{i}^2}{\omega^2}\left(\frac{m_\mathrm{M}}{2m_\mathrm{m}}\eta_\mathrm{m}\zeta_0 Z(\zeta_1) + \frac{\eta_\mathrm{M}\omega^2}{\omega^2-\Omega_\mathrm{M}^2}\right).$$

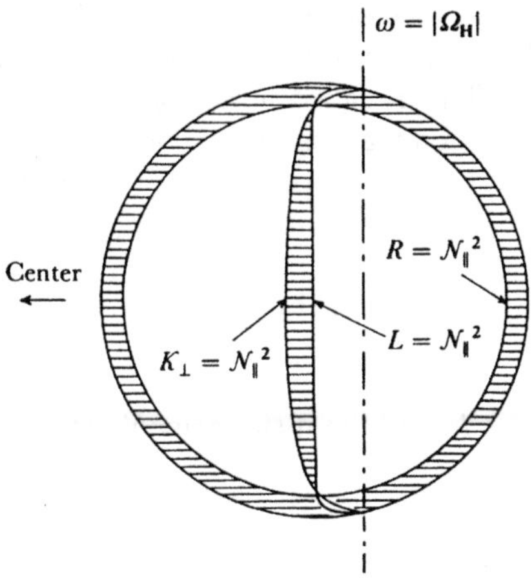

Figure 12.3 L cutoff layer($L = N_\parallel^2$), R cutoff layer ($R = N_\parallel^2$), and the ion-ion hybrid resonance layer ($K_\perp = N_\parallel^2$) of ICRF wave in a tokamak with two ion components D^+, H^+. The shaded area is the region of $N_\perp^2 < 0$.

The resonance condition is $K_{xx} = N_\parallel^2$. The value of $Z(\zeta_1)$ that appears in the dispersion equation is finite and $0 > Z(\zeta_1) > -1.08$. The condition

$$\eta_m \geq \eta_{cr} \equiv \frac{2}{1.08}\frac{m_m}{m_M}2^{1/2}N_\parallel \frac{v_{Ti}}{c}\left(\frac{\eta_M \omega^2}{\omega^2 - \Omega_M^2} + N_\parallel^2 \frac{\omega^2}{\Pi_i^2}\right)$$

is necessary to obtain the resonance condition. This point is different from the cold plasma dispersion equation (note the difference between K_{xx} and $K_\perp$).

It is deduced from the dispersion equation (12.25) that the mode conversion [12.2] from the fast wave to the ion Bernstein wave occurs at the resonance layer when $\eta_m \geq \eta_{cr}$. When the L cutoff layer and the ion-ion hybrid resonance layer are close to each other, as shown in Figure 12.3, the fast wave propagating from the outside torus penetrates the L cutoff layer partly by the tunneling effect and is converted to the ion Bernstein wave. The mode converted wave is absorbed by ion cyclotron damping or electron Landau damping. The theory of mode conversion is described in Chapter 10 of [12.1]. ICRF experiments related to this topic were carried out in TFR.

When $\eta_m < \eta_{cr}$, $K_\perp = N_\parallel^2$ cannot be satisfied and the ion-ion hybrid resonance layer disappears. In this case, a fast wave excited by the loop antenna outside the torus can pass through the R cutoff region (because the width is small) and is reflected by the L cutoff layer and bounced back and forth in the region surrounded by $R = N_\parallel^2$ and $L = N_\parallel^2$. In this region, there is a layer satisfying $\omega = |\Omega_m|$, and the minority m ions are heated by the fundamental ion cyclotron damping. The majority M ions are heated by the Coulomb collisions with m ions. If the mass of M ions is l times as large as the mass of m ions, the M ions are also heated by the lth harmonic ion cyclotron damping. This type of experiment was carried out in PLT with good heating efficiency. This is called *minority heating*. The absorption power P_{e0} due to electron Landau damping per unit volume is given by (12.14), and it is important only in the case $\zeta_0 \le 1$. In this case we have $E_y/E_z \approx K_{zz}/K_{yz} \approx 2\zeta_0^2/(2^{1/2}b^{1/2}\zeta_0(-i))$ and P_{e0} is [12.3]

$$P_{e0} = \frac{\omega\epsilon_0}{4}|E_y|^2\left(\frac{\Pi_e}{\omega}\right)^2\left(\frac{k_\perp v_{Te}}{\Omega_e}\right)^2 2\zeta_{0e}\pi^{1/2}\exp(-\zeta_{0e}^2). \tag{12.26}$$

The absorption power P_{in} by the n-th harmonic ion cyclotron damping is given by (12.15) as follows:

$$P_{in} = \frac{\omega\epsilon_0}{2}|E_x + iE_y|^2\left(\frac{\Pi_i}{\omega}\right)^2\left(\frac{n^2}{2\times n!}\right)\left(\frac{b}{2}\right)^{n-1}$$
$$\times\frac{\omega}{2^{1/2}k_z v_{Ti}}\pi^{1/2}\exp\left(-\frac{(\omega - n|\Omega_i|)^2}{2(k_z v_{Ti})^2}\right). \tag{12.27}$$

The absorption power due to the second harmonic cyclotron damping is proportional to the beta value of the plasma. In order to evaluate the absorption power by (12.26) and (12.27), we need the spatial distributions of E_x and E_y and it is possible to calculate these distributions numerically [12.4].

In the range of the higher harmonic ion cyclotron frequencies ($\omega \sim 2\Omega_i, 3\Omega_i$), the direct excitation of the ion Bernstein wave has been studied by an external antenna or waveguide, which generates a high-frequency electric field parallel to the magnetic field [12.5].

12.3 Lower Hybrid Heating

Since $|\Omega_i| \ll \Pi_i$ in a tokamak plasma ($n_e \ge 10^{13}\text{cm}^{-3}$), the lower hybrid resonance frequency becomes

$$\omega_{LH}^2 = \frac{\Pi_i^2 + \Omega_i^2}{1 + \Pi_e^2/\Omega_e^2 + Zm_e/m_i} \approx \frac{\Pi_i^2}{1 + \Pi_e^2/\Omega_e^2}.$$

There are relations $\Omega_e \gg \omega_{LH} \gg |\Omega_i|$, $\Pi_i^2/\Pi_e^2 = |\Omega_i|/\Omega_e$. For a given frequency ω, lower hybrid resonance $\omega = \omega_{LH}$ occurs at the position where the electron density satisfies the following condition:

$$\frac{\Pi_e^2(x)}{\Omega_e^2} = \frac{\Pi_{res}^2}{\Omega_e^2} \equiv p, \qquad p = \frac{\omega^2}{\Omega_e|\Omega_i| - \omega^2}.$$

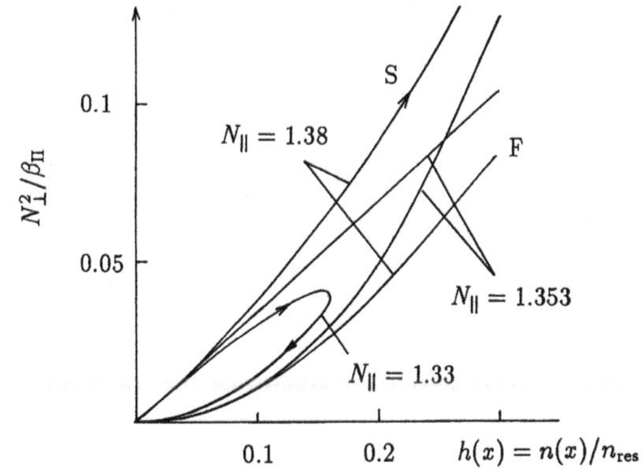

Figure 12.4 Trace of lower hybrid wave in $N_\perp^2 - h(x)\ (= \Pi_{\rm e}^2(x)/\Pi_{\rm res}^2)$ diagram for the case of $p = 0.353$, $N_{\|{\rm cr}}^2 = 1 + p = 1.353$. This corresponds to the case of H^+ plasma in $B = 3\,{\rm T}$, and $f = \omega/2\pi = 10^9$ Hz. The electron density for the parameter $\beta_{\rm II} = 7.06 \times 10^3\ (= \Pi_{\rm res}^2/\omega^2)$ is $n_{\rm res} = 0.31 \times 10^{20}\ {\rm m}^{-3}$.

When the dispersion equation (12.20) of cold plasma is solved about $N_\perp^2$ using $N^2 = N_\|^2 + N_\perp^2$, we have

$$N_\perp^2 = \frac{K_\perp \widetilde{K}_\perp - K_\times^2 + K_\| \widetilde{K}_\perp}{2K_\perp} \pm \left[\left(\frac{K_\perp \widetilde{K}_\perp - K_\times^2 + K_\| \widetilde{K}_\perp}{2K_\perp} \right)^2 + \frac{K_\|}{K_\perp}(K_\times^2 - \widetilde{K}_\perp^2) \right]^{1/2},$$

where $\widetilde{K}_\perp = K_\perp - N_\|^2$. The relations $h(x) \equiv \Pi_{\rm e}^2(x)/\Pi_{\rm res}^2$, $K_\perp = 1 - h(x)$, $K_\times = ph(x)\Omega_{\rm e}/\omega$, $K_\| = 1 - \beta_{\rm II} h(x)$, $\beta_{\rm II} \equiv \Pi_{\rm res}^2/\omega^2 \sim O(m_{\rm i}/m_{\rm e})$, $\alpha \equiv \Pi_{\rm res}^2/(\omega\Omega_{\rm e}) \sim O(m_{\rm i}/m_{\rm e})^{1/2}$, and $\beta_{\rm II} h \gg 1$ reduce this to

$$N_\perp^2(x) = \frac{\beta_{\rm II} h}{2(1-h)} \left(N_\|^2 - (1 - h + ph) \pm \left[(N_\|^2 - (1 - h + ph))^2 - 4(1-h)ph \right]^{1/2} \right). \tag{12.28}$$

The slow wave corresponds to the case of the plus sign in (12.28). In order for the slow wave to propagate from the plasma edge with low density ($h \ll 1$) to the plasma center with high density ($\Pi_{\rm e}^2 = \Pi_{\rm res}^2$, $h = 1$), $N_\perp(x)$ must be real. Therefore, following condition

$$N_\| > (1 - h)^{1/2} + (ph)^{1/2}$$

is necessary. The right-hand side of the inequality has the maximum value $(1 + p)^{1/2}$ in the range $0 < h < 1$, so that the accessibility condition of the resonant region to the lower hybrid wave becomes

$$N_{\parallel}^2 > N_{\parallel,\mathrm{cr}}^2 = 1 + p = 1 + \frac{\Pi_{\mathrm{res}}^2}{\Omega_{\mathrm{e}}^2}. \tag{12.29}$$

If this condition is not satisfied, the externally excited slow wave propagates into the position where the square root term in (12.28) becomes zero and transforms to the fast wave there. Then the fast wave returns to the low-density region (see Figure 12.4). The slow wave that satisfies the accessibility condition can approach the resonance region and $N_{\perp}$ can become large, so that the dispersion relation of hot plasma must be used to examine the behavior of this wave. Near the lower hybrid resonance region, the approximation of the electrostatic wave, (12.131) is applicable. Since $|\Omega_{\mathrm{i}}| \ll \omega \ll \Omega_{\mathrm{e}}$, the terms of ion contribution and electron contribution are given by (12.133) and (12.135), respectively, that is,

$$1 + \frac{\Pi_{\mathrm{e}}^2}{k^2}\frac{m_{\mathrm{e}}}{T_{\mathrm{e}}}(1 + I_0 e^{-b}\zeta_0 Z(\zeta_0)) + \frac{\Pi_{\mathrm{i}}^2}{k^2}\frac{m_{\mathrm{i}}}{T_{\mathrm{i}}}(1 + \zeta Z(\zeta)) = 0,$$

where $\zeta_0 = \omega/(2^{1/2}k_z v_{\mathrm{Te}})$, and $\zeta = \omega/(2^{1/2}k v_{\mathrm{Ti}}) \approx \omega/(2^{1/2}k_{\perp} v_{\mathrm{Ti}})$. Since $I_0 e^{-b} \approx 1 - b + (3/4)b^2$, $\zeta_0 \gg 1$, $\zeta \gg 1$, $1 + \zeta Z(\zeta) \approx -(1/2)\zeta^{-2} - (3/4)\zeta^{-4}$, we have

$$\left(\frac{3\Pi_{\mathrm{i}}^2}{\omega^4}\frac{\kappa T_{\mathrm{i}}}{m_{\mathrm{i}}} + \frac{3}{4}\frac{\Pi_{\mathrm{e}}^2}{\Omega_{\mathrm{e}}^4}\frac{\kappa T_{\mathrm{e}}}{m_{\mathrm{e}}}\right)k_{\perp}^4 - \left(1 + \frac{\Pi_{\mathrm{e}}^2}{\Omega_{\mathrm{e}}^2} - \frac{\Pi_{\mathrm{i}}^2}{\omega^2}\right)k_{\perp}^2 - \left(1 - \frac{\Pi_{\mathrm{e}}^2}{\omega^2}\right)k_z^2 = 0. \tag{12.30}$$

Using the notations $\rho_{\mathrm{i}} = v_{\mathrm{Ti}}/|\Omega_{\mathrm{i}}|$ and

$$s^2 \equiv 3\left(\frac{|\Omega_{\mathrm{e}}\Omega_{\mathrm{i}}|}{\omega^2} + \frac{1}{4}\frac{T_{\mathrm{e}}}{T_{\mathrm{i}}}\frac{\omega^2}{|\Omega_{\mathrm{e}}\Omega_{\mathrm{i}}|}\right) = 3\left(\frac{1+p}{p} + \frac{1}{4}\frac{T_{\mathrm{e}}}{T_{\mathrm{i}}}\frac{p}{1+p}\right),$$

we have

$$\left(\frac{3\Pi_{\mathrm{i}}^2}{\omega^4}\frac{\kappa T_{\mathrm{i}}}{m_{\mathrm{i}}} + \frac{3}{4}\frac{\Pi_{\mathrm{e}}^2}{\Omega_{\mathrm{e}}^4}\frac{\kappa T_{\mathrm{e}}}{m_{\mathrm{e}}}\right) = \frac{\Pi_{\mathrm{i}}^2}{\omega^2}\frac{m_{\mathrm{e}}}{m_{\mathrm{i}}}\frac{v_{\mathrm{Ti}}^2 s^2}{\Omega_{\mathrm{i}}},$$

$$\left(1 + \frac{\Pi_{\mathrm{e}}^2}{\Omega_{\mathrm{e}}^2} - \frac{\Pi_{\mathrm{i}}^2}{\omega^2}\right) = \frac{1}{1+p}\frac{1-h}{h}\frac{\Pi_{\mathrm{i}}^2}{\omega^2}.$$

Then the dimensionless form of (12.30) is

$$(k_{\perp}\rho_{\mathrm{i}})^4 - \frac{1-h}{h}\frac{m_{\mathrm{i}}}{m_{\mathrm{e}}}\frac{1}{(1+p)s^2}(k_{\perp}\rho_{\mathrm{i}})^2 + \left(\frac{m_{\mathrm{i}}}{m_{\mathrm{e}}}\right)^2\frac{1}{s^2}(k_z\rho_{\mathrm{i}})^2 = 0. \tag{12.31}$$

This dispersion equation has two solutions. One corresponds to the slow wave in a cold plasma and the other to the plasma wave in a hot plasma.

The slow wave transforms to the plasma wave at the location where (12.30) or (12.31) has equal roots [12.6]~[12.8]. The condition of zero discriminant is $1/h = 1 + 2k_z\rho_{\rm i}(1+p)s$ and

$$\frac{\Pi_{\rm e}^2(x)}{\Omega_{\rm e}^2} = \frac{\Pi_{\rm M.C.}^2}{\Omega_{\rm e}^2} \equiv \frac{p}{1+2k_z\rho_{\rm i}(1+p)s}.$$

Accordingly, the mode conversion occurs at the position satisfying

$$\frac{\omega^2}{\Pi_{\rm i}^2} = \left(1 - \frac{\omega^2}{|\Omega_{\rm i}|\Omega_{\rm e}}\right) + \frac{N_\parallel v_{\rm Te} 2\sqrt{3}}{c}\left(\frac{T_{\rm i}}{T_{\rm e}} + \frac{1}{4}\left(\frac{\omega^2}{\Omega_{\rm i}\Omega_{\rm e}}\right)^2\right)^{1/2},$$

and the value of $k_\perp^2\rho_{\rm i}^2$ at this position becomes

$$k_\perp^2\rho_{\rm i}^2|_{\rm M.C.} = \frac{m_{\rm i}}{m_{\rm e}}\frac{k_z\rho_{\rm i}}{s}.$$

If the electron temperature is high enough at the plasma center to satisfy $v_{\rm Te} > (1/3)c/N_\parallel$, the wave is absorbed by electrons due to electron Landau damping.

After the mode conversion, the value $N_\perp$ becomes large so that $c/N_\perp$ becomes comparable to the ion thermal velocity ($c/N_\perp \sim v_{\rm Ti}$). Since $\omega \gg |\Omega_{\rm i}|$, the ion motion is not affected by the magnetic field within the time scale of ω^{-1}. Therefore, the wave with phase velocity c/N is absorbed by ions due to ion Landau damping. When ions have velocity $v_{\rm i}$ larger than $c/N_\perp$ ($v_{\rm i} > c/N_\perp$), the ions are accelerated or decelerated at each time satisfying $v_{\rm i}\cos(\Omega_{\rm i}t) \approx c/N_\perp$ and are subjected to stochastic heating.

The wave is excited by the array of waveguides, as shown in Figure 12.5, with an appropriate phase difference to provide the necessary parallel index $N_\parallel = k_z c/\omega = 2\pi c/(\lambda_z\omega)$. In the low-density region at the plasma boundary, the component of the electric field parallel to the magnetic field is larger for the slow wave than for the fast wave. Therefore, the direction of wave-guides is arranged to excite the electric field parallel to the line of magnetic force. The coupling of waves to plasmas is discussed in detail in [12.9] and the experiments of LHH are reviewed in [12.10].

For the current drive by lower hybrid wave, the accessibility condition (12.29) and $c/N_\parallel \gg v_{\rm Te}$ are necessary. If the electron temperature is high and $\kappa T_{\rm e} \sim 10\,{\rm keV}$, then $v_{\rm Te}/c$ is already $\sim 1/7$. Even if $N_\parallel$ is chosen to be small under the accessibility condition, (12.29), the wave is subjected to absorbtion by electron damping in the outer part of the plasma, and it cannot be expected that the wave can propagate into the central part of the plasma.

When the value of $N_\parallel$ is chosen to be $N_\parallel \sim (1/3)(c/v_{\rm Te})$, electron heating can be expected and has been observed experimentally. Under the condition that the mode conversion can occur, ion heating can be expected. However, the experimental results are less clear than those for electron heating.

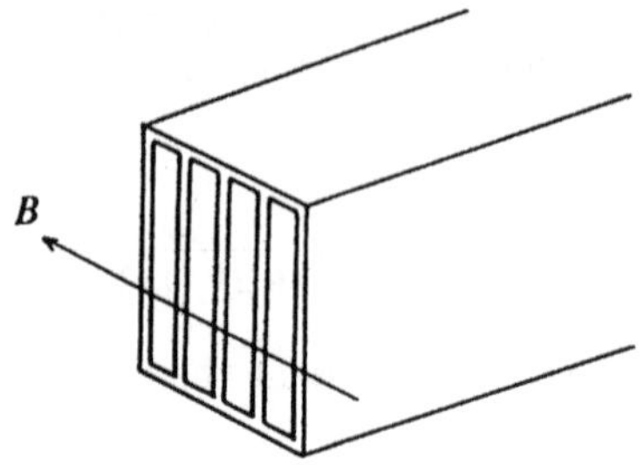

Figure 12.5 Array of waveguides to excite a lower hybrid wave (slow wave).

12.4 Electron Cyclotron Heating

The dispersion relation of waves in the electron cyclotron range of frequency in a cold plasma is given by (11.79). The plus and minus signs in (11.79) correspond to ordinary and extraordinary waves, respectively. The ordinary wave can propagate only when $\omega^2 > \Pi_e^2$ as is clear from (11.86) (in the case of $\theta = \pi/2$). This wave can be excited by an array of waveguides, like that used for lower hybrid waves (Figure 12.5), which emit an electric field parallel to the magnetic field. The phase of each waveguide is selected to provide the appropriate value of the parallel index $N_{\parallel} = k_z c/\omega$ $= 2\pi c/(\omega\lambda_z)$.

The dispersion relation of the extraordinary wave is given by (11.87). When $\theta = \pi/2$, it is given by (11.52). It is necessary to satisfy $\omega_{\rm UH}^2 > \omega^2$ $> \omega_{\rm L}^2, \omega_{\rm LH}^2$. As is seen from the CMA diagram of Figure 12.5, the extraordinary wave can access the plasma center from the high magnetic field side (see Figure 12.6) but cannot access from the low field side because of $\omega = \omega_{\rm R}$ cutoff. The extraordinary wave can be excited by the waveguide, which emits an electric field perpendicular to the magnetic field (see Section 12.2.1). The ion's contribution to the dielectric tensor is negligible. When relations $b \ll 1$, $\zeta_0 \gg 1$ are satisfied for the electron, the dielectric tensor of a hot plasma is

$$K_{xx} = K_{yy} = 1 + X\zeta_0 Z_{-1}/2, \qquad K_{zz} = 1 - X + N_\perp^2 \chi_{zz},$$

$$K_{xy} = -iX\zeta_0 Z_{-1}/2, \qquad K_{xz} = N_\perp \chi_{xz}, \qquad K_{yz} = iN_\perp \chi_{yz},$$

$$\chi_{xz} \approx \chi_{yz} \approx 2^{-1/2} X Y^{-1} \frac{v_{\rm T}}{c} \zeta_0 (1 + \zeta_{-1} Z_{-1}),$$

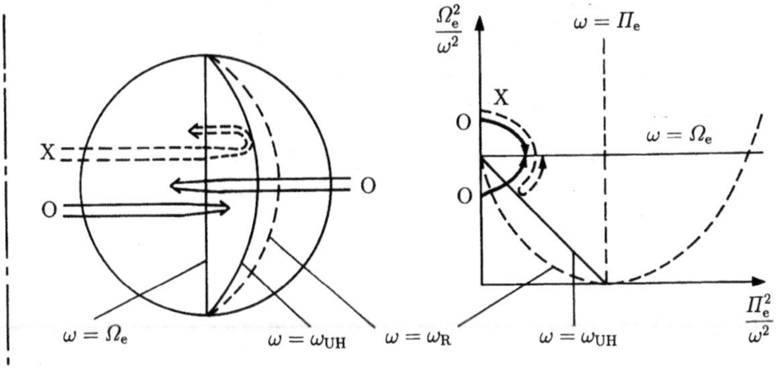

Figure 12.6 The locations of electron cyclotron resonance ($\omega = \Omega_e$), upper hybrid resonance ($\omega = \omega_{LH}$), and R cut off ($\omega = \omega_R$) in the case of $\Omega_{e0} > \Pi_{e0}$ in tokamak configuration, where Ω_{e0} and Π_{e0} are electron cyclotron resonance frequency and plasma frequency at the plasma center, respectively (left figure). The right figure is the CMA diagram near the electron cyclotron frequency region.

$$\chi_{zz} \approx XY^{-2}\left(\frac{v_T}{c}\right)^2 \zeta_0\zeta_{-1}(1+\zeta_{-1}Z_{-1}),$$

$$X \equiv \frac{\Pi_e^2}{\omega^2}, \qquad Y \equiv \frac{\Omega_e}{\omega}, \qquad \zeta_{-1} = \frac{\omega - \Omega_e}{2^{1/2}k_z v_T}, \qquad N_\perp = \frac{k_\perp c}{\omega}.$$

Maxwell equation is

$$(K_{xx} - N_\parallel^2)E_x + K_{xy}E_y + N_\perp(N_\parallel + \chi_{xz})E_z = 0,$$

$$-K_{xy}E_x + (K_{yy} - N_\parallel^2 - N_\perp^2)E_y + iN_\perp\chi_{yz}E_z = 0,$$

$$N_\perp(N_\parallel + \chi_{xz})E_x - iN_\perp\chi_{yz}E_y + (1 - X - N_\perp^2(1-\chi_{zz}))E_z = 0.$$

The solution is

$$\frac{E_x}{E_z} = -\frac{iN_\perp^2\chi_{xz}(N_\parallel + \chi_{xz}) + K_{xy}(1 - X - N_\perp^2(1-\chi_{zz}))}{N_\perp(i\chi_{xz}(K_{xx} - N_\parallel^2) + K_{xy}(N_\parallel + \chi_{xz}))},$$

$$\frac{E_y}{E_z} = -\frac{N_\perp^2(N_\parallel + \chi_{xz})^2 - (K_{xx} - N_\parallel^2)(1 - X - N_\perp^2(1-\chi_{zz}))}{N_\perp(i\chi_{xz}(K_{xx} - N_\parallel^2) + K_{xy}(N_\parallel + \chi_{xz}))}.$$

The absorption power P_{-1} per unit volume is given by (12.15) as follows:

$$P_{-1} = \omega X\zeta_0 \frac{\pi^{1/2}}{2} \exp\left(-\frac{(\omega - \Omega_e)^2}{2k_z^2 v_{Te}^2}\right) \frac{\epsilon_0}{2}|E_x - iE_y|^2.$$

When $\omega = \Omega_e$, then $\zeta_{-1} = 0$, $Z_{-1} = i\pi^{1/2}$, $K_{xx} = 1 + ih$, $K_{xy} = h$, $\chi_{yz} = \chi_{xz} = 2^{1/2}X(v_{Te}/c)\zeta_0 = X/(2N_\parallel)$, $\chi_{zz} = 0$, $h \equiv \pi^{1/2}\zeta_0 X/2$. Therefore, the dielectric tensor $\boldsymbol{K}$ becomes

$$\boldsymbol{K} = \begin{bmatrix} 1+ih & h & N_\perp\chi_{xz} \\ -h & 1+ih & iN_\perp\chi_{xz} \\ N_\perp\chi_{xz} & -iN_\perp\chi_{xz} & 1-X \end{bmatrix}.$$

For the ordinary wave (O wave), we have

$$\frac{E_x - iE_y}{E_z} = \frac{iN_\perp^2(\mathrm{O})N_\parallel(N_\parallel + \chi_{xz}) - i(1-N_\parallel^2)(1 - X - N_\perp^2(\mathrm{O}))}{N_\perp(\mathrm{O})(N_\parallel h + i\chi_{xz}(1-N_\parallel^2))}.$$

When $N_\parallel \ll 1$ and the incident angle is nearly perpendicular, (11.82) gives $1 - X - N_\perp^2(\mathrm{O}) = (1-X)N_\parallel^2$. Since $\chi_{xz} = X/2N_\parallel$, then $\chi_{xz} \gg N_\parallel$. Therefore, the foregoing equation reduces to

$$\frac{E_x - iE_y}{E_z} = \frac{iN_\perp(\mathrm{O})N_\parallel\chi_{xz}}{N_\parallel h + i\chi_{xz}}.$$

For the extraordinary wave (X wave), we have

$$\frac{E_x - iE_y}{E_y} = -\frac{iN_\perp^2(\mathrm{X})N_\parallel(N_\parallel + \chi_{xz}) - i(1-N_\parallel^2)(1 - X - N_\perp^2(\mathrm{X}))}{N_\perp^2(\mathrm{X})(N_\parallel + \chi_{xz})^2 - (K_{xx} - N_\parallel^2)(1 - X - N_\perp^2(\mathrm{X}))}.$$

When $N_\parallel \ll 1$ and $\omega = \Omega_e$, (11.83) gives $1 - X - N_\perp^2(\mathrm{X}) \approx -1 + N_\parallel^2$. Since $\chi_{xz}^2 = (2\pi)^{-1/2}(v_{Te}/cN_\parallel)Xh \ll h$, the foregoing equation reduces to

$$\frac{E_x - iE_y}{E_y} = \frac{-(1 + N_\perp^2(\mathrm{X})N_\parallel(N_\parallel + \chi_{xz}))}{h - i(1 + N_\perp^2(\mathrm{X})(N_\parallel + \chi_{xz})^2)} \sim \frac{-1}{h}.$$

The absorption power per unit volume [12.11] at $\omega = \Omega_e$ for an ordinary wave is

$$P_{-1}(\mathrm{O}) \approx \frac{\omega\epsilon_0}{2}|E_z|^2 \frac{hN_\perp^2(\mathrm{O})N_\parallel^2\chi_{xz}^2}{(N_\parallel h)^2 + \chi_{xz}^2}\exp(-\zeta_{-1}^2)$$

$$\approx \frac{\omega\epsilon_0}{2}|E_z|^2 \frac{1}{(2\pi)^{1/2}}\left(\frac{\Pi_e}{\omega}\right)^2\left(\frac{v_{Te}}{cN_\parallel}\right)\frac{N_\perp^2(\mathrm{O})N_\parallel^2}{N_\parallel^2 + (v_{Te}/c)^2(2/\pi)}, \tag{12.32}$$

and the absorption power per unit volume for an extraordinary wave is

$$P_{-1}(\mathrm{X}) \sim \frac{\omega\epsilon_0}{2}|\mathrm{E_y}|^2\frac{1}{\mathrm{h}} = \frac{\omega\epsilon_0}{2}|\mathrm{E_y}|^2 2\left(\frac{2}{\pi}\right)^{1/2}\left(\frac{\Pi_e}{\omega}\right)^{-2}\left(\frac{\mathrm{N_\parallel v_{Te}}}{\mathrm{c}}\right). \tag{12.33}$$

Since $P(\mathrm{O}) \propto n_e T_e^{1/2}/N_\parallel$, $P(\mathrm{X}) \propto N_\parallel T_e^{1/2}/n_e$, the ordinary wave is absorbed more in the case of higher density and perpendicular incidence, but the extraordinary wave has the opposite tendency.

Heating and current drive by electron cyclotron waves are reviewed in [12.12].

Remark

As long as the plasma current is driven by electromagnetic induction of the current transformer in a tokamak device, the discharge is a necessarily pulsed operation with finite duration. If the plasma current can be driven in a noninductive way, a steady-state tokamak reactor is possible in principle. Current drive by neutral beam injection has been proposed by Ohkawa [12.13] and current drive by traveling wave has been proposed by Wort [12.14]. The momenta of particles injected by NBI or of traveling waves are transferred to the charged particles of the plasma, and the resultant charged particle flow produces the plasma current. Current drive by NBI was demonstrated by DITE, TFTR, etc. Current drive by LHW, proposed by Fisch, was demonstrated by JFT-2, JIPPT-II, WT-2, PLT, Alcator C, Versator 2, T-7, Wega, JT-60 and so on. Current drive by electron cyclotron wave was demonstrated by Cleo, T-10, WT-3, Compass-D, DIII-D, TCV and so on.

12.5 Lower Hybrid Current Drive

The theory of current drive by waves is described here according to Fisch and Karney [12.15]. When a wave is traveling along the line of magnetic force, the velocity distribution function near the phase velocity of the wave is flattened by the diffusion in velocity space. Denote the diffusion coefficient in velocity space by the wave by $D_{\rm rf}$ (refer to Section 11.3); then the Fokker-Planck equation is given by

$$\frac{\partial f}{\partial t} + \boldsymbol{v}\cdot\nabla_{\rm r} f + \left(\frac{\boldsymbol{F}}{m}\right)\cdot\nabla_{\rm v} f = \frac{\partial}{\partial v_z}\left(D_{\rm rf}\frac{\partial f}{\partial v_z}\right) + \left(\frac{\delta f}{\delta t}\right)_{\rm F.P.} \tag{12.34}$$

where $(\delta f/\delta t)_{\rm F.P.}$ is the Fokker-Planck collision term (refer to Section 11.1.3)

$$\left(\frac{\delta f}{\delta t}\right)_{\rm F.P.} = -\sum_{\rm i,e}\left(\frac{1}{v^2}\frac{\partial}{\partial v}(v^2 J_{\rm v}) + \frac{1}{v\sin\theta}\frac{\partial}{\partial\theta}(\sin\theta J_\theta)\right), \tag{12.35}$$

$$J_{\rm v} = -D_\parallel\frac{\partial f}{\partial v} + Af, \quad J_\theta = -D_\perp\frac{1}{v}\frac{\partial f}{\partial\theta}. \tag{12.36}$$

When the velocity v of a test particle is larger than the thermal velocity $v_{\rm T}^*$ of field particles ($v > v_{\rm T}^*$), the diffusion tensor in velocity space $D_\parallel, D_\perp$ and the coefficient of dynamic friction A are reduced to (refer to Section 11.1.3)

$$D_{\|} = \frac{v_{\mathrm{T}}^{*2}\nu_0}{2}\left(\frac{v_{\mathrm{T}}^*}{v}\right)^3, \quad D_{\perp} = \frac{v_{\mathrm{T}}^{*2}\nu_0}{2}\frac{v_{\mathrm{T}}^*}{2v},$$

$$A = -D_{\|}\frac{m}{m^*}\frac{v}{v_{\mathrm{T}}^{*2}},$$

where v_{T}^* and ν_0 are

$$v_{\mathrm{T}}^{*2} = \frac{T^*}{m^*}, \qquad \nu_0 = \left(\frac{qq^*}{\epsilon_0}\right)^2 \frac{n^* \ln \Lambda}{2\pi v_{\mathrm{T}}^{*3} m^2}. = \Pi^{*4}\frac{\ln \Lambda}{2\pi v_{\mathrm{T}}^{*3} n^*},$$

where $\Pi^{*2} \equiv qq^* n^*/(\epsilon_0 m)$. (v, θ, ψ) are spherical coordinates in velocity space. v_{T}^*, q^*, n^* are the thermal velocity, charge, and density of field particles, respectively, and v, q, n are quantities of test particles. Let us consider the electron distribution function in a homogeneous case in space without external force ($\boldsymbol{F} = 0$). Collision terms of electron-electron and electron-ion (charge number $= Z$) are taken into account. When dimensionless quantities $\tau = \nu_{0\mathrm{e}} t$, $u = v/v_{\mathrm{Te}}^*$, $w = v_z/v_{\mathrm{Te}}^*$, $D(w) = D_{\mathrm{rf}}/v_{\mathrm{Te}}^{*2}\nu_{0\mathrm{e}}$ are introduced, the Fokker-Planck equation reduces to

$$\frac{\partial f}{\partial \tau} = \frac{\partial}{\partial w}\left(D(w)\frac{\partial f}{\partial w}\right) + \frac{1}{2u^2}\frac{\partial}{\partial u}\left(\frac{1}{u}\frac{\partial f}{\partial u} + f\right) + \frac{1+Z}{4u^3}\frac{1}{\sin\theta}\frac{\partial}{\partial \theta}\left(\sin\theta\frac{\partial f}{\partial \theta}\right).$$

When Cartesian coordinates in velocity space $(v_x, v_y, v_z) = (v_1, v_2, v_3)$ are used instead of spherical coordinates in velocity space, the Fokker-Planck collision term in Cartesian coordinates is given as follows ($v > v_{\mathrm{T}}^*$ is assumed):

$$A_i = -D_0 v_{\mathrm{T}}^* \frac{m}{m^*}\frac{v_i}{v^3}, \tag{12.37}$$

$$D_{ij} = \frac{D_0}{2}\frac{v_{\mathrm{T}}^*}{v^3}\left((v^2\delta_{ij} - v_i v_j) + \frac{v_{\mathrm{T}}^{*2}}{v^2}(3v_i v_j - v^2\delta_{ij})\right), \tag{12.38}$$

$$J_i = A_i f - \sum_j D_{ij}\frac{\partial f}{\partial v_j}, \tag{12.39}$$

$$D_0 \equiv \frac{(qq^*)^2 n^* \ln \Lambda}{4\pi\epsilon_0^2 m^2 v_{\mathrm{T}}^*} = \frac{v_{\mathrm{T}}^{*2}\nu_0}{2}, \tag{12.40}$$

$$\left(\frac{\delta f}{\delta t}\right)_{\mathrm{F.P.}} = -\nabla_{\mathrm{v}} \cdot \boldsymbol{J}.$$

A_i is the *coefficient of dynamic friction* and D_{ij} is the component of *diffusion tensor*. Let us assume that the distribution function of the perpendicular velocities v_x, v_y to the line of magnetic force is Maxwellian. Then the one-dimensional Fokker-Planck equation on the distribution

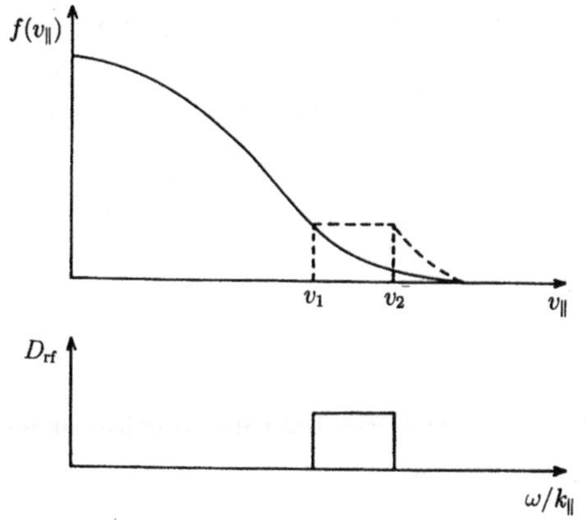

Figure 12.7 Distribution function $f(v_\parallel)$ of electrons is flattened in the region from $v_1 = c/N_1$ to $v_2 = c/N_2$ due to the interaction with the lower hybrid wave whose spectra of parallel index $N_\parallel$ ranges from N_1 to N_2.

function $F(w) = \int f \, dv_x \, dv_y$ of parallel velocity $w = v_z/v_{\rm Te}^*$ can be deduced by $(v_x, \ v_y)$ integration:

$$\iint \left(\frac{\delta f}{\delta t}\right)_{\rm F.P.} dv_x dv_y = \iint (-\nabla_{\rm v} \cdot \boldsymbol{J}) \, dv_x dv_y$$

$$= \iint \frac{\partial}{\partial v_z}\left(-A_z f + \sum_j D_{zj}\frac{\partial f}{\partial v_j}\right) dv_x dv_y.$$

When $|v_z| \gg |v_x|, \ |v_y|$, the approximation $v \approx |v_z|$ can be used. The resultant one-dimensional Fokker-Planck equation on $F(w)$ is

$$\frac{\partial F}{\partial \tau} = \frac{\partial}{\partial w}\left(D(w)\frac{\partial F}{\partial w}\right) + \left(1 + \frac{Z}{2}\right)\frac{\partial}{\partial w}\left(\frac{1}{w^3}\frac{\partial}{\partial w} + \frac{1}{w^2}\right) F(w),$$

and the steady-state solution is

$$F(w) = C \ \exp \int^w \frac{-w dw}{1 + w^3 D(w)/(1 + Z/2)}.$$

$F(w)$ is shown in Figure 12.7 schematically (when $D(w) = 0$, this solution is Maxwellian). $F(w)$ is asymmetric with respect to $w = 0$, so that the current is induced. The induced current density J is

$$J = env_{\rm Te}^* j,$$

where $j = \int wF(w) dw$, and

$$j \approx \frac{w_1 + w_2}{2} F(w_1)(w_2 - w_1). \tag{12.41}$$

On the other hand, this current tends to dissipate by Coulomb collision. Dissipated energy must be supplied by the input energy from the wave in order to sustain the current. Necessary input power $P_{\rm d}$ is

$$P_{\rm d} = -\int \frac{nmv^2}{2}\left(\frac{\delta f}{\delta t}\right)_{\rm F.P.} {\rm d}\boldsymbol{v} = \int \frac{nmv^2}{2}\frac{\partial}{\partial v_z}\left(D_{\rm rf}\frac{\partial f}{\partial v_z}\right){\rm d}\boldsymbol{v}$$

$$= nmv_{\rm Te}^{*2}\nu_0 \int \frac{w^2}{2}\frac{\partial}{\partial w}\left(D(w)\frac{\partial F}{\partial w}\right){\rm d}w = nmv_{\rm Te}^{*2}\nu_0 p_{\rm d},$$

where $p_{\rm d}$ is given by use of the steady-state solution of $F(w)$, under the assumption of $w^3 D(w) \gg 1$, as follows:

$$p_{\rm d} = \left(1+\frac{Z}{2}\right)F(w_1)\ln\left(\frac{w_2}{w_1}\right) \approx \left(1+\frac{Z}{2}\right)F(w_1)\frac{w_2-w_1}{w_1},$$

$$\frac{j}{p_{\rm d}} = \frac{1.5}{1+0.5Z_{\rm i}}\frac{2}{3}w^2. \qquad (12.42)$$

More accurately, this ratio is [12.15]

$$\frac{j}{p_{\rm d}} = \frac{1.12}{1+0.12Z_{\rm i}}1.7w^2. \qquad (12.43)$$

The ratio of the current density J and the necessary input power $P_{\rm d}$ per unit volume to sustain the current is given by

$$\frac{J}{P_{\rm d}} = \frac{env_{\rm Te}^{*}j}{nT_{\rm e}\nu_0 p_{\rm d}} = 0.16\frac{T_{\rm e\,keV}}{n_{19}}\langle w^2\rangle\frac{1.12}{1+0.12Z_{\rm i}}\left(\frac{{\rm A/m^2}}{{\rm W/m^3}}\right), \qquad (12.44)$$

where $T_{\rm e\,keV}$ is the electron temperature in 1 keV units and

$$\frac{I_{\rm CD}}{W_{\rm LH}} = \frac{1}{2\pi R}\frac{\int J 2\pi r{\rm d}r}{\int P_{\rm d}2\pi r{\rm d}r},$$

and the current drive efficiency of LHCD $\eta_{\rm LH}^{\rm T}$ is

$$\eta_{\rm LH}^{\rm T} \equiv \frac{Rn_{19}I_{\rm CD}}{W_{\rm LH}} = \frac{\int \eta_{\rm LH}(r)P_{\rm d}(r)2\pi r{\rm d}r}{\int P_{\rm d}(r)2\pi r{\rm d}r}\left(10^{19}\frac{\rm A}{\rm Wm^2}\right),$$

where $\eta_{\rm LH}(r)$ is local current drive efficiency given by

$$\eta_{\rm LH}(r) = \frac{Rn_{19}J(r)}{2\pi RP_{\rm d}(r)} = 0.026(\kappa T_{\rm e})_{\rm keV}\langle w^2\rangle\frac{1.12}{1+0.12Z_{\rm i}}\left(10^{19}\frac{\rm A}{\rm Wm^2}\right). \qquad (12.45)$$

(R is the major radius in meter.) The square average $\langle w^2\rangle$ of the ratio of the phase velocity (in the direction of the magnetic field) of traveling waves to

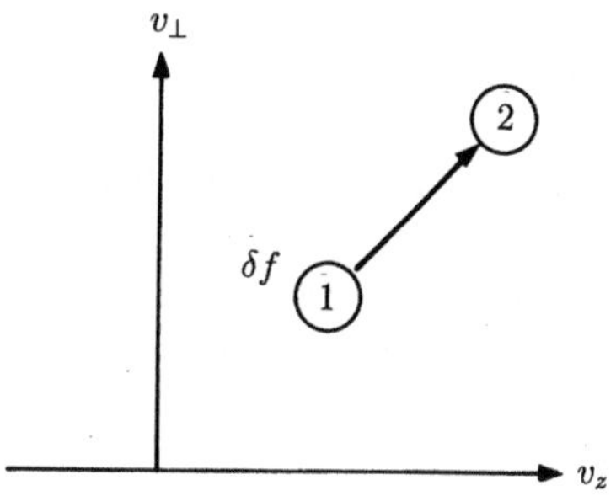

Figure 12.8 The displacement in velocity space of small number, δf, of electrons from coordinates to be subscripted 1 to those to be subscripted 2.

the electron thermal velocity is of the order of $20 \sim 50$. In the experiment (1994) of JT60U, a plasma current of $I_p = 3\,\mathrm{MA}$ was driven by a lower hybrid wave with $W_{\mathrm{LH}} = 4.8\,\mathrm{MW}$ when $n = 1.2 \times 10^{19}\,\mathrm{m}^{-3}$, $\langle T_{\mathrm{e\,keV}} \rangle \sim 2\,\mathrm{keV}$, R=3.5 m and B_t=4 T ($\eta_{\mathrm{LH}} \sim 3$). An empirical scaling of $\eta_{\mathrm{LH}} = 12\langle T_{\mathrm{e\,keV}}\rangle/(5+Z_{\mathrm{eff}})10^{19}\ \mathrm{A/(m^2W)}$ is proposed. These results are consistent with the theoretical results.

There are some constraints on the application of LHCD because of the accessibility condition (refer to Section 12.3). The lower hybrid wave cannot drive the current beyond the threshold density. As the electron temperature of central fusion core plasma is high, the lower hybrid wave is absorbed by Landau damping before accessing the core plasma (Section 12.3). However, the experimental data of current drive efficiency is relatively high compared with the other methods and LHCD plays a role of off-axis current drive of half the outer region of plasma.

12.6 Electron Cyclotron Current Drive

Electron cyclotron current drive (ECCD) relies on the generation of an asymmetric resistivity due to the selective heating of electrons moving in a particular toroidal direction. Fisch et al. [12.16] proposed that the collisionality of plasma is somehow altered so that, for example, electrons moving to the left collide less frequently with ions than do electrons moving to the right. The result would be a net electric current with electrons moving, on average, to the left and ions moving to the right.

Consider the displacement in velocity space of a small number, δf, of electrons from coordinates to be subscripted 1 to those to be subscripted 2 as is shown in Figure 12.8. The energy expended to produce this displacement is given by

$$\Delta E = (E_2 - E_1)\delta f,$$

where E_i is the kinetic energy associated with velocity-space location i. Electrons at different coordinates will lose their momentum parallel to the magnetic field, which is in the z direction, at a rate of ν_1, but now lose it at a rate of ν_2. The z directed current density is then given by

$$j(t) = -e\delta f\big(v_{z2}\exp(-\nu_2 t) - v_{z1}\exp(-\nu_1 t)\big). \tag{12.46}$$

Consider the time-smoothed current J over a time interval δt which is large compared with both $1/\nu_1$ and $1/\nu_2$ so that

$$J = \frac{1}{\Delta t}\int_0^{\Delta t} j(t)\mathrm{d}t = -\frac{e\delta f}{\Delta t}\left(\frac{v_{z2}}{\nu_2} - \frac{v_{z1}}{\nu_1}\right).$$

Therefore, the necessary input power density P_d to induce the current density is

$$P_\mathrm{d} = \frac{\Delta E}{\Delta t} = \frac{E_2 - E_1}{\Delta t}\delta f.$$

The ratio of J/P_d becomes

$$\frac{J}{P_\mathrm{d}} = -e\frac{v_{z2}/\nu_2 - v_{z1}/\nu_1}{E_2 - E_1} \Rightarrow -e\frac{\boldsymbol{s}\cdot\nabla(v_z/\nu)}{\boldsymbol{s}\cdot\nabla E}, \tag{12.47}$$

where $\boldsymbol{s}$ is the unit vector in the direction of the displacement in velocity space. Let us estimate ν of (12.47). The deceleration rate of momentum of a test electron by collision with electrons and ions is expressed by (refer to (2.23) and (2.15))

$$\frac{\mathrm{d}p}{\mathrm{d}t} = -\frac{p}{\tau_{\mathrm{ee}\|}} - \frac{p}{\tau_{\mathrm{ei}\|}} = -\left(1 + \frac{Z_\mathrm{i}}{2}\right)\frac{\nu_0}{u^3}p,$$

where

$$\nu_0 = \left(\frac{e^2 n_\mathrm{e}}{\epsilon_0 m_\mathrm{e}}\right)^2 \frac{\ln\Lambda}{2\pi n_\mathrm{e} v_\mathrm{Te}^3}, \qquad u \equiv \frac{v}{v_\mathrm{Te}}.$$

$v_\mathrm{Te} = (\kappa T_\mathrm{e}/m_\mathrm{e})^{1/2}$ is electron thermal velocity. Therefore, we have

$$\frac{\mathrm{d}p}{\mathrm{d}t} = -\nu_\mathrm{M} p, \qquad \nu_\mathrm{M} \equiv (2 + Z_\mathrm{i})\frac{\nu_0}{2u^3}.$$

In order to estimate $\mathrm{d}u/\mathrm{d}t$, we must use the energy relaxation time $\tau_\mathrm{ee}^\epsilon$ (refer to (2.24))

$$\frac{\mathrm{d}E}{\mathrm{d}t} = -\frac{E}{\tau_\mathrm{ee}^\epsilon}, \qquad E = \frac{m_\mathrm{e}}{2}u^2 v_\mathrm{Te}^2,$$

that is,

$$\frac{du}{dt} = -\frac{u}{2\tau_{ee}^{\epsilon}} = -\frac{\nu_0}{2u^3}u.$$

Each term in (12.46) of $j(t)$ must be modified as follows:

$$j(t) = j_0 \exp(-\int \nu_{\rm M} {\rm d}t) = j_0 \left(\frac{u(t)}{u_0}\right)^{2+Z_{\rm i}}, \tag{12.48}$$

because of

$$-\int \nu_{\rm M}{\rm d}t = -\int \nu_{\rm M}\frac{{\rm d}t}{{\rm d}u}{\rm d}u = (2+Z_{\rm i})\int \frac{{\rm d}u}{u} = (2+Z_{\rm i})\ln\frac{u(t)}{u_0}.$$

Then the integral of $j(t)$ of (12.48) reduces to

$$\int_0^{\infty} j(t){\rm d}t = j_0\int_{u_0}^{0}\left(\frac{u(t)}{u_0}\right)^{2+Z_{\rm i}}\frac{{\rm d}t}{{\rm d}u}{\rm d}u = \frac{j_0}{\nu_0}\frac{2u_0^3}{5+Z_{\rm i}}.$$

Accordingly ν in (12.47) is

$$\nu = \nu_0\frac{5+Z_{\rm i}}{2u^3}, \tag{12.49}$$

and

$$\frac{J}{P_{\rm d}} = \frac{en_{\rm e}v_{\rm Te}}{n_{\rm e}T_{\rm e}\nu_0}\frac{j}{p_{\rm d}}, \qquad \frac{j}{p_{\rm d}} \equiv \frac{4}{5+Z_{\rm i}}\frac{\boldsymbol{s}\cdot\nabla(u^3w)}{\boldsymbol{s}\cdot\nabla u^2},$$

where $w \equiv v_z/v_{\rm Te}$. In the case of ECCD we have $j/p_{\rm d} \approx 6wu/(5+Z_{\rm i})$ and

$$\frac{J}{P_{\rm d}} = \frac{en_{\rm e}v_{\rm Te}}{n_{\rm e}T_{\rm e}\nu_0}\frac{\langle 6wu\rangle}{5+Z_{\rm i}} = 0.096\frac{T_{\rm e\,keV}}{n_{19}}\frac{\langle 6wu\rangle}{5+Z_{\rm i}}. \tag{12.50}$$

The ratio of driven current $I_{\rm CD}$ to ECCD power $W_{\rm EC}$ is

$$\frac{I_{\rm CD}}{W_{\rm EC}} = \frac{1}{2\pi R}\frac{\int J2\pi r{\rm d}r}{\int P_{\rm d}2\pi r{\rm d}r},$$

and the current drive efficiency $\eta_{\rm EC}^{\rm T}$ of ECCD is

$$\eta_{\rm EC}^{\rm T} \equiv \frac{Rn_{19}I_{\rm CD}}{W_{\rm CD}} = \frac{\int \eta_{\rm EC}(r)P_{\rm d}(r)2\pi r{\rm d}r}{\int P_{\rm d}2\pi r{\rm d}r},$$

where $\eta_{\rm EC}(r)$ is the local current drive efficiency given by

$$\eta_{\rm EC}(r) = \frac{Rn_{19}J(r)}{2\pi RP_{\rm d}} = 0.015T_{\rm e\,keV}\frac{\langle 6wu\rangle}{5+Z_{\rm i}}\left(10^{19}\frac{\rm A}{\rm Wm^2}\right). \tag{12.51}$$

Experimental results of current drive efficiency are $\eta_{\rm EC} = 0.4 \sim 0.8 \times 10^{19}$ $\rm Am^{-2}W^{-1}$ for $T_{\rm e\,0} = 7 \sim 20$keV. The range of experimental parameters is still limited at the present time. However, electron cyclotron waves can be launched in vacuum and propagate directly into plasma without interactions near the plasma boundary. Because of the localized absorption propertity of EC wave, ECCD is a very effective tool to control the current profiles.

12.7 Neutral Beam Current Drive

When a fast neutral beam is injected into a plasma, it changes to a fast ion beam by charge exchange or ionization processes. When the fast ions have higher energy than $E_{\rm cr} = m_{\rm b}v_{\rm cr}^2/2$ given by (2.31), they are decelerated mainly by electrons in the plasma and the fast ions with $E < E_{\rm cr}$ are decelerated mainly by ions in the plasma. The distribution function of the ion beam can be obtained by solving the Fokker-Planck equations. The Fokker-Planck collision term (12.35) of the fast ions with $E \gg E_{\rm cr}$ is dominated by the dynamic friction term in (12.36) due to electrons. The dynamic friction term of electrons on the fast ion in the case of $v < v_{\rm T}^*$ is given by [11.1]

$$A = -\frac{v}{2\tau_{\rm be}^{\epsilon}}.$$

Then the Fokker-Planck equation is reduced to

$$\frac{\partial f_{\rm b}}{\partial t} + \frac{\partial}{\partial v}\left(\frac{-vf_{\rm b}}{2\tau_{\rm be}^{\epsilon}}\right) = \phi\delta(v - v_{\rm b}), \qquad (12.52)$$

where $v_{\rm b}$ is the initial injection velocity and $\tau_{\rm be}^{\epsilon}$ is the energy relaxation time of beam ions and electrons as described by (2.32). The right-hand side is the source term of beam ions. The steady-state solution of the Fokker-Planck equation is

$$f_{\rm b} \propto 1/v.$$

However, the dynamic friction term due to ions or the diffusion term dominates the collision term in the region of $v < v_{\rm cr}$. Therefore, the approximate distribution function of the ion beam is given by $f_{\rm b} \propto v^2/(v^3 + v_{\rm cr}^3)$, that is,

$$f_{\rm b}(v) = \frac{n_{\rm b}}{\ln(1 + (v_{\rm b}/v_{\rm cr})^3)^{1/3}} \frac{v^2}{v^3 + v_{\rm cr}^3} \qquad (v \le v_{\rm b}), \qquad (12.53)$$

$$f_{\rm b}(v) = 0 \qquad (v > v_{\rm b}). \qquad (12.54)$$

The necessary ion injection rate ϕ per unit time per unit volume to keep the steady-state condition of the beam is derived by substitution of the solved $f_{\rm b}(v)$ into the Fokker-Planck equation

$$\phi = \frac{n_{\rm b}}{2\tau_{\rm be}^{\epsilon}} \frac{(1 + (v_{\rm cr}/v_{\rm b})^3)^{-1}}{(\ln(1 + (v_{\rm b}/v_{\rm cr})^3))^{1/3}},$$

and necessary power is

$$P_{\rm b} = \frac{m_{\rm b}v_{\rm b}^2}{2}\phi \approx \frac{m_{\rm b}v_{\rm b}^2 n_{\rm b}}{4\,\ln(v_{\rm b}/v_{\rm cr})\tau_{\rm be}^{\epsilon}}. \tag{12.55}$$

The average velocity of the decelerating ion beam is

$$\bar{v}_{\rm b} = v_{\rm b}(ln(v_{\rm b}/v_{\rm cr}))^{-1}. \tag{12.56}$$

Then the current density J driven by the fast ion's beam consists of terms due to fast ions and bulk ions and electrons of the plasma:

$$J = Z_{\rm i}en_{\rm i}\bar{v}_{\rm i} + Z_{\rm b}en_{\rm b}\bar{v}_{\rm b} - en_{\rm e}\bar{v}_{\rm e},$$

$$n_{\rm e} = Z_{\rm i}n_{\rm i} + Z_{\rm b}n_{\rm b},$$

where $\bar{v}_{\rm i}$ and $\bar{v}_{\rm e}$ are the average velocities of ions with density $n_{\rm i}$ and electrons with density $n_{\rm e}$, respectively. The electrons of the plasma receive momentum by collision with fast ions and lose it by collision with plasma ions; that is,

$$m_{\rm e}n_{\rm e}\frac{{\rm d}\bar{v}_{\rm e}}{{\rm d}t} = m_{\rm e}n_{\rm e}(\bar{v}_{\rm b} - \bar{v}_{\rm e})\nu_{\rm eb\|} + m_{\rm e}n_{\rm e}(\bar{v}_{\rm i} - \bar{v}_{\rm e})\nu_{\rm ei\|} = 0,$$

so that

$$(Z_{\rm i}^2 n_{\rm i} + Z_{\rm b}^2 n_{\rm b})\bar{v}_{\rm e} = Z_{\rm b}^2 n_{\rm b}\bar{v}_{\rm b} + Z_{\rm i}^2 n_{\rm i}\bar{v}_{\rm i}.$$

Since $n_{\rm b} \ll n_{\rm i}$, then

$$n_{\rm e}\bar{v}_{\rm e} = \frac{Z_{\rm b}^2}{Z_{\rm i}}n_{\rm b}\bar{v}_{\rm b} + Z_{\rm i}n_{\rm i}\bar{v}_{\rm i},$$

so that [12.13]

$$J = \left(1 - \frac{Z_{\rm b}}{Z_{\rm i}}\right) Z_{\rm b}en_{\rm b}\bar{v}_{\rm b}. \tag{12.57}$$

The driven current density consists of the fast ion beam term $Z_{\rm b}en_{\rm b}\bar{v}_{\rm b}$ and the term of dragged electrons by the fast ion beam, $-Z_{\rm b}^2 en_{\rm b}\bar{v}_{\rm b}/Z_{\rm i}$. Then the ratio of $J/P_{\rm d}$ becomes

$$\frac{J}{P_{\rm d}} = (1 - Z_{\rm b}/Z_{\rm i})\frac{Z_{\rm b}en_{\rm b}\bar{v}_{\rm b}}{m_{\rm b}n_{\rm b}v_{\rm b}\bar{v}_{\rm b}/4\tau_{\rm be}^{\epsilon}} = \frac{2eZ_{\rm b}(2\tau_{\rm be}^{\epsilon})}{m_{\rm b}v_{\rm b}}\left(1 - \frac{Z_{\rm b}}{Z_{\rm i}}\right). \tag{12.58}$$

When the charge number of beam ions is equal to that of the plasma ions, that is, when $Z_{\rm b} = Z_{\rm i}$, the current density becomes zero for linear (cylindrical) plasmas. For toroidal plasmas, the motion of circulating electrons is disturbed by collision with the trapped electrons (banana electrons), and the term of the dragged electrons is reduced. Thus, $J/P_{\rm d}$ becomes [12.17]

$$\frac{J}{P_{\rm d}} = \frac{2eZ_{\rm b}(2\tau_{\rm be}^{\epsilon})}{m_{\rm b}v_{\rm b}}\left(1 - \frac{Z_{\rm b}}{Z_{\rm i}}(1 - G(Z_{\rm eff}, \epsilon))\right), \tag{12.59}$$

$$G(Z_{\rm eff},\epsilon) = \left(1.55 + \frac{0.85}{Z_{\rm eff}}\right)\epsilon^{1/2} - \left(0.2 + \frac{1.55}{Z_{\rm eff}}\right)\epsilon, \qquad (12.60)$$

where ϵ is the inverse aspect ratio. When the effect of pitch angle of ionized beam is taken into account, the factor $\xi \equiv v_{\parallel}/v = R_{\rm tang}/R_{\rm ion}$ must be multiplied to (12.59), where $R_{\rm tang}$ is the minimum value of R along the neutral beam path and $R_{\rm ion}$ is R of the ionization position.

The driving efficiency calculated by the bounce average Fokker-Planck equation [12.17] becomes

$$\frac{J}{P_{\rm d}} = \frac{2eZ_{\rm b}(2\tau_{\rm be}^{\epsilon})}{m_{\rm b}v_{\rm b}}\left(1 - \frac{Z_{\rm b}}{Z_{\rm i}}(1 - G(Z_{\rm eff},\epsilon))\right)\xi_0 F_{\rm nc} x_{\rm b} J_0(x_{\rm b}, y),$$

$$\frac{J}{P_{\rm d}} = \frac{2eZ_{\rm b}(2\tau_{\rm be}^{\epsilon})}{m_{\rm b}v_{\rm cr}}\left(1 - \frac{Z_{\rm b}}{Z_{\rm i}}(1 - G(Z_{\rm eff},\epsilon))\right)\xi_0 F_{\rm nc} J_0(x_{\rm b}, y), \qquad (12.61)$$

where

$$x_{\rm b} \equiv \frac{v_{\rm b}}{v_{\rm cr}}, \qquad y = 0.8\frac{Z_{\rm eff}}{A_{\rm b}},$$

$$J_0(x,y) = \frac{x^2}{x^3 + (1.39 + 0.61y^{0.7})x^2 + (4 + 3y)},$$

and $F_{\rm nc} = 1 - b\epsilon^{\sigma}$ is the correction factor [12.18]. Finally, we have

$$\frac{J}{P_{\rm d}}\left(\frac{\rm Am}{\rm W}\right) = \frac{15.8T_{\rm e\,keV}\xi_0}{Z_{\rm b}n_{\rm e19}}\left(1 - \frac{Z_{\rm b}}{Z_{\rm i}}(1 - G)\right)(1 - b\epsilon^{\sigma})J_0(x_{\rm b}, y). \qquad (12.62)$$

The local current drive efficiency $\eta_{\rm NB}$ of neutral beam current drive (NBCD) is

$$\eta_{\rm NB} \equiv \frac{Rn_{\rm e19}J}{2\pi RP_{\rm d}}\left(10^{19}\frac{\rm A}{\rm Wm^2}\right)$$

$$= 2.52T_{\rm e\,keV}\xi_0\left(1 - \frac{Z_{\rm b}}{Z_{\rm i}}(1 - G)\right)(1 - b\epsilon^{\sigma})J_0(x_{\rm b}, y). \qquad (12.63)$$

When $Z_{\rm b} = 1, Z_{\rm eff} = 1.5, A_{\rm b} = 2, x_{\rm b}^2 = 4$, then $((1 - b\epsilon^{\sigma})J_0) \sim 0.2$. When $\langle\epsilon\rangle \sim 0.15$, then $\eta_{\rm NB} \sim 0.29T_{\rm e\,keV}(10^{19}{\rm A/Wm^2})$. Negative ion-based neutral beam injection with 3.5 MeV beam energy is carried out in JT60-U and experimental data of the current drive efficiency is $\eta_{\rm NB} \sim 0.6 \times 10^{19}{\rm A/Wm^2}$ for $T_{\rm e\,0} \sim 4\,{\rm keV}$. When the beam energy is 1 MeV, it is possible to on-axis current drive by NBCD. A neutral beam can be launched in vacuum. It is advantageous that the launching structure does not have to be in close proximity to the plasma.

When the application of a current drive to the fusion grade plasma with $n_{\rm e} \sim 10^{20}{\rm m^{-3}}$ is considered, the necessary input power for any current drive of full plasma current occupies a considerable amount of the fusion output. Therefore, substantial part of plasma current must be driven by bootstrap current as was described in Section 4.5.3 (also refer to Section 4.7 and Problem 4 in this chapter).

Problems

1. Necessary Conditions for Wave Heating What are necessary conditions to heat the plasma center by wave heating?

2. Group Velocity The phase velocity is $\boldsymbol{v}_{\rm ph} = (\omega(\boldsymbol{k})/k)\hat{\boldsymbol{k}}$, where $\hat{\boldsymbol{k}}$ is the unit vector in the direction of $\boldsymbol{k}$, and group velocity is $\boldsymbol{v}_{\rm g} = (\partial\omega(\boldsymbol{k})/\partial k_x, \partial\omega(\boldsymbol{k})/\partial k_y, \partial\omega(\boldsymbol{k})/\partial k_z)$. Describe $\boldsymbol{v}_{\rm ph}$ and $\boldsymbol{v}_{\rm g}$ of compressional and torsional Alfvén waves (refer to Section 11.4.1) and electron plasma wave (refer to Problem 1 in Chapter 10). Derive $v_{\rm ph}$ and $v_{\rm g}$ of electron plasma wave.

3. Minority Heating in ICRF Experiment of deuterium minority heating in D-T plasma was carried out in JET (JET Team: 17th IAEA Fusion Energy Conf. (Yokahama), 1998, CD1/2). Deutrium concentration is $\eta_{\rm m} = 0.09$ and tritium concentration is $\eta_{\rm M} = 0.91$. Wave frequency is $f = 28\,{\rm MHz}$ and the toroidal field is $B_{\rm t} = 3.7{\rm T}$, so that the frequency is equal to deuteron cyclotron frequency at plasma center. The core electron density is $n_{\rm e} \sim 5 \times 10^{19}{\rm m}^{-3}$ and ion temperature in the center is $T_{\rm i} \sim 7\,{\rm keV}$. The parallel index $N_{\parallel} = c/(\omega/k_{\parallel})$ can be calculated from $k_{\parallel} \sim 8{\rm m}^{-1}$. The upper limit of the minority concentration for minority heating can be estimated from the dispersion relation in Section 12.2. Is the result of this simple analysis consistent with the experimental results of JET?

4. Current Drive Non-inductive operational scenario of ITER was introduced in Section 4.8. An example listed in Table 4.5 is as follows: total plasma current is $I_{\rm p} = 9{\rm MA}$, in which the bootstrap current is $I_{\rm bs} \approx 5{\rm MA}$ and the driven current is $I_{\rm CD} = 4{\rm MA}$. The powers of neutral beam and lower hybrid wave are $P_{\rm NB} = 34{\rm MW}$ and $P_{\rm LH} \approx 30{\rm MW}$, respectively, and $R = 6.35{\rm m}$, $\langle n_{\rm e}\rangle = 0.67 \times 10^{20}{\rm m}^{-3}$, $\langle T_{\rm e}\rangle = 11\,{\rm keV}$. Assuming that the current $I_{\rm NB} = 2{\rm MA}$ in the central region is driven by NBCD and the current $I_{\rm LH} = 2{\rm MA}$ in the outer region is driven by LHCD, estimate the necessary current drive efficiencies of NBCD and LHCD.

13

Plasma Transport by Turbulence

13.1 Fluctuation Loss, Bohm, GyroBohm Diffusion, and Convective Loss

A plasma will be, in many cases, more or less unstable, and fluctuations in the density, temperature, electric field, and magnetic field will induce collective motions of particles and bring about anomalous losses. We will study such losses in this section.

Assume the plasma density $n(\boldsymbol{r}, t)$ consists of the zeroth order term $n_0(\boldsymbol{r}, t)$ and first order perturbation terms $\tilde{n}_k(\boldsymbol{r}, t) = n_k \exp i(\boldsymbol{k} \cdot \boldsymbol{r} - \omega_k t)$ and

$$n = n_0 + \sum_k \tilde{n}_k. \tag{13.1}$$

Since n and n_0 are real, there are the following relations:

$$\tilde{n}_{-k} = (\tilde{n}_k)^*, \qquad n_{-k} = n_k^*, \qquad \omega_{-k} = -\omega_k^*,$$

where $*$ denotes the complex conjugate. ω_k is generally complex and $\omega_k = \omega_{k\mathrm{r}} + i\gamma_k$ and

$$\omega_{-k\mathrm{r}} = -\omega_{k\mathrm{r}}, \qquad \gamma_{-k} = \gamma_k.$$

The plasma is forced to move by perturbation. When the velocity is expressed by

$$\boldsymbol{V}(\boldsymbol{r}, t) = \sum_k \tilde{\boldsymbol{V}}_k = \sum_k \boldsymbol{V}_k \exp i(\boldsymbol{k} \cdot \boldsymbol{r} - \omega_k t), \tag{13.2}$$

then $\boldsymbol{V}_{-k} = \boldsymbol{V}_k^*$ and the equation of continuity

$$\frac{\partial n}{\partial t} + \nabla \cdot (n\boldsymbol{V}) = 0$$

may be written as

$$\frac{\partial n_0}{\partial t} + \sum_k \frac{\partial \tilde{n}_k}{\partial t} + \nabla \cdot \left(\sum_k n_0 \tilde{\boldsymbol{V}}_k + \sum_{k,k'} \tilde{n}_k \tilde{\boldsymbol{V}}_{k'} \right) = 0.$$

When the first- and the second-order terms are separated, then

$$\sum_k \frac{\partial \tilde{n}_k}{\partial t} + \nabla \cdot \sum_k n_0 \tilde{\boldsymbol{V}}_k = 0, \tag{13.3}$$

$$\frac{\partial n_0}{\partial t} + \nabla \cdot \left(\sum_{k,k'} \tilde{n}_k \tilde{\boldsymbol{V}}_{k'} \right) = 0. \tag{13.4}$$

Here we have assumed that the time derivative of n_0 is second order. The time average of the product of (13.3) and $\tilde{n}_{-k}$ becomes

$$\left. \begin{aligned} \gamma_k |n_k|^2 + \nabla n_0 \cdot \mathrm{Re}(n_k \boldsymbol{V}_{-k}) + n_0 \boldsymbol{k} \cdot \mathrm{Im}(n_k \boldsymbol{V}_{-k}) = 0, \\ \omega_{k\mathrm{r}} |n_k|^2 + \nabla n_0 \cdot \mathrm{Im}(n_k \boldsymbol{V}_{-k}) - n_0 \boldsymbol{k} \cdot \mathrm{Re}(n_k \boldsymbol{V}_{-k}) = 0. \end{aligned} \right\} \tag{13.5}$$

If the time average of (13.4) is taken, we find that

$$\frac{\partial n_0}{\partial t} + \nabla \cdot \left(\sum_k \mathrm{Re}(n_k \boldsymbol{V}_{-k}) \exp(2\gamma_k t) \right) = 0. \tag{13.6}$$

The diffusion equation is

$$\frac{\partial n_0}{\partial t} = \nabla \cdot (D \nabla n_0),$$

and the particle flux $\boldsymbol{\Gamma}$ is

$$\boldsymbol{\Gamma} = -D \nabla n_0 = \sum_k \mathrm{Re}(n_k \boldsymbol{V}_{-k}) \exp 2\gamma_k t. \tag{13.7}$$

Equation (13.5) alone is not enough to determine the quantity $\nabla n_0 \cdot \mathrm{Re}(n_k \boldsymbol{V}_{-k}) \exp 2\gamma_k t$. Denote $\beta_k = n_0 \boldsymbol{k} \cdot \mathrm{Im}(n_k \boldsymbol{V}_{-k}) / \nabla n_0 \cdot (\mathrm{Re}(n_k \boldsymbol{V}_{-k}))$; then (13.7) is reduced to

$$D |\nabla n_0|^2 = \sum_k \frac{\gamma_k |n_k|^2 \exp 2\gamma_k t}{1 + \beta_k},$$

and

$$D = \sum_k \gamma_k \frac{|\tilde{n}_k|^2}{|\nabla n_0|^2} \frac{1}{1 + \beta_k}. \tag{13.8}$$

This is the anomalous diffusion coefficient due to fluctuation loss.

Let us consider the case in which the fluctuation $\tilde{\boldsymbol{E}}_k$ of the electric field is electrostatic and can be expressed by a potential $\tilde{\phi}_k$. Then the perturbed electric field is expressed by

$$\tilde{\boldsymbol{E}}_k = -\nabla \tilde{\phi}_k = -i\boldsymbol{k} \cdot \phi_k \exp i(\boldsymbol{k} \cdot \boldsymbol{r} - \omega_k t).$$

The electric field results in an $\tilde{\boldsymbol{E}}_k \times \boldsymbol{B}$ drift, i.e.,

$$\tilde{\boldsymbol{V}}_k = (\tilde{\boldsymbol{E}}_k \times \boldsymbol{B})/B^2 = -i(\boldsymbol{k} \times \boldsymbol{b})\tilde{\phi}_k/B. \qquad (13.9)$$

where $\boldsymbol{b} = \boldsymbol{B}/B$. Equation (13.9) gives the perpendicular component of fluctuating motion. The substitution of (13.9) into (13.3) yields

$$\tilde{n}_k = \nabla n_0 \cdot \left(\frac{\boldsymbol{b} \times \boldsymbol{k}}{B}\right)\frac{\tilde{\phi}_k}{\omega_k}. \qquad (13.10)$$

In general ∇n_0 and $\boldsymbol{b}$ are orthogonal. Take the z axis in the direction of $\boldsymbol{b}$ and the x axis in the direction of $-\nabla n$, i.e., let $\nabla n = -\kappa_n n_0 \hat{\boldsymbol{x}}$, where κ_n is the inverse of the scale of the density gradient and $\hat{\boldsymbol{x}}$ is the unit vector in the x direction. Then Equation (13.10) gives

$$\frac{\tilde{n}_k}{n_0} = \frac{\kappa_n}{B}\frac{k_y}{\omega_k}\tilde{\phi}_k = k_y \kappa_n \frac{T_e}{eB\omega_k}\frac{e\tilde{\phi}_k}{T_e} = \frac{\omega_k^*}{\omega_k}\frac{e\tilde{\phi}_k}{T_e},$$

where k_y is y (poloidal) component of the propagation vector $\boldsymbol{k}$. The quantity

$$\omega_k^* \equiv k_y \kappa_n \frac{T_e}{eB}$$

is the *electron drift frequency*. If the frequency ω_k is real (i.e., if $\gamma_k = 0$), $\tilde{n}_k$ and $\tilde{\phi}_k$ have the same phase, and the fluctuation does not contribute to anomalous diffusion as is clear from (13.8). When $\gamma_k > 0$, so that ω is complex, there is a phase difference between $\tilde{n}_k$ and $\tilde{\phi}_k$ and the fluctuation in the electric field contributes to anomalous diffusion. (When $\gamma_k < 0$, the amplitude of the fluctuation is damped and does not contribute to diffusion.) $\tilde{\boldsymbol{V}}_k$ is expressed by

$$\tilde{\boldsymbol{V}}_k = -i(\boldsymbol{k} \times \boldsymbol{b})\frac{T_e}{eB}\frac{\tilde{\phi}_k}{T_e} = -ik_y \frac{T_e}{eB}\frac{\tilde{n}_k}{n_0}\frac{\omega_{kr} + \gamma_k i}{\omega_k^*}\hat{\boldsymbol{x}}.$$

Then the diffusion particle flux may be obtained from (13.7) as follows:

$$\Gamma = D\kappa_n n_0 = \mathrm{Re}(\tilde{n}_{-k}\tilde{V}_{kx}) = \left(\sum_k \frac{k_y \gamma_k}{\omega_k^*}\left|\frac{\tilde{n}_k}{n_0}\right|^2\right)\frac{T_e}{eB}n_0,$$

and

$$D = \left(\sum_k \frac{k_y \gamma_k}{\kappa_n \omega_k^*}\left|\frac{\tilde{n}_k}{n_0}\right|^2\right)\frac{T_e}{eB} = \sum_k \left|\frac{\tilde{n}_k}{n_0}\right|^2 \frac{\gamma_k}{\kappa_n^2}. \qquad (13.11)$$

The anomalous diffusion coefficient due to fluctuation loss increases with time and eventually the term with the maximum growth rate $\gamma_k > 0$ becomes dominant. However, the amplitude $|\tilde{n}_k|$ will saturate due to nonlinear effects; the saturated amplitude will be of the order of

$$|\tilde{n}_k| \approx |\nabla n_0| \Delta x \approx \frac{\kappa_n}{k_x} n_0.$$

Δx is the correlation length of the fluctuation and the inverse of the propagation constant k_x in the x (radial) direction. Then Equation (13.11) yields

$$D = \frac{\gamma_k}{\kappa_n^2} \left| \frac{\tilde{n}_k}{n_0} \right|^2 \approx \frac{\gamma_k}{k_x^2}. \tag{13.12}$$

When the nondimensional coefficient inside the parentheses in (13.11) is assumed to be at its maximum of 1/16, we have the *Bohm diffusion coefficient*

$$D_{\mathrm{B}} = \frac{1}{16} \frac{T_{\mathrm{e}}}{eB}. \tag{13.13}$$

It appears that (13.13) gives the largest possible diffusion coefficient. When the density and potential fluctuations $\tilde{n}_k$, $\tilde{\phi}_k$ are measured, $\tilde{\boldsymbol{V}}_k$ can be calculated by (13.9), and the estimated outward particle flux Γ by (13.7) and diffusion coefficient D can be compared to the values obtained by experiment. As the relation of $\tilde{n}_k$ and $\tilde{\phi}_k$ is given by (13.10), the phase difference will indicate whether ω_k is real (oscillatory mode) or $\gamma_k > 0$ (growing mode), so that this equation is very useful in interpreting experimental results.

Let us take an example of the fluctuation driven by ion temperature gradient drift instability (refer to Section 9.8). The mode is described by

$$\phi(r, \theta, z) = \sum \phi_{mn}(r) \exp(-im\theta + inz/R).$$

The growth rate of the fluctuation has the maximum at around $k_\theta = (-i/r)\partial/\partial\theta = -m/r$ of [13.1]

$$|k_\theta| = \frac{m}{r} \sim \frac{\alpha_\theta}{\rho_{\mathrm{i}}}, \qquad \alpha_\theta = 0.7 \sim 0.8.$$

Then the correlation length Δ_θ in θ direction is $\Delta_\theta \sim \rho_{\mathrm{i}}/\alpha_\theta$ (ρ_{i} is ion Larmor radius).

The propagation constant $k_\parallel$ along the line of magnetic force near the rational surface $q(r_m) = m/n$ is

$$k_\parallel = -i\boldsymbol{b} \cdot \nabla = \frac{B_\theta}{B} \left(\frac{-m}{r} \right) + \frac{B_{\mathrm{t}}}{B} \left(\frac{n}{R} \right) \approx \frac{1}{R} \left(n - \frac{m}{q(r)} \right)$$

$$= \frac{m}{rR} \frac{rq'}{q^2} (r - r_m) = \frac{s}{Rq} k_\theta (r - r_m)$$

where $q(r) \equiv (r/R)(B_{\mathrm{t}}/B_\theta)$ is the safety factor (B_θ and B_{t} are poloidal and toroidal fields, respectively) and s is the shear parameter (refer to Section

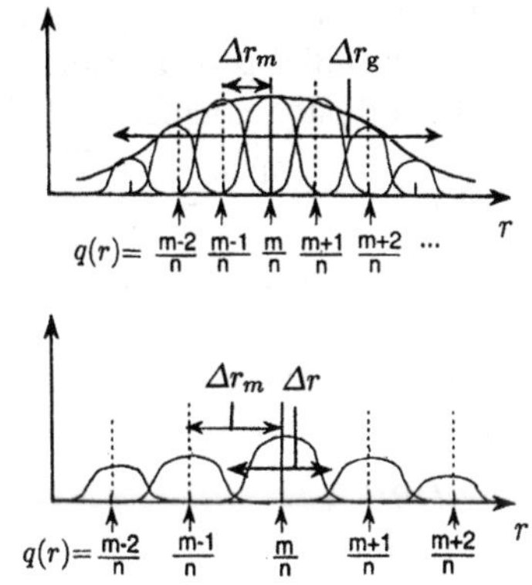

Figure 13.1 In the upper figure, the radial width of eigenmode Δr is larger than the radial separation of the rational surfaces Δr_m. A semi-global eigenmode structure $\Delta r_{\rm g}$ takes place due to the mode couplings. In the lower figure, the radial width of eigenmode Δr is smaller than the radial separation of the rational surfaces Δr_m. The modes with the radial width Δr are independent of each other.

9.5.2) $s \equiv rq'/q$. $|k_\parallel|$ is larger than the inverse of the connection length qR of torus and is less than the inverse of, say, the pressure gradient scale $L_{\rm p}$, that is

$$\frac{1}{qR} < |k_\parallel| < \frac{1}{L_{\rm p}}.$$

The radial width $\Delta r = |r - r_m|$ of the mode near the rational surface $r = r_m$ is roughly expected to be $\Delta r = |r - r_m| = (Rq/s)(k_\parallel/k_\theta) = (\rho_{\rm i}/s\alpha_\theta)$ $\sim O(\rho_{\rm i}/s)$. The more accurate radial width of the eigenmode of ion temperature gradient driven drift turbulence is given by [13.2, 13.3]

$$\Delta r = \rho_{\rm i}\left(\frac{qR}{sL_{\rm p}}\right)^{1/2}\left(\frac{\gamma_k}{\omega_{kr}}\right)^{1/2}.$$

The radial separation length Δr_m of the adjacent rational surface r_m and r_{m+1} is

$$q'\Delta r_m = q(r_{m+1}) - q(r_m) = \frac{m+1}{n} - \frac{m}{n} = \frac{1}{n}, \qquad \Delta r_m = \frac{1}{nq'} = \frac{m/n}{rq'}\frac{r}{m} \sim \frac{1}{sk_\theta}.$$

When the mode width Δr is larger than the radial separation of the rational surface Δr_m, the different modes are overlapped and the toroidal mode coupling takes place (see Figure 13.1). The half width $\Delta r_{\rm g}$ of the envelope of coupled modes is estimated to be [13.4, 13.5]

$$\Delta r_{\rm g} = \left(\frac{\rho_{\rm i} L_{\rm p}}{s}\right)^{1/2}.$$

The radial correlation length becomes the large value of $\Delta r_{\rm g}$ ($\Delta r_{\rm g}/\Delta r \sim (L_{\rm p}/\rho_{\rm i})^{1/2}$) and the radial propagation constant becomes $k_r \sim 1/\Delta r_{\rm g}$. In this case, the diffusion coefficient D is

$$D = (\Delta r_{\rm g})^2 \gamma_k \sim \frac{\rho_{\rm i} L_{\rm p}}{s} \omega_k^* \sim \frac{T}{eB}\frac{\alpha_\theta}{s}.$$

where ω_k^* is the drift frequency (Section 9.8, Section 12.6). This coefficient is of the Bohm type.

When the mode width Δr is less than Δr_m (weak shear case), there is no coupling between different modes and the radial correlation length is

$$\Delta r = \rho_{\rm i}\left(\frac{qR}{sL_{\rm p}}\right)^{1/2}.$$

The diffusion coefficient D in this case is

$$D \sim (\Delta r)^2 \omega_k^* \sim \rho_{\rm i}^2 \left(\frac{qR}{sL_{\rm p}}\right)\left(\frac{k_\theta T}{eBL_{\rm p}}\right) \sim \frac{T}{eB}\frac{\rho_{\rm i}}{L_{\rm p}}\left(\frac{\alpha_\theta qR}{sL_{\rm p}}\right) \propto \frac{T}{eB}\frac{\rho_{\rm i}}{L_{\rm p}}. \tag{13.14}$$

This is called *gyro-Bohm type diffusion coefficient.* It may be expected that the transport in toroidal systems becomes small in the weak shear region of negative shear configuration near the minimum q position (refer to Section 4.6).

Next, let us consider *stationary convective losses* across the magnetic flux. Even if fluctuations in the density and electric field are not observed at a fixed position, it is possible that the plasma can move across the magnetic field and continuously escape. When a stationary electric field exists and the equipotential surfaces do not coincide with the magnetic surfaces $\phi =$ const., the $\boldsymbol{E} \times \boldsymbol{B}$ drift is normal to the electric field $\boldsymbol{E}$, which itself is normal to the equipotential surface. Consequently, the plasma drifts along the equipotential surfaces (see Figure 13.2) which cross the magnetic surfaces. The resultant loss is called *stationary convective loss.* The particle flux is given by

$$\Gamma_k = n_0 \frac{E_y}{B}. \tag{13.15}$$

The losses due to diffusion by binary collision are proportional to B^{-2}, but fluctuation or convective losses are proportional to B^{-1}. Even if the magnetic field is increased, the loss due to fluctuations does not decrease rapidly.

13.2 Loss by Magnetic Fluctuation

When the magnetic field in a plasma fluctuates, the lines of magnetic force will wander radially. Denote the radial shift of the field line by Δr and the

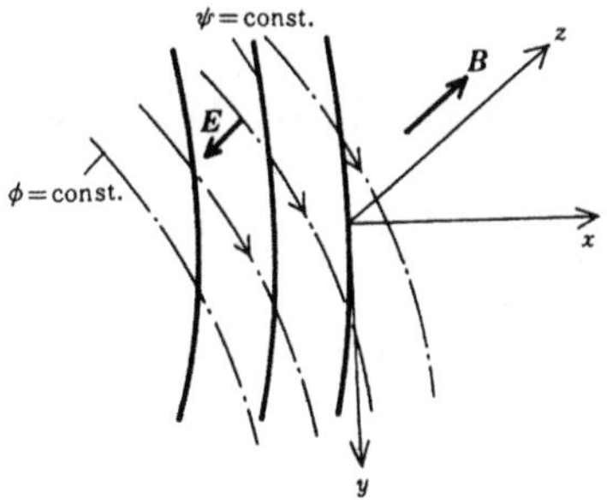

Figure 13.2 Magnetic surface $\psi = \text{const.}$ and electric-field equipotential $\phi = \text{const.}$ The plasma moves along the equipotential surfaces by virtue of $\boldsymbol{E} \times \boldsymbol{B}$.

radial component of magnetic fluctuation $\delta \boldsymbol{B}$ by δB_r, respectively. Then we find

$$\Delta r = \int_0^L b_r \mathrm{d}l,$$

where $b_r = \delta B_r / B$ and l is the length along the line of magnetic force. The ensemble average of $(\Delta r)^2$ is given by

$$\langle (\Delta r)^2 \rangle = \left\langle \int_0^L b_r \, \mathrm{d}l \int_0^L b_r \, \mathrm{d}l' \right\rangle = \left\langle \int_0^L \mathrm{d}l \int_0^L \mathrm{d}l' \, b_r(l) \, b_r(l') \right\rangle$$
$$= \left\langle \int_0^L \mathrm{d}l \int_{-l}^{L-l} \mathrm{d}s \, b_r(l) \, b_r(l+s) \right\rangle \approx L \langle b_r^2 \rangle \, l_{\mathrm{corr}},$$

where l_{corr} is

$$l_{\mathrm{corr}} = \frac{\left\langle \int_{-\infty}^{\infty} b_r(l) \, b_r(l+s) \, \mathrm{d}s \right\rangle}{\langle b_r^2 \rangle}.$$

If electrons run along the lines of magnetic force with the velocity v_{Te}, the diffusion coefficient D_{e} of electrons becomes [13.6]

$$D_{\mathrm{e}} = \frac{\langle (\Delta r)^2 \rangle}{\Delta t} = \frac{L}{\Delta t} \langle b_r^2 \rangle l_{\mathrm{corr}} = v_{\mathrm{Te}} l_{\mathrm{corr}} \left\langle \left(\frac{\delta B_r}{B} \right)^2 \right\rangle. \qquad (13.16)$$

We may take $l_{\mathrm{corr}} \sim R$ in the case of tokamak and $l_{\mathrm{corr}} \sim a$ in the case of reverse field pinch (RFP) (refer to Section 5.2.2).

13.3 Dimensional Analysis of Transport

The determination of scaling law between the overall energy confinement time $\tau_{\rm E}$ and the parameters of apparatus such as

$$\tau_{\rm E} = f(n, T, B, a, q, a/R)$$

is one of the main objectives of large experimental devices and the scaling law is usally in the form of a power law

$$\tau_{\rm E} = n^p T^q B^r a^s.$$

Dimensional analysis of energy confinement time is discussed here according to Connor and Taylor [13.7].

A: Collisionless Vlasov equation in electrostatic limit (collisionless, low beta)
We first consider the collisionless Vlasov model, in which plasma distribution function for each species is described by

$$\frac{\partial f_{\rm i}}{\partial t} + (\boldsymbol{v}\cdot\nabla)f_{\rm i} + \frac{e_{\rm i}}{m_{\rm i}}(\boldsymbol{E}+\boldsymbol{v}\times\boldsymbol{B})\cdot\frac{\partial f_{\rm i}}{\partial \boldsymbol{v}} = 0, \tag{13.17}$$

$$\sum e_{\rm i}\int f_{\rm i}(\boldsymbol{x},\boldsymbol{v}){\rm d}\boldsymbol{v} = 0. \tag{13.18}$$

The energy loss per unit area and unit time is given by

$$Q = \sum\int \boldsymbol{v}\frac{m_{\rm i}v^2}{2}f_{\rm i}{\rm d}\boldsymbol{v} = Q(n, T, B, a).$$

We now seek all the linear transformations of the independent and dependent variables

$$f\rightarrow\alpha f,\quad v\rightarrow\beta v,\quad x\rightarrow\gamma x,\quad B\rightarrow\delta B,\quad t\rightarrow\epsilon t,\quad E\rightarrow\eta E,$$

which leave the basic equations (13.17) and (13.18) invariant. There are three such transformations:

$$\mathrm{A}_1:\quad f\rightarrow\alpha f,$$

$$\mathrm{A}_2:\quad v\rightarrow\beta v,\quad B\rightarrow\beta B,\quad t\rightarrow\beta^{-1}t,\quad E\rightarrow\beta^2 E,$$

$$\mathrm{A}_3:\quad x\rightarrow\gamma x,\quad B\rightarrow\gamma^{-1}B,\quad t\rightarrow\gamma t,\quad E\rightarrow\gamma^{-1}E.$$

Under these combined transformations, the heat flux transformation as $Q\rightarrow\alpha\beta^6 Q$, temperature as $T\rightarrow\beta^2 T$, density as $n\rightarrow\alpha\beta^3 n$. Consequently, if the heat flux is as

$$Q = \sum c_{pqrs}n^p T^q B^r a^s,$$

the requirement that it remains invariant under the transformations $A_1 - A_3$ imposes the following restritions on the exponents:

$$p = 1, \qquad 3p + 2q + r = 6, \qquad s - r = 0,$$

so that the general expression for Q is restricted to

$$Q = \sum c_q n a^3 B^3 \left(\frac{T}{a^2 B^2}\right)^q = n a^3 B^3 F\left(\frac{T}{a^2 B^2}\right),$$

where F is some unknown function. The corresponding energy confinement time is proportional to nTa/Q and so is restricted to the form

$$B\tau_{\mathrm{E}} = F\left(\frac{T}{a^2 B^2}\right).$$

This scaling law is the exact consequences of the model as long as the boundary conditions do not introduce any dominant additional physical effects. Consequently, if the scaling is assumed to follow a power law, then it must be

$$B\tau_{\mathrm{E}} = \left(\frac{T}{a^2 B^2}\right)^q.$$

Further, we made the stronger assumption that a local transport coefficient exists. In such a case, the confinement time would be proportional to a^2. Then exponet q must be equal to -1; that is,

$$\tau_{\mathrm{E}} \propto a^2 \frac{B}{T}.$$

This represents the ubiqitous Bohm diffusion coefficient.

B: Collisional Vlasov equation in the electrostatic limit (collisioal, low beta) The basic equations are

$$\frac{\partial f_{\mathrm{i}}}{\partial t} + (\boldsymbol{v} \cdot \nabla) f_{\mathrm{i}} + \frac{e_{\mathrm{i}}}{m_{\mathrm{i}}} (\boldsymbol{E} + \boldsymbol{v} \times \boldsymbol{B}) \cdot \frac{\partial f_{\mathrm{i}}}{\partial \boldsymbol{v}} = C(f, f), \tag{13.19}$$

where $C(f, f)$ is the Coulomb collisional term together with charge neutrality and electrostatic approximation (13.18). There are two transformations which leave the basic equations (13.19) and (13.18) invariant:

$$\mathrm{B}_1: \quad f \to \beta f, \quad v \to \beta v, \quad B \to \beta B, \quad t \to \beta^{-1} t, \quad E \to \beta^2 E,$$

$$\mathrm{B}_2: \quad f \to \gamma^{-1} f, \quad x \to \gamma x, \quad B \to \gamma^{-1} B, \quad t \to \gamma t, \quad E \to \gamma^{-1} E.$$

Then we have $Q \to \beta^7 \gamma^{-1} Q, \ T \to \beta^2 T, \ n \to \beta^4 \gamma^{-1} n$. The constraint on exponents is

$$4p + 2q + r = 7, \qquad s - r - p = -1,$$

so that

$$Q = na^3B^3F\left(\frac{n}{B^4a^3}, \frac{T}{a^2B^2}\right), \qquad B\tau_{\mathrm{E}} = F\left(\frac{n}{B^4a^3}, \frac{T}{a^2B^2}\right),$$

$$B\tau_{\mathrm{E}} = \left(\frac{n}{B^4a^3}\right)^p \left(\frac{T}{a^2B^2}\right)^q .$$

C: Collisionless Vlasov equation at high beta (collisionless, high beta)
The basic equations are (13.17) and Maxwell equation:

$$\nabla \times \boldsymbol{E} = -\frac{\partial B}{\partial t}, \quad \nabla \times \boldsymbol{B} = \mu_0 \boldsymbol{j}, \quad \boldsymbol{j} = \sum e_i \boldsymbol{v}_i f_i \mathrm{d}\boldsymbol{v}. \tag{13.20}$$

There are two transformations:

$$\mathrm{C}_1: \ f \to \beta^{-3} f, \ v \to \beta v, \ B \to \beta B, \ t \to \beta^{-1} t, \ E \to \beta^2 E, \ j \to \beta j,$$

$$\mathrm{C}_2: \ f \to \gamma^{-2} f, \ x \to \gamma x, \ B \to \gamma^{-1} B, \ t \to \gamma t, \ E \to \gamma^{-1} E. \ j \to \gamma^{-2} j.$$

The constraint on the exponents is

$$2p + r - s = 2, \qquad r + 2q = 3,$$

and the energy confinement time is

$$B\tau_{\mathrm{E}} = F\left(na^2, \frac{T}{a^2B^2}\right) = F(N, \beta), \quad N \equiv na^2.$$

D: Collisional Vlasov equation, high beta (collisional, high beta)
The basic equations are Vlasov equation including the collisional term (13.19) and Maxwell equation (13.20) and charge neutrality (if the Debye length is neglegible). The transformation is

$$\mathrm{D}_1: \ f \to \beta^5 f, \ v \to \beta v, \ x \to \beta^{-4} x, \ B \to \beta^5 B, \ t \to \beta^{-5} t, \ E \to \beta^6 E, \ j \to \beta^9 j,$$

and the constraint on the exponents is

$$2p + \frac{q}{2} + \frac{5r}{4} - s = \frac{11}{4}.$$

Then the energy confinement time is

$$B\tau_{\mathrm{E}} = F\left(na^2, Ta^{1/2}, Ba^{5/4}\right).$$

E: MHD fluid models
MHD equations are

$$\frac{\partial \rho_{\mathrm{m}}}{\partial t} + \nabla \cdot (\rho_{\mathrm{m}}) \boldsymbol{v} = 0,$$

$$\rho_{\rm m}\left(\frac{\partial}{\partial t}+(\boldsymbol{v}\cdot\nabla)\right)\boldsymbol{v}+\nabla p-\boldsymbol{j}\times\boldsymbol{B}=0,$$

$$\frac{\gamma\rho_{\rm m}}{\gamma-1}\left(\frac{\partial}{\partial t}+(\boldsymbol{v}\cdot\nabla)\right)(p\rho_{\rm m}^{\gamma})=\eta j^2$$

$$\boldsymbol{E}+\boldsymbol{v}\times\boldsymbol{B}=\eta\boldsymbol{j},$$

and Maxwell equation (13.20). In the case of ideal MHD fluid, there are three transformations:

$$\mathrm{E}_1:\ \ n\to\alpha n,\ \ B\to\alpha^{1/2}B,\ \ E\to\alpha^{1/2}E,\ \ p\to\alpha p,\ \ j\to\alpha^{1/2}j,$$

$$\mathrm{E}_2:\ \ v\to\beta v,\ \ t=\beta^{-1}t,\ \ B\to\beta B,\ \ E\to\beta^2 E,\ \ p=\beta^2 p,\ \ j\to\beta j,\ \ T\to\beta^2 T,$$

$$\mathrm{E}_3:\ \ x\to\gamma x,\ \ t\to\gamma t,\ \ j\to\gamma^{-1}j.$$

The constraint on the exponents is

$$p+r/2=1,\qquad 2q+r=3,\quad s=0,$$

and the energy confinement time in ideal MHD case is

$$B\tau_{\rm E}=(na^2)^{1/2}F\left(\frac{nT}{B^2}\right)=N^{1/2}F(\beta).$$

F: Resistive MHD fluid model
In the case of resistive MHD fluid, there are two transformations:

$$\mathrm{F}_1:\ \ n\to\alpha n,\ \ B\to\alpha^{1/2}B,\ \ E\to\alpha^{1/2}E,\ \ p\to\alpha p,\ \ j\to\alpha^{1/2}j,$$

$$\mathrm{F}_2:\ \ v\to\beta v,\ \ x=\beta^{-4}x,\ \ t=\beta^{-5}t,\ \ B\to\beta B,\ \ E\to\beta^2E,\ \ p=\beta^2p,$$
$$j\to\beta^5 j,\ \ T\to\beta^2T,$$

and the constraint on the exponents is

$$p+r/2=1,\qquad 2q+r-4s=3.$$

The energy confinement time in the case of resistive MHD is

$$\frac{B\tau_{\rm E}}{n^{1/2}a}=F\left(\frac{n}{B^2a^{1/2}},Ta^{1/2}\right)=F_1(\beta,Ta^{1/2})=F_2\left(\beta,\frac{\tau_{\rm A}}{\tau_{\rm R}}\right).$$

It is useful to discuss dimensional analysis of scaling law from Kadomtsev's viewpoint [13.8]. From the variables (n,T,B,a), one can construct four independent dimensionless parameters. If we select the set of

$$(n,T,B,a)\to\left(\beta,\frac{\rho_{\rm i}}{a},\frac{\nu_{\rm ei}}{\Omega_{\rm e}},\frac{\lambda_{\rm D}}{a}\right),$$

where $\rho_{\rm i}$ and $\lambda_{\rm D}$ are ion Larmor radius and Debye length, respectively, then the confinement time can be written by

$$\Omega_{\rm e}\tau_{\rm E} = F\left(\beta, \frac{\rho_{\rm i}}{a}, \frac{\nu_{\rm ei}}{\Omega_{\rm e}}, \frac{\lambda_{\rm D}}{a}\right)$$

When the Debye length is negligible or charge neutrality can be assumed, we can drop $(\lambda_{\rm D}/a)$ and

$$\Omega_{\rm e}\tau_{\rm E} = F\left(\beta, \frac{\rho_{\rm i}}{a}, \frac{\nu_{\rm ei}}{\Omega_{\rm e}}\right).$$

In the case of MHD fluid models, which do not refer to particle aspects, we have only the dimensionless parameters β and $\tau_{\rm A}/\tau_{\rm R}$, where $\tau_{\rm A} = (an^{1/2}/B)(2\mu_0/m_{\rm i})^{1/2}$ and $\tau_{\rm R} = \mu_0 a^2/\eta \propto a^2 T^{3/2}$. Thus, we can write

$$\frac{\tau_{\rm E}}{\tau_{\rm A}} = F\left(\beta, \frac{\tau_{\rm A}}{\tau_{\rm R}}\right).$$

This confinement scaling is equivalent to ones derived in ideal (**E**) and resistive (**F**) MHD fluids models.

There are reviews on dimensional analysis of energy confinement time [13.9].

13.4 Computer Simulation by Gyrokinetic Particle and Full Orbit Particle Models

The use of computers as an aid to understanding the complex nonlinear behavior of turbulence, such as mode-mode coupling, nonlinear saturation, and self-organization, has continued to grow and has become a dominant factor in the theoretical component of plasma research. The evaluation of the saturation level of pertubations is essential to study the plasma transport, which is very difficult to estimate analytically.

However, computer simulation of plasma presents many difficulties. There is a numerically stiff problem, since plasma behavior contains mutilevel wide ranges of length and time scales (refer to Section 2.3). (Resistive diffusion time $\tau_{\rm R} \sim 10^3$ s, Coulomb collision time $\tau_{\rm ei} \sim 0.1$ ms, MHD Alfvén transit time $\tau_{\rm H} \sim 0.1\,\mu$s, ion and electron cyclotron periods $\tau_{\rm ci} \sim 30$ ns, $\tau_{\rm ce} \sim 7$ ps, electron plasma oscillation period $\tau_{\rm p} \sim 10$ ps; mean free path $\lambda_{\rm ei} \sim 10$ km, plasma radius $a \sim 1$ m, Larmor radii $\rho_{\rm ci} \sim 3$ mm, $\rho_{\rm ce} \sim 50\,\mu$m, Debye length $\lambda_{\rm D} \sim 70\,\mu$m for typical fusion grade plasma). Because of these difficulties, simulations including all the relevant physics are not possible even for use of the most advanced computers in the near future. Therefore, it is necessary to use numerically feasible approximate models while retaining the essential features of aimed plasma dynamics.

In this section, a brief introduction of gyrokinetic particle model and full orbit particle model is presented. Numerical algorithm of simulation is not discussed here. Readers should refer to the excellent textbooks [13.10]~[13.12] and cited references.

13.4.1 Gyrokinetic Particle Model

Vlasov equation of a distribution function $F(\boldsymbol{x}, \boldsymbol{v}, t)$ in the space $(\boldsymbol{x}, \boldsymbol{v})$ is given by

$$\frac{\partial F}{\partial t} + \boldsymbol{v} \cdot \frac{\partial F}{\partial \boldsymbol{x}} + \frac{q}{m}(\boldsymbol{E} + \boldsymbol{v} \times \boldsymbol{B}) \cdot \frac{\partial F}{\partial \boldsymbol{v}} = 0.$$

Let us apply gyrokinetic change of variables from $(\boldsymbol{x}, \boldsymbol{v})$ to $(\boldsymbol{R}, \mu, v_\parallel, \varphi)$. ($\boldsymbol{R}$: gyrocenter coordinates, $\mu \equiv v_\perp^2/2B$: magnetic moment, $v_\parallel$: parallel component of velocity, φ: phase angle), where

$$\boldsymbol{x} = \boldsymbol{R} + \boldsymbol{\rho}, \quad \boldsymbol{\rho} = (\boldsymbol{b} \times \boldsymbol{v}_\perp)/\Omega, \quad \Omega \equiv qB/m, \quad \boldsymbol{b} \equiv \boldsymbol{B}/B.$$

Distribution function $F(\boldsymbol{x}, \boldsymbol{v}, t)$ can be expressed by the sum of the zeroth order term f, which is independent of gyrophase and the first order term g, which is dependent of gyrophase,

$$F(\boldsymbol{x}, \boldsymbol{v}, t) = f(\boldsymbol{R}, \mu, v_\parallel, t) + g(\boldsymbol{R}, \mu, v_\parallel, \varphi, t).$$

In the case of electrostatic perturbation $\boldsymbol{E} = -\nabla\Phi$, f and g are given by [13.13]

$$\frac{\partial f}{\partial t} + \left(v_\parallel \boldsymbol{b} + \boldsymbol{v}_{\rm d} - \frac{1}{B}\frac{\partial \Psi}{\partial \boldsymbol{R}} \times \boldsymbol{b} \right) \cdot \frac{\partial f}{\partial \boldsymbol{R}} + \left(-\frac{q}{m}\frac{\partial \Psi}{\partial \boldsymbol{R}} \cdot \boldsymbol{b} - \mu \boldsymbol{b} \cdot \frac{\partial B}{\partial \boldsymbol{R}} \right) \frac{\partial f}{\partial v_\parallel} = 0,$$

(13.21)

$$g = \frac{q}{mB}\frac{\partial f}{\partial \mu}(\Phi(\boldsymbol{x}) - \langle \Phi \rangle).$$

$\boldsymbol{v}_{\rm d}$ is curvature and ∇B drift ($\boldsymbol{v}_{\rm d} = \boldsymbol{b} \times (\mu \cdot \nabla B + v_\parallel^2(\boldsymbol{b} \cdot \nabla)\boldsymbol{b})/\Omega$). $\langle\ \rangle = \oint {\rm d}\varphi/2\pi$ means gyrophase average. $\Phi(\boldsymbol{x})$ and $\langle \Phi \rangle$ are expressed by

$$\Phi(\boldsymbol{x}) = \sum_k \phi(\boldsymbol{k}) \exp i\boldsymbol{k} \cdot \boldsymbol{x} = \sum_k \phi(\boldsymbol{k}) \exp(\boldsymbol{k} \cdot \boldsymbol{R}) \exp i\boldsymbol{k} \cdot \boldsymbol{\rho},$$

$$\langle \Phi \rangle \equiv \langle \Phi(\boldsymbol{R} + \boldsymbol{\rho}) \rangle = \sum_k \phi(\boldsymbol{k}) J_0 \left(\frac{\boldsymbol{k}_\perp v_\perp}{\Omega} \right) \exp(i\boldsymbol{k} \cdot \boldsymbol{R}).$$

When f is Maxwellian in $v_\perp$ ($f \propto \exp(-v_\perp^2/2v_{\rm T}^2)/2\pi v_{\rm T}^2$), we find

$$\frac{q}{mB}\frac{\partial f}{\partial \mu} = -\frac{q}{T}, \qquad \frac{T}{m} = v_{\rm T}^2,$$

and Ψ is given by

$$\Psi(\boldsymbol{R}) \equiv \langle\Phi\rangle + \frac{1}{2}\frac{q}{T}(\langle\Phi\rangle^2 - \langle\Phi^2\rangle) \simeq \langle\Phi\rangle - \frac{1}{2}\frac{q}{T}\frac{v_\perp^2}{\Omega^2}\left|\frac{\partial\langle\Phi\rangle}{\partial\boldsymbol{R}}\right|^2 .$$

We used the relation $\langle\Phi^2\rangle = \sum_k(\sum_{k'}\phi(\boldsymbol{k}')\phi(\boldsymbol{k}-\boldsymbol{k}'))J_0(k_\perp v_\perp/\Omega)\exp i\boldsymbol{k}\cdot\boldsymbol{R}$. Then the distribution function F in $(\boldsymbol{x},\boldsymbol{v})$ space becomes

$$F(\boldsymbol{x},\boldsymbol{v},t) = \langle f(\boldsymbol{R},\mu,v_\parallel,t)\rangle\Big(1-(q/T)\big(\Phi(\boldsymbol{x})-\langle\langle\Phi\rangle(\boldsymbol{R})\rangle\big)\Big), \qquad \boldsymbol{R} = \boldsymbol{x}-\boldsymbol{\rho},$$

and Poisson's equation is given by

$$\nabla^2\Phi(\boldsymbol{x}) = -(e/\epsilon_0)\int(F_{\rm i} - F_{\rm e})\mathrm{d}\boldsymbol{v}. \tag{13.22}$$

As f is expressed by

$$f(\boldsymbol{R},\boldsymbol{v}_\perp,v_\parallel,t) = \sum_k \bar{f}(\boldsymbol{k},v_\parallel,t)\frac{1}{2\pi v_{\rm T}^2}\exp\left(-\frac{v_\perp^2}{2v_{\rm T}^2}\right)\exp i\boldsymbol{k}\cdot(\boldsymbol{x}-\boldsymbol{\rho}),$$

$\langle f\rangle$ is

$$\langle f(\boldsymbol{R},\boldsymbol{v}_\perp,v_\parallel,t)\rangle = \sum_k \bar{f}(\boldsymbol{k},v_\parallel,t)\frac{1}{2\pi v_{\rm T}^2}\exp\left(-\frac{v_\perp^2}{2v_{\rm T}^2}\right)\exp(i\boldsymbol{k}\cdot\boldsymbol{x})J_0(k_\perp v_\perp/\Omega).$$

We define $\tilde{f}(\boldsymbol{x},v_\parallel,t)$ by

$$\tilde{f}(\boldsymbol{x},v_\parallel,t) \equiv \int\langle f(\boldsymbol{R},\boldsymbol{v}_\perp,v_\parallel,t)\rangle 2\pi v_\perp \mathrm{d}v_\perp = \sum_k \bar{f}(\boldsymbol{k},v_\parallel,t)\exp(-b/2)\exp i\boldsymbol{k}\cdot\boldsymbol{x},$$

where $\rho_{\rm T} = v_{\rm T}/\Omega,\quad b \equiv (k_\perp\rho_{\rm T})^2$. Furthermore, $\tilde{\Phi}(\boldsymbol{x})$ is defined by

$$\tilde{f}(\boldsymbol{x},v_\parallel,t)\tilde{\Phi}(\boldsymbol{x}) \equiv \int\langle\langle\Phi\rangle\rangle\langle f\rangle 2\pi v_\perp\mathrm{d}v_\perp.$$

Then we obtain

$$\tilde{\Phi}(\boldsymbol{x}) \simeq \sum_k \phi(\boldsymbol{k})\Gamma_{\rm o}(b)\exp i\boldsymbol{k}\cdot\boldsymbol{x}, \qquad \Gamma_{\rm o}(b) \equiv I_{\rm o}(b)\exp(-b).$$

The terms $\int F\mathrm{d}\boldsymbol{v}$, appearing in the right-hand side of Poisson's equation (13.22), are expressed by

$$\int F\mathrm{d}\boldsymbol{v} = \hat{n}(\boldsymbol{x}) - q/T(\Phi - \tilde{\Phi})\hat{n}(\boldsymbol{x}),$$

$$\hat{n}(\boldsymbol{x}) \equiv \int\tilde{f}(\boldsymbol{x},v_\parallel,t)\mathrm{d}v_\parallel.$$

Therefore, (13.22) is reduced to

$$\nabla^2\Phi(\boldsymbol{x}) = -\frac{e}{\epsilon_0}(\hat{n}_\mathrm{i} - \hat{n}_\mathrm{e}) + \frac{T_\mathrm{e}}{T_\mathrm{i}}\frac{1}{\lambda_\mathrm{D}^2}\frac{\hat{n}_\mathrm{i}}{n_0}(\Phi - \tilde{\Phi}). \tag{13.23}$$

Here we assumed zero electron Larmor radius. $\lambda_\mathrm{D} = (\epsilon_0 T_\mathrm{e}/n_0 e^2)^{1/2}$ is Debye length. The second term in the right-hand side of (13.23) is

$$\frac{T_\mathrm{e}}{T_\mathrm{i}}\frac{1}{\lambda_\mathrm{D}^2}(\Phi - \tilde{\Phi}) \simeq \frac{T_\mathrm{e}}{T_\mathrm{i}}\frac{(k_\perp \rho_\mathrm{i})^2}{\lambda_\mathrm{D}^2}\Phi \approx -\frac{\Pi_\mathrm{i}^2}{\Omega_\mathrm{i}^2}\nabla_\perp^2\Phi.$$

Usually $\Pi_\mathrm{i}^2/\Omega_\mathrm{i}^2 = (T_\mathrm{e}/T_\mathrm{i})(\rho_\mathrm{i}/\lambda_\mathrm{D})^2$ is much larger than 1.

Now we discuss the numerical schemes for solving the gyrokinetic Vlasov-Poisson systems (13.21) and (13.23). Applying the discrete representation for the distribution function of N particles

$$f(\boldsymbol{R}, \mu, v_\parallel, t) = \sum_{j=1}^{N} \delta(\boldsymbol{R} - \boldsymbol{R}_j(t))\delta(\mu - \mu_j)\delta(v_\perp - v_{\perp j}(t)),$$

to (13.21), the equations of motion in the gyrocenter coordinates for the j-th gyrokinetic particles must satisfiy

$$\frac{\mathrm{d}\boldsymbol{R}_j}{\mathrm{d}t} = v_{\parallel j}\boldsymbol{b} + \boldsymbol{v}_\mathrm{d} - \frac{1}{B}\left.\left(\frac{\partial \Psi}{\partial \boldsymbol{R}} \times \boldsymbol{b}\right)\right|_{R_j \mu_j} \tag{13.24}$$

$$\frac{\mathrm{d}v_{\parallel j}}{\mathrm{d}t} = -\frac{q}{m}\left.\left(\frac{\partial \Psi}{\partial \boldsymbol{R}} \cdot \boldsymbol{b}\right)\right|_{R_j \mu_j} - \mu_j \boldsymbol{B}\cdot\nabla B. \tag{13.25}$$

Then the function $\tilde{f}$ is given by

$$\tilde{f}(\boldsymbol{x}, v_\parallel, t) = \int \langle f(\boldsymbol{R}, \boldsymbol{v}_\perp, v_\parallel, t)\rangle 2\pi v_\perp \mathrm{d}v_\perp = \sum_{j=1}^{N}\langle \delta(\boldsymbol{x} - \boldsymbol{R}_j - \boldsymbol{\rho}_j)\rangle \delta(v_\parallel - v_{\parallel j}).$$

$\hat{n}_\mathrm{i}$, $\hat{n}_\mathrm{e}$ in the first term of the right-hand side of (13.23) are

$$\hat{n}_\mathrm{i} = \int \tilde{f}(\boldsymbol{x}, v_\parallel, t)\mathrm{d}v_\parallel = \sum_{j=1}^{N}\langle \delta(\boldsymbol{x} - \boldsymbol{R}_j^\mathrm{i} - \boldsymbol{\rho}_j^\mathrm{i})\rangle,$$

$$\hat{n}_\mathrm{e} = \sum_{j=1}^{N}\delta(\boldsymbol{x} - \boldsymbol{R}_j^\mathrm{e}).$$

When $k_\perp \rho_\mathrm{i} < 1$, the four points average by $\varphi = 0, \pi/2, \pi, 3\pi/2$ is a good approximation of the gyrophase average $\langle\ \rangle$.

Drift wave trapped electron mode in tokamak is studied in [13.14]. The simulation parameters used include the system size $L_x \times L_y \times L_z = 64\Delta \times 64\Delta \times 32\Delta_z$ with $\Delta = \rho_\mathrm{i}$ and $\Delta_z = 25\rho_\mathrm{i}$. The aspect ratio $R_0/a = 4$, $n_0 = \langle n\rangle = 4$ particles/cell (total ions $\sim 0.5 \times 10^6$). $\langle T_\mathrm{e}\rangle/T_\mathrm{i} = 4$, $m_\mathrm{i}/m_\mathrm{e} = 1836$, $\Omega_\mathrm{i}\Delta t = 0.1$, $\omega_\mathrm{e}^*/\Omega_\mathrm{i} = 0.1m$ (m is the poloidal

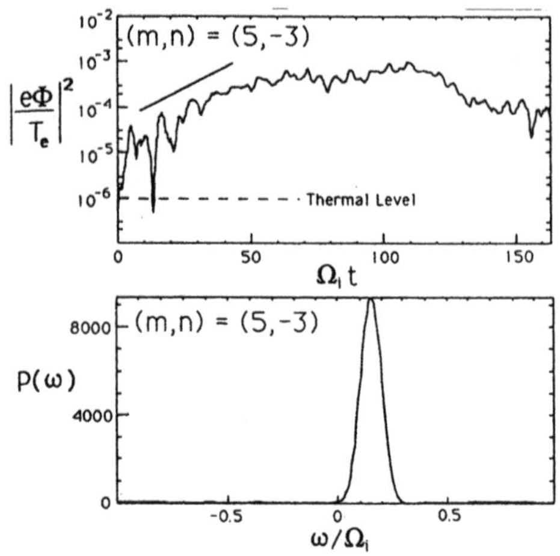

Figure 13.3 Upper figure: Temporal evolution of mode amplitude of drift wave trapped electron mode (m,n) = (5,-3). The solid line is the linear theory growth rate. Lower figure: Power spectrum density versus frequency. After [13.14].

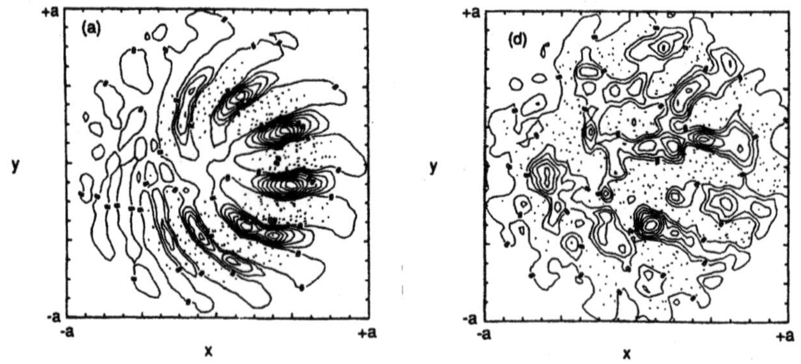

Figure 13.4 Plots of the electrostatic potential of ITG turbulence in a poloidal cross-section during the linear phase (left-hand side) and nonlinearly saturated steady-state (right-hand side). After [13.15].

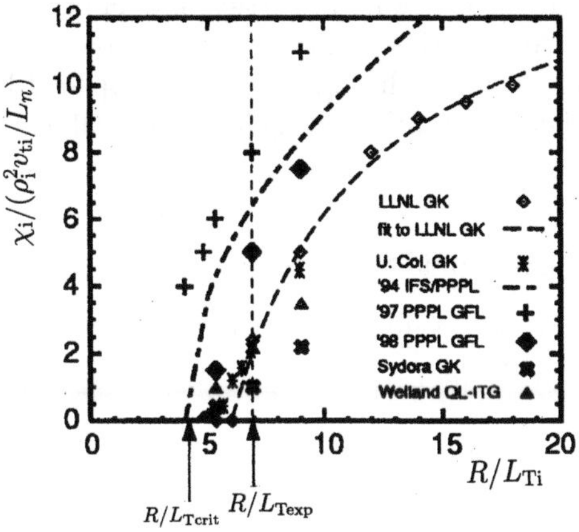

Figure 13.5 Ion thermal diffusion coefficient χ_i of ITG turbulence versus R/L_T from the gyrofluid codes 94IFS/PPPl, 97PPPL GFL, 98PPPL GFL and from the gyrokinetic particle codes of LLNL and University of Colorado flux-tube and UCLA (Sydora) global codes. After Ref.[13.18]. The point ($R/L_{Texp} = 7$, $\chi_i/(\rho_i^2 v_{ti}/L_n) = 0.16$) is the experimental value of DIII-D base data. The values of the threshold R/L_{Tcrit} of gyrofluid code is somewhat different from that of LLNL gyrokinetic results.

mode number) where $\omega_e^* = k_\theta T_e/L_n e B_0$. Results from the simulation of drift wave trapped electron mode with $\eta_e \equiv d(\ln T_e)/d(\ln n_e)=1$ are shown in Figure 13.3. The time evolution of the mode amplitude for $(m, n) = (5, -3)$ is shown in the upper figure. The saturation amplitude reaches $e\Phi/T_e \approx 0.035$. The linear growth rate in the initial phase and real frequency agree well with the theoretical linear eigenmode analysis.

Ion temperature gradient (ITG) mode is studied in [13.15]. In this analysis, $\delta f/f$ method [13.16] is used. The electrons are assumed to be adiabatic ($\delta n_e/n_0 = e\phi/T_e$) and the number of total ions is $\sim 10^6$. Figure 13.4 plots the electrostatic potential during the linear phase and nonlinearly saturated steady-state.

The ion thermal diffusivity of ITG turbulence in tokamak is studied in [13.17, 13.18]. The number of particles in the simulation is in the range of 5×10^5 to 1.34×10^8. For 10^6 or more (two particles per cell), χ_i at the later time does not appear to change with increasing particle number. The thermal diffusivity is defined formally as $\chi_i = 1.5 L_n \langle \tilde{v}_r \tilde{T}_i \rangle / T_i$ where $\tilde{v}_r$ and $\tilde{T}_i$ are the fluctuating components of radial ion velocity and ion temperature. The dependence of χ_i on R/L_T is scanned and the fit can be expressed by an offset linear dependence on R/L_T (refer to Figure 13.5)

$$\frac{\chi_i}{\rho_i^2 v_{ti}/L_n} = 15.4\big(1.0 - 6.0(L_T/R)\big). \tag{13.26}$$

The second term in the right-hand side of (13.26) is called *Dimit shift.*

We have discussed the gyrokinetic particle model of electrostatic perturbations. The formulation of the gyrokinetic particle model of electromagnetic perturbations has also been developed [13.19], in which the effect of magnetic field fluctuations is included as well as the effect of electric field fluctuations.

13.4.2 Full Orbit Particle Model

Fundamental equations of full orbit particle model are simple and are described as follows [13.10]:

$$\frac{\mathrm{d}\boldsymbol{r}_{\mathrm{s}j}(t)}{\mathrm{d}t} = \boldsymbol{v}_{\mathrm{s}j}(t),$$

$$\frac{\mathrm{d}\boldsymbol{v}_{\mathrm{s}j}(t)}{\mathrm{d}t} = \frac{q_{\mathrm{s}}}{m_{\mathrm{s}}}\left(\boldsymbol{E}(\boldsymbol{r}_{\mathrm{s}j}(t),t) + \boldsymbol{v}_{\mathrm{s}j}(t) \times \boldsymbol{B}(\boldsymbol{r}_{\mathrm{s}j}(t),t)\right),$$

$$\rho = \sum_{\mathrm{s=e,i}} q_{\mathrm{s}} \sum_j \delta(\boldsymbol{r} - \boldsymbol{r}_{\mathrm{s}j}(t)),$$

$$\boldsymbol{j} = \sum_{\mathrm{s=e,i}} q_{\mathrm{s}} \sum_j \boldsymbol{v}_{\mathrm{s},j}(t)\delta(\boldsymbol{r} - \boldsymbol{r}_{\mathrm{s}j}(t)),$$

$$\epsilon_0 \nabla \cdot \boldsymbol{E} = \rho, \qquad \nabla \cdot \boldsymbol{B} = 0,$$

$$\nabla \times \boldsymbol{E} = -\frac{\partial \boldsymbol{B}}{\partial t}, \qquad \nabla \times \boldsymbol{B} = \mu_0 \boldsymbol{j} + \frac{1}{c^2}\frac{\partial \boldsymbol{E}}{\partial t}.$$

m_{s} and q_{s} are the mass and charge of s species. The other notations are as usual. However, the number of particles of real plasma is far beyond $10^7 \sim 10^8$, which is the limit of most advanced supercomputers, at least in the near future. In full orbit particle model, the concept of superparticle with finite size is introduced instead. Let us consider the system with the volume of $V = L_x L_y L_z$ which contains N_{e} electrons and N_{i} ions. Λ number of particls are put together to one superparticle with the mass and the charge of ($\Lambda \gg 1$)

$$m_{\mathrm{s}}^{\mathrm{sp}} = \Lambda m_{\mathrm{s}}, \qquad q_{\mathrm{s}}^{\mathrm{sp}} = \Lambda q_{\mathrm{s}}.$$

Then the average values of the density and temperature of superparticles are

$$n_{\mathrm{s}}^{\mathrm{sp}} = n_{\mathrm{s}}/\Lambda, \qquad T_{\mathrm{s}}^{\mathrm{sp}} = \Lambda T_{\mathrm{s}}.$$

Then the plasma frequency $\Pi_{\mathrm{s}}^{\mathrm{sp}}$, cyclotron frequency $\Omega_{\mathrm{s}}^{\mathrm{sp}}$, thermal velocity $v_{\mathrm{T}}^{\mathrm{sp}}$, Alfvén velocity $v_{\mathrm{A}}^{\mathrm{sp}}$, Debye length $\lambda_{\mathrm{Ds}}^{\mathrm{sp}}$, Larmor radius $\rho_{\mathrm{cs}}^{\mathrm{sp}}$, and beta ratio β^{sp} are the same as the original system. However, Coulomb collision frequency is greatly enhanced; that is

$$\nu_{\rm ei}^{\rm sp} \sim 0.4\frac{\Pi_{\rm e}^{\rm sp}}{n_{\rm e}^{\rm sp}(\lambda_{\rm De}^{\rm sp})^3} = 0.4\frac{\Lambda\Pi_{\rm e}}{n_{\rm e}(\lambda_{\rm De})^3} = \Lambda\nu_{\rm ei}.$$

Coulomb collision frequency of superparticles with zero size can be comparable to $1/10 \sim 1/100$ of the electron plasma frequency when $N_{\rm e}^{\rm sp} \equiv n_{\rm e}^{\rm sp}(\lambda_{\rm De}^{\rm sp})^3$ is selected to be from several to several times tens. Therefore, low frequency waves will be masked by Coulomb collision. To avoid this effect, the charge density distribution of finite size with shape factor of $S(\boldsymbol{r})$ is introduced.

$$q_{\rm s}^{\rm sp}\delta(\boldsymbol{r} - \boldsymbol{r}_{{\rm s}j}) \rightarrow q_{\rm s}^{\rm sp}S(\boldsymbol{r} - \boldsymbol{r}_{{\rm s}j})$$

$$\int S(\boldsymbol{r} - \boldsymbol{r}_{{\rm s}j}){\rm d}\boldsymbol{r} = 1.$$

When the effective radius of the shape factor is R, the effective Coulomb collision frequency is reduced by one to three orders of magnitude depending on the size $R \sim (1 \sim 5)\lambda_{\rm D}$ and $n_{\rm e}(\lambda_{\rm De})^3 \sim (10 \sim 10^3)$ [13.20]. The dispersion relation with $k < R^{-1}$ is not affected much. Therefore, the simulation with use of superparticles with a proper finite size can reproduce the wave phenomena with wavelength larger than R. The equations of motion of superparticles are

$$\frac{{\rm d}\boldsymbol{r}_{{\rm s}j}^{\rm sp}(t)}{{\rm d}t} = \boldsymbol{v}_{{\rm s}j}^{\rm sp}(t),$$

$$\frac{{\rm d}\boldsymbol{v}_{{\rm s}j}^{\rm sp}(t)}{{\rm d}t} = \frac{q_{\rm s}^{\rm sp}}{m_{\rm s}^{\rm sp}}\left(\boldsymbol{E}^*(\boldsymbol{r}_{{\rm s}j}(t), t) + \boldsymbol{v}_{{\rm s}j}^{\rm sp}(t) \times \boldsymbol{B}^*(\boldsymbol{r}_{{\rm s}j}(t), t)\right),$$

where $\boldsymbol{E}^*$, $\boldsymbol{B}^*$ are the fields that finite size superparticles feel; that is,

$$\boldsymbol{E}^*(\boldsymbol{r}, t) \equiv \int \boldsymbol{E}(\boldsymbol{r}', t)S(\boldsymbol{r}' - \boldsymbol{r}){\rm d}\boldsymbol{r}',$$

$$\boldsymbol{B}^*(\boldsymbol{r}, t) \equiv \int \boldsymbol{B}(\boldsymbol{r}', t)S(\boldsymbol{r}' - \boldsymbol{r}){\rm d}\boldsymbol{r}',$$

$$\rho = \sum_{\rm s=e,i} q_{\rm s}^{\rm sp}\sum_{j=1}^{N_{\rm s}^{\rm sp}} S(\boldsymbol{r} - \boldsymbol{r}_{{\rm s}j}(t)),$$

$$\boldsymbol{j} = \sum_{\rm s=e,i} q_{\rm s}^{\rm sp}\sum_{j=1}^{N_{\rm s}^{\rm sp}} \boldsymbol{v}_{{\rm s},j}(t)S(\boldsymbol{r} - \boldsymbol{r}_{{\rm s}j}(t)),$$

$$\epsilon_0\nabla\cdot\boldsymbol{E} = \rho, \qquad \nabla\cdot\boldsymbol{B} = 0,$$

$$\nabla\times\boldsymbol{E} = -\frac{\partial\boldsymbol{B}}{\partial t}, \qquad \nabla\times\boldsymbol{B} = \mu_0\boldsymbol{j} + \frac{1}{c^2}\frac{\partial\boldsymbol{E}}{\partial t}.$$

When time and length are normalized by

$$\tilde{t} = t\Pi_{\rm e}, \qquad \tilde{\boldsymbol{r}} = \boldsymbol{r}\Delta^{-1},$$

and $\boldsymbol{E}$, $\boldsymbol{B}$, ρ, and $\boldsymbol{j}$ are normalized by

$$\tilde{\boldsymbol{E}} = \frac{\boldsymbol{E}}{m_{\rm e}\Delta\Pi_{\rm e}^2/e}, \qquad \tilde{\boldsymbol{B}} = \frac{\boldsymbol{B}}{m_{\rm e}\Pi_{\rm e}/e},$$

$$\tilde{\rho} = \frac{\rho}{en_{\rm e}}, \qquad \tilde{\boldsymbol{j}} = \frac{\boldsymbol{j}}{en_{\rm e}\Delta\Pi_{\rm e}} \quad (\Delta: \text{ grid size}),$$

then the dimensionless forms of equations are [13.21]

$$\frac{\mathrm{d}\tilde{\boldsymbol{r}}_{\mathrm{s}j}(\tilde{t})}{\mathrm{d}\tilde{t}} = \tilde{\boldsymbol{v}}_{\mathrm{s}j}(\tilde{t}), \tag{13.27}$$

$$\frac{\mathrm{d}\tilde{\boldsymbol{v}}_{\mathrm{s}j}(\tilde{t})}{\mathrm{d}\tilde{t}} = \frac{Q_{\rm s}}{M_{\rm s}}\left(\tilde{\boldsymbol{E}}^*(\tilde{\boldsymbol{r}}_{\mathrm{s}j}(\tilde{t}),\tilde{t}) + \tilde{\boldsymbol{v}}_{\mathrm{s}j}(\tilde{t}) \times \tilde{\boldsymbol{B}}^*(\tilde{\boldsymbol{r}}_{\mathrm{s}j}(\tilde{t}),\tilde{t})\right), \tag{13.28}$$

$$\tilde{\rho} = \frac{\tilde{L}_x\tilde{L}_y\tilde{L}_z}{N_{\rm e}^{\rm sp}}\sum_{\rm s=e,i} Q_{\rm s}\sum_{j=1}^{N_{\rm s}^{\rm sp}} \tilde{S}(\tilde{\boldsymbol{r}} - \tilde{\boldsymbol{r}}_{\mathrm{s}j}(\tilde{t})), \tag{13.29}$$

$$\tilde{\boldsymbol{j}} = \frac{\tilde{L}_x\tilde{L}_y\tilde{L}_z}{N_{\rm e}^{\rm sp}}\sum_{\rm s=e,i} Q_{\rm s}\sum_{j=1}^{N_{\rm s}^{\rm sp}} \tilde{\boldsymbol{v}}_{\mathrm{s},j}(\tilde{t})\tilde{S}(\tilde{\boldsymbol{r}} - \tilde{\boldsymbol{r}}_{\mathrm{s}j}(\tilde{t})), \tag{13.30}$$

$$\tilde{\nabla}\cdot\tilde{\boldsymbol{E}} = \tilde{\rho}, \qquad \tilde{\nabla}\cdot\tilde{\boldsymbol{B}} = 0, \tag{13.31}$$

$$\tilde{\nabla}\times\tilde{\boldsymbol{E}} = -\frac{\partial\tilde{\boldsymbol{B}}}{\partial\tilde{t}}, \qquad \tilde{c}^2\tilde{\nabla}\times\tilde{\boldsymbol{B}} = \tilde{\boldsymbol{j}} + \frac{\partial\tilde{\boldsymbol{E}}}{\partial\tilde{t}}, \tag{13.32}$$

$$\tilde{c} = \frac{c}{\Delta\Pi_{\rm e}}, \qquad \tilde{S}(\tilde{\boldsymbol{r}} - \tilde{\boldsymbol{r}}_{\mathrm{s}j}(\tilde{t})) = \Delta^3 S(\boldsymbol{r} - \boldsymbol{r}_{\mathrm{s}j}(t)). \tag{13.33}$$

where c is light velocity and $Q_{\rm s} \equiv q_{\rm s}/e$ $(Q_{\rm e} = -1, Q_{\rm i} = q_{\rm i}/e)$, $M_{\rm s} \equiv m_{\rm s}/m_{\rm e}$ $(M_{\rm e} = 1, M_{\rm i} = m_{\rm i}/m_{\rm e})$.

Tilt stability of a field reversed configuration (FRC) is studied by full orbit particle ions and electrons in [13.22]. The grid number is $49 \times 49 \times 32$ and the total number of superparticle is 10^6. $\Pi_{\rm e}\Delta t = 1.5$, $c\Delta t/\Delta < 1$. The ion and electron mass ratio is set to be $m_{\rm i}/m_{\rm e} = 50$ and $\Omega_{\rm e} \sim \Pi_{\rm e}/5$. Figure 13.6 shows contour plots of mass density in the poloidal cross-section at the periods of $t/\tau_{\rm A}$=0.0, 1.0, 3.0, and 5.0. $\tau_{\rm A}$ is Alfvén transit time. In the case of this simulation, $|\Omega_{\rm i}| < \tau_{\rm A}^{-1}$ holds. Stabilizing effect on the tilting by cycling ions which cross the separatrix is discussed. Tilt stability of FRC is also discussed by hybrid model of full orbit particle ions and fluid electrons [13.23]. The stabilizing effect due to finite ion Larmor radius is analyzed.

In the case of $|\Omega_{\rm i}| \gg \tau_{\rm A}^{-1}$, full orbit particle simulation of electromagnetic perturbation is very difficult due to the excess amount of computer run-time. The toroidal particle code (TPC) is developed for the electrostatic turbulance [13.24]. TPC solves Poisson's equation

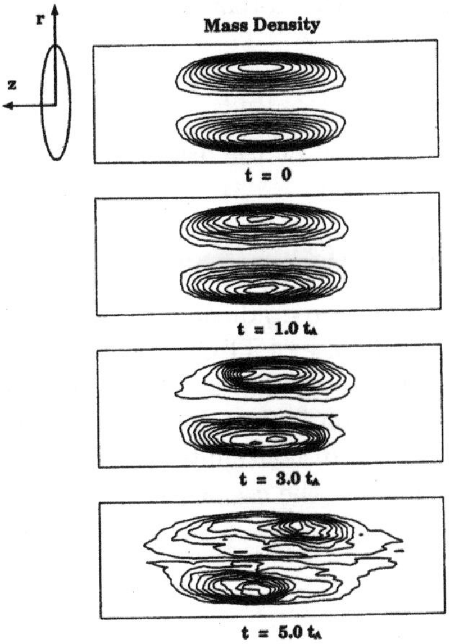

Figure 13.6 Contour plots of mass density in the poloidal cross-section of field reversed configuration (FRC) at the periods of $t/t_{\mathrm{A}} = 0.0, 1.0, 3.0,$ and 5.0. After [13.22].

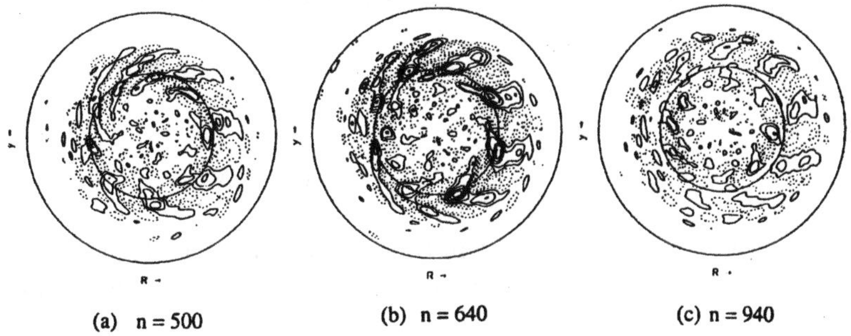

Figure 13.7 Potential structure in poloidal cross-section due to ITG at three different times for a reversed magnetic shear configuration of tokamak. After [13.25]. In the quasi-steady state (c), discontinuity of the potential structure across the $q_{\min}$ surface is recovered.

$$\nabla^2 \Phi = -\frac{1}{\epsilon_0}\rho = -\frac{e}{\epsilon_0}\sum_j ((q_{\rm i}/e)S(\boldsymbol{r}-\boldsymbol{r}^{\rm i}_j) - S(\boldsymbol{r}-\boldsymbol{r}^{\rm e}_j)). \tag{13.34}$$

The ion motion is given by the Lorentz equation

$$\frac{d\boldsymbol{r}^{\rm i}}{dt} = \boldsymbol{v}^{\rm i}, \qquad \frac{d\boldsymbol{v}^{\rm i}}{dt} = \frac{q^{\rm i}}{m^{\rm i}}\left(\boldsymbol{E} + \boldsymbol{v}^{\rm i}\times\boldsymbol{B}\right), \tag{13.35}$$

and the electron motion is given by the drift equation

$$\frac{dv^{\rm e}_{\parallel}}{dt} = -\frac{e}{m_{\rm e}}E_{\parallel} - \mu(\boldsymbol{b}\cdot\nabla)B. \tag{13.36}$$

In [13.25], the ion temperature gradient (ITG) turbulences in tokamak configuration are analyzed by use of TPC. In these simulations, electrons are treated as an adiabatic fluid ($\delta n_{\rm e}/n_{\rm e} = e\Phi/T_{\rm e}$). The effects of reversed magnetic shear configuration on ITG turbulence are studied and it is observed that the existence of discontinuity and/or gap in the structure of ITG perturbed potential across the $q_{\rm min}$ surface as is shown in Figure 13.7 (refer to Section 4.6).

13.5 Zonal Flow

13.5.1 Hasegawa-Mima Equation for Drift Turbulence

A model equation is the equation of continuity for ions, in which the parallel ion inertia is neglected. Ions are assumed to be cold:

$$\frac{\partial n}{\partial t} + \nabla\cdot(n\boldsymbol{v}_\perp) = \frac{\partial n}{\partial t} + (\boldsymbol{v}_\perp\cdot\nabla)n + n\nabla\cdot\boldsymbol{v}_\perp = 0. \tag{13.37}$$

The ion motion consists of $\boldsymbol{E}\times\boldsymbol{B}$ drift and polarization drift (refer to Section 2.1.3):

$$\boldsymbol{v}_\perp = -\frac{1}{B}\nabla\phi\times\hat{z} - \frac{1}{\Omega_{\rm i}B}\frac{\rm d}{{\rm d}t}\nabla\phi, \qquad \frac{\rm d}{{\rm d}t} = \frac{\partial}{\partial t} - \frac{1}{B}(\nabla\phi\times\hat{z})\cdot\nabla. \tag{13.38}$$

A slow variation of the potential ϕ in the parallel direction along the magnetic field line allows the electrons to obey the Boltzmann distribution, that is,

$$n = n_0 + \delta n, \qquad \frac{\delta n}{n_0} = \frac{e\phi}{T_{\rm e}} \equiv \tilde{\phi}. \tag{13.39}$$

We assume the following ordering:

$$\frac{1}{\Omega_{\rm i}}\frac{\rm d}{{\rm d}t} \sim \delta, \quad \tilde{\phi}\sim\delta, \quad L_n\nabla\sim\delta^{-1}.$$

Denoting $c_s^2 \equiv T_e/m_i$, $\rho_s \equiv c_s/\Omega_i$ and $\Omega_i = eB/m_i$, (13.37), (13.38), and (13.39) reduce to

$$\boldsymbol{v}_\perp = -\rho_s c_s(\nabla\tilde{\phi} \times \hat{\boldsymbol{z}}) - \rho_s^2 \left(\frac{\partial}{\partial t} - \rho_s c_s (\nabla\tilde{\phi} \times \hat{z}) \cdot \nabla \right) \nabla\tilde{\phi}, \tag{13.40}$$

$$\frac{\partial\tilde{\phi}}{\partial t} - \rho_s c_s(\nabla\tilde{\phi} \times \hat{\boldsymbol{z}}) \left(\nabla\tilde{\phi} + \frac{\nabla n_0}{n_0} \right)$$

$$+(1+\tilde{\phi})\nabla \cdot \left(-\rho_s c_s(\nabla\tilde{\phi} \times \hat{\boldsymbol{z}}) - \rho_s^2 \left(\frac{\partial}{\partial t} - \rho_s c_s(\nabla\tilde{\phi} \times \hat{z}) \cdot \nabla \right) \nabla\tilde{\phi} \right) = 0,$$

$$\frac{\partial\tilde{\phi}}{\partial t} - \rho_s c_s \left(\frac{\partial\tilde{\phi}}{\partial y} \frac{\mathrm{d}n_0}{\mathrm{d}x} \frac{1}{n_0} \right) - \rho_s^2 \frac{\partial\nabla^2\tilde{\phi}}{\partial t} + \rho_s^3 c_s \nabla \cdot ((\nabla\tilde{\phi} \times \hat{z}) \cdot \nabla)\nabla\tilde{\phi} = 0,$$

$$(1-\rho_s^2\nabla^2)\frac{\partial\tilde{\phi}}{\partial t} + \rho_s^3 c_s \left(\frac{\partial\nabla^2\tilde{\phi}}{\partial x}\frac{\partial\tilde{\phi}}{\partial y} - \frac{\partial\nabla^2\tilde{\phi}}{\partial y}\frac{\partial\tilde{\phi}}{\partial x} \right) + v_d^* \frac{\partial\tilde{\phi}}{\partial y} = 0, \tag{13.41}$$

where v_d^* is the drift velocity of the electron and

$$v_d^* = \frac{\kappa_n T_e}{eB} = c_s(\kappa_n \rho_s).$$

(13.41) is called the *Hasegawa-Mima-Charney equation* [13.26]. Here we used the following relation:

$$((\nabla\tilde{\phi} \times \hat{z}) \cdot \nabla)\nabla^2\tilde{\phi} = \left(\frac{\partial\nabla^2\tilde{\phi}}{\partial x}\frac{\partial\tilde{\phi}}{\partial y} - \frac{\partial\nabla^2\tilde{\phi}}{\partial y}\frac{\partial\tilde{\phi}}{\partial x} \right).$$

When the density gradient is negligible, (13.41) becomes

$$(1-\rho_s^2\nabla^2)\frac{\partial\tilde{\phi}}{\partial t} + \rho_s^4 \Omega_i \left(\frac{\partial\nabla^2\tilde{\phi}}{\partial x}\frac{\partial\tilde{\phi}}{\partial y} - \frac{\partial\nabla^2\tilde{\phi}}{\partial y}\frac{\partial\tilde{\phi}}{\partial x} \right) = 0. \tag{13.42}$$

(13.42) is called the *Hasegawa-Mima equation* [13.27].
A solution of (13.41) is

$$\tilde{\phi} = A \exp i(k_x x + k_y y) \exp(-i\omega_k^1 t), \qquad \omega_k^1 = \frac{1}{1+\rho_s^2 k^2} k_y v_d^*. \tag{13.43}$$

There are two constants of motion:

$$\frac{\partial}{\partial t} \int (\tilde{\phi}^2 + \rho_s^2 (\nabla\tilde{\phi})^2) \mathrm{d}V = 0, \tag{13.44}$$

$$\frac{1}{2}\frac{\partial}{\partial t} \int \left((\nabla\tilde{\phi})^2 + \rho_s^2 (\nabla^2\tilde{\phi})^2 \right) \mathrm{d}V - \int v_d^* \nabla^2\tilde{\phi} \frac{\partial\tilde{\phi}}{\partial y} \mathrm{d}V = 0. \tag{13.45}$$

If (13.45) is multiplied by ϕ and integrated over the entire volume, the nonlinear equation becomes

$$\int \tilde{\phi}((\nabla\tilde{\phi}\times\hat{z})\cdot\nabla)\nabla^2\tilde{\phi}\mathrm{d}V = \int \tilde{\phi}\nabla\cdot((\nabla\tilde{\phi}\times\hat{z})\nabla^2\tilde{\phi})\mathrm{d}V = \int \nabla\cdot(\tilde{\phi}\nabla^2\tilde{\phi}(\nabla\tilde{\phi}\times\hat{z}))\mathrm{d}V$$

$$= \int \boldsymbol{J}_1 \cdot \boldsymbol{n}\mathrm{d}S, \qquad \boldsymbol{J}_1 \equiv \tilde{\phi}\nabla^2\tilde{\phi}(\nabla\tilde{\phi}\times\hat{z}),$$

$$\frac{1}{2}\frac{\partial}{\partial t}\int (\tilde{\phi}^2 + \rho_\mathrm{s}^2(\nabla\tilde{\phi})^2)\mathrm{d}V = -\rho_\mathrm{s}^3 c_\mathrm{s}\int \boldsymbol{J}_1\cdot\boldsymbol{n}\mathrm{d}S - \frac{1}{2}\int v_\mathrm{d}^*\frac{\partial\tilde{\phi}^2}{\partial y}\mathrm{d}V \rightarrow 0.$$

Similarly, if we multiply (13.41) by $\nabla^2\tilde{\phi}$, we have

$$\int \left(\nabla^2\tilde{\phi}((\nabla\tilde{\phi}\times\hat{\boldsymbol{z}})\cdot)\nabla^2\tilde{\phi} + \nabla^2\tilde{\phi}\left(\frac{\partial\tilde{\phi}}{\partial t} - \rho_\mathrm{s}\nabla^2\frac{\partial\tilde{\phi}}{\partial t}\right) + v_\mathrm{d}^*\nabla^2\tilde{\phi}\frac{\partial\tilde{\phi}}{\partial y}\right)\mathrm{d}V = 0,$$

$$-\nabla\cdot\boldsymbol{J}_2 \equiv \nabla\cdot\left(\frac{\partial\tilde{\phi}}{\partial t}\nabla\tilde{\phi} + \frac{1}{2}(\nabla^2\tilde{\phi})^2(\nabla\tilde{\phi}\times\hat{\boldsymbol{z}})\right)$$

$$= \frac{1}{2}\frac{\partial}{\partial t}(\nabla\tilde{\phi})^2 + \frac{\partial\tilde{\phi}}{\partial t}\nabla^2\tilde{\phi} + \frac{1}{2}((\nabla\tilde{\phi}\times\hat{\boldsymbol{z}})\cdot\nabla)(\nabla^2\tilde{\phi})^2,$$

$$-\nabla\cdot\boldsymbol{J}_2 - \frac{1}{2}\frac{\partial}{\partial t}\left((\nabla\tilde{\phi})^2 + \rho_\mathrm{s}^2(\nabla^2\tilde{\phi})^2\right) + v_\mathrm{d}^*\nabla^2\tilde{\phi}\frac{\partial\tilde{\phi}}{\partial y} = 0,$$

$$\frac{1}{2}\frac{\partial}{\partial t}\int\left((\nabla\tilde{\phi})^2 + \rho_\mathrm{s}^2(\nabla^2\tilde{\phi})^2\right)\mathrm{d}V - \int v_\mathrm{d}^*\nabla^2\tilde{\phi}\frac{\partial\tilde{\phi}}{\partial y}\mathrm{d}V = -\nabla\cdot\boldsymbol{J}_2 \rightarrow 0.$$

When $\boldsymbol{E}\times\boldsymbol{B}$ drift velocity is denoted by $\boldsymbol{v}_E$, we have $v_E = c_\mathrm{s}\rho_\mathrm{s}|\nabla\tilde{\phi}|$ and $\nabla\times\boldsymbol{v}_E|_z = c_\mathrm{s}\rho_\mathrm{s}\nabla^2\tilde{\phi}$. Equations (13.44) and (13.45) reduce to

$$\int\left(\left(\frac{\delta n}{n_0}\right)^2 + \frac{m_\mathrm{i}v_E^2}{T_\mathrm{e}}\right)\mathrm{d}V = \mathrm{const.}$$

$$\frac{1}{2}\frac{\partial}{\partial t}\int\left(\frac{1}{\rho_\mathrm{s}^2}\frac{v_E^2}{c_\mathrm{s}^2} + \frac{1}{c_\mathrm{s}^2}(\nabla\times v_E)^2\right)\mathrm{d}V + \Omega_\mathrm{i}\frac{1}{c_\mathrm{s}^2}(\nabla\times v_E)_z\kappa_n v_{Ex}\mathrm{d}V = 0.$$

Let us normalize the coordinates (x, y) and time t as follows:

$$x = \rho_\mathrm{s}\hat{x}, \quad y = \rho_\mathrm{s}\hat{y}, \quad t = \Omega^{-1}\hat{t}, \quad \boldsymbol{k} = \rho_\mathrm{s}^{-1}\hat{\boldsymbol{k}}, \quad \omega = \Omega_\mathrm{i}\hat{\omega}.$$

Then (13.41) and (13.43) reduce to

$$\partial_{\hat{t}}(\hat{\nabla}^2\tilde{\phi} - \tilde{\phi}) - (\kappa_n\rho_\mathrm{s})\partial_{\hat{y}}\tilde{\phi} + (\partial_{\hat{y}}\hat{\nabla}^2\tilde{\phi})\partial_{\hat{x}}\tilde{\phi} - (\partial_{\hat{x}}\hat{\nabla}^2\tilde{\phi})\partial_{\hat{y}}\tilde{\phi} = 0, \tag{13.46}$$

$$\hat{\omega}_k^1 = \frac{(\kappa_n\rho_\mathrm{s})\hat{k}_{\hat{y}}}{1+\hat{k}^2}.$$

From now on the notation ˆ is omitted. When we expand $\tilde{\phi}(x, t)$ in a spatial Fourier series

$$\tilde{\phi}(\boldsymbol{x}, t) = \sum \tilde{\phi}_k(t) \exp(i\boldsymbol{k} \cdot \boldsymbol{x}), \qquad \tilde{\phi}_k^* = \tilde{\phi}_{-k}, \tag{13.47}$$

where the asterisk means conjugate complex.

Let us consider three waves with wavenumbers $\boldsymbol{k}_1$, $\boldsymbol{k}_2$, and $\boldsymbol{k}_3$ such that $\boldsymbol{k}_1 + \boldsymbol{k}_2 + \boldsymbol{k}_3 = 0$. Let us suppose that these waves have amplitudes larger than other waves in the summation of (13.47) and study the energy flow among these three waves. Equations (13.41) and (13.47) are reduced to

$$\frac{\mathrm{d}\tilde{\phi}_{k1}}{\mathrm{d}t} + i\omega_{k1}\tilde{\phi}_{k1} = \sum_{\boldsymbol{k}1+\boldsymbol{k}2+\boldsymbol{k}3=0} \Lambda^{k1}_{k2,k3}\tilde{\phi}^*_{k2}\tilde{\phi}^*_{k3}, \tag{13.48}$$

$$\Lambda^{k1}_{k2,k3} = \frac{1}{2}\frac{1}{1+k_1^2}((\boldsymbol{k}_2 \times \boldsymbol{k}_3)\cdot\hat{\boldsymbol{z}})(k_3^2 - k_2^2). \tag{13.49}$$

Denoting $\tilde{\phi}_j = \tilde{\phi}_{kj}$, (13.48) is

$$\frac{\mathrm{d}\tilde{\phi}_1}{\mathrm{d}t} + i\omega_1\tilde{\phi}_1 = \Lambda^1_{2,3}\tilde{\phi}^*_2\tilde{\phi}^*_3, \tag{13.50}$$

$$\frac{\mathrm{d}\tilde{\phi}_2}{\mathrm{d}t} + i\omega_2\tilde{\phi}_2 = \Lambda^2_{3,1}\tilde{\phi}^*_3\tilde{\phi}^*_1, \tag{13.51}$$

$$\frac{\mathrm{d}\tilde{\phi}_3}{\mathrm{d}t} + i\omega_3\tilde{\phi}_3 = \Lambda^3_{1,2}\tilde{\phi}^*_1\tilde{\phi}^*_2. \tag{13.52}$$

Without loss of generality, we can assume that

$$k_1 < k_2 < k_3.$$

First we consider a case in which the k_2 mode is highly populated, so that $|\tilde{\phi}_2| \gg |\tilde{\phi}_1|, |\tilde{\phi}_3|$. Then (13.50)∼(13.52) are reduced to

$$\tilde{\phi}_i = A_i \exp(-i\omega_i t),$$

$$A_2 = \text{const.} \qquad \frac{\mathrm{d}A_1}{\mathrm{d}t} = \Lambda^1_{2,3}A^*_2A^*_3\exp(i\theta t), \qquad \frac{\mathrm{d}A_3}{\mathrm{d}t} = \Lambda^3_{1,2}A^*_1A^*_2\exp(i\theta t), \tag{13.53}$$

where $\theta \equiv (\omega_1 + \omega_2 + \omega_3)$ is frequency mismatch. From (13.53), we have

$$\frac{\mathrm{d}^2A_1}{\mathrm{d}t^2} - i\theta\frac{\mathrm{d}A_1}{\mathrm{d}t} - \Lambda^1_{2,3}\Lambda^3_{1,2}|A_2|^2A_1 = 0. \tag{13.54}$$

Hence, instability occurs when

$$\theta^2 - 4\Lambda^1_{2,3}\Lambda^3_{1,2}|A_2|^2 < 0.$$

and the growth rate is

$$\gamma = \left(\Lambda^1_{2,3} \Lambda^3_{1,2} |A_2|^2 - \frac{1}{4}\theta^2 \right)^{1/2} . \tag{13.55}$$

Because of the assumption $k_1 < k_2 < k_3$, then $\Lambda^1_{2,3}\Lambda^3_{1,2} > 0$ and the system can be unstable. The cascade in $\boldsymbol{k}$ space occurs from the wave with wave number k_2 such that $k_1 < k_2 < k_3$ to the waves with wavenumbers k_1 k_3. If the frequency mismatch $\theta = (\omega_1 + \omega_2 + \omega_3)$ is zero, the case occurs from the wave with the highest frequency $\omega_2 = -(\omega_2 + \omega_3)$ to the waves with lower frequencies ω_1 and ω_3.

On the other hand, if modes 1 or 3 are highly populated, the system is stable, since $\Lambda^2_{3,1}, \Lambda^3_{1,2}$ are always negative.

When we introduce a number

$$N_p \equiv \frac{(1 + k_p^2)|\tilde{\phi}_p|^2}{|k_q^2 - k_r^2|}, \tag{13.56}$$

we have the following relations from (13.50)∼(13.52),

$$N_3 - N_1 = \text{const.}, \qquad N_2 + N_3 = \text{const.} \qquad N_1 + N_2 = \text{const.}. \tag{13.57}$$

These equations mean that a loss of 1 in N_2 appears as a gain of 1 in N_1, N_3. N_k mode has the energy $W_k = (1 + k^2)|\tilde{\phi}_k|$ as is seen in (13.44). Hence, from (13.56) and (13.57) the partition of energy of mode N_1 and N_3 is

$$\frac{\Delta W_1}{\Delta W_2} = \frac{k_3^2 - k_2^2}{k_3^2 - k_1^2}, \qquad \frac{\Delta W_3}{\Delta W_2} = \frac{k_2^2 - k_1^2}{k_3^2 - k_1^2}.$$

Computer experiments of $\boldsymbol{k}$ spectrum cascade yield [13.26] that the cascade in k_x plane tends to stop at a critical value of $k_x = k_c$ where the nonlinear term and the linear term in (13.41) are comparable, that is, $(k_x \rho_s)^3 \tilde{\phi} = \kappa_n \rho_s$, $k_c \rho_s \equiv (\kappa_n \rho_s / \tilde{\phi})^{1/3}$ and the energy spectrum tends to condense near $k_x \approx k_c$. With respect to the k_y dependence of the spectrum, energy spectrum tends to condense at $k_y \approx 0$. These results suggest the appearance of zonal flows in drift turbulence.

Evolution of k Spectrum Power Density for $\tilde{\phi}$

Let $\langle\tilde{\phi}\rangle_{pi}$ and $\langle\tilde{\phi}\rangle_{ki}$ be the functions of 2D wave vectors $\boldsymbol{p}_i = (p_{ix}, p_{iy})$, $\boldsymbol{k}_i = (k_{ix}, k_{iy})$ corresponding to the large spatial scales with low freqencies and small spatial scales with high frequencies of $\tilde{\phi}$ and $|\boldsymbol{p}_i| \ll |\boldsymbol{k}_i|$ Then we have

$$\tilde{\phi} = \langle\tilde{\phi}_p\rangle + \tilde{\phi}_k.$$

The bracket means average over the fast time scales. Then an average of $\tilde{\phi}_1\tilde{\phi}_2$ over the fast time scales is given by

$$\langle\tilde{\phi}_1\tilde{\phi}_2\rangle = \langle\tilde{\phi}_{1p}\rangle\langle\tilde{\phi}_{2p}\rangle + \langle\tilde{\phi}_{1k}\tilde{\phi}_{2k}\rangle.$$

To obtain the equation for the evolution of large scale components, let us Fourier transform (13.46) and average over the characteristic times of small scale [13.28]:

$$\partial_t(p^2+1)\langle\tilde{\phi}_p\rangle - i(\kappa_n\rho_s)\partial_y\langle\tilde{\phi}_p\rangle - \int [\boldsymbol{p}_2,\boldsymbol{p}_1]p_2^2\langle\tilde{\phi}_{p1}\rangle\langle\tilde{\phi}_{p2}\rangle\delta(\boldsymbol{p}_1+\boldsymbol{p}_2-\boldsymbol{p})\mathrm{d}\boldsymbol{p}_1\mathrm{d}\boldsymbol{p}_2$$

$$-\int [\boldsymbol{k}_2,\boldsymbol{k}_1]k_2^2\langle\tilde{\phi}_{k1}\tilde{\phi}_{k2}\rangle\delta(\boldsymbol{k}_1+\boldsymbol{k}_2-\boldsymbol{p})\mathrm{d}\boldsymbol{k}_1\mathrm{d}\boldsymbol{k}_2 = 0, \qquad (13.58)$$

where [ ] means the z component of vector multiplication $[\boldsymbol{a},\boldsymbol{b}] = a_xb_y - a_yb_x$ and $\partial_t = \partial/\partial t$. By introducing $\boldsymbol{k}$ and $\boldsymbol{p}$ as follows

$$\boldsymbol{k}_1 = -\boldsymbol{k} + \frac{1}{2}\boldsymbol{p}, \qquad \boldsymbol{k}_2 = \boldsymbol{k} + \frac{1}{2}\boldsymbol{p},$$

$$[\boldsymbol{k}_2,\boldsymbol{k}_1]k_2^2 = \big((k_x+p_x/2)(-k_y+p_y/2) - (k_y+p_y/2)(-k_x+p_x/2)\big)(\boldsymbol{k}+\boldsymbol{p}/2)^2$$
$$= [\boldsymbol{k},\boldsymbol{p}](\boldsymbol{k}\cdot\boldsymbol{p}+k^2+p^2/4),$$

the second integral in (13.58) can be rewritten in the form of

$$\int [\boldsymbol{k},\boldsymbol{p}](\boldsymbol{k}+\boldsymbol{p}/2)^2\langle\tilde{\phi}_{p/2-k}\tilde{\phi}_{p/2+k}\rangle\mathrm{d}\boldsymbol{k} = \int [\boldsymbol{k},\boldsymbol{p}](\boldsymbol{k}\cdot\boldsymbol{p})\langle\tilde{\phi}_{p/2-k}\tilde{\phi}_{p/2+k}\rangle\mathrm{d}\boldsymbol{k}$$

$$= \int (-k_xk_y(p_x^2-p_y^2) - (k_y^2-k_x^2)p_xp_y)\langle\tilde{\phi}_{p/2-k}\tilde{\phi}_{p/2+k}\rangle\mathrm{d}\boldsymbol{k},$$

where $[\boldsymbol{k},\boldsymbol{p}](k^2+p^2/4) = -[\boldsymbol{k}_1,\boldsymbol{k}_2](k_1^2+k_2^2)/2$. Performing an inverse Fourier transform on (13.58), we get the following equation for the evolution of large scales:

$$\partial_t(\nabla^2\tilde{\phi}_{\mathrm{L}} - \tilde{\phi}_{\mathrm{L}}) - (\kappa_n\rho_s)\partial_y\tilde{\phi}_{\mathrm{L}} + (\partial_y\nabla^2\tilde{\phi}_{\mathrm{L}})\partial_x\tilde{\phi}_{\mathrm{L}} - (\partial_x\nabla^2\tilde{\phi}_{\mathrm{L}})\partial_y\tilde{\phi}_{\mathrm{L}}$$
$$= -\partial_{xx}A - \partial_x\partial_yB + \partial_{yy}A, \qquad (13.59)$$

where

$$A(\boldsymbol{x},t) = 2\int \frac{k_xk_y}{k^2(1+k^2)}n_k\mathrm{d}\boldsymbol{k},$$

$$B(\boldsymbol{x},t) = 2\int \frac{k_y^2-k_x^2}{k^2(1+k^2)}n_k\mathrm{d}\boldsymbol{k},$$

$$n_k(\boldsymbol{k},\boldsymbol{x},t) \equiv \frac{1}{2}k^2(1+k^2)\int \langle\tilde{\phi}_{p/2-k}\tilde{\phi}_{p/2+k}\rangle\exp(i\boldsymbol{p}\cdot\boldsymbol{x})\frac{\mathrm{d}\boldsymbol{p}}{(2\pi)^2}. \qquad (13.60)$$

n_k is the power density of high-frequency spectrum of $\tilde{\phi}$.

To derive the evolution equation for the density of high-frequency n_k, the evolution equation for the Fourier component $\tilde{\phi}_k$ is described as follows:

$$-\partial_t(1+k^2)\tilde{\phi}_k - (\kappa_n\rho_s)ik_y\tilde{\phi}_k$$

$$-\int(-k'_y k'^2\tilde{\phi}_{k'}q_x\tilde{\phi}_q)+k'_x k'^2\tilde{\phi}_{k'}q_y\tilde{\phi}_q)\delta(\boldsymbol{q}+\boldsymbol{k}'-\boldsymbol{k})\mathrm{d}\boldsymbol{k}'\mathrm{d}\boldsymbol{q}=0.$$

Integrating by $\boldsymbol{k}'$ and taking account of

$$k'^2(k'_x q_y-k'_y q_x)=[\boldsymbol{k},\boldsymbol{q}](\boldsymbol{k}-\boldsymbol{q})^2,\qquad(\boldsymbol{k}'=\boldsymbol{k}-\boldsymbol{q}),$$

we have

$$\partial_t\tilde{\phi}_k+\frac{i(\kappa_n\rho_s)k_y}{1+k^2}\tilde{\phi}_k+\int\frac{[\boldsymbol{k},\boldsymbol{q}](\boldsymbol{k}-\boldsymbol{q})^2}{1+k^2}\tilde{\phi}_{k-q}\tilde{\phi}_q\mathrm{d}\boldsymbol{q}=0.$$

Let us take the sum of the evolution equation for $\tilde{\phi}_k$ muliplied by $\tilde{\phi}_{k'}$ and the similar equation for $\tilde{\phi}_{k'}$ muliplied by $\tilde{\phi}_k$; as a result we get

$$\partial_t(\tilde{\phi}_k\tilde{\phi}_{k'})+i(\omega_k^1+\omega_{k'}^1)(\tilde{\phi}_k\tilde{\phi}_{k'})+R_{k\,k'}=0,\tag{13.61}$$

where

$$R_{k\,k'}\equiv\int\left(\frac{[\boldsymbol{k},\boldsymbol{q}](\boldsymbol{k}-\boldsymbol{q})^2}{1+k^2}\tilde{\phi}_{k-q}\tilde{\phi}_{k'}+\frac{[\boldsymbol{k}',\boldsymbol{q}](\boldsymbol{k}'-\boldsymbol{q})^2}{1+k^2}\psi_{k'-q}\tilde{\phi}_k\right)\tilde{\phi}_q\mathrm{d}\boldsymbol{q}.$$

and $\omega_k^1=(\kappa_n\rho_s)k_y/(1+k^2)$.

Let us suppose that the concentration of turbulence spectrum at large scales is high enough to neglect the interaction of small scales among themselves in comparison with their interaction with large scales. The correlator of the small-scale field $\langle\tilde{\phi}_k\tilde{\phi}_{k'}\rangle$ is of an appreciable value only for the small values of $\boldsymbol{k}+\boldsymbol{k}'$, which are comparable to the characteristic wave number of large-scale motions. Taking account of these arguments, $\boldsymbol{k}$ and $\boldsymbol{k}'$ are replaced by $-\boldsymbol{k}+\boldsymbol{p}/2$ and $\boldsymbol{k}+\boldsymbol{p}/2$ as follows:

$$\boldsymbol{k}\to-\boldsymbol{k}+\boldsymbol{p}/2,\qquad\boldsymbol{k}'\to\boldsymbol{k}+\boldsymbol{p}/2,\qquad\tilde{\phi}_k\tilde{\phi}_{k'}\to\tilde{\phi}_{p/2-k}\tilde{\phi}_{p/2+k}.$$

Then we have

$$(\kappa_n\rho_s)^{-1}i(\omega_k^1+\omega_{k'}^1)=i\frac{-(k_y-p_y/2)}{1+(\boldsymbol{k}-\boldsymbol{p}/2)^2}+\frac{(k_y+p_y/2)}{1+(\boldsymbol{k}+\boldsymbol{p}/2)^2}=\frac{\partial}{\partial\boldsymbol{k}}\left(\frac{k_y}{1+k^2}\right)\cdot i\boldsymbol{p},$$

as $|\boldsymbol{p}|\ll|\boldsymbol{k}|$, and the first two terms of (13.61) reduce to

$$\partial_t n_k+(\kappa_n\rho_s)\frac{\partial}{\partial\boldsymbol{k}}\left(\frac{k_y}{1+k^2}\right)\cdot\frac{\partial n_k}{\partial\boldsymbol{x}}.$$

The first integral of $R_{kk'}$ reduces to

$$\text{first term of }R_{p/2-k,p/2+k}=$$

$$\int\left(\frac{-[\boldsymbol{k}-\boldsymbol{p}/2,\boldsymbol{q}](\boldsymbol{k}-\boldsymbol{p}/2+\boldsymbol{q})^2}{1+(\boldsymbol{k}-\boldsymbol{p}/2)^2}\tilde{\phi}_{-k+p/2-q}\tilde{\phi}_{k+p/2}\right)\tilde{\phi}_q\mathrm{d}\boldsymbol{q}$$

$$= \int \frac{-[\boldsymbol{k}-\boldsymbol{p}'/2,\boldsymbol{q}](\boldsymbol{k}-\boldsymbol{p}'/2+\boldsymbol{q}/2)^2}{1+(\boldsymbol{k}-\boldsymbol{p}'/2-\boldsymbol{q}/2)^2}\frac{2}{(\boldsymbol{k}+\boldsymbol{q}/2)^2(1+(\boldsymbol{k}+\boldsymbol{q}/2)^2)^2)}$$

$$\times\left(\tilde{\phi}_{p'/2-(k+q/2)}\tilde{\phi}_{p'/2+(k+q/2)}\frac{(\boldsymbol{k}+\boldsymbol{q}/2)^2(1+(\boldsymbol{k}+\boldsymbol{q}/2)^2)^2)}{2}\right)\tilde{\phi}_q \mathrm{d}\boldsymbol{q},$$

where $\boldsymbol{p}=\boldsymbol{p}'+\boldsymbol{q}$. Finally we have

$$\frac{1}{2}k^2(1+k^2)\int R_{p/2-k,p/2+k}\exp(i\boldsymbol{p}\cdot\boldsymbol{x})\frac{\mathrm{d}\boldsymbol{p}}{(2\pi)^2}$$

$$=\frac{1}{2}k^2(1+k^2)\int R_{p/2-k,p/2+k}\exp((i\boldsymbol{p}'\cdot\boldsymbol{x})+i\boldsymbol{q}\cdot\boldsymbol{x}))\frac{\mathrm{d}\boldsymbol{p}'}{(2\pi)^2}$$

$$=\frac{-k^2}{1+k^2}[\boldsymbol{k},\boldsymbol{q}](n_{k+q/2}-n_{k-q/2})\tilde{\phi}_{\mathrm{L}}+\frac{2[\boldsymbol{k},\boldsymbol{q}]}{(1+k^2)}(\boldsymbol{k}\cdot\boldsymbol{p}')n_k\tilde{\phi}_{\mathrm{L}}+\frac{k^2}{(1+k^2)^2}[\boldsymbol{p}',\boldsymbol{q}]n_k\tilde{\phi}_{\mathrm{L}}$$

$$=\frac{-k^2}{1+k^2}[\boldsymbol{k},\boldsymbol{q}]\frac{\partial n_k}{\partial\boldsymbol{k}}\cdot\boldsymbol{q}\tilde{\phi}_{\mathrm{L}}+\frac{2[\boldsymbol{k},\boldsymbol{q}]}{(1+k^2)}(\boldsymbol{k}\cdot\boldsymbol{p}')n_k\tilde{\phi}_{\mathrm{L}}+\frac{k^2}{(1+k^2)^2}[\boldsymbol{p}',\boldsymbol{q}]n_k\tilde{\phi}_{\mathrm{L}}.$$

Here $|\boldsymbol{q}|,|\boldsymbol{p}'|\ll|\boldsymbol{k}|$ is assumed and the integrand of $R_{kk'}$ was expanded. Equation (13.61) is reduced to

$$\partial_t n_k+(\kappa_n\rho_{\mathrm{s}})\frac{\partial}{\partial\boldsymbol{k}}\left(\frac{k_y}{1+k^2}\right)\cdot\nabla_x n_k-\frac{k^2}{1+k^2}[\boldsymbol{k},\boldsymbol{q}]\frac{\partial n_k}{\partial\boldsymbol{k}}\cdot\boldsymbol{q}\tilde{\phi}_{\mathrm{L}}+\frac{2[\boldsymbol{k},\boldsymbol{q}]}{(1+k^2)}(\boldsymbol{k}\cdot\boldsymbol{p}')n_k\tilde{\phi}_{\mathrm{L}}$$

$$+\frac{k^2}{(1+k^2)^2}[\boldsymbol{p}',\boldsymbol{q}]n_k\tilde{\phi}_{\mathrm{L}}=0. \tag{13.62}$$

If ω_k is newly introduced,

$$\omega_k=\omega_k^{\mathrm{l}}+\omega_k^{\mathrm{nl}}\equiv\frac{(\kappa_n\rho_{\mathrm{s}})k_y+(\hat{\boldsymbol{v}}_E\cdot\boldsymbol{k})k^2}{1+k^2},\qquad \hat{\boldsymbol{v}}_E\equiv\frac{\boldsymbol{v}_E}{c_{\mathrm{s}}}=\left(-\frac{\partial\tilde{\phi}}{\partial\hat{y}},\frac{\partial\tilde{\phi}}{\partial\hat{x}}\right),$$

(13.62) is reduced to [13.28]

$$\left(\partial_t+\frac{\partial\omega_k}{\partial\boldsymbol{k}}\partial_x-\frac{\partial\omega_k}{\partial\boldsymbol{x}}\partial_k\right)n_k=0. \tag{13.63}$$

When the ordinary coordinates and time are used instead of normalized dimensionless coordinates and time, (13.63) becomes

$$\left(\frac{\partial}{\partial t}+\frac{\partial\omega_k}{\partial\boldsymbol{k}}\frac{\partial}{\partial x}-\frac{\partial\omega_k}{\partial\boldsymbol{x}}\frac{\partial}{\partial k}\right)n_k=0,\qquad \omega_k=\frac{k_y v_{\mathrm{d}}^*+(\boldsymbol{v}_E\cdot\boldsymbol{k})\rho_{\mathrm{s}}^2k^2}{1+\rho_{\mathrm{s}}k^2}. \tag{13.64}$$

13.5.2 Generation of Zonal Flow

The scale of sheared mean flow is macroscopic and the stabilizing effect of sheared mean flow on drift turbulence was discussed in Section 4.6. On the other hand, the scale of zonal flow is mesoscopic scale. The difference of sheared mean flow and zonal flow is illustrated in Figure 13.8.

Numerical simulation studies of drift turbulence have played a crucial role in the developement of research on zonal flow. The synergy between the theory and numerical simulation has been the key promotor of interest in the physics of zonal flow. A result of numerical simulation based on the gyrokinetic particle model (refer to Section 13.4.1) is shown in Figure 13.9 [13.29]. It is clearly seen in this figure that radial size of turbulent eddies is greatly reduced by the shearing of zonal flows.

The potential ϕ_z of zonal field has the structure of $k_y = k_z = 0$ on the magnetic surface with finite k_x (radial direction), so that the phase velocity of ϕ_z along the magnetic field line is infinity. Therefore, adiabatic response of the potential is modified to

$$\frac{\delta n}{n_0} = \frac{e}{T_{\rm e}}(\phi - \bar{\phi}), \qquad \phi = \bar{\phi} + \delta\phi_{\rm d},$$

where $\bar{\phi}$ is a spatial average of ϕ on a magnetic surface. Therefore, adiabatic response is

$$\frac{\delta n}{n_0} = \frac{e\delta\phi_{\rm d}}{T_{\rm e}} \equiv \tilde{\phi}_{\rm d}.$$

We assume the following ordering:

$$\frac{e\bar{\phi}}{T_{\rm e}} \equiv \bar{\phi}_{\rm z}, \qquad \tilde{\phi}_{\rm d} \sim \delta, \qquad \bar{\phi}_{\rm z} \sim \delta, \qquad \nabla\tilde{\phi}_{\rm d} \sim O(1), \qquad \nabla\bar{\phi}_{\rm z} \sim O(1).$$

The equation of continuity for ions is derived [13.30] by modifying (13.41) to

$$(1 - \rho_{\rm s}^2\nabla^2)\frac{\partial\tilde{\phi}_{\rm d}}{\partial t} - \rho_{\rm s}c_{\rm s}(\nabla\tilde{\phi}_{\rm d} \times \hat{\boldsymbol{z}}) \cdot \frac{\nabla n_0}{n_0} - \rho_{\rm s}c_{\rm s}(\nabla\bar{\phi}_{\rm z} \times \hat{\boldsymbol{z}}) \cdot \nabla\tilde{\phi}_{\rm d}$$

$$+\rho_{\rm s}^3 c_{\rm s}\nabla \cdot \Big(((\nabla\bar{\phi}_{\rm z} \times \hat{\boldsymbol{z}}) \cdot \nabla)\nabla\tilde{\phi}_{\rm d} + ((\nabla\tilde{\phi}_{\rm d} \times \hat{\boldsymbol{z}}) \cdot \nabla)\nabla\bar{\phi}_{\rm z}\Big) = 0, \qquad (13.65)$$

$$-\frac{\partial}{\partial t}\rho_{\rm s}^2\nabla^2\bar{\phi}_{\rm z} + \rho_{\rm s}^3 c_{\rm s}\nabla \cdot \Big((\nabla\tilde{\phi}_{\rm d} \times \hat{\boldsymbol{z}} \cdot \nabla)\nabla\tilde{\phi}_{\rm d}\Big) = 0. \qquad (13.66)$$

A low-dimensional model for generation of zonal flow is developed under the assumption of

$$\tilde{\phi}_{\rm d} = \tilde{\phi}_{\rm d0}(t)\cos(k_y y - \omega_0 t) + \tilde{\phi}_{\rm d\,c}(t)\sin(k_x x - \omega_{\rm z} t)\cos(k_y y - \omega_0 t)$$

$$+\tilde{\phi}_{\rm d\,s}(t)\sin(k_x x - \omega_{\rm z} t)\sin(k_y - \omega_0 t), \qquad (13.67)$$

$$\bar{\phi}_{\rm z} = \bar{\phi}_{\rm z}(t)\cos(k_x x - \omega_{\rm z} t), \qquad (13.68)$$

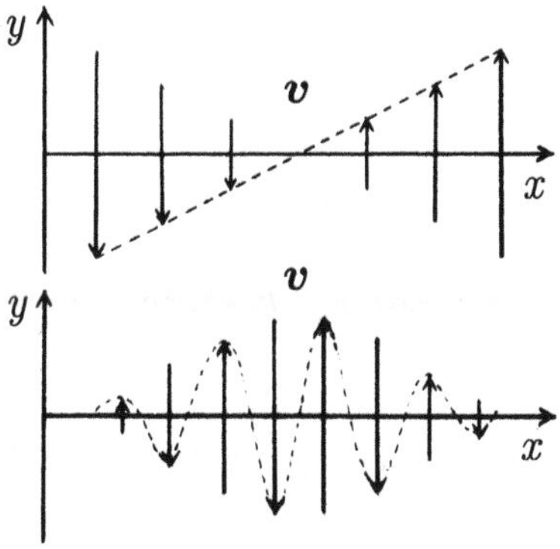

Figure 13.8 Upper figure: Sheared mean flow. Lower figure: Zonal flow.

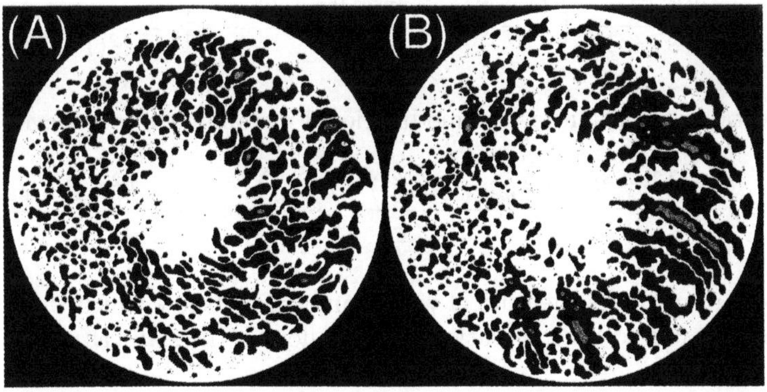

Figure 13.9 Radial size of turbulent eddies shown in poloidal contour plots of ambient potential fluctuation $e\phi/T_e$ gets reduced due to random shearing by self-regulated $\boldsymbol{E} \times \boldsymbol{B}$ zonal flow from gyrokinetic particle simulation with zonal flow (A) and with zonal flow suppressed (B). After [13.29].

with

$$\omega_0 = \frac{k_y v_{\rm d}^*}{1+\rho_{\rm s}^2 k^2}.$$

When (13.67) and (13.68) are substituted in (13.65) and (13.66), components of $\cos(k_y y - \omega_0 t)$, $\sin(k_x x - \omega_{\rm z} t)\cos(k_y y - \omega_0 t)$, $\sin(k_x x - \omega_{\rm z} t)\sin(k_y - \omega_0 t)$ and $\cos(k_x x - \omega_{\rm z} t)$ reduce to

$$(1+\rho_{\rm s}^2 k^2)\frac{{\rm d}\tilde{\phi}_{\rm d\,0}}{{\rm d}t} - \frac{\rho_{\rm s} c_{\rm s}}{2} k_x k_y \bar{\phi}_{\rm z} \tilde{\phi}_{\rm d\,s} = 0,$$

$$(1+\rho_{\rm s}^2 k^2)\frac{{\rm d}\tilde{\phi}_{\rm d\,c}}{{\rm d}t} - \frac{k_y k_x^2 \rho_{\rm s}^2 v_{\rm d}^*}{1+k_y^2\rho_{\rm s}^2}\tilde{\phi}_{\rm d\,s} = 0,$$

$$(1+\rho_{\rm s}^2 k^2)\frac{{\rm d}\tilde{\phi}_{\rm d\,s}}{{\rm d}t} + \frac{k_y k_x^2 \rho_{\rm s}^2 v_{\rm d}^*}{1+k_y^2\rho_{\rm s}^2}\tilde{\phi}_{\rm d\,c} + \rho_{\rm s} c_{\rm s} k_x k_y (1 + k_y^2\rho_{\rm s}^2 - k_x^2\rho_{\rm s}^2)\bar{\phi}_{\rm z}\tilde{\phi}_{\rm d\,0} = 0,$$

$$\frac{{\rm d}\bar{\phi}_{\rm z}}{{\rm d}t} + \frac{\rho_{\rm s} c_{\rm s}}{2} k_x k_y \tilde{\phi}_{\rm d\,0}\tilde{\phi}_{\rm d\,s} = 0.$$

If we assume that the drift wave $\tilde{\phi}_{\rm d\,0}$ is pumped and the following conditions are satisfied,

$$|\tilde{\phi}_{\rm d\,c}|, |\tilde{\phi}_{\rm d\,s}|, |\bar{\phi}_{\rm z}| \ll |\tilde{\phi}_{\rm d\,0}|,$$

and that the perturbed state variables have a time dependence given as $\exp(\gamma t)$, then the linearized system of equations under the assumption $\tilde{\phi}_{\rm d\,0} =$ const. yields the following dispersion relation:

$$\gamma^2 = \frac{c_{\rm s}^4 k_x^2 k_y^2}{\Omega_{\rm i}^2(1+k^2\rho_{\rm s}^2)^2}\left(\frac{(1+k_y^2\rho_{\rm s} - k_x^2\rho_{\rm s}^2)(1+k^2\rho_{\rm s}^2)}{2}\tilde{\phi}_{\rm d\,0}^2 - \frac{k_x^2\rho_s^2 v_{\rm d}^{*2}}{c_{\rm s}^2(1+k^2\rho_{\rm s}^2)^2}\right). \tag{13.69}$$

The unstable condition for the modulational instability of four waves is

$$k_{x{\rm c}}^2\rho_{\rm s}^2 < \frac{(1+k^2\rho_{\rm s}^2)^3((1+k_y^2\rho_{\rm s} - k_x^2\rho_{\rm s}^2))}{2}\left(\frac{c_{\rm s}\tilde{\phi}_{\rm d0}}{v_{\rm d}^*}\right)^2 \approx \frac{(1+k_y^2\rho_{\rm s})^4}{2}\left(\frac{c_{\rm s}\tilde{\phi}_{\rm d0}}{v_{\rm d}^*}\right)^2. \tag{13.70}$$

This condition shows the critical wave number $k_{x{\rm c}}$ below which the drift wave is unstable to the generation of zonal flow in the slab case.

There is the other intuitive way to analyze the driving mechanism of zonal flow [13.31]. Let us assume that $\boldsymbol{v} = \bar{\boldsymbol{v}} + \tilde{\boldsymbol{v}}$, where $\bar{\boldsymbol{v}}$ is the component of large spatial scale with low frequency and $\tilde{\boldsymbol{v}}$ is the component of small spatial scale with high frequency. Since the potential of zonal flow has $k_\theta = k_z = 0$ and $k_r \neq 0$, a component of poloidal zonal flow can be extracted by average mean of Reynold stress over a magnetic surface. The second order of

$$\frac{\partial \boldsymbol{v}}{\partial t} = -(\boldsymbol{v}\cdot\nabla)\boldsymbol{v} + \cdots$$

becomes

$$\frac{\partial \bar{v}_\theta^{\mathrm{z}}}{\partial t} = -\langle((\tilde{\boldsymbol{v}}\cdot\nabla)\tilde{\boldsymbol{v}})_\theta\rangle + \cdots = -\frac{1}{r^2}\frac{\partial}{\partial r}\langle(r^2\tilde{v}_r\tilde{v}_\theta)\rangle + \cdots \approx \frac{\partial}{\partial r}\langle(\tilde{v}_r\tilde{v}_\theta)\rangle + \cdots . \tag{13.71}$$

with use of the following relation:

$$((\tilde{\boldsymbol{v}}\cdot\nabla)\tilde{\boldsymbol{v}})_\theta = (\tilde{\boldsymbol{v}}\cdot\nabla)\tilde{v}_\theta + \frac{\tilde{v}_r\tilde{v}_\theta}{r} = \frac{1}{r^2}\frac{\partial}{\partial r}(r^2\tilde{v}_r\tilde{v}_\theta) + \frac{1}{r}\frac{\partial\tilde{v}_\theta^2}{\partial\theta} + \frac{\partial}{\partial z}(\tilde{v}_\theta\tilde{v}_z).$$

The notation $\langle\ \rangle$ means the average over a magnetic surface (θ, z). Since

$$\langle(\tilde{v}_r\tilde{v}_\theta)\rangle = \frac{k_r k_\theta}{B^2}\langle\phi^2\rangle = c_{\mathrm{s}}^2(k_r k_\theta \rho_{\mathrm{s}}^2)\tilde{\phi}^2,$$

$$\tilde{\phi}^2 = \int\int(\tilde{\phi}_{k'}\tilde{\phi}_{p-k'})\mathrm{d}\boldsymbol{k}'\exp(i\boldsymbol{p}\cdot\boldsymbol{x})\mathrm{d}\boldsymbol{p} = \int\int(\tilde{\phi}_{p/2+k}\tilde{\phi}_{p/2-k})\exp(i\boldsymbol{p}\cdot\boldsymbol{x})\mathrm{d}\boldsymbol{p}\mathrm{d}\boldsymbol{k}$$

$$= \int\frac{2n_k}{\rho_{\mathrm{s}}^2k^2(1+\rho_{\mathrm{s}}^2k^2)}\mathrm{d}\boldsymbol{k}\rho_{\mathrm{s}}^2,$$

(13.71) becomes [13.31]

$$\frac{\partial \bar{v}_\theta^{\mathrm{z}}}{\partial t} = -\frac{\partial}{\partial r}\int c_{\mathrm{s}}^2\frac{(k_r k_\theta\rho_{\mathrm{s}}^2)2n_k}{\rho_{\mathrm{s}}^2k^2(1+\rho_{\mathrm{s}}^2k^2)}\mathrm{d}k_r\mathrm{d}k_\theta\rho_{\mathrm{s}}^2 - \gamma_{\mathrm{damp}}\bar{v}_\theta^{\mathrm{z}}, \tag{13.72}$$

where n_k is the power density of high-frequency spectrum of $\tilde{\phi}$ and the equation of n_k was described by (13.64). n_k can be obtained by linearizing (13.64) for small perturbations $n_k^{(1)}$ with the frequency Ω and the radial wave number K,

$$n_k^{(1)} \propto \exp(iKr - i\Omega t),$$

and

$$i\left(-\Omega + \frac{\partial\omega_k}{\partial k_r}K + i\gamma\right)n_k^{(1)} - \frac{\partial\omega_k}{\partial r}\frac{\partial n_k^{(0)}}{\partial k_r} = 0,$$

and

$$n_k^{(1)} = \frac{\frac{\partial\omega_k}{\partial r}\frac{\partial n_k^{(0)}}{\partial k_r}}{-i\left(\Omega - \frac{\partial\omega_k}{\partial k_r}K - i\gamma\right)} = R(K,\Omega)\frac{\partial}{\partial r}(k_\theta\bar{v}_\theta^{\mathrm{z}})\frac{\partial n_k^{(0)}}{\partial k_r},$$

$$R(K,\Omega) = \frac{i}{\Omega - \frac{\partial\omega_k}{\partial k_r}K - i\gamma} = \frac{-\gamma + i(\Omega - (\partial\omega_k/\partial k_r)K)}{(\Omega - (\partial\omega_k/\partial k_r)K)^2 + \gamma^2}.$$

Finally, we obtain

$$\frac{\partial \bar{v}_\theta^{\mathrm{z}}}{\partial t} = -\frac{\partial}{\partial r}\int c_{\mathrm{s}}^2\frac{(k_r k_\theta\rho_{\mathrm{s}}^2)2}{(1+\rho_{\mathrm{s}}^2k^2)}Rk_\theta\frac{\partial\bar{v}_\theta^{\mathrm{z}}}{\partial r}\frac{\partial n_k^{(0)}}{\partial k_r}\mathrm{d}k_r\mathrm{d}k_\theta\rho_{\mathrm{s}}^2 - \gamma_{\mathrm{damp}}\bar{v}_\theta^{\mathrm{z}}, \tag{13.73}$$

(13.73) is the equation of diffusion

$$\frac{\partial \bar{v}_\theta^{\rm z}}{\partial t} = \frac{\partial}{\partial r} D_{rr} \frac{\partial \bar{v}_\theta^{\rm z}}{\partial r} - \gamma_{\rm damp} \bar{v}_\theta^{\rm z}, \tag{13.74}$$

if the following diffusion coefficient is introduced:

$$D_{rr} = -c_{\rm s}^2 \int \frac{2(k_\theta \rho_{\rm s})^2}{(1+\rho_{\rm s}^2 k^2)} R(K,\Omega) k_r \frac{\partial n_k^{(0)}}{\partial k_r} \rho_{\rm s}^2 {\rm d}k_r {\rm d}k_\theta. \tag{13.75}$$

The growth rate $\gamma_{\rm z}$ of zonal flow due to the first term of (13.75) is

$$\gamma_{\rm z} = -K^2 {\rm Re}({\rm D_{rr}}) = -K^2 c_{\rm s}^2 \int \frac{2(k_\theta \rho_{\rm s})^2}{(1+\rho_{\rm s}^2 k^2)} \frac{\gamma k_r (\partial n_k^{(0)}/\partial k_r)}{(\Omega - (\partial \omega_k/\partial k_r)K)^2 + \gamma^2} \rho_{\rm s}^2 {\rm d}k_r {\rm d}k_\theta. \tag{13.76}$$

$k_r(\partial n_k/\partial r)$ is negative in typical drift turbulence, so that $\gamma_{\rm z}$ is positive and zonal flows are driven, if the growth rate is larger than damping rate $\gamma_{\rm damp}$ of zonal flow, that is,

$$k_r \frac{\partial n_{(0)}}{\partial k_r} < 0, \qquad \gamma_{\rm z} - \gamma_{\rm damp} > 0. \tag{13.77}$$

In other words, when the diffusion coefficient is negative, the system can be unstale.

13.5.3 Geodesic Acoustic Mode (GAM)

When one constructs an eigenmode in the regime of time scale of $|\partial/\partial t| \sim \omega_{\rm t} = v_{T_{\rm i}}/qR$, one finds the geodesic acoustic mode (GAM) [13.32]. GAM is a perturbation for which $m = n = 0$ electrostatic potential is linearly couped by toroidal effect to $m = 1/n = 0$ side band density perturation. The equation of continuity and the equation of motion are

$$\frac{\partial n}{\partial t} + \nabla \cdot (n \boldsymbol{v}_\perp) + \nabla_\parallel (n v_\parallel) = 0, \tag{13.78}$$

$$n m_{\rm i} \left(\frac{\partial}{\partial t} \boldsymbol{v} + (\boldsymbol{v} \cdot \nabla) \boldsymbol{v} \right) = -\nabla p + \boldsymbol{j} \times \boldsymbol{B}, \tag{13.79}$$

$$\nabla \cdot \boldsymbol{j} = 0, \qquad \boldsymbol{E} + \boldsymbol{v} \times \boldsymbol{B} = 0.$$

Since $\nabla \cdot (\boldsymbol{E} \times \boldsymbol{B}/B^2) = -\boldsymbol{E} \cdot \nabla \times (\boldsymbol{B}/B^2) = 2\boldsymbol{E} \cdot (\nabla B \times \boldsymbol{b})/B^2$ $= 2(E_r/B)\sin\theta/R$ due to the geodesic curvature of magnetic field line on the magnetic surface, a density accumulation is caused due to the equation of continuity ($\tilde{n}$ is normalized density fluctuation):

$$\frac{\partial \tilde{n}}{\partial t} - \frac{2}{R} \sin\theta \tilde{v}_{E\times B} + \nabla_\parallel \tilde{v}_\parallel = 0. \tag{13.80}$$

Since the axisymmetric toroidal field is given by (refer to Section 4.2)

$$\boldsymbol{B} = B_{\rm t}\boldsymbol{i}_\varphi + \frac{1}{R}\nabla\psi \times \boldsymbol{i}_\varphi,$$

the poloidal component of the equation of motion is given by

$$\boldsymbol{B}_{\rm p} \cdot \left(nm_{\rm i}\frac{\rm d}{{\rm d}t}\boldsymbol{v} + (T_{\rm i} + T_{\rm e})\nabla n \right) = B_{\rm t}R^{-1}\boldsymbol{j} \cdot \nabla\psi.$$

The density accumulation generates radial current across the magnetic surface and acts to reverse $\tilde{\boldsymbol{E}}$ [13.32]. Since $\int \boldsymbol{j} \cdot \boldsymbol{n}{\rm d}S = 0$ and $RB_{\rm t}$ =const, we have

$$\int \frac{R^2}{|\nabla\psi|}\frac{B_{\rm t}(\boldsymbol{j} \cdot \nabla\psi)}{R}{\rm d}S = \int \frac{R^2}{|\nabla\psi|}\boldsymbol{B}_{\rm p} \cdot \left(nm_{\rm i}\frac{\rm d}{{\rm d}t}\boldsymbol{v} + (T_{\rm i} + T_{\rm e})\nabla n \right) {\rm d}S = 0,$$

and the poloidal components of the equation of motion is reduced to [13.33] $({\rm d}S = 2\pi(R_0 + r\cos\theta)r{\rm d}\theta,\ c_{\rm s}^2 \equiv (T_{\rm i} + T_{\rm e})/m_{\rm i})$

$$\frac{\partial \tilde{v}_{E\times B}}{\partial t} + \frac{2c_{\rm s}^2}{2\pi R}\oint \tilde{n}\sin\theta{\rm d}\theta = 0. \tag{13.81}$$

The parallel components of equation of motion are

$$\frac{\partial \tilde{v}_\parallel}{\partial t} + c_{\rm s}^2\nabla_\parallel \tilde{n} = 0. \tag{13.82}$$

In the case in which density perturbaton is in the form of $\tilde{n} = \tilde{n}_1\sin\theta$, we have

$$\frac{\partial \tilde{v}_{E\times B}}{\partial t} = -\frac{c_{\rm s}^2}{R}\tilde{n}, \tag{13.83}$$

$$\frac{\partial^2 \tilde{n}}{\partial t^2} + \frac{2}{R^2}\sin\theta\frac{c_{\rm s}^2}{R}\tilde{n}_1 - c_{\rm s}^2\nabla_\parallel^2\tilde{n} = 0$$

and the dispersion relation is $(\tilde{n} \propto \exp(ik_r r + ik_\parallel l - i\Omega t)\sin\theta)$

$$-\Omega^2 + \frac{2c_{\rm s}^2}{R^2} + k_\parallel^2 c_{\rm s}^2 = 0.$$

The frequency of GAM is given by $(k_\parallel = 1/q_{\rm s}R)$

$$\omega_{\rm GAM}^2 = 2c_{\rm s}^2R^{-2}\left(1 + \frac{k_\parallel^2R^2}{2}\right) = c_{\rm s}^2\left((k_\parallel^2 + \frac{2}{R^2}\right) = 2c_{\rm s}^2R^{-2}(1+q_{\rm s}^{-2}/2) \approx 2c_{\rm s}^2R^{-2}, \tag{13.84}$$

$$v_{{\rm ph}\parallel} = \omega_{\rm GAM} \approx 2^{1/2}q_{\rm s}c_{\rm s}.$$

When $q_{\rm s} \gg 1$ or $T_{\rm e} \gg T_{\rm i}$, GAM is not damped by ion Landau damping. $\tilde{n}$ of GAM is $k_r\rho_{\rm s}$ times smaller than normalized electrostatic potential as is reduced from (13.83) and (13.84) $(E_r = -ik_r\phi)$,

$$\tilde{n} = 2^{1/2}k_r\rho_{\rm s}(1 + q_{\rm s}^{-2}/2)^{1/2}\frac{e\phi}{T_{\rm e}}.$$

In the case in which density perturbation has dependence other than $\sin\theta$, we have

$$-\Omega i\tilde{n} - \frac{2}{R}\sin\theta\tilde{v}_{E\times B} + \nabla\tilde{v}_{\parallel} = 0, \quad ik_{\parallel}c_{\rm s}^2\tilde{n} - i\Omega\tilde{v}_{\parallel} = 0, \quad -i\Omega\tilde{v}_{E\times B} = 0.$$

One solution is zonal flow branch ($\nabla_{\parallel} = (\partial\theta/\partial l)\partial\partial\theta = (1/q_{\rm s}R)\partial/\partial\theta$);

$$\Omega = 0, \qquad \tilde{v}_{\parallel} = -2q_{\rm s}\cos\theta\tilde{v}_{E\times B},$$

where $\tilde{v}_{\parallel}$ is return flow along the magnetic field line. The other one is ion sound wave ($\nabla_{\parallel} = ik_{\parallel}$);

$$\Omega^2 = k_{\parallel}^2 c_{\rm s}^2.$$

13.5.4 Zonal Flow in ETG Turbulence

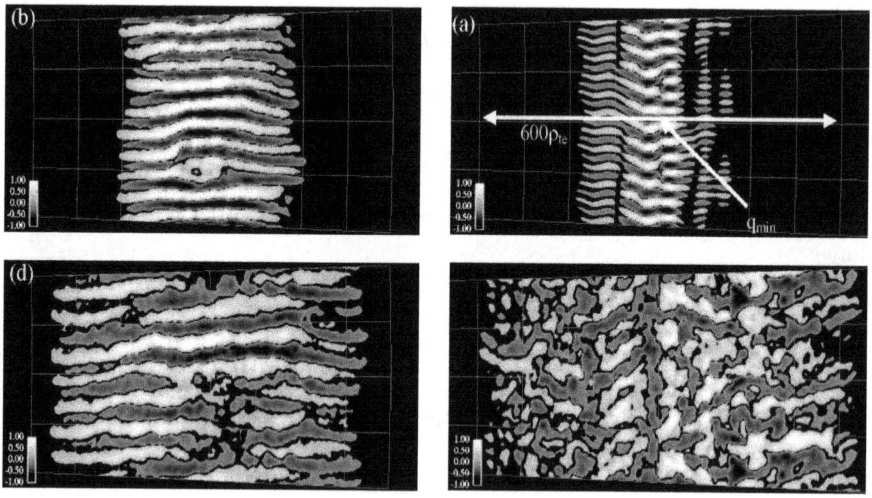

Figure 13.10 Electrostatic potential contours of ETG turbulence. Figures in the left-hand side are the case of positive shear configuration of tokamak. The upper one is the eletrostatic potential at initial saturation phase $t = 172L_n/v_{T\rm e}$ and the lower one is at quasi-steady state $t = 250L_n/v_{T\rm e}$. At quasi-steady state positive shear configuration, large quasi-linear streamers are seen. Figures in the right-hand side are the case of reversed shear configuration. The upper one is the potential contour at linear phase $t = 110L_n/v_{T\rm e}$. The lower one is at quasi-steady state $t = 380L_n/v_{T\rm e}$, at which quasi-steady zonal flows are sustained. The length of the arrow is $600\rho_{T\rm e}$. After [13.35].

The electron temperature gradient driven (ETG) turbulence is considered to be one candidate for causing anomalous electron thermal transport. ETG is almost isomorphic to ITG in electrostatic limit, with the role of electrons and ions reversed. If the isomorphism could be assumed, ETG turbulence at electron gyroradius scale would produce electron thermal diffusion coefficient $\chi_e^{ETG} \sim \rho_e^2 v_{Te}/L_T$ and $\chi_e^{ETG} \sim (m_e/m_i)^{1/2}\chi_i^{ITG}$, which is too small to be relevant to tokamak experimental results. This isomorphism is broken if one considers zonal flow in the scale of Debye shielding length. As stated in Section 13.5.3 for ITG turbulence, the proper electron response with $n_e/n = e(\phi - \bar{\phi})/T_e$ was essential to obtaining an enhanced zonal flow amplitude [13.30]. On the other hand, for ETG turbulence, the ion dynamics asymptotes to $n_i/n = -e\phi/T_i$, as it is unmagnetized for $k_\perp \rho_i \gg 1$. Because of this, ETG-driven zonal flows are expected to be weaker than that for ITG-driven zonal flows in regulating turbulence.

Recently several gyrokinetic particle simulations have provided substantial progress in understanding ETG turbulence. However, results from different simulation models provided different estimations of electron heat conductivity χ_e and streamers at the present time. Nevertheless, because of important and interesting topics, results of ETG turbulence by toroidal particle code GT3D [13.34] based on the gyrokinetic particle model (electrostatic) are shown in Figure 13.10 in the cases of positive magnetic shear and reversed shear configurations of tokamak [13.35]. The minor radius a is $200 \sim 10^5$ times the electron Larmor radius ρ_e. Figures in the left-hand side are the case of positive shear configuration. At quasi-steady state, large quasi-linear streamers are seen and provide $\chi_e \sim 10\chi_{GB}$, $\chi_{GB} = \rho_e^2 v_{Te}/L_T$. Figures in the right-hand side are the case of reversed shear configuration. ETG mode with $k_\theta \rho_{Te}$ is excited around the q_{min} surface. At quasi-steady state $t = 380L_n/v_{Te}$, at which quasi-steady zonal flows are sustained and ETG turbulence is suppressed. As a possible cause to drive zonal flow of ETG turbulence, self-organization process (inverse cascade of $\boldsymbol{k}$ spectrum) is examined.

13.5.5 Remarks

Zonal flows are universal phenomena in microturbulence. Zonal flows have been observed in the experiments of CHS [13.36] and ASDEX-U [13.37]. There are extensive reviews on zonal flows [13.38]. The summary of Section 13.5 is as follows:

Zonal flows in ITG are pumped by modulational instability of the drift waves.

Magnetic field structures with $m = 0,\ n = 0$ and finite k_r, called zonal field, can be generated by drift wave turbulence. Zonal flows can modify the onset criterion for turbulence and turbulent transport.

A variant of zonal flows, refered to as GAM, is likely important in regulating turbulence.

In low collisonality regimes, zonal flows damp by scale-independent friction between trapped and circulating ions in ITG induced turbulence [13.38].

Understanding of the zonal flow drive and damping has suggested several routes to improving confinement via externally driven flow shear amplification or by tuning the configuration design to lower the zonal flow damping.

Further studies of zonal flows driven by ETG are required.

Problems

1. Energy Confinement Time Estimate the energy confinement times of fusion core plasma of tokamak reactor with parameters of $n_e = 10^{20}\mathrm{m}^{-3}$, $T_i = T_e = 10\,\mathrm{keV}$, atomic weight of D-T fuel $= 2.5$, effective $Z = 1.0$, $I_p = 15\,\mathrm{MA}$, $B_t = 5\mathrm{T}$, $a = 2\,\mathrm{m}$, $A = 3$ and the safety factor $q_s = 3$ by the formula of $\tau_E \approx a^2/(4\chi_T)$ with different χ_T, where $\chi_T = T/(16eB)$ (Bohm), $\chi_T = (T/eB)(\rho_i/a)$ (GyroBohm), $\chi_T = \epsilon^{-3/2} q_s^2 \rho_i^2 \nu_{ii}$ (neoclassical ion thermal diffusion).

2. Electrostatic Fluctuation Loss Assume that fluctuations of density $\tilde{n}$ and potential $\tilde{\phi}$ are electrostatic and their time fluctuations are measured. Do these fluctuations contribute anomalous transport when the phase difference between $\tilde{n}$ and $\tilde{\phi}$ is zero?

3. Dimensional Analysis of Transport The power law scaling for thermal energy confinement time $\tau_E^{\mathrm{scaling}}$ is expressed by

$$\tau_E^{\mathrm{scaling}} = C I^{\alpha_I} B_t^{\alpha_B} P^{\alpha_P} R^{\alpha_R} n^{\alpha_n} M^{\alpha_M} \epsilon^{\alpha_\epsilon} \kappa^{\alpha_\kappa}.$$

Dimensional analysis of scaling law (Section 13.3) suggests that it should be expressed by dimensionless form (Kadomtsev constraint); that is,

$$\frac{\tau_E^{\mathrm{fit}}}{\tau_B} = C_f \rho_*^{\gamma_\rho} \nu_*^{\gamma_\nu} \beta^{\gamma_\beta} q_I^{\gamma_q} M^{\gamma_M} \epsilon^{\gamma_\epsilon} \kappa^{\gamma_\kappa},$$

where ρ_* is the ratio of ion Larmor radius to minor radius, ν_* is the ratio of ion-ion collision frequency to ion banana excusion frequency, and M is the ion's atomic mass unit. They are given by

$$\tau_B \equiv \frac{a^2}{T/eB}, \quad \rho_* \equiv \frac{\rho_i}{a} \propto \frac{M^{1/2}}{\epsilon}\frac{T^{1/2}}{RB}, \quad \beta \propto \frac{nT}{B^2}, \quad q_I = \frac{Ka}{R}\frac{B_t}{B_\theta} \propto \epsilon^2 \kappa \frac{RB}{I},$$

$$\nu_* \equiv \frac{\nu_{ii}}{\nu_{bi}} = \frac{\nu_{ii}}{\epsilon^{2/3} v_{Ti}/(q_I R)} \propto \frac{q_I}{\epsilon^{3/2}}\frac{nR}{T^2}.$$

For the dimensional parameters I, B_t, R, P, n of $\tau_E^{\mathrm{scaling}}$, there are four dimensionless parameters ρ_*, ν_*, β, q_I in τ_{fit}. There must be a constraint among α^I, α^B, α^P, α^R, α^n in order to express τ_{scaling} in dimensionless form. Derive this constraint. Does $\tau_E^{\mathrm{IPB98y2}}$ given by (4.96) in Section 4.6 satisfy this constraint? Heating power P should be replaced by use of power balance; that is, $P = 3nT2\pi R\pi a^2\kappa/\tau_E \propto \epsilon^2 \kappa n T R^3/\tau_E$. Is $S \equiv \tau_R/\tau_A$ an independent non-dimensional parameter?

References

[Ch1] Introduction to Plasmas

[1.1] R. F. Post: Chapter 22, "Plasma Physics in the 20th Century", in *Twentieth Century Physics*, vol. III, edited by L. M. Brown, A. Pais and B. Pippard: IOP Publishing, Bristol and American Institute of Physics Press, New York (1995).

[1.2] W. R. Arnold, J. A. Phillips, G. A. Sawyer, E. J. Stovall, Jr. and J. C. Tuck: Cross Sections for the Reactions $D(d,p)T$, $D(d,n)He^3$, $T(d,n)He^4$, and $He^3(d,p)He^4$ below 120kV. *Phys. Rev.* **93**, 483 (1954).

[1.3] C. F. Wandel, T. Hesselberg Jensen and O. Kofoed-Hansen: A Compilation of Some Rates and Cross Sections of Interest in Controlled Thermonuclear Research. *Nucl. Instr. and Methods* **4**, 249 (1959).

[1.4] J. L. Tuck: Thermonuclear Reaction Rates. *Nucl. Fusion* **1**, 201 (1961).

[1.5] T. Takizuka and M. Yamagiwa: Reactivity for DT Plasma with Beam-Induced Tail in Magnetic Confinement System. *JAERI-M* 87-066 (1987), Japan Atomic Energy Research Institute.

[Ch2] Particle Orbit and Magnetic Configuration

[2.1] M. Tabak, J. Hammer, M. E. Glinsky, W. L. Kruer, S. C. Wilks et al.: Ignition and Hain Gain with Ultrapowerful Lasers. *Phys. Plasmas* **1**, 1626 (1994).

[2.2] D. V. Sivukhin: *Reviews of Plasma Physics* **4**, p. 93, (edited by M. A. Leontovich), Consultant Bureau, New York (1966).

[2.3] K. Miyamoto: *Plasma Physics for Nuclear Fusion* (Revised Edition), The MIT Press, Cambridge, Mass. (1989).

[2.4] L. Spitzer, Jr.: *Physics of Fully Ionized Gases*, Interscience, New York (1962).

[2.5] T. H. Stix: Heating of Toroidal Plasmas by Neutral Injection. *Plasma Phys.* **14**, 367 (1972).

[2.6] A. A. Ware: Pinch Effect for Trapped Particles in a Tokamak. *Phys. Rev. Lett.* **25**, 15 (1970).

[2.7] A. I. Morozov and L. S. Solovev: *Rev. of Plasma Physics* **2**, 201, (edited by M. A. Leontovich), Consultant Bureau, New York (1966).

[2.8] R. G. Littlejohn: Hamiltonian Formulation of Guiding Center Motion. *Phys. Fluids* **24**, 1730 (1981).
R. G. Littlejohn: Variational Principles of Guiding Center Motion. *J. Plasma Phys.* **29**, 111 (1983).

[Ch3] Magnetohydrodynamics

[3.1] K. Miyamoto: *Plasma Physics for Nuclear Fusion* (Revised Edition), ch. 6. The MIT Press, Cambridge, Mass. (1989).

[Ch4] Tokamak

[4.1] L. A. Artsimovich: Tokamak Devices. *Nucl. Fusion* **12**, 215 (1972). H. P. Furth: Tokamak Research. *Nucl. Fusion* **15**, 487 (1975).
[4.2] J. Wesson: *Tokamaks*, 2nd ed., Clarendon Press, Oxford (1997).
[4.3] V. S. Mukhovatov and V. D. Shafranov: Plasma Equilibrium. *Nucl. Fusion* **11**, 605 (1971).
[4.4] Equip TFR: Tokamak Plasma Diagnostics. *Nucl. Fusion* **18**, 647 (1978).
[4.5] L. A. Artsimovich, G. A. Bobrovsky, E. P. Gorbunov, D. P. Ivanov, V. D. Kirillov et al.: Experiments in Tokamak Devices. *Plasma Phys. Contr. Nucl. Fusion Res.* (Conf. Proceedings, Novosibirsk 1969) vol.**1**, 157 (1969) IAEA, Vienna. English Translation in *Nucl. Fusion Special Suppl.* p.17 (1969).
[4.6] M. A. Gashev, G. K. Gustov, K. K. D'yachenko, E. G. Komar, I. F. Malyshev et al.: Fundamental Technical Characteristics of the Experimental Thermonuclear System "Tokamak-3". *Sov. Atomic Energy* **17**, 1017 (1964).
[4.7] L. S. Solovev: The Theory of Hydromagnetic Stabilitiy of Toroidal Plasma. *Sov. Physics JETP* **26**, 400 (1968).
[4.8] N. M. Zueva and L. S. Solovev: Equilibrium and Stability of Plasma in Axially Symmetric Toroidal Systems. *Atomnaya Energia* **24** 453 (1968).
[4.9] R. H. Weening: Analytical Spherical Torus Plasma Equilibrium Model. *Phys. Plasmas* **7**, 3654 (2000).
[4.10] V. D. Shafranov: Equilibrium of a Plasma Toroid in a Magnetic Field. *Sov. Phys. JETP* **37**, 775 (1960).
[4.11] D. Pfirsch and A. Schlüter: Der Einfluss der Elektrischen Leitfähigkeit auf das Gleichgewichtsverhalten von Plasmen Niedrigen Drucks in Stelleratoren. *MPI/PA/7/62*, Max-Planck Institut für Physik und Astrophysik, München (1962).
[4.12] J. Wesson: *Tokamaks*, 2nd ed., sec. 3.7, Clarendon Press, Oxford (1997).
[4.13] B. B. Kadomtsev: Disruptive Instability in Tokamaks. *Sov. J. Plasma Phys.* **1**, 389 (1975).
[4.14] M. Greenwald: Density Limits in Toroidal Plasma. *Plasma Phys. Control. Fusion* **44**, R27 (2002).
[4.15] ASDEX-U Team: Overview of ASDEX Upgrade Results. *16th IAEA Fusion Energy Conf.* (Conf. Proceedings, Montreal 1996) vol.**1**, 79 (1996) IAEA, Vienna.
[4.16] J. A. Wesson: Hydromagnetic Stability of Tokamaks. *Nucl. Fusion* **18**, 87 (1978).
[4.17] B. B. Kadomtsev and O. P. Pogutse: Trapped Particles in Toroidal Magnetic Systems. *Nucl. Fusion* **11**, 67 (1971).
[4.18] J. W. Connor and H. R. Wilson: Survey of Theories of Anomalous Transport. *Plasma Phys. Contr. Fusion* **36**, 719 (1994).
[4.19] F. Wagner and U. Stroth: Transport in Toroidal Devices - The Experimentalist's View. *Plasma Phys. Contr. Fusion* **35**, 1321 (1993).
[4.20] DIII-D team: DIII-D Research Program Progress. *Plasma Phys. Contr. Nucl. Fusion Res.* (Conf. Proceedings, Washington D. C. 1990) vol.**1**, 69 (1991) IAEA, Vienna.
[4.21] F. Troyon, R. Gruber, H. Saurenmann, S. Semenzato and S. Succi: MHD-Limits to Plasma Confinement. *Plasma Phys. Contr. Fusion* **26**, 209 (1984).

[4.22] A. Sykes, M. F. Turner and S. Patel: Beta Limits in Tokamaks due to High-n Ballooning Modes. *Proc. 11th European Conf. on Contr. Fusion Plasma Phys.*, Part II, 363 (1983) Aachen.
T. Tuda, M. Azumi, K. Itoh, G. Kurita, T. Takeda et al.: Accessible Beta Value of Tokamaks. *Plasma Phys. Contr. Nucl. Fusion Res.* (Conf. Proceedings, London 1984) vol.**2**, 173 (1985) IAEA Vienna.

[4.23] ITER Physics Basis: *Nucl. Fusion* **39**, No. 12 (1999).

[4.24] K. Borrass: Disruptive Tokamak Density Limit as Scrape-Off Layer/Divertor Phenomenon. *Nucl. Fusion* **31**, 1035 (1991)
K. Borrass, R. Farengo, G. C. Vlases: Access Conditions for H Modes with Detached Divertor Plasmas. *Nucl. Fusion* **36**, 1389, (1996).
B. LaBombard, J. A. Goetz, I. Hutchinson, D. Jablonski, J. Kesner et al.: Experimental Investigation of Transport Phenomena in the Scrap-Off Layer and Diverter. *Nucl. Materials* **241-243**, 149 (1997).

[4.25] K. Miyamoto: *Plasma Physics for Nuclear Fusion* (Revised Edition), The MIT Press, Cambridge, Mass. (1989).

[4.26] A. A. Galeev and R. Z. Sagdeev: Transport Phenomena in a Collisionless Plasma in a Toroidal Magnetic System. *Sov. Phys. JETP* **26**, 233 (1968).

[4.27] F. L. Hinton and R. D. Hazeltine: The Theory of Plasma Transport in Toroidal Confinement System. *Rev. Modern Phys.* **48**, 239 (1976).

[4.28] R. J. Bickerton, J. W. Connor and J. B. Taylor: Diffusion Driven Plasma Currents and Bootstrap Tokamak. *Nature Phys. Sci.* **229**, 110 (1971).

[4.29] A. A. Galeev: Diffusion-Electrical Phenomena in a Plasma Confined in a Tokamak. *Sov. Phys. JETP* **32**, 752 (1971).

[4.30] M. N. Rosenbluth, R. D. Hazeltine and F. L. Hinton: Plasma Transport in Toroidal Confinement Systems. *Phys. Fluids* **15**, 116 (1972).

[4.31] D. J. Sigmar: Bootstrap Current in Tokamaks with Neutral Injection. *Nucl. Fusion* **13**, 17 (1973).

[4.32] T. Ozeki, M. Azumi, S. Tokuda, and S. Ishida: Effects of a Hollow Current Profile in the Idial MHD Stability of High β_p Plasmas in a Tokamak. *Nucl. Fusion* **33**, 1025 (1993).

[4.33] R. J. Goldston: Energy Confinement Scaling in Tokamaks. *Plasma Phys. Contr. Fusion* **26**, 87 (1984).
S. M. Kaye: A Review of Energy Confinement and Local Transport Scaling Results in Neutral- Beam-Heated Tokamaks. *Phys. Fluids* **28**, 2327 (1985).

[4.34] P. N. Yushimanov, T. Takizuka, K. S. Riedel, D. J. W. F. Kardaun, J. G. Cordey, S. M. Kaye, and D. E. Post: Scalings for Tokamak Energy Confinement. *Nucl. Fusion* **30**, 1999, (1990).
N. A. Uckan, P. N. Yushimanov, T. Takizuka, K. Borras, J. D. Callen, et al.: Energy and Particle Confinement in ITER. *Plasma Phys. Contr. Nucl. Fusion Res.* (Conf. Proceedings, Washington D. C. 1990) vol. **3**, 307 (1991) IAEA, Vienna.

[4.35] F. Wagner, G. Becker, K. Behringer et. al.: Regime of Improved Confinement and High Beta in Neutral-Beam-Heated Divertor Discharges of the ASDEX Tokamaks. *Phys. Rev. Lett.* **49**, 1408 (1982).
F. Wagner, G. Becker, K. Behringer, D. Campbell, M. Keilhacker et al.: Confinement and β_p-Studies in Neutral-Beam-Heated ASDEX Plasmas.

Plasma Phys. Contr. Nucl. Fusion Research (Conf. Proceedings, Baltimore 1982) vol. **1**, 43 (1983) IAEA, Vienna.

[4.36] ASDEX Team: The H Mode of ASDEX. *Nucl. Fusion* **29**, 1959 (1989).

[4.37] R. J. Groebner: An Emerging Understanding of H Mode Discharges in Tokamaks. *Phys. Fluids* **B5**, 2343 (1993).

[4.38] E. J. Doyle, C. L. Rettig, K. H. Burrell, P. Gohil, R. J. Groebner et. al.: Turbulence and Transport Reduction Mechanisms in the Edge and Interior of DIII-D H Mode Plasmas. *Plasma Phys. Contr. Nucl. Fusion Res.* (Conf. Proceedings, Würzburg 1992) vol. **1**, 235 (1992) IAEA, Vienna.

[4.39] S. I. Itoh and K. Itoh: Model of L- to H- Mode Transition in Tokamak. *Phys. Rev. Lett.* **60**, 2276 (1988).
K. C. Shaing and E. C. Crume: Bifurcation Theory of Poloidal Rotation in Tokamaks. *Phy. Rev. Lett.* **63**, 2369 (1989).

[4.40] K. Ida: Experimental Studies of the Physical Mechanism Determining the Radial Electric Field and Its Radial Structure in a Toroidal Plasma. *Plasma Phys. Contr. Fusion* **40**, 1429 (1998).

[4.41] H. Bigrali, D. H. Diamond, Y.-B. Kim, B. A. Carreras, V. E. Lynch, F. L. Hinton et al.: Influence of Sheared $\boldsymbol{E} \times \boldsymbol{B}$ Rotation on Edge Turbulence Dynamics and Access to Enhanced Confinement Regimes. *Plasma Phys. Contr. Nucl. Fusion Res.* (Conf. Proceedings, Washington D.C. 1990) vol. **2**, 191 (1991) IAEA, Vienna.

[4.42] T. H. Dupree: Theory of Phase Space Density Granulation in Plasma. *Phys. Fluids*: **15**, 334 (1972).
T. Boutros-Ghali and T. H. Dupree: Theory of Two-Ponit Correlation Function in a Vlasov Plasma. *Phys. Fluids*: **24**, 1839 (1981).

[4.43] TFTR Team, Recent TFTR Results. *Plasma Phys. Contr. Nucl. Fusion Res.* (Conf. Proceedings, Washington D.C. 1990) vol. **1**, 9, (1991) IAEA, Vienna.

[4.44] T. S. Taylor, T. H. Osborne, K. H. Burrel et al.: Confinement and Stability of VH-Mode Discharges in the DIII-D Tokamaks. *Plasma Phys. and Contr. Nucl. Fusion Res.* (Conf. Proceedings, Würzburg 1992) vol. **1**, 167 (1992) IAEA, Vienna.

[4.45] JT60U Team: Improved Confinement and Transport Barrier in the JT-60U High β_p H Mode. *Plasma Phys. Contr. Nucl. Fusion Research* (Conf. Proceedings, Seville 1994) vol. **1**, 199 (1995) IAEA Vienna.

[4.46] F. L. Hinton and G. M. Staebner: Particle and Energy Confinement Bifurcation in Tokamaks. *Phys. Fluids* **B5**, 1281 (1993).

[4.47] O1-6 (DIII-D), O1-2 (TFTR), O1-3 (JT-60U), A5-5 (JET), O2-2 (Tore Supra) in *16th IAEA Fusion Energy Conf.* (Conf. Proceedings, Montreal 1996) IAEA, Vienna.

[4.48] H. Zohn: Edge Localized Models (ELM). *Plasma Phys. Contr. Fusion* **38**, 105 (1996).
J. W. Connor: A Review of Models for ELM. *Plasma Phys. Contr. Fusion* **40** 191 (1998).
M. Becoulet, G. Huysmans, Y. Sarazin, X. Garbet, Ph Ghendrih et al.: Edge Localized Mode Physics and Operational Aspects in Tokamaks. *Plasma Phys. Control. Fusion* **45**, A93 (2003).

[4.49] JET Team: Fusion Energy Production from a Deuterium-Tritium Plasma in the JET Tokamak. *Nucl. Fusion* **32**, 187 (1992).

[4.50] JT60U Team: Recent JT-60U Results toward Steadt State Operations of Tokamaks. *Plasma Phys. Contr. Nucl. Fusion Research* (Conf. Proceedings, Seville 1994) vol. **1**, 31 (1995) IAEA, Vienna.

[4.51] TFTR Team: Review of Recent D-T Experiments from TFTR. *Plasma Phys. Contr. Nucl. Fusion Research* (Conf. Proceedings, Seville 1994) vol. **1**, 11 (1995) IAEA, Vienna.

[4.52] The JET Team: Physics of High Performance JET Plasmas in D-T. *17th Fusion Energy Conf.* (Conf. Proceedings, Yokohama 1998) vol. **1**, 29 (1998) IAEA, Vienna.

[4.53] T. Fujita, S. Ide, Y. Kamada, T. Suzuki, T. Oikawa et al.: Quasisteady High-Confinement Reversed Shear Plasma with Large Bootstrap Current Fraction under Full Noninductive Current Drive Condition in JT-60U. *Phys. Rev. Lett.* **87**, 085001 (2001).

[4.54] P. A. Politzer, A. W. Hyatt, T. C. Luce, F. W. Perkins, R. Prater et al.: Stationary, High Bootstrap Fractions in Dlll-D without Inductive Current Control. *Nucl. Fusion* **45**, 417 (2005).

[4.55] J. L. Luxon: A Design Retrospective of the Dlll-D Tokamak. *Nucl. Fusion* **42**, 614 (2002).

[4.56] M. Okabayashi, J. Bialek, A. Bodeson, M. S. Chance, M. S. Chu et al.: Control of the Resistive Wall Mode with Internal Coils in the Dlll-D Tokamak. *20th IAEA Fusion Energy Conf.* (Vilamoura 2004) EX/3-1Ra.
H. Reimerdes, J. Bialek, M. S. Chance, M. S. Chu, A. M. Garofalo et al.: Measurement of Resistive Wall Mode Stability in Rotating High Beta Plasma. *20th IAEA Fusion Energy Conf.* (Vilamoura 2004) EX/3-1Rb.

[4.57] Y. Kamada, T. Oikawa, L. Lao, T. Takizuka, T. Hatae et al.: Disappearance of Giant ELMs and Appearance of Minute Grassy ELMs in JT60U High Triangularity Discharge. *Plasma Phys. Contrl. Fusion* **42**, A247 (2000).

[4.58] P. T. Lang, A. Kallenbach, J. Bucalossi, G. D. Conway, A. Degeling et al. (ASDEX-U Team): Integrated Exhaust Senarios with Actively Controlled ELMs. *20th IAEA Fusion Energy Conf.* (Vilamoura 2004) EX/2-6.

[4.59] R. Yoshino, Y. Neyatani, N. Isei, A. Tanga, D. J.Cambell et al.: Disruptive Amelioration Experiments in JT60U and JET. *Plasma Phys. Contr. Nucl. Fusion Res.* (Conf. Proceedings, Seville 1994) vol. **1**, 685, (1995) IAEA, Vienna.
R. Yoshino, T. Kondoh, Y. Neyatani, K. Itami, Y. Kawano, and N. Isei: Fast Plasma Shutdown by Killer Pellet Injection in JT60U with Reduced Heat Flux on the Divertor Plate and Avoiding Runaway Electron Genration. *Plasma Phys. Contr. Fusion* **39**, 313 (1997).

[4.60] G. Pautasso, K. Büchl, J. C. Fuchs, O. Gruber, A. Herrmann et al.: Use of Impurity Pellets to Control Energy Dissipation during Disruption. *Nucl. Fusion* **36**, 1291 (1996).

[4.61] R. Yoshino: Neural-Net Predictor for Beta Limit Disruptions in JT60U. *Nucl. Fusion* **45**, 1232 (2005).

[4.62] ITER Team: ITER: System Studies and Design Space Analysis. *Plasma Phys. Contr. Nucl. Fusion Res.* (Conf. Proceedings, Washington D.C. 1990) vol. **3**, 413, (1991) IAEA, Vienna.

[4.63] K. Miyamoto: Parameter Sensitivity of ITER Type Experimental Tokamak Reactor toward Compactness ll. *J. Plasma Fusion Res.* **76**, 166 (2000).

[4.64] R. Aymar, V. Chuyanov, M. Huguet, and Y. Shimonura: ITER-FEAT - The Future International Burning Plasma Experiment Overview. *18th IAEA Fusion Energy Conf.* (Sorrento 2000) OV/1.
Technical Bases for the ITER-FEAT Outline Design Dec. 1999 (Draft for TAC Review).
M. Shimada, V. Mukhavatov, G. Federici, Y. Gribov, A. Kukushkin et et: Performance of ITER as Burning Plasma Experiment. *The 19th IAEA Fusion Energy Conf.* (Lyon in 2002) CT-2.

[4.65] INTOR Team: International Tokamak Reactor - Phase Two A Part 1. *Nucl Fusion* **23**, 1513 (1983).

[4.66] Y. K.M. Peng, and D. J. Strickler: Features of Spherical Torus Plasma. *Nucl. Fusion* **26**, 769 (1986).

[4.67] D. A. Gates, R. Akers, L. Appel, P. G. Carolan, N. Conway et al.: High-Performance Discharges in the Small Tight Aspect Ratio Tokamak (START). *Phys. Plasmas* **5**, 1775 (1988).

[4.68] D. C. Robinson: The Physics of Sperical Confinement Systems. *Plasma Phys. Contrl. Fusion* **41**, A143 (1999).

[4.69] B. P. LeBlanc, R. E. Bell, S. M. Kaye, D. Stutman, M. G. Bell et al.: Confinement Studies of Auxiliary Heated NSTS Plasmas. *Nucl. Fusion* **44**, 513 (2004).
M. Ono, M. G. Bell, R. E. Bell, T. Bigelow, M. Bitter et al.: Progress toward High-Performance, Steady-State Spherical Torus. *Plasma Phys. Control. Fusion* **45**, A335 (2003).

[4.70] R. J. Akers, J. W. Ahn, G Y Antar, L. C. Appel, D. Applegate *et. al.*: Transport and Confinement in the Mega Ampere Spherical Tokamak (MAST) Plasma. *Plasma Phys. Contr. Fusion* **45**, A175 (2003).

[4.71] T. Hayashi, N. Mizuguchi, T. H. Watanabe, Y. Todo and T. Sato: Non-linear Simulation of Internal Reconnection Events in Sperical Tokamaks. *Nucl. Fusion* **40**, 721 (2000).

[4.72] H. R. Wilson, G. Voss, J-W Ahn, R. J. Akers, L. Appel et al.: The Spherical Tokamak Fusion Power Plant. *19th IAEA Fusion Energy Conf.* (Lyon 2002) FT/1-5.
R. J. Akers, A. Bond, R. J. Buttery, P. C. Carolan, G. F. Counsell et al.: The Spherical Tokamak Fusion Plant. *Nucl. Fusion* **40**, 1223 (2000).

[4.73] F. Najmabadi, S. Jardin, M. Tillack, R. Miller, T. K. Mau et. al.: The ARIES-ST Study. *17th IAEA Fusion Energy Conf.* (Yokohama 1998) FTP-08.

[4.74] S. Nishio, K. Tobita, S. Konishi, T. Ando, S. Hiroki et al.: Tight Aspect Ratio Tokamak Power Reactor with Superconducting TF Coils. *19th IAEA Fusion Energy Conf.* (Lyon 2002) FT/P1-21.
S. Nishio, K. Tobita, K. Tokimatsu, K. Shinya, I. Senda et al.: Technological and Environmental Prospects of Low Aspect Ratio Tokamak Reactor (VECTOR). *20th IAEA Fusion Energy Conf.* (Vilamoura 2004) FT/P7-35.

[4.75] K. Tobita, S. Nishio, M. Enoeda, M. Sato, T. Isono et al.: Design Study of Fusion Demo Plant at JAERI. *Fusion Eng. Des.* **81**, 1151 (2006).

[4.76] S. Shiraiwa, S. Ide et al.: Formation of Advanced Tokamak Plasmas without the Use of an Ohmic-Heating Solenoid. *Phys. Rev. Lett.* **92**, 035001 (2004).
Y. Takase, T. Fukuda, G. Xiang, M. Gyaznevich, S. Ide et al. Plasma Current Start-up, Ramp-up, and Achievement of Advanced Tokamak Plasma without the Use of Ohic Heating Solenoid in JT-60U. J. Plasma Fusion Res. **78**, 719 (2002).

[4.77] R. L. Miller, Y. R. Lin-Liu, A. D. Turnbull and V. S. Chan et. al.: Stable Equilibria for Bootstrap-Current-Driven Low Aspect Ratio Tokamaks. *Phys. Plasmas* **4**, 1062 (1997).
C. P. C. Wong, and R. D. Stambaugh: Tokamak Reactor Design as a Function of Aspect Ratio. *Fusion Eng. Des.* **51-52**, 387 (2000).
Y. R. Lin-Liu and R. D. Stambaugh: Optimum Equilibria for High Performance, Steady State Tokamaks. *Nucl. Fusion* **44**, 548 (2004).

[4.78] M. Ono, M. Peng, C. Kessel, C. Neumeyer, J. Schmidt et al.: Next-Step Spherical Torus Experiment and Spherical Torus Strategy in the Course of Development of Fusion Energy. *Nucl. Fusion* **44**, 452 (2004).

[4.79] M. Fujiwara, S. Itoh, K. Matsuoka, K. Matsuura, K. Miyamoto, and A. Ogata: Conceptional Design of a Verical Field Control System in JIPP Tll. *Jpn. J. Appl. Phys.* **14**, 675 (1975).

[Ch5] Reversed Field Pinch

[5.1] D. C. Robinson and R. E. King: Factors Influencing the Period of Improved Stability in Zeta. *Plasma Phys. Contr. Nucl. Fusion Res.* (Conf. Proceedings, Novosibirsk 1968) vol. **1**, 263 (1969) IAEA, Vienna.

[5.2] H. A. B. Bodin and A. A. Newton: Reversed-Field-Pinch Research. *Nucl. Fusion* **20**, 1255 (1980).

[5.3] H. A. B. Bodin: Reversed Field Pinch: Status and Trends. *Plasma Phys. Contr. Fusion* **29**, 1297 (1987).

[5.4] MST Team: Confinement and Fluctuations in the MST Reversed Field Pinch. *Plasma Phys. Contr. Nucl. Fusion Res.* (Conf. Proceedings, Washington D. C. 1990) vol. **2**, 519 (1991) IAEA, Vienna.
TPE-1RM20 Team: Improved High Theta Mode and Dynamo Activity in a Reversed Field Pinch on TPE-1RM20. *19th IAEA Fusion Energy Conference* (Conf. Proceedings, Montreal 1996) vol. **2**, 95 (1997) IAEA, Vienna.

[5.5] EX4/3(RFX), EX4/4(TPE-RX): *17th IAEA Fusion Energy Conf.* (Conf. Proceedings, Yokohama 1998) vol. **1** 367, 375 (1998) IAEA, Vienna.

[5.6] J. B. Taylor: Relaxtion of Toroidal Plasma and Generation of Reversed Magnetic Fields. *Phys. Rev. Lett.* **33**, 1139 (1974).

[5.7] T. H. Jensen and M. S. Chu: Current Drive and Helicity Injection. *Phys. Fluids* **27**, 2881 (1984).

[5.8] V. D. Shafranov and E. I. Yurchenko: *Sov. Phys. JETP* **26**, 682 (1968).

[5.9] D. A. Backer, M. D. Bausman, C. J. Buchenauer, L. C. Burkhardt, G. Chandler, J. N. Dimarco et al.: Performance of the ZT-40M Reversed-Field Pinch with an Inconel Liner. *Plasma Phys. Contr. Nucl. Fusion Res.* (Conf. Proceeding, Baltimore 1982) vol. **1**, 587 (1983) IAEA, Vienna.

[5.10] K. Miyamoto: Dynamic Processes in Relaxation Phenomena of RFP Plasma. *Plasma Phys. Contr. Fusion* **30**, 1493 (1988).
Y. Ueda, N. Asakura, S. Matsuzuka, K. Yamagishi, S. Shinohara, K. Miyamoto et al.: Change of the Magnetic Field Structure during Relaxation in High Pinch Parameter Regime of the Reversed Field Pinch REPUTE-1. *Nucl. Fusion* **27**, 1453 (1987).

[5.11] K. Kusano and T. Sato: Simulation Study of the Self-Reversal Process in the the Reversed-Field Pinch Based on a Non-Linearly Driven Reconnection Model. *Nucl. Fusion* **26**, 1051 (1986).

[5.12] D. D. Schnack, E. J. Caramana, and R. A. Nebel: Three Dimensional Magnetohydrodynamic Studies of the Reversed-Field Pinch. *Phys. Fluids* **28**, 321 (1985).

[5.13] A. R. Jacobson, and R. W. Moses: Nonlocal DC Electric Conductivity of a Lorentz Plasma in a Stochastic Magnetic Field. *Phys. Rev.* **A29**, 3335 (1984).
R. W. Moses, K. F. Schoenberg, and D. A. Baker: Empirical Modeling and Dependence of Reversed Field Pinch Loop Voltage on Edge Plasma Conditions. *Phys. Fluids* **31**, 3152 (1988).

[5.14] K. A. Werley, J. N. Dimarco, R. A. Krakowski, and C. G. Bathke: Energy Confinement and Future Reversed Field Pinches. *Nucl. Fuison* **36**, 629 (1996).

[5.15] M. Valisa, F. Auriemma, A. Canton, L. Carraro, R. Lorenzini et al.: Density Limit in the Reversed Field Pinch. *20th IAEA Fusion Energy Conf.* (Vilamoura 2004) EX/P4-13.

[5.16] J. S. Sarff, S. A. Hokin, H. Ji, S. C. Prager, and C. R. Sovinec: Fluctuation and Transport Reduction in a Reversed Field Pinch by Inductive Poloidal Current Drive. *Phys. Rev. Lett.* **72**, 3670 (1994).

[5.17] S. C. Prager, J. Adney, A. Almagri, J. Anderson, A. Blair et al.: Overview of Results in the MST Reversed Field Pinch Experiment. *20th IAEA Fusion Energy Conf.* (Vilamoura 2004) OV/4-2.

[5.18] J. S. Sarff, N. E. Lanier, S. C. Prager, and M. R. Stoneking: Increased Confinement and β by Inductive Poloidal Current Drive in the Reversed Field Pinch. *Phys. Rev. Lett.* **78**, 62 (1997).

[5.19] B. E. Chapman, J. K. Anderson, T. M. Biewer, D. L. Brower, S. Castillo it et al.: Reduced Edge Instability and Improved Confinement in the MST Reversed-Field Pinch. *Phys. Rev. Lett.* **87**, 205001 (2001).
B. E. Chapman, A. F. Almagri, J. K. Anderson, T. M. Biewer, P. K. Chattopadhyay et al.: High Confinement Plasmas in the Madison Symmetric Torus Reversed-Field Pinch. *Phys. Plasmas* **9**, 2061 (2002).

[5.20] J. S. Sarff, A. F. Almagri, J. K. Anderson, T. M. Biewer, D. L. Brower et al.: Overview of Improved Confinement and Plasma Control in the MST Reversed Field Pinch. *19th IAEA Fusion Energy Conf.* (Lyon 2002) OV/4-3.

[5.21] R. Bartiromo, P. Martin, S. Martini, T. Bolzonella, A. Canton, and P. Innocente: Core Transport Improvement during Poloidal Current Drive in the RFX Reversed Field Pinch. *Phys. Rev. Lett.* **82**, 1462 (1999).

[5.22] Y. Yagi, H. Koguchi, Y. Hirano, T. Shimada, H. Sakakita, and S. Sekine: Increased Confinement Improvement in a Reversed-Field Pinch Using Double-Pulsed Poloidal Current Drive. *Phys. Plasmas* **10**, 2925 (2003).

[5.23] H. Sakakita, Y. Yagi, T. Asai, G. Fiksel, L. Frassinetti, K. Hayase et al.: Characteristics of the TPE Reversed-Field Pinch Plasmas in Conventional and Improved Confinement in Reversed-Field Pinch Plasmas. *20th IAEA Fusion Energy Conf.* (Vilamoura 2004) EX/P2-17.

[5.24] Y. Yagi, H. Koguchi, Y. Hirano, H. Sakakita, and L. Frassinetti: An empirical Scaling Law for Improved Confinement in Reversed-Field Pinch Plasmas. *Nucl. Fusion* **45**, 138 (2005).

[5.25] M. K. Bevir and J. W. Gray: Relaxation, Flux Consumption and Quasi Steady State Pinches. *Proc. of Reversed Field Pinch Theory Workshop* (LANL Los Alamos, 1980)
M. K. Bevir, C. G. Gimblett, and G. Miller: Quasi-Steady-State Toroidal Discharges. *Phys. Fluids* **28**, 1826 (1985).

[5.26] K. F. Schoenberg, J. C. Ingraham, C. P. Munson, P.G. Weber et al.: Oscillating Field Current Drive Experiments in a Reversed Field Pinch. *Phys. Fluids* **31**, 2285 (1988).

[Ch6] Stellarator

[6.1] L. Spitzer, Jr.: The Stellarator Concept. *Phys. Fluids* **1**, 253 (1958).

[6.2] A. I. Morozov and L. S. Solovév: *Rev. of Plasma Physics* **2**, 1 (edited by M. A. Leontovich) Consultants Bureau, New York (1966).

[6.3] K. Nagasaki, K. Itoh, M. Wakatani and A. Iiyoshi: Vacuum Magnetic Surface of $l = 2$ Helical Systems. *J. Phys. Soc. Japan* **57**, 2000 (1988).

[6.4] J. L. Johnson, C. R. Oberman, R. M. Kulsrud, and E. A. Frieman: Some Stable Hydromagnetic Equililibria. *Phys. Fluids* **1**, 281 (1958).

[6.5] S. Chandrasekhar and P. C. Kendal: On Force-Free Magnetic Fields. *Astrophys. J.* **126**, 457 (1957).
P. Barberio-Corsetti: Helical Equilibria. *Plasma Phys.* **15**, 1131 (1973).
D. Corresa and D. Lortz: A Class of Helically Symmetric MHD-Equation. *Nucl. Fusion* **13**, 127 (1973).

[6.6] W VIIA Team: Neutral-Injection Heating in the Wendelstein Vll-A Stellarator. *Plasma Phys. Contr. Nucl. Fusion Res.* (Conf. Proceedings, Baltimore 1982) **2**, 241 (1983) IAEA, Vienna.

[6.7] E. D. Andryukhina, G. M. Batanov, M. S. Berezhetshij, M. A. Blokh, G. S. Vorosov et al.: First Results on Currentless Plasma Production and Heating by ECR Waves in the L-2 Stellarator. *Plasma Phys. Contr. Nucl. Fusion Res.* (Conf. Proceedings, London 1984) vol. **2**, 409 (1985) IAEA, Vienna.

[6.8] K. Uo, A. Iiyoshi, T. Obiki, O. Motojima, S. Morimoto, M. Wakatani et al.: NBI Heating of Currentless Plasma in Heliotron E. *Plasma Phys. Contr. Nucl. Fusion Res.* (Conf. Proceedings, London 1984) vol. **2**, 383 (1985) IAEA, Vienna.

[6.9] L. Garcia, B. A. Carreras, J. H. Harris, H. R. Hicks, and V. E. Lynch: Equilibrium Studies for Low-Aspect-Ratio Torsatron. *Nucl. Fusion* **24**, 115 (1984).

[6.10] C. Gourdon, D. Marty, E. K. Maschke and J. P. Dumont: Configurations of the Stellarator Type with Mean Well and Shear. *Plasma Phys. Contr. Nucl. Fusion Res.* (Conf. Proceedings, Novosibirsk 1968) vol. **1**, 847 (1969) IAEA, Vienna.

[6.11] K. Uo: The Helical Heliotron Field for Plasma Confinement. *Plasma Phys.* **13**, 243 (1971).

[6.12] A. Mohri: A Negative V'' System without Use of Toroidal Solenoid Coils. *J. Phys. Soc. Japan* **28**, 1549 (1970).

[6.13] C. Gourdon, D. Marty, E. K. Maschke, and J. Touche: The Torsatron without Toroidal Field Coils as a Solution of the Divertor Problem. *Nucl. Fusion* **11**, 161 (1971).

[6.14] Yu. N. Petrenco and A. P. Popryadukhin: Magnetic Field Properties of a Stellarator with Elliptical Coils. *3rd International Symp. on Toroidal Plasma Confinements*, D 8, (1973, Garching).

[6.15] H. Wobig and S. Rehker: A Stellarator Coil System without Helical Windings. *Proc. 7th Symp. on Fusion Technol.* 345 (1971, Grenoble).

[6.16] A. Iiyoshi, A. Komori, A. Ejiri, M. Emoto, H. Funaba et al.: Overview of the Large Helical Device Project. *Nucl Fusion* **39**, 1245 (1999).
M. Fujiwara, O. Motojima, Y. Yamada, T. Watari, M. Okamoto et al.: Progress in LHD (Large Helical Device) Program. *Plasma Phys. Contr. Fusion* **39**, A261 (1997).

[6.17] F. Wagner and the W7-X Team: Physics, Technologies, and Status of Wendelstein 7-X Device. *20th IAEA Fusion Energy Conf.* (Vilamoura 2004), FT/3-5.
H. Wobig: Theory of Advanced Stellarators. *Plasma Phys. Contr. Fusion* **41**, A159 (1999).

[6.18] B. B. Kadomtsev and O. P. Pogutse: Trapped Particles in Toroidal Magnetic Systems. *Nucl. Fusion* **11**, 67 (1971).

[6.19] J. A. Derr and J. L. Shohet: Superbanana Orbits in Stellarator Geometries. *Phys. Rev. Lett.* **44**, 1730 (1979).

[6.20] M. Wakatani, S. Kodama, M. Nakasuga, and K. Hanatani: Velocity-Space Loss Regions in Toroidal Helical Systems. *Nucl. Fusion* **21**, 175 (1981).

[6.21] J. W. Connor and R. J. Hastie: Neoclassical Diffusion in an $l = 3$ Stellarator. *Phys. Fluids* **17**, 114 (1974).

[6.22] L. M. Kovrizhnykh: Neoclassical Theory of Transport Processes in Toroidal Magnetic Confinement Systems, with Emphasis on Non-Axisymmetric Configurations. *Nucl. Fusion* **24**, 851 (1984).

[6.23] D. E. Hastings, W. A. Houlberg, and K. C. Shaing: The Ambipolar Electic Field in Stellarators. *Nucl. Fusion* **25**, 445 (1985).

[6.24] K. M. Young: The C-Stellarator - A Review of Containment. *Plasma Phys.* **16**, 119 (1974).

[6.25] K. Miyamoto: Recent Stellarator Research. *Nucl. Fusion* **18**, 243 (1978).

[6.26] B. A. Carreras, G. Grieger, J. H. Harris, J. L. Johnson, J. F. Lyon, O. Motojima, F. Rau, H. Renner, J. A. Rome, K. Uo, M. Wakatani, and H. Wobig: Progress in Stellarator/Heliotron Research: 1981-1986. *Nucl. Fusion* **28**, 1613 (1988).

[6.27] S. Sudo, Y. Takeiri, Z. Zushi, F. Sano, K. Itoh, K. Kondo, and A. Iiyoshi: Scalings of Energy Confinement and Density Limit in Stellarator/Heliotron Devices. *Nucl. Fusion* **30**, 11 (1990).

[6.28] U. Stroth, M. Murakami, R. A. Dory, H. Yamada, S. Okamura, F. Sano, and T. Obiki: Energy Confinement Scaling from the International Stellarator Database. *Nucl. Fusion* **36**, 1063 (1996).

[6.29] A. Fujisawa, H. Iguchi, T. Minami, Y. Yoshimura, H. Sanuki, K. Itoh et al.: Electron Thermal Transport Barrier and Density Fluctuation Reduction in a Toroidal Helical Plasma. *Phys. Rev. Lett.* **82**, 2669 (1999).

[6.30] H. Massberg, C. D. Beidler, U. Gasparino, M. Rome, K. S. Dyailin et al.: The Neoclassical "Electron Root" Feature in Wendelstein-7-AS Stellarator. *Phys. Plasmas* **7**, 295 (2000).

[6.31] O. Motojima, N. Ohyabu, A. Komori, O. Kaneko, H. Yamada et al.: Recent Advance in LHD Experiment. *19th Fusion Energy Conf.* (Lyon 2002), OV/1-6.

[6.32] K. Ida, S. Inagaki, T. Shimozuka, T. Tamura, H. Funaba et al.: Characteristics of Transport in Electron Internal Transport Barriers and in the Vicinity of Rational Surfaces in the Large Helical Device. *Phys. Plasmas* **11**, 2551 (2004).

[6.33] U. Stroth, J. Baldzuhm, J. Geiger, T. Geist, L. Giannone et al.: High-Confinement NBI Discharges in the W7-AS Stellarator. *Plasma Phys. Contr. Fusion* **40**, 1551 (1998).

[6.34] K. McCormick, P. Grigull, R. Burhenn, R. Brakel, H. Ehmler et al.: New Advanced Operational Regime in the W7-AS Stellarator. *Phys. Rev. Lett.* **89**, 015001 (2001).

[6.35] F. Wagner, S. Bäuml, J. Baldzuhn, N. Basse, R. Brakel et al.: W7-AS: One Step of the Wendelstein Stellarator Line. *Phys. Plasmas* **12**, 072509 (2005).

[6.36] K. Kawahata, N. Ohyabu, O. Kaneko, A. Komori, H. Yamada et al.: Experimental Studies on NBI and ICRF Heated Plasmas in the Large Helical Device. *Plasma Phys. Contr. Fusion* **42**, B51 (2000).

[6.37] O. Motojima, K. Ida, K. Y. Watababe, Y. Nagayama, A. Komori et al.: Confinement and MHD Stability in the Large Helical Device. *20th Fusion Energy Conf.* (Vilamoura 2004), OV/1-4.

[6.38] S. Hamada: Hydromagnetic Equilibria and Their Proper Coordinates. *Nucl. Fusion* **2**, 23 (1962).

[6.39] A. H. Boozer: Guiding Center Drift Equations. *Phys. Fluids* **23**, 904 (1980).

[6.40] N. Nakajima, J. Todoroki, and M. Okamoto: On Relation between Hamada and Boozer Magnetic Coordinates Systems. *J. Plasma Fusion Res.* **68**, 395 (1992).

[6.41] R. B. White, A. H. Boozer, and R. Hay: Drift Hamiltonian in Magnetic Coordinates. *Phys. Fluids* **25**, 575 (1982).

[6.42] R. B. White and M. S. Chance: Hamiltonian Guiding Center Drift Orbit Calculation for Plasmas of Arbitrary Cross Section. *Phys. Fluids* **27**, 2455 (1984).

[6.43] A. H. Boozer: Time-Dependent Drift Hamiltonian. *Phys. Fluids* **27**, 2441 (1984).

[6.44] P. J. Heitzenroeder, T. G. Brown, J. H. Chrzanowski, M. J. Cole, P. L. Goranson et al.: Component Manufacturing Development for the National Compact Stellarator Experiment (NCSX). *20th Fusion Energy Conf.* (Vilamoura 2004), FT/P7-22.
M. C. Zarnstorff, L. A. Berry, A. Brooks, F. Fredrickson, G-Y. Fu et al.: Physics of Compact Advanced Stellarator NCSX. *Plasma Phys. Contr. Fusion* **43**, A237 (2001).

[6.45] S. Okamura, K. Matsuoka, S. Nishimura, M. Isobe, C. Suzuki et al.: Confinement Characteristics of the Quasi-Axisymmetric Stellarator CHS-qa. *Nucl. Fusion* **44**, 575 (2004).

[6.46] F. S. B. Anderson, A. F. Almagri, D. T. Anderson, P. G. Mathews, J. N. Talmadge, and J. L. Shohet: Helically Symmetric Experiment (HSX), Goals, Design and Status. *Fusion Technol.* **27**, 273 (1995).
S. P. Gerhardt, A. F. Almagri, D. T. Anderson, F. S. B. Anderson, D. Brower et al.: Measurement and Modeling of Electrode Biased Discharges in HSX Stellarator. *20th Fusion Energy Conf.* (Vilamoura 2004), EX/8-3.

[6.47] D. A. Spong, S. P. Hirshman, J. F. Lyon, L. A. Berry, D. J. Strickler et al.: Recent Advances in Quasi-Poloidal Stellarator Physics Issues. *20th Fusion Energy Conf.* (Vilamoura 2004), IC/P6-51.
B. E. Nelson, A. Brooks, R. D. Benson, L. A. Berry, T. G. Brown et al.: Engineering Aspects of Compact Stellarators. *19th Fusion Energy Conf.* (Lyon 2002), FT/2-4.

[6.48] F. Wagner, T. Andreeva, J. Baldzuhn, A. Benndorf, H. Bolt et al.: Physics, Technologies, and Status of Wendelstein 7-X Device. *20th Fusion Energy Conf.* (Vilamoura 2004), FT/3-5.
J. Nührenbberg et. al.: Overview on Wendelstein 7-X Theory. *Fusion Technology* **27**, 71 (1995)
W. Lotz, P. Merkl, J. Nührenbberg, and E. Strumberger: Collisionless α Particle Confinement in Stellarators. *Plasma Phys. Contr. Fusion* **34**, 1037 (1992).

[6.49] D. A. Spong, S. P. Hirshman, J. C. Whitson, D. B. Batchelor, R. Sanchez, B. A. Carreras et al.: Design Studies of Low Aspect Ratio Quasi-Omigenous Stellarators. *Nucl. Fusion* **40**, 563 (2000).
J. R. Cary and S. G. Shasharina: Omnigenous Stellarator. *Plasma Phys. Reports* **23**, 509 (1997).

[6.50] M Okamoto, M. Yokoyama, and N. Nakajima: New Concepts of Helical Magnetic Configurations. *J. Plasma Fusion Res.* **74**, 1107 (1998)

[6.51] C. D. Beidler, E. Harmeyer, F. Herrngger, Yu. Igitkhanov, A. Kendl et al.: The Helias Reactor HSR4/18. *Nucl. Fusion* **41**, 1759 (2001).
H. Wobig, C. D. Beider, J. Kisslinger, E. Harmeyer, F. Herrnegger et al.: Power Balance in Stellarator Reactors. *17th Fusion Energy Conf.* (Yokohama 1998), FTP/01.

[6.52] A. Sagara, S. Imagawa, O. Mitarai, T. Dolan, T. Tanaka et al.: Improved Structure and Long-Life Blanket Concepts for Heliotron Reactors. *Nucl. Fusion* **45**, 258 (2005).

[Ch7] Mirror and Tandem Mirror

[7.1] D. V. Sivkhim: *Rev. of Plasma Physics* **4**, 93 (edited by M. A. Leontovich) Consultants Bureau, New York, 1966

[7.2] Yu. V. Gott, M. S. Ioffe and V. G. Tel'kovskii: Some New Results for Plasma Containment in Magnetic Trap. Nucl. Fusion Suppl. Pt.3, 1045 (1962) (Conf. Proceedings, Salzburg 1961) IAEA, Vienna. (Translation of Russian Papers: U.S. Atomic Energy Commission, Division of Tecnical Information Office of Technical Service, Depart. of Commerce, Washington 25 D.C. 1963).

Yu. T. Baiborodov, M. S. Ioffe, V. M. Petrov, and R. I. Sobolev: An Adiabatic Trap with Combined Magnetic Field. *Sov. Atomic Energy* **14**, 459 (1963).

[7.3] Yu. T. Baiborodov, M. S. Ioffe, B. I. Kanaev, R. I. Sobolev, and E. E. Yushimanov: Investigation of Plasma Decay in the PR-6 Adiabatic Trap. *Plasma Phys. Contr. Nucl. Fusion Res.* (Conf. Proceedings, Madison 1971)**2**, 647 (1971) IAEA, Vienna.

[7.4] F. H. Coensgen, W. F. Cummins, V. A. Finlayson, W. E. Nexsen, Jr., and T. C. Simonen: Plasma Containment in the LRL-2X-Experiment. *Plasma Phys. Contr. Nucl. Fusion Res.* (Conf. Proceedings, Madison 1971) **2**, 721 (1971).

[7.5] F. H. Coensgen, W. F. Cummins, B. G. Logan, A. W. Halvik, W. E. Nexsen, T. C. Simonen, B. W. Stallard, and W. C. Turne: Stabilization of a Neutral-Beam-Sustained Mirror-Confined Plasma. *Phys. Rev. Lett.* **35**, 1501 (1975).

[7.6] T. K. Fowler, Ed.: Fusion Research in Open-Ended Systems. *Nucl. Fusion* **9**, 3 (1969).

[7.7] M. S. Ioffe and B. B. Kadomotsev: Plasma Confinement in Adiabatic Traps. *Sov. Phys. Uspekhi* **13**, 225 (1970).

[7.8] R. F. Post: The Magnetic Mirror Approach to Fusion. *Nucl. Fusion* **27**, 1579 (1987).

[7.9] E. G. Harris: Unstable Plasma Oscillation in a Magnetic Field. *Phys. Rev. Lett.* **2**, 34 (1959).

[7.10] R. A. Dory, G. E. Guest, and E. G. Harris: Unstable Electrostatic Plasma Waves Propagating Perpendicular to a Magnetic Field. *Phys. Rev. Lett.* **14**, 131 (1965).

[7.11] G. E. Guest and R. A. Dory: Microinstability of a Mirror-Confined Plasma. *Phys. Fluids* **8**, 1853 (1965).

[7.12] J. Gordey, G. Kuo-Petravic, E. Murphy, M. Petravie, D. Sweetman, and E. Thompson: Ion Cyclotron Instabilities and the Effect of Microwave Heating in the Phoenix II Magnetic Well. *Plasma Phys. Contr. Nucl. Fusion Res.* (Conf. Proceedings, Novosibirsk 1968) **2**, 267 (1969) IAEA, Vienna.

[7.13] R. F. Post and M. N. Rosenbluth: Electrostatic Instability of a Magnetic Field. *Phys. Fluids* **9**, 730 (1966).

[7.14] H. Postman, H. Dunlap, R. Dory, G. Haste, and R. Young: Observation of Negative-Mass Instability in an Energetic Proton Plasma. *Phys. Rev. Lett.* **16**, 265 (1966).

[7.15] B. B. Kadomtsev and O. P. Pogutse: Cyclotron Instability and Abrupt Plasma Density Decreases in Adiabatic Traps. *Plasma Phys. Contr. Nucl. Fusion Res.* (Conf. Proceedings, Novosibirsk 1968) **2**, 125 (1969) IAEA, Vienna.

[7.16] R. Z. Sagdeev and V. D. Shafranov: On the Instability of a Plasma with an Anisotropic Distribution of Velocities in a Magnetic Field. *Sov. Phys. JETP* **12**, 130 (1961).

[7.17] J. Sharer and A. Trivelpiece: Cycloton Wave Instabilities in Finite Mirror-Confined Plasmas. *Phys. Fluids* **10**, 591 (1967).

[7.18] H. Ikegami, H. Ikezi, T. Kawamura, H. Momota, K. Takayama, and Y. Terashima: Characteristics of Microinstabilities in a Hot Electron Plasma. *Plasma Phys. Contr. Nucl. Fusion Res.* (Conf. Proceedings, Novosibirsk 1968) **2**, 423 (1969) IAEA, Vienna.

[7.19] G. I. Dimov, V. V. Zakaidakov, and M. E. Kishinevskii: Thermonuclear Confinement System with Twin Mirror Systems. *Sov. J. Plasma Phys.* **2**, 326 (1976).

[7.20] T. K. Fowler and B. G. Logan: The Tandem Mirror Reactor. *Comments Plasma Phys. Contr. Fusion Res.* **2**, 167 (1977).

[7.21] V. P. Pastukhov: Collisional Losses of Electrons from an Adiabatic Trap in a Plasma with a Positive Potential. *Nucl. Fusion* **14**, 3 (1974).

[7.22] D. E. Baldwin and B. G. Logan: Improved Tandem Mirror Fusion Reactor. *Phys. Rev. Lett.* **43**, 1318 (1979).

[7.23] R. H. Cohen, I. B. Bernstein, J. J. Dorning, and G. Rowland: Particle and Energy Exchange between Untrapped and Electrically Confined Populations in Magnetic Mirrors. *Nucl. Fusion* **20**, 1421 (1980)

[7.24] F. H. Coensgen, C. A. Anderson, T. A. Casper, J. F. Clauser, W. C. Condit et al.: Electrostatic Plasma-Confinement Experiments in a Tandem Mirror System. *Phys. Rev. Lett.* **44**, 1132 (1980).
T. C. Simonen, C. A. Anderson, T. A. Casper, J. F. Clauser, F. H. Coesgen et al.: Plasma Confinement Experiments in the TMX Tandem Mirror. *Plasma Phys. Contr. Nucl. Fusion Res.* (Conf. Proceedings, Brussel 1980) **1**, 97 (1981) IAEA, Vienna.

[7.25] T. C. Simonen, S. L. Allen, D. E. Baldwin, T. A. Casper, J. F. Clauser et al.: Thermal Barrier Confinement Experiments in the TMX-U Tandem Mirror. *Plasma Phys. Contr. Nucl. Fusion Res.* (Conf. Proceedings, London 1984) **2**, 255 (1985) IAEA, Vienna.
D. P. Grubb, S. L. Allen, T. A. Casper, J. F. Clauser, F. H. Coesgen et al.: Thermal-Barrier Prediction and Identification in a Tandem Mirror. *Phys. Rev. Lett.* **53**, 783 (1984).

[7.26] T. Cho, M. Ichimura, M. Inutake, K. Ishii, S. Miyoshi et al.: Studies of Potential Formation and Transport in the Tandem Mirror GAMMA 10. *Plasma Phys. Contr. Nucl. Fusion Res.* (Conf. Proceedings, London 1984) **2**, 275 (1985) IAEA, Vienna.

[7.27] K. Yatsu, L. G. Bruskin, T. Cho, M. Hamada, M. Hirata et al.: Plasma Confinement in the GAMMA 10 Tandem Mirror. *17th Fusion Energy Conf.* (Yokohama 1998), EX4/6.

[7.28] T. Cho, H. Higaki, M. Hirata, H. Hojo, M. Ichimura et al.: Advances in Potential Formation and Finding in Sheared Radial Electric-Field Effects on Turbulence and Loss Suppression in GAMMA 10. *20th Fusion Energy Conf.* (Vilamoura 2004), EX/9-6Rd.

[Ch8] Magnetohydrodynamic (MHD) Instabilities

[8.1] G. Bateman: *MHD Instabilities,* The MIT Press, Cambridge, Mass. (1978).

[8.2] M. Kruskal and M. Schwarzschield: Some Instabilities of a Completely Ionized Plasma. *Proc. Roy. Soc.* **A223**, 348 (1954).

[8.3] M. N. Rosenbluth, N. A. Krall, and N. Rostoker: Finite Larmor Radius Stabilization of "Weakly" Unstable Confined Plasmas. *Nucl. Fusion Suppl.* pt. **1** 143 (1962).

[8.4] M. N. Rosenbluth and C. L. Longmire: Stability of Plasma Confined by Magnetic Fields. *Ann. Phys.* **1**, 120 (1957).

[8.5] M. D. Kruskal, J. L. Johnson, M. B. Gottlieb, and L. M. Goldman: Hydromagnetic Instability in a Stellarator. *Phys. Fluids* **1**, 421 (1958).

[8.6] V. D. Shafranov: Hydromagnetic Instability in a Stellarator. *Sov. Phys. JETP* **6**, 545 (1958).

[8.7] I. B. Berstein, E. A. Frieman, M. D. Kruskal, and R. M. Kulsrud: An Energy Principle for Hydromagnetic Stability Problems. *Proc. Roy. Soc.* **A244**, 17 (1958).

[8.8] B. B. Kadmotsev: *Rev. of Plasma Phys.* **2**, 153 (edited by M. A. Loentovich) Consultant Bureau, New York (1966).

[8.9] J. M. Greene and J. L. Johnson: Interchange Instabilities in Ideal Hydromagnetic Theory. *Plasma Phys.* **10**, 729 (1968).

[8.10] K. Miyamoto: *Plasma Physics and Controlled Nuclear Fusion.* Springer, Berlin (2005).

[8.11] J. W. Connor, R. J. Hastie, and J. B. Talor: High Mode Number Stability of an Axisymmetric Toroidal Plasma. *Proc. Roy. Soc.* **A365** 1, (1979).
J. W. Connor, R. J. Hastie, and J. B. Talor: Shear, Periodicity, and Plasma Ballooning Modes. *Phys. Rev. Lett.* **40**, 396 (1978).

[8.12] B. R. Suydam: Stability of a Linear Pinch. *Proc. 2nd U. N. International Conf. on Peaceful Uses of Atomic Energy*, Geneva, vol. **31**, 157 (1958).

[8.13] W. A. Newcomb: Hydromagnetic Stability of a Diffuse Linear Pinch. *Annal. Phys.* **10**, 232 (1960).

[8.14] D. C. Robinson: High-β Diffuse Pinch Configurations. *Plasma Phys.* **13**, 439 (1971).

[8.15] K. Hain and R. Lüst: Zur Stabilität Zylindrersymmetyrischer Plasma Konfigurationen mit Volumenstrőmen. *Z. Naturforsh.* **13a**, 936 (1958).

[8.16] K. Matsuoka and K. Miyamoto: Stabilization of Low-m Kink Modes in Tokamaks. *Jpn. J. Appl. Phys.* **18**, 817 (1979).

[8.17] R. M. Kulsrud: Stability of Negative V'' Systems. *Plasma Phys. Contr. Nucl. Fusion Res.*, (Conf. Proceedings, Culham 1965) vol. **1**, 127 (1966) IAEA, Vienna.
H. P. Furth, J. Killeen, and M. N. Rosenbluth: Stabilization by Shear and Negative V''. *Plasma Phys. Contr. Nucl. Fusion Res.*, (Conf. Proceedings, Culham 1965) vol. **1**, 103 (1966) IAEA, Vienna.

[8.18] J. A. Wesson and A. Sykes: Tokamak Beta Limit. *Nucl. Fusion* **25** 85 (1985).

[8.19] J. M. Greene and M. S. Chance: The Second Region of Stability against Ballooning Modes. *Nucl. Fusion* **21**, 453 (1981).

[8.20] T. Ozeki, M. Azumi, S. Tokuda, and S. Ishida: Effects of a Hollow Current Profile on the Ideal MHD Stability of High β_p Plasmas in a Tokamak. *Nucl. Fusion* **33**, 1025 (1993).

[8.21] B. B. Kadomtsev and O. P. Pogutse: *Rev. of Plasma Physics* **5**, 304 (ed. by M. A. Leontovich) Consultant Bureau, New York (1970).

[8.22] A. Hasegawa and Liu Chen: Kinetic Processes in Plasma Heating by Resonant Mode Conversion of Alfvén Wave. *Phys. Fluids* **19**, 1924 (1976).

[8.23] H. L. Berk, J. W. Van Dam, Z. Guo, and D.M. Lindberg: Continuum Damping of Low-n Toroidicity-Induced Shear Alfvén Eigenmodes. *Phys. Fluids* **B4** 1806 (1992).

[8.24] R. Betti and J. P. Freidberg: Stability of Alfvén Gap Modes in Burning Plasmas. *Phys. Fluids* **B4**, 1465 (1992).

[8.25] Y. Todo, T. -H. Watanabe, H.-B. Park, and T. Sato: Fokker-Planck Simulation Study of Alfvén Eigenmode Bursts. *Nucl. Fusion* **41**, 1153 (2001).
Y. Todo, T. Sato, K. Watanabe, T. H. Watanabe, and R. Horiuchi: Magnetohydrodynamic Vlasov Simulation of the Toroidal Alfvén Eigenmode. *Phys. Plasma* **2**, 2711 (1995).

[8.26] M. N. Rosenbruth, H. L. Berek, J. W. Van Dam, and D. M. Lindberg: Mode Structure and Continuum Damping of High-n Toroidal Alfvén Eigenmodes. *Phys. Fluids* **B4**, 2189 (1992).

[8.27] J. Candy and N. M. Rosenbluth: Alpha-Particle-Driven Nonideal Toroidal Alfvén Eigenmodes. *Plasma Phys. Contr. Fusion* **35**, 957 (1993).
J. Candy and N. M. Rosenbluth: Nonideal Theory of Toroidal Alfvén Eigenmodes. *Phys. Plasmas* **1**, 356 (1994).
R. R. Mett and S. M. Mahajan: Kinetic Theory of a High-n Toroidicity-Induced Alfvén Eigenmodes. *Phys. Fluids* **B4**, 2885 (1992).

[8.28] G. Y. Fu and C. Z. Cheng: Theory of a High-n Toroidicity-Induced Shear Alfvén Eigenmode in Tokamaks. *Phys. Fluids* **B2**, 985 (1990).
H. L. Berk, J. W. Van Dam, D. Borba, J. Candy, G. T. A. Huysmans, and S. Sharapov: More on Core-Localized Toroidal Alfvén Eigenmodes. *Phys. Plasmas* **2**, 3401 (1995).

[8.29] F. Zonca and L.Chen: Theory of Continuum Damping of Toroidal Alfvén Eigenmodes in Finite-β Tokamaks. *Phys. Fluids* **B5**, 3668 (1993).
F. Zonca and L. Chen: Theory of Toroidal Alfvén Modes Excited by Energetic Particles in Tokamaks. *Phys. Plasmas* **3**, 323 (1996).

[8.30] ITER Physics Basis: *Nucl. Fusion* **39**, No. 12, 2471 (1999).

[8.31] E. J. Strait, W. W. Heidbrik, A. D. Turnbull, M. S. Chu, and H. H. Duong: Stability of Neutral Beam Driven TAE Modes in DIII-D. *Nucl. Fusion* **33**, 1849 (1993).
K.-L. Wong: A Review of Alfvén Eigenmode Observations in Toroidal Plasmas. *Plasma Phys. Contr. Fusion* **41**, R1 (1999).
A. Fukuyama and T. Ozeki: Linear Theory of Alfvén Eigenmodes in Tokamak Plasmas. *J. Plasma Fusion Res.* **75**, 537 (1999) (in Japanese).

[8.32] L. Chen, R. B. White, and M. N. Rosenbluth: Excitation of Internal Kink Modes by Trapped Energetic Beam Ions. *Phys. Rev. Lett.* **52**, 1122 (1984).

[8.33] V. D. Shafranov: Hydromagnetic Stability of a Current-Carrying Pinch in a Strong Logitudinal Magnetic Field. *Sov. Phys. Tech. Phys.* **15**, 175 (1970).

[Ch9] Resistive Instabilities

[9.1] H. P. Furth, J. Killeen, and M. Rosenbluth: Finite-Resistivity Instabilities of a Sheet Pinch. *Phys. Fluids* **6**, 459 (1963).

[9.2] H. P. Furth, P. H. Rutherford, and H. Selberg: Tearing Mode in the Cylindrical Tokamak. *Phys. Fluids* **16**, 1054 (1973).

[9.3] R. J. Goldston and R. H. Rutherford: *Introduction to Plasma Physics* Institute of Physics Publishing, London (1995).

[9.4] P. H. Rutherford: Nonlinear Growth of the Tearing Mode. *Phys. Fluids* **16**, 1903 (1973).

[9.5] R. Fitzpatrick: Helical Temperature Perturbations Associated with Tearing Modes in Tokamak Plasmas. *Phys. Plasmas* **2**, 825 (1995).

[9.6] H. R. Wilson, J. W. Connor, R. H. Hastie, and C. C. Hegna: Threshold for Neoclassical Magnetic Islands in a Low Collision Frequency Tokamak. *Phys. Plasma* **3**, 248 (1996).
A. I. Smolyakov, A. Hirose, E. Lazzaro, G. B. Re, and J. D. Callen: Rotating Nonlinear Magnetic Islands in a Tokamak Plasma. *Phys. Plasma* **2**, 1581 (1995).
F. L. Hinton and J. R. Robertson: Neoclassical Dielectric Property of a Tokamak Plasma. *Phys. Fluids* **27**, 1243 (1984).

[9.7] D. A. Gates, B. Lloyd, A. W. Morris, G. McArdle, M. R. O'Brien et al.: Neoclassical Islands on Compass-D. *Nucl. Fusion* **37**, 1593, (1997).

[9.8] F. F. Chen: *Introduction to Plasma Physics and Controlled Fusion* 2nd ed. Plenum Press, New York, London (1984).

[9.9] S. S. Moiseev and R. Z. Sagdeev: On the Bohm Diffusion Coefficient. *Sov. Phys. JETP* **17**, 515 (1963).

[9.10] F. F. Chen: Resistive Oberstabilities and Anomalous "Diffusion". *Phys. Fluids* **8**, 912 (1965).
F. F. Chen: "Universal" Overstability of a Resistive Inhomogeous Plasma. *Phys. Fluids* **8**, (1965)

[9.11] F. Troyon, R. Gruber, H. Saurenmann, S. Semenzato and S. Succi: MHD-Limits to Plasma Cnfinement. *Plasma Phys. Contr. Fusion* **26**, 209 (1984).

[9.12] J. M. Finn: Resistive Wall Stabilization of Kink and Tearing Mode. *Phys. Plasmas* **2**, 198, (1995).

[9.13] D. J. Ward and A. Bondeson: Stabilization of Ideal Modes by Resistive Walls in Tokamaks with Plasma Rotation and Its Effect on the Beta Limit. *Phys. Plasmas* **2**, 1570, (1995).

[9.14] M. Okabayashi, N. Pomphrey, and R. E. Hatcher: Circuit Equation Formulation of Resistive Wall Mode Feedback Stabilization Schemes. *Nucl. Fusion* **38**, 1607 (1998).

[Ch10] Boltzmann Equation and Landau Damping

[10.1] D. V. Sivukhin: *Reviews of Plasma Physics* **4**, 93 (edited by M. A. Leontobich) Consultants Bureau, New York (1966).

[10.2] T. H. Stix: *The Theory of Plasma Waves*, McGraw Hill, New York (1962).
T. H. Stix: *Waves in Plasmas*, American Institute of Physics, New York, (1992).

[10.3] L. D. Landau: On the Vibrations of the Electronic Plasma. *J. Phys.* (USSR) **10**, 45 (1946).

[10.4] J. H. Malmberg, C. B. Wharton, and W. E. Drummond: Landau Damping of Electron Plasma Waves. *Plasma Phys. Contr. Nucl. Fusion Res.* (Conf. Proceedings, Culham 1965) vol. **1**, 485 (1966) IAEA, Vienna.

[Ch11] Plasma as Medium of Waves

[11.1] T. H. Stix: *The Theory of Plasma Waves*, McGraw-Hill, New York (1962).
T. H. Stix: *Waves in Plasmas*, American Institute of Physics, New York, (1992).

[11.2] W. P. Allis, S. J. Buchsbanm, and A. Bers: *Waves in Anisotropic Plasmas*, The MIT Press, Cambrige, Mass. (1963).

[11.3] B. D. Fried and S. D. Conte: *The Plasma Dispersion Function*, Academic Press, New York (1961).

[11.4] N. A. Krall and M. N. Rosenbluth: Universal Instability in Complex Field Geometries. *Phys. Fluids* **8**, 1488 (1965).

[11.5] M. N. Rosenbluth and R. F. Post: High-Frequency Electrostatic Plasma Instability Inherent to "Loss Cone" Particle Distribution. *Phys. Fluids* **8**, 547 (1965).

[11.6] E. G. Harris: Unstable Plasma Oscillations in a Magnetic Field. *Phys. Rev. Lett.* **2**, 34 (1959).

[11.7] E. G. Harris: *Physics of Hot Plasma*, 145 (edited by B. J. Rye and J. B. Taylar) Oliver & Boyd, Edinburgh (1970).

[11.8] B. B. Kadomtsev and O. P. Pogutse: Trapped Particles in Toroidal Magnetic Systems. *Nucl. Fusion* **11**, 67 (1971).

[11.9] W. M. Tang, G. Rewolt, and Liu Chen: Microinstabilities in Weak Density Gradient Tokamak Systems. *Phys. Fluids* **29**, 3715 (1986).

[Ch12] Wave Heating and Non-Inductive Current Drive

[12.1] M. Porkolab: *Fusion* (edited by E. Taylor) vol. **1** Part B, 151, Academic Press, New York (1981).

[12.2] J. E. Scharer, B. D. McVey, and T. K. Mau: Fast-Wave Ion-Cyclotron and First-Harmonic Heating of Large Tokamak. *Nucl. Fusion* **17**, 297 (1977).

[12.3] T. H. Stix: Fast-Wave Heating of a Two-Component Plasma. *Nucl. Fusion* **15**, 737 (1975).

[12.4] A. Fukuyama, S. Nishiyama, K. Itoh, and S.I. Itoh: Kinetic Description of Propagation and Absorption Structures of ICRF Waves. *Nucl. Fusion* **23** 1005 (1983).

[12.5] M. Ono, T. Watari, R. Ando, J. Fujita et al.: Ion-Bernstein-Wave Heating in the JIPP TIIU Tokamak Plasma. *Phys. Rev. Lett.* **54**, 2339 (1985).

[12.6] T. H. Stix: Radiation and Absorption via Mode Conversion in an Inhomogeneous Collision-Free Plasma. *Phys. Rev. Lett.* **15**, 878 (1965).

[12.7] V. M. Glagolev: Propagation and Absorption of Ion Hybrid Waves in a Weakly Inhomogeneous Plasma Layer - I. *Plasma Phys.* **14**, 301 and 315 (1972).

[12.8] M. Brambilla: Propagation and Absorption of Waves at the Lower Hybrid Resonance. *Plasma Phys.* **18**, 669 (1976).

[12.9] S. Bernabei, M. A. Heald, W. M. Hooke, R. W. Motley, F. J. Paoloni, M. Brambilla, and W.D. Getty: Plasma-Wave Coupling and Propagation Using Phased Waveguid Arrays. *Nucl. Fusion* **17**, 929 (1977).

[12.10] S. Takamura: *Fundamentals of Plasma Heatings* (in Japanese), Nagoya University Press, Nagoya (1986).

[12.11] I. Fidone, G. Granata, and G. Ramponi: Wave Absorption near the Electron Cyclotron Frequency. *Phys. Fluids* **21**, 645 (1978).

[12.12] R. Prator: Heating and Current Drive by Electron Cyclotron Waves. *Phys. Plasmas* **11**, 2349 (2004).

[12.13] T. Ohkawa: New Methods of Driving Plasma Current in Fusion Devices. *Nucl. Fusion* **10**, 185 (1970).

[12.14] D. J. H. Wort: The Peristaltic Tokamak. *Plasma Phys.* **13**, 258 (1971).

[12.15] N. J. Fisch: Confining a Tokamak Plasma with RF-Driven Currents. *Phys. Rev. Lett.* **41**, 873 (1978).
C. F. F. Karney and N. J. Fisch: Numerical Studies of Current Generation by Radio-Frequency Traveling Waves. *Phys. Fluids* **22**, 1817 (1979).

[12.16] N. J. Fisch and A. H. Boozer: Creating an Asymmetric Plasma Resistivity with Waves. *Phys. Rev. Lett.* **45**, 720 (1980).

[12.17] N. J. Fisch: Theory of Current Drive in Plasmas. *Rev. Modern Phys.* **59**, 175 (1987).

[12.18] D. F. H. Start, J. G. Cordey, and E. M. Jones: Effect of Trapped Electrons on Bean Driven Currents in Toroidal Plasmas. *Plasma Phys.* **22**, 303 (1980).

[12.19] K. Okano: Neoclassical Formula for Neutral Beam Current Drive. *Nucl. Fusion* **30**, 423 (1990).

[Ch13] Plasma Transport by Turbulence

[13.1] W. Horton: Spectral Distribution of Drift-Wave Fluctuations in Tokamaks. *Phys. Rev. Lett.* **37**, 1269 (1976).
W. Horton: Drift Waves and Transport. *Rev. Modern Phys.* **71**, 735 (1999)

[13.2] S. Hamaguchi and W. Horton: Effects of Sheared Flows on Ion-Temperature-Gradient-Driven Turbulent Transport. *Phys. Fluids* **B4**, 319 (1992).

[13.3] W. Horton, Jr., R. Esres, H. Kwak, and D.-I. Choi: Toroidal Mode Coupling Effects on Drift Wave Stability. *Phys. Fluids* **21**, 1366 (1978).

[13.4] J. Y. Kim and M. Wakatani: Radial Structue of High-Mode-Number Toroidal Modes in General Equilibrium Profiles. *Phys. Rev. Lett.* **73**, 2200 (1994).

[13.5] Y. Kishimoto, J. Y. Kim, T. Fukuda, S. Ishida, T. Fujita, T. Tajima, W. Horton, G. Furnish, and M. J. LeBrun: Effect of Weak/Negative Magnetic Shear and Plasma Shear Rotation on Self-Organized Critical Gradient Transport in Toroidal Plasmas - Formation of Internal Transport Barrier. *6th IAEA Fusion Energy Conf.* (Conf. Proceedings, Motreal 1996) vol. **2**, 581 (1997) IAEA, Vienna.
Y. Kisihimoto: Anomalous Transport. *Jour. Plasma Fusion Res.* **76**, 1280 (2000) (in Japanese).

[13.6] A. B. Rechester and M. N. Rosenbluth: Electron Heat Transport in a Tokama with Destroyed Magnetic Surface. *Phys. Rev. Lett.* **40**, 38 (1978).

[13.7] J. W. Connor, and J. B. Taylor: Scaling Laws for Plasma Confinement. *Nuclear Fusion* **17**, 1047 (1977).

[13.8] B. B. Kadomtsev: Tokamaks and Dimensional Analysis. *Sov. J. Plasma Phys.* **1**, 295 (1975).

[13.9] J. W. Connor: Invariance Principles and Plasma Confinement. *Plasma Phys. Control. Fusion* **30**, 619 (1988)
J. Cordy, K. Thompsen, A. Chudnovskiy, O. J. W. F. Kardaun, T. Takizuka et al.: Scaling of the Energy Confinement Time with β and Collisionality Approaching ITER Conditions. *Nucl. Fusion* **45**, 1078 (2005).

[13.10] R. W. Hockney and J. W. Eastwood: *Computer Simulation Using Particles*, McGraw-Hill, NewYork (1981).

[13.11] T. Tajima: *Computational Plasma Physics: With Application to Fusion and Astrophysics*, Addison-Wesley, Redwood City, Calif. (1989).

[13.12] C. K. Birdsall and A. B. Langdon: *Plasma Physics via Computer Simulation*, IOP Pubishing Ltd., Bristol (1991).

[13.13] W. W. Lee: Gyrokinetic Approach in Particle Simulation. *Phys. Fluids* **26**, 556 (1983).
W. W. Lee: Gyrokinetic Particle Simulation Model. *J. Comput. Physics* **72**, 243 (1987).

[13.14] R. D. Sydora: Three Dimensional Gyrokinetic Simulation of Low-Frequency Drift Instabilities. *Phys. Fluids* **B2**, 1455 (1990).

[13.15] S. E. Parker, W. W. Lee, and R. A. Santoro: Gyrokinetic Simulation of Ion Temperature Gradient Driven Turbulence in 3D Toroidal Geometry. *Phys. Rev. Lett.* **71**, 2042 (1993).

[13.16] S. E. Parker, and W. W. Lee: A Fully Nonlinear Characteristic Method for Gyrokinetic Simulation. *Phys. Fluids* **B5**, 77 (1993).

[13.17] A. M. Dimits, B. I. Cohen, N. Mattor, W. M. Nevins, and D. E. Shumaker: Simulation of Ion Temperature Gradient Turbulence in Tokamaks. *Nucl. Fusion* **40**, 661 (2000).

[13.18] A. M. Dimits, G. Bateman, M. A. Beer, B. I. Cohen, W. Dorland et al.: Comparisons and Physics Basis of Tokamak Transport Models and Turbulence Simulation. *Phys. Plasmas* **7**, 969 (2000).

[13.19] R. G. Littlejohn: Hamiltonian Formulation of Guiding Center Motion. *Phys. Fluids* **24**, 1730 (1981).
R. G. Littlejohn: Variational Principles of Guiding Center Motion. *J. Plasma Phys.* **29**, 111 (1983).
A. Brizard: Nonlinear Gyrokinetic Maxwell-Vlasov Equations Using Magnetic Co-ordinates. *J. Plasma Phys.* **41**, 541 (1989).
H. Sugama: Gyrokinetic Field Theory. *Phys. Plasmas* **7**, 466 (2000).

[13.20] A. B. Langdon and C. K. Birdsall: Theory of Plasma Simulation Using Finite-Size Particles. *Phys. Fluids* **13**, 2115 (1970).
H. Okuda and C. K. Birdsall: Collisions in a Plasma of Finite-Size Particles. *Phys. Fluids* **13**, 2123 (1970).
H. Okuda: Verification of Theory for Plasma of Finite-Size Particles. *Phys. Fluids* **15**, 1268 (1972).

[13.21] H. Naitou: Basic Theory of Particle Simulation. *J. Plasma Fusion Res.* **74**, 470 (1998) (in Japanese).

[13.22] K. Nishimura, R. Horiuchi, and T. Sato: Tilt Stabilization by Cycling Ions Crossing Magnetic Separatrix in a Field Reversed Configuration. *Phys. Plasmas* **4**, 4035 (1997).
R. Horiuchi, and T. Sato: The Meandering Orbit Effect in Stabilization of the Tilting Instability in a Field Reversed Configuration. *Phys. Fluids* **B 2**, 2652 (1990).

[13.23] E. V. Belova, S. C. Jardin, H. Ji, M. Yamada, and R. Kulsrud: Numerical Study of Tilting Stability of Prolate Field Reversed Configurations. *Phys. Plasma* **7**, 4996 (2000).

[13.24] M. J. LeBrun, T. Tajima, M. G. Gray, G. Furnish, and W. Horton: Toroidal Effects on Drift Wave Turbulence. *Phys. Fluids* **B5**, 752 (1993).

[13.25] Y. Kishimoto, J-Y Kim, W. Horton, T. Tajima, M. J. LeBrun, and H. Shirai: Toroidal Mode Structure in Weak and Reversed Magnetic Shear

Plasma and Its Role in the Internal Transport Barrier. *Plasma Phys. Contr. Fusion* **40**, A663 (1998).

[13.26] A. Hasegawa, C. G. Maclennan, and Y. Kodama: Nonlinear Behavior and Turbulence Spectra of Drift Waves and Rossy Waves. *Phys. Fluids* **22**, 2122 (1979).

[13.27] A. Hasegawa, and K. Mima: Pseudo-Three-Dimensional Turbulence in Magnetized Nonuniform Plasma. *Phys. Fluids* **21**, 87 (1978).

[13.28] A. I. Dyachenko, S. V. Nazarenko, and V. E. Zakharov: Wave-Vortex Dynamics in Drift and β-Plane Turbulence. *Phys. Lett.* **A165**, 330 (1992).

[13.29] Z. Lin, T. S. Hahm, W. W. Lee, W. M. Tang, and R. B. White: Turbulent Transport Reduction by Zonal Flows: Massively Parallel Simulations. *Science* **281**, 1835 (1998).

[13.30] P. N. Guzdar, R. G. Kleva, and Liu Chen: Shear Flow Generation by Drift Waves Revisited. *Phys. Plasmas* **87**, 459 (2001).

[13.31] A. I. Smolyakov, P. H. Diamond, and M. Malkov: Coherent Structure Phenomenoa in Drift Wave-Zonal Flow Turbulence. *Phys. Rev. Lett.* **84**, 491 (2000).

[13.32] N. Winsor, J. L. Johnson, and J. M. Dawson: Geodesic Acoustic Waves in Hydromagnetic System. *Phys. Fluids* **11**, 2448 (1968).

[13.33] K. Itoh, K. Hallatschek, and S.-I. Itoh: Excitation of Geodesic Acoustic Mode in Toroidal Plasmas. *Plasma Phys. Contr. Fusion* **47**, 451 (2005).

[13.34] Y. Idomura, S. Tokuda, and Y. Kishimoto: Global Gyrokinetic Simulation of Ion Temperature Gradient Driven Turbulence in Plasmas Using a Canonical Maxwellian Distribution. *Nucl. Fusion* **43**, 234 (2003).

[13.35] Y.Idomura, S. Totsuka, and Y. Kishimoto: Global Gyrokinetic Simulation of Toroidal Electron Temperature Gradient Driven Mode in Reversed Shear Tokamaks. *20th Fusion Energy Conf.* (Vilamoura in 2004) TH/8-1.
Y. Idomura, M. Wakatani, and S. Tokuda: Stability of $E \times B$ Zonal Flow in Electron Temperature Gradient Driven Turbulence. *Phys. Plasmas* **7**, 3551 (2000).

[13.36] A. Fujisawa, K. Itoh, H. Iguchi, K. Matsuoka, S. Okamomura et al.: Identification of Zonal Flows in a Toroidal Plasma. *Phys. Rev. Lett.* **93**, 165002 (2004).

[13.37] G. D. Conway, B. Scott, J. Schirmer, M. Reich, A. Kendl and ASDEX-U Team: Direct Measurement of Zonal Flows and Geodesic Acoustic Mode Oscillations in ASDEX Upgrade Using Doppler Reflectometry. *Plasma Phys. Contr. Fusion* **47**, 1165 (2005).

[13.38] P. H. Diamond, S.-L. Itoh, K. Itoh, and T. S. Hahm: Zonal Flows in Plasma - A Review. *Plasma Phys. Contr. Fusion* **47**, R35 (2005).
P. H. Diamond, K. Itoh, S.-L. Itoh, and T. S. Hahm: Overview of Zonal Flow Physics. *20th Fusion Energy Conf.* (Vilamoura 2004) OV/2-1.

Index

Milton Keynes UK
Ingram Content Group UK Ltd.
UKHW021824071024
449327UK00021B/1422

9 780367 390082